STUDY GUIDE

to accompany

Asking About Life

second edition

TOBIN & DUSHECK

LORI K. GARRETT
Danville Area Community College

Harcourt College Publishers
Fort Worth Philadelphia San Diego New York Orlando Austin
San Antonio Toronto Montreal London Sydney Tokyo

to process will be too great. "Cramming" *may* get you by in the short-term, but you will not retain the information for the future.

When you finish reading a chapter, answer the Review and Thought Questions in the text, then work through this study guide. Read the Key Concepts for a quick review of major points, then review the material in the Extended Chapter Outline. If any part is not clear to you, return to the textbook and read that section again. If it remains unclear, jot down the problem area and ask your instructor for clarification.

Once you have reviewed the material, fill in the Vocabulary Building section. For the most benefit, try to define the terms from memory after reading the chapter, and always use your own words—do not copy from the text. Explain the term and, when possible, provide an example. Then check your definitions against the text and make corrections as needed for future reference. Be sure to review this vocabulary list before an exam.

Now you can complete the Chapter Test. Work through the entire test once, then go back to any questions you previously skipped. Mark any questions you are still unable to answer. Next, check your answers against the material in the textbook or in the Answer section at the end of this study guide. Mark any items you answered incorrectly, make corrections as needed, then review that material in the text.

Keep your study guide handy to help you prepare for quizzes and exams, and review the material often. Repetition is essential for thorough learning.

Biology is a fascinating area of study, but any science can be challenging. Use your study time wisely—there are no shortcuts to learning. The more effort you put forth, the more knowledge you will gain. As you study, constantly look for common themes and analogies that can help you comprehend and retain the complex material. Think of analogies that will help you remember what you learn, and think of how each concept relates to your life. Try to avoid just memorizing by always asking "Why" things are as they are. Once you wade past the complicated jargon, most science is delightfully simple and logical.

As your study skills improve and your knowledge increases, you will find that your eyes are opening and you are seeing the world around you differently than you ever have before. You will start to see that simplicity and balance abound, and you will realize that, simply by living, you already are a biologist.

Lori K. Garrett
May 2000

Acknowledgments

I thank Ms. Lee Marcott for giving me the opportunity to again be involved with this book, and for all her hard work coordinating this project. Lee—I've been on your end and know how crazy it can get! You do a great job and working with you is a pleasure. Thanks also go to all the other Harcourt staff who helped this project to fruition.

I thank my parents, Pat and Harry Garrett, for their unending support of all of my endeavors, and for knowing that education is a necessity, not a luxury.

And a final thank you to my students, who are my continuing education.

I dedicate this book to my nephew, Jeremy "Spike" Garrett, who I hope will discover the joy of learning for the sake of the soul and the spirit.

Contents

Preface		iii
Acknowledgments		v
Chapter 1	The Unity and Diversity of Life	1
Chapter 2	The Chemical Foundations of Life	13
Chapter 3	Biological Molecules Great and Small	27
Chapter 4	Why Are All Organisms Made of Cells?	41
Chapter 5	Directions and Rates of Biochemical Processes	55
Chapter 6	How Do Organisms Supply Themselves with Energy?	67
Chapter 7	Photosynthesis: How Do Organisms Get Energy from the Sun?	79
Chapter 8	Cell Reproduction	91
Chapter 9	From Meiosis to Mendel	103
Chapter 10	The Structure, Replication, and Repair of DNA	117
Chapter 11	How Are Genes Expressed?	129
Chapter 12	Jumping Genes and Other Unconventional Genetic Systems	143
Chapter 13	Genetic Engineering and Recombinant DNA	155
Chapter 14	Human Genetics	167
Chapter 15	What is the Evidence for Evolution?	179
Chapter 16	Microevolution: How Does a Population Evolve?	193
Chapter 17	Macroevolution: How Do Species Evolve?	207
Chapter 18	How Did the First Organisms Evolve?	221
Chapter 19	Classification: What's In a Name?	233
Chapter 20	Prokaryotes: How Does the Other Half Live?	243
Chapter 21	Classifying the Protists and Multicellular Fungi	255
Chapter 22	How Did Plants Adapt to Dry Land?	269
Chapter 23	Protostome Animals: Most Animals Form Mouth First	283
Chapter 24	Deuterostome Animals: Echinoderms and Chordates	299
Chapter 25	Ecosystems	311
Chapter 26	Biomes and Aquatic Communities	323
Chapter 27	Communities: How Do Species Interact?	337
Chapter 28	Populations: Extinctions and Explosions	349
Chapter 29	The Ecology of Animal Behavior	363
Chapter 30	Structural and Chemical Adaptations of Plants	373
Chapter 31	What Drives Water Up and Sugars Down?	387

13. You are testing a new diet drug. Your test subjects are to follow a prescribed diet and exercise program while taking the new medication. Which of the following would serve as an appropriate control group for your experiment?
    a) overweight individuals who are given a different diet drug
    b) normal weight individuals who receive the test drug
    c) overweight individuals who follow the prescribed diet and exercise but take a placebo
    d) overweight individuals who take the test drug but follow no diet or exercise program

14. Which of the following is not attributed to Charles Darwin?
    a) the theory of natural selection
    b) the genetic basis of heredity
    c) existence of a common ancestor
    d) All of these came from Darwin.

15. Which of the following is not typical of a prokaryotic organism?
    a) contains genetic material
    b) has membrane-bound organelles
    c) is single-celled
    d) contains no nucleus

Part 2: **Matching**

For each of the following, match the correct term with its definition or example. More than one answer may be appropriate.

hypothesis **model** **theory**

1. ____________________ A group of generally accepted, related explanations that have been repeatedly supported by experimental testing.

2. ____________________ An informed guess that attempts to answer the question under study.

3. ____________________ A simplified view of a process, usually resulting from observation.

homeostasis **adaptation** **negative feedback**

4. ____________________ When you get too hot, you sweat more, but as you cool down, sweating decreases.

5. ____________________ Although the air outside you is relatively dry, you maintain a wet world inside of you.

6. ____________________ Many plants living in the desert can store water.

7. ____________________ Body temperature is relatively stable from day to day.

eukaryotic **prokaryotic** **multicellular**

8. ____________________ Composed of many cells.
9. ____________________ Contains membrane-bound organelles.
10. ____________________ A small cell that has no membrane-bound organelles.
11. ____________________ A scissor-tailed flycatcher (bird).
12. ____________________ A staphylococcus bacterium.

mean **standard deviation** **normal distribution**

13. ____________________ Data are equally distributed on each side of the mean.
14. ____________________ Reveals how much the data deviate from the mean.
15. ____________________ Arithemetic average.
16. ____________________ This can help determine what sample size is needed.
17. ____________________ 68% of the data are within one standard deviation.

Part 3: Short Answer

Write your answers in the space provided or on a separate piece of paper.

1. What are the steps used when following the scientific method? Explain all steps and how they are dependent upon each other.

2. Why is a control necessary in all experiments? Be able to design one.

3. Why can a hypothesis be disproven but never be proven.

4. What are the characteristics of a good experiment?

5. What is meant by a testable hypothesis, and why is that important?

6. List the three domains into which organisms are categorized.

7. List the seven characteristics shared by all organisms that were discussed in this chapter. Think of examples of each of these.

8. Differentiate between eukaryotic and prokaryotic organisms, and give an example of each.

9. Why is it better to conduct experiments on relatively large groups, rather than on individuals?

10. Why must statistical analysis be part of a good experiment?

11. How can the mean, standard deviation, and a normal distribution be used to evaluate data?

12. What is meant by saying that organisms are more alike at the molecular level than at the level of the whole organism?

13. Define natural selection. How it is related to adaptations and how does it lead to evolution?

14. How do genes allow continuity of traits from one generation to the next?

15. Define homeostasis and how negative feedback is used to maintain it. Provide an example.

Part 4: Critical Thinking—Using Your Knowledge

Answer each of these in essay form, using complete sentences and paragraphs. Provide as much information as you can. (For extra essay practice, write out answers to the Review and Thought Questions in your textbook.)

1. Think of an example of a time when you tried to figure out how something (perhaps a machine) worked. How did you do it? List the specific actions you took. Can you correlate these actions to steps in the scientific process? Do you often use some version of the scientific method without realizing it?

2. What were the flaws in Barry Marshall's experiment in which he ingested bacteria to test his hypothesis that those bacteria cause ulcers? Did he follow the scientific method?

3. Human experimentation such as Marshall did (although he performed it on himself) is generally not conducted. As a more recent example, some researchers have been willing to inject themselves with HIV (the virus that causes AIDS) to test new drugs designed to prevent AIDS. Members of the public have also offered to volunteer for these injections. What are the advantages of this type of human research? What are the potential disadvantages? What are some reasons why this type of research is often considered unethical and not permitted?

4. Keeping your response to the previous question in mind, design a good experiment or series of experiments to test Marshall's hypothesis about certain bacteria being the cause of ulcers. What variables would you need to consider? Who would you include in your sample? What would you use for a control?

CHAPTER TEST

The following test has four parts. Complete as much of the exam as you can from memory. If you cannot answer a question, skip it. Once you complete all that you can, try to answer any questions you skipped. If you still cannot answer them, consult your textbook for the answers. Once you have completed all sections of the test, check your answers for Parts 1 - 3 against those in the back of this book. Highlight any incorrect answers then review that material in your textbook. Correct your answers for future reference.

Part 1: Multiple Choice

For each of the following, select all correct responses—more than one may be correct.

1. Which of the following are located within the nucleus?
   a) electrons and protons
   b) electrons and neutrons
   c) protons and neutrons
   d) neutrons only

2. Which of the following results from sharing electrons?
   a) covalent bond
   b) ionic bond
   c) hydrogen bond
   d) all of these

3. Which of the following is an ion?
   a) C
   b) NaCl
   c) H_2O
   d) OH^-

4. Which of the following contributes little to the mass of an atom?
   a) electrons
   b) protons
   c) neutrons
   d) both *b* and *c*

5. Isotopes of an element differ from each other in:
   a) the number of electrons in their outer shell
   b) the total number of electrons
   c) the number of neutrons
   d) the number of protons

6. All elements in the same column of the Periodic Table contain the same number of:
   a) total electrons
   b) electrons in the outer shell
   c) protons
   d) neutrons in the nucleus

7. Water makes up approximately what percentage of the material in living organisms?
   a) 30%
   b) 50%
   c) 70%
   d) 90%

8. How does increased temperature affect molecular activity?
   a) slows molecular movement
   b) increases molecular movement
   c) depends on the molecules involved
   d) no effect

9. Water has a high amount of surface tension. What characteristic of water causes this?
   a) cohesion
   b) high density
   c) adhesion
   d) high heat capacity

10. Many of the special characteristics of water can be attributed to its:
   a) ionic nature
   b) nonpolar molecules
   c) hydrogen bonds
   d) hydrophilic interactions

11. Which of the following is a way in which water is important to life?
   a) Most biochemical reactions occur in or near water.
   b) Many biochemical reactions use water.
   c) Many biochemical reactions produce water.
   d) Water can split into H^+ and OH^- ions that determine pH.

12. The balance between which of the following determines pH?
   a) water and hydronium ions
   b) acids and ions
   c) hydrogen ion and hydroxide ions
   d) water and hydrogen

13. In humans, what is the normal blood pH?
   a) 5.0
   b) 7.0
   c) 7.4
   d) 8.6

14. A buffer is anything that:
   a) releases hydrogen ions
   b) combines with hydrogen ions
   c) stops changes in pH
   d) minimizes normal fluctuations in pH

15. An element's chemical reactivity is primarily determined by:
   a) how many electrons are in its outer shell
   b) how many atoms it contains
   c) how many neutrons it has
   d) the mass of its subatomic particles

Chapter 3
Biological Molecules Great and Small

KEY CONCEPTS

1. Other than water, almost all molecules found in living organisms are made of carbon chains. All such compounds are called organic compounds.
2. All living organisms are made of the same few small molecules.
3. A small number of functional groups determine the properties of the many diverse biological molecules.
4. Lipids are oily substances that are usually insoluble in water (hydrophobic) although some are amphipathic (such as fatty acids). Lipids are high in energy. They include fats, oils, phospholipids, steroids, and cholesterol.
5. Sugars are the building blocks of carbohydrates. They contain many hydroxyl groups, typically having a hydrogen to oxygen ratio of 2:1.
6. Nucleotides are the building blocks of DNA, which carries our genes; RNA, which is involved in protein synthesis; and energy-storing molecules such as ATP. Nucleotides are composed of a sugar, phosphate group(s), and one of five nitrogenous bases.
7. Amino acids are the building blocks of proteins. They contain carboxyl groups, amino groups, and a side chain referred to as the R group.
8. Small molecules usually link small organic molecules together by removing the equivalent of one water molecule. This process is a dehydration synthesis, or condensation, reaction. The reverse process, which breaks larger molecules into smaller ones, adds water and is called hydrolysis.
9. DNA and RNA are composed of long chains of nucleotides linked together by dehydration synthesis reactions.
10. Amino acids are linked together by peptide bonds to form polypeptide chains. One or more polypeptide chains fold into unique shapes to form proteins.
11. A protein's unique three-dimensional structure is determined by four different levels of organization:
    - primary—determined by the amino acid sequence;
    - secondary—determined by interactions between side chains on the amino acids in localized areas, or domains, of the polypeptide;
    - tertiary—determined by the interactions along the full length of the polypeptide; and
    - quaternary—determined by the folding together of multiple polypeptides to form a functional protein.

EXTENDED CHAPTER OUTLINE
HOW DO ORGANISMS BUILD BIOLOGICAL MOLECULES?

How big are biological molecules?

1. Cells mostly have very small or very large molecules, with few of intermediate size.
2. Macromolecules are large molecules that are made of small building-block molecules linked together in long chains.
3. With the exception of water, almost all molecules found in living organisms are made of chains of carbon atoms.
4. Organic compound refers to all carbon- and hydrogen-containing compounds.
5. Organic chemistry is the study of carbon-and-hydrogen compounds. Biochemistry is a somewhat broader field—the study of the chemistry of life.
6. The carbon-based building blocks are simple and universal.
7. Most biological structures are made from only about 35 small molecules.
8. The small building-block molecules are generally categorized as sugars, amino acids, nucleotides, and lipids.
9. Sugars build macromolecules called polysaccharides; amino acids build proteins; and nucleotides build nucleic acids. Lipids do not form long polymers, but they do form sheets.

Why are biological structures made from so few building blocks?

1. All living organisms are made of the same few small molecules.
2. Every organism on Earth uses the same 20 amino acids and almost all use glucose as their main source of energy.
3. The small number of molecules in organisms supports the idea that life on our planet shares a common origin—that we are all related.

What kinds of structures do carbon atoms form?

1. Each carbon atom can bind with up to four other atoms.
2. A hydrocarbon chain is a chain of carbon atoms with hydrogen atoms attached to each carbon.
3. Hydrocarbon chains can also form rings. Some are flat and are called aromatic compounds.
4. Some carbon rings can fold into odd shapes instead of lying flat.
5. Carbon atoms typically form either long chains or five- or six-carbon rings.

Functional groups: How do small molecules differ from one another?

1. A molecule's structure determines its biological function.
2. Small groupings of atoms—functional groups—play the most important role in determining a molecule's chemical properties. A molecule may have few or many functional groups.
3. The hydroxyl group ($-OH^-$) is a very common functional group. Because it is polar, it readily forms hydrogen bonds. Compounds with hydroxyl groups tend to be very hydrophilic and dissolve easily in water.
4. The carboxyl group ($-COOH$) is common in organic acids because it readily releases its hydrogen, leaving behind a COO^- ion.
5. The amino group ($-NH_2$) makes a molecule basic because it tends to pick up hydrogen ions, forming NH_3^+ ions. Amines are a class of compounds that all have amino groups and almost all smell somewhat to very unpleasant.
6. A very small number of functional groups determine the properties of a wide variety of biological compounds.

3. The characteristic that most determines a polypeptide's folding is hydrophobic interactions of nonpolar side chains.
4. Each protein usually maintains its characteristic shape, but small changes in the environment around a protein can alter its shape and thus its function.

Are the interactions among amino acids enough to determine protein structure?

1. Despite all the possible ways an individual protein could fold, proteins quite reliably fold to their "native" functional shape.
2. Until recently, it has been generally believed that in a normal cell environment, amino acid sequence alone is enough to determine the structure of a protein.

What factors determine how proteins fold?

1. Several obstacles are present that could prevent normal protein folding.
2. Promiscuous interactions refer to bonding that occurs when individual polypeptides bind to other polypeptides instead of to themselves.
3. Chaperones are a group of enzymes and special proteins that bind to an emerging polypeptide to block promiscuous interactions so it can fold correctly. They also help proteins maintain their shapes and change shapes to take on different roles.

VOCABULARY BUILDING

In your own words, first write a brief definition, then a full explanation for each of the following terms. Include examples where appropriate. Complete this section from your memory—you will not learn the terms by simply copying definitions from the textbook. Once you have finished, check your responses against the information in the chapter and make any necessary corrections.

macromolecule —

organic —

organic chemistry —

biochemistry —

hydrocarbon chain —

aromatic compound —

functional group —

hydroxyl (OH) group —

carboxyl (COOH) group —

amino (NH_2) group —

lipid —

sugar —

amino acid —

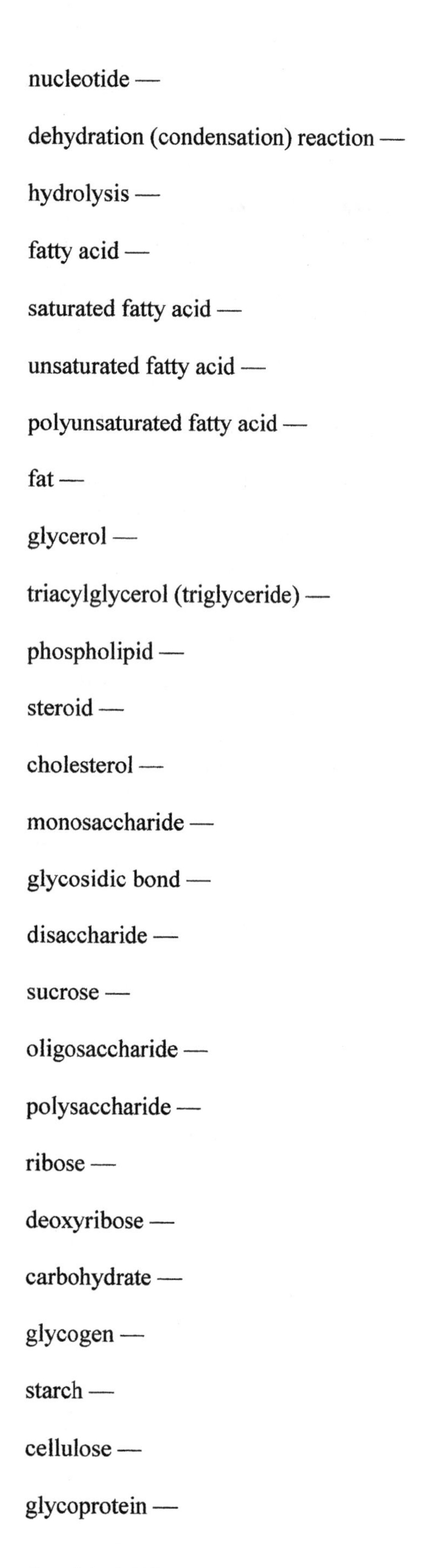

nucleotide —

dehydration (condensation) reaction —

hydrolysis —

fatty acid —

saturated fatty acid —

unsaturated fatty acid —

polyunsaturated fatty acid —

fat —

glycerol —

triacylglycerol (triglyceride) —

phospholipid —

steroid —

cholesterol —

monosaccharide —

glycosidic bond —

disaccharide —

sucrose —

oligosaccharide —

polysaccharide —

ribose —

deoxyribose —

carbohydrate —

glycogen —

starch —

cellulose —

glycoprotein —

Part 3: Short Answer

Write your answers in the space provided or on a separate piece of paper.

1. What are the four major categories of organic macromolecules? Provide examples.

2. Explain the general structure of an amino acid.

3. Explain the general structure of a nucleotide.

4. What is the structure of a phospholipid and how does that effect how a phospholipid molecule behaves in water?

5. Why do most carbohydrates dissolve easily in water?

6. Explain the process of dehydration reactions.

7. What kind of compound is cellulose? Why don't most organisms use it?

8. List five different proteins and their general functions.

9. What type of change in the structure of a protein is likely to have the greatest impact on the protein's function?

10. Most organic compounds have a hydrocarbon backbone, so what is mostly responsible for their diverse properties?

11. Explain the process of hydrolysis.

12. Explain the different functional groups found in organic molecules and understand how these groups determine the chemical properties of the molecules in which they are found.

13. What kind of compound is starch?

14. What functional group promotes hydrogen bond formation?

15. How do fatty acids interact with water?

Part 4: **Critical Thinking: Using Your Knowledge**

Answer each of these in essay form, using complete sentences and paragraphs. Provide as much information as you can. (For extra essay practice, write out answers to the Review and Thought Questions in your textbook.)

1. Start at the level of an amino acid and build a protein. Then explain the four levels of structural organization in a protein.

2. Explain saturated and unsaturated fats. What are their structural similarities and differences? What are their characteristics and where are they found? What aspects of their chemical structure account for their different effects on humans' cardiovascular health?

3. How can only 35 small building blocks make almost all of the molecules needed to sustain life? How does this support the idea of a common origin for all life?

4. How does the simplicity of the organic molecules of life aid homeostasis?

5. What are the advantages of having glucose as the universal energy source?

6. Most differences between any two individuals are a result of differences in their genes. Knowing that genes carry the instructions for making proteins, explain how proteins alone can account for all the variation seen between two people.

7. The cell membrane is composed primarily of two layers of phospholipid molecules. How must these be arranged? Based on your answer, now explain what types of molecules you think might pass directly through the cell membrane and which could not.

8. Although most animals cannot digest cellulose, cows and termites can, but the actual digestion is done by microorganisms that live in the cows' and termites' digestive system. How can this relationship benefit the cow or termite? How does it benefit the microorganism? Chemically, why is this relationship possible?

9. Proteins may begin to denature as temperatures climb over 105°. Knowing what you do about protein functions, explain why a very high fever can be harmful or even fatal.

10. If you are a gardener, you probably apply fertilizers that contain nitrogen and phosphorous. Specifically, why do plants need these nutrients, and how many ways can you think of in which each is used?

Chapter 4
Why Are All Organisms Made of Cells?

KEY CONCEPTS

1. All organisms are made of one or more cells.
2. Cells are the basic living unit of organization of all organisms.
3. All cells come from other cells.
4. Each cell is enclosed by a plasma membrane and keeps its genetic information (DNA) within a nucleus (eukaryotes) or nucleoid (prokaryotes). Eukaryotes have other organelles in their cytoplasm.
5. Cell size is limited because as a cell gets larger, its surface-to-volume ratio decreases, impairing the cell's ability to adequately carry out all vital processes, such as absorbing nutrients and excreting wastes.
6. Cellular organization allows a division of labor among specialized cells within the organism and allows the organism to outlive most of the cells from which it is composed.
7. Eukaryotic cells contain a variety of membrane-bound organelles that have specialized functions. Prokaryotic cells contain no internal membranes.
8. Organisms from all six kingdoms share common characteristics but have distinctive features.
9. The plasma membrane consists of a phospholipid bilayer that has proteins located throughout and between the layers, and carbohydrates on the outer layer only.
10. Simple diffusion is the process by which substances move from an area of high concentration to an area of low concentration.
11. Osmosis is the movement of water across a selectively permeable membrane in response to a concentration gradient. The pressure exerted is called osmotic pressure.
12. Tonicity refers to the concentration of solutes on one side of a selectively permeable membrane compared to the solute concentration on the other side. A hypertonic solution has a higher solute concentration, a hypotonic solution has a lower solute concentration, and isotonic solutions have the same solute concentrations.
13. Molecules pass through plasma membranes by passive transport, facilitated diffusion, and active transport. Active transport uses energy to move a substance against a concentration or charge gradient.
14. Plasma membranes can fuse easily and spontaneously form small sacs called vesicles, which can be used to take in or expel materials.
15. Cells have special mechanisms that allow them to communicate with each other, such as chemical signals, gap junctions, adhering junctions, tight junctions, and plasmodesmata.

EXTENDED CHAPTER OUTLINE

WHY ARE ALL ORGANISMS MADE OF CELLS?

All organisms are made of cells

1. All organisms are composed of cells.
2. All cells are alive and are the basic living unit of organization of all organisms.
3. New cells come from division of preexisting cells.

Every cell consists of a boundary, a cell body, and a set of genes

1. A plasma membrane is a cell's boundary, defining the cell's limits and regulating the inside.
2. The genetic information for each cell is stored in DNA molecules.
3. Eukaryotes house DNA inside the membrane-bound nucleus. In prokaryotes, the DNA is in a particular region of the cell called the nucleoid, which is not surrounded by a membrane.
4. In eukaryotic cells, outside of the nucleus are several organelles that have specialized functions. Most of these are membrane-bound structures.
5. Prokaryotes do not have organelles because they have no internal membranes.
6. The cytoplasm is the cell body, which is all the cellular material outside of the nucleus but inside of the plasma membrane.
7. Cytosol is the part of the cytoplasm that is not contained within membrane-bound organelles.
8. The cytoskeleton is a complex network of protein fibers throughout the cytosol that helps give the cell its shape, holds organelles in place, and participates in cell movement.

How are cells alive?

1. Cells are the fundamental living units of life, and they have all the characteristics of life discussed in Chapter 1. They:
   - are made of organized parts,
   - perform chemical reactions,
   - obtain energy from their surroundings,
   - respond to their environments,
   - change over time,
   - reproduce, and
   - share an evolutionary history.

What are the advantages of cellular organization?

1. Subdivision into tiny cells has many advantages.
2. Cells must maintain a relatively constant internal environment (homeostasis).
3. The plasma membrane's ability to admit or excrete molecules is limited by its size.
4. As cells get bigger, their surface-to-volume ratio decreases. They have proportionately less membrane and cannot absorb nutrients or expel wastes adequately.
5. Many eukaryotic cells have convoluted surface membranes and almost all have elaborate internal membrane systems.
6. In multicellular organisms, individual cells often perform separate and specialized tasks, allowing a division of labor among numerous cells.
7. In multicellular organisms, cells often live and die independently of the whole organism.
8. Multicellular organisms' cells are continuously replaced—the organism outlives its cells.

WHAT'S IN A CELL?

1. Vesicles, in both plant and animal cells, and vacuoles in plant cells are membranous saclike structures with no apparent internal organization.

What role does the nucleus play in the life of a cell?

1. In cells that are not dividing, DNA forms a diffuse mass within the nucleus, called chromatin—a DNA and protein complex.
2. In dividing cells, the DNA forms distinct rodlike structures called chromosomes.
3. The nucleus is enclosed by the *nuclear envelope,* which is a double membrane.
4. The two layers of the nuclear envelope come together periodically and form nuclear pores, which form channels connecting the nucleoplasm (nuclear contents) and the cytoplasm.
5. The nucleus' main function is storing the DNA—the genetic information that is passed on.
6. The DNA contains the instructions for building every polypeptide the body will ever use.

The cytosol is the cytoplasm that lies outside the organelles

1. Although the cytosol is aqueous, about 20% of its weight is from proteins, giving it a consistency similar to gelatin.
2. Cytosol contains organelles, vesicles, vacuoles, and structures not contained in membranes.
3. Ribosomes are complexes of RNA and protein. They provide a platform on which proteins are built. Ribosomes may be free in the cytosol or "membrane-bound," meaning attached to the cell's internal membranes.
4. Proteasomes, themselves made of protein, degrade old proteins and recycle their amino acids.

The endoplasmic reticulum is a folded membrane

1. The endoplasmic reticulum (ER) is a single sheet of membrane enclosing a complex network of cavities and channels (the lumen) that are all interconnected.
2. The ER membrane is continuous with the outer membrane of the nucleus.
3. Rough endoplasmic reticulum (rough ER) is studded with ribosomes and is mostly devoted to modifying recently made proteins, especially ones being exported to organelles or out of the cell.
4. Smooth endoplasmic reticulum (smooth ER) has no ribosomes and is involved primarily with the synthesis and metabolism of lipids. It also detoxifies some substances, such as alcohol.
5. Most eukaryotic cells contain both rough ER and smooth ER in varying amounts, depending on the functions of the cell.

The Golgi complex directs the flow of newly made proteins

1. The Golgi complex is usually near the nucleus and composed of stacks of flattened membranous disks.
2. Most proteins exported out of cells are glycoproteins.
3. The Golgi complex modifies glycoproteins, which are proteins with carbohydrates attached. The Golgi complex packages them in membrane-bound vesicles, then "labels" them by modifying the carbohydrate.
4. The labels added by the Golgi complex help direct the glycoproteins so they will either be sent to other parts of the cell or expelled from the cell.

The lysosomes function as digestive vats

1. Lysosomes are a specialized type of vesicle found in all eukaryotic cells.
2. Lysosomes contain enzymes that break down proteins, nucleic acids, sugars, lipids, and other complex molecules.
3. The large vacuole in a plant cell is essentially a giant lysosome.
4. Lysosomes are especially numerous in cells that perform phagocytosis, the process by which cells take in and consume large particles of solid food.
5. A lysosome's membrane prevents it from digesting the cell's own structures.

Peroxisomes produce peroxide and metabolize small organic molecules

1. Peroxisomes are also membrane-bound vesicles.
2. Peroxisomes contain enzymes that use oxygen to break down various compounds through reactions that produce hydrogen peroxide (H_2O_2).
3. Hydrogen peroxide is toxic to cells, so peroxisomes also contain an enzyme called catalase that breaks down the peroxide once it is formed, protecting the cell.

Mitochondria capture the energy from small organic molecules in the form of ATP

1. Mitochondria are the eukaryotic cell's powerhouses—in cells that do not get energy directly from sunlight, mitochondria produce nearly all of the ATP the cells need to do their work, such as chemical reactions.
2. Mitochondria are among the most numerous organelles in eukaryotic cells.
3. The mitochondrion has two membranes. The outer layer is smooth, but the inner layer is highly folded into inner partitions called cristae.
4. Mitochondria have their own DNA and make some of their own proteins.
5. In the evolutionary process, mitochondria are thought to have arisen by incorporation of bacteria into the cells.

A plant cell's chloroplast is just one kind of plastid

1. Plastids, found in nearly all plant cells, are also enclosed by double membranes.
2. Chloroplasts perform photosynthesis—the process through which organisms capture energy from sunlight and use it to build energy-rich sugar molecules.
3. Sugars are synthesized and temporarily stored in the chloroplasts.
4. Chloroplasts are large, rounded, and green. The color comes from the pigment chlorophyll.
5. Chloroplasts have separate internal membranes arranged in flattened disklike sacs called thylakoids. Thylakoids are stacked into numerous structures called grana.
6. Chromoplasts arise from chloroplasts and contain pigments that produce yellow, orange, and red colors in plants.
7. Amyloplasts store starches.

What does the cytoskeleton do?

1. The cytoskeleton gives structure and support to the cytoplasm and its contents, and provides the force for most cell movements, such as transporting materials and changing cell shape.
2. The cytoskeleton is made of at least three types of protein filaments: microtubules, actin filaments, and intermediate filaments.
3. Microtubules are hollow cylinders that originate in structures called microtubule organizing centers (MTOCs).
4. Microtubules are especially important in moving the DNA during cell division.
5. In most eukaryotic cells, the major MTOCs are near the nucleus, in the centrosome—a region that contains an organelle called the centriole. This region has a major role in cell division.
6. Microtubules are made of two globular protein molecules called tubulins, but up to 50 other proteins, called microtubule-associated proteins, or MAPs, may be associated with them.
7. Some MAPs act as motors that use ATP to move subcellular structures along the microtubule tracks. Dyneins move away from the MTOC; kinesins move toward it.
8. Actin filaments (microfilaments) provide the main force for movement and shape changes. Actin is a protein also found in muscles.
9. Intermediate filaments are fibrous proteins, such as keratin, and provide strength in areas of cells that are subjected to mechanical stress.

WHAT DO MEMBRANES DO?

1. Membranes are essential boundaries that separate the inside of a cell from its outside.
2. Membranes regulate the contents of the spaces they enclose.
3. Membranes serve as a "workbench" for biochemical reactions.
4. Membranes participate in energy conversions.

What kinds of molecules do membranes contain?

1. Plasma membranes of all cells look the same.
2. Plasma membranes consist mostly of lipids and proteins, with some carbohydrates.
3. Phospholipids are the most important lipids in plasma membranes. Due to their amphipathic nature, they arrange in two sheets, each with its hydrophobic tails pointed toward the inside of the membrane—away from the water—and its hydrophilic head pointing outward.
4. Proteins are located in both layers of the membrane and in the spaces between the layers.
5. Membrane proteins are also amphipathic; their hydrophobic parts interact with hydrophobic parts of the membrane and the hydrophilic areas interact with hydrophilic parts of the membrane or with the aqueous solutions inside or outside of the cell.

How does the stucture of a membrane establish its functional properties?

1. The basic structure of the plasma membrane is a lipid bilayer, with two sheets of phospholipids arranged tail-to-tail.
2. Proteins are dispersed throughout the membrane, contributing to the membrane's structure and function.
3. Some proteins span the membrane, while others are confined to the inner or outer surface.
4. The membrane is fluid—both protein and lipid molecules move freely.
5. The lipid bilayer serves as a hydrophobic barrier that limits movement of hydrophilic molecules across the membrane.
6. Some membrane proteins help transport specific molecules across the membrane.

HOW DO MEMBRANES REGULATE THE SPACES THEY ENCLOSE?

1. Plasma membranes are *selectively permeable*, meaning some substances (especially water) are allowed to pass through while others are not.

Why does water move across membranes?

1. Concentration refers to the number of molecules of a substance in a given volume.
2. Simple diffusion is the random movement of molecules or ions from an area of higher concentration to an area of lower concentration.
3. A concentration gradient exists whenever there is a difference in the concentration of a certain substance in two adjacent areas.
4. The rate at which diffusion occurs depends on different factors, such as:
   - temperature—heat increases molecular motion which increases the rate of diffusion;
   - size—smaller molecules generally diffuse faster than larger ones; and
   - concentration gradient—a greater concentration difference hastens diffusion.
5. Diffusion occurs through open solutions and across membranes.
6. Osmosis is the movement of water across any selectively permeable membrane in response to a concentration gradient.
7. The pressure exerted by osmosis is called osmotic pressure.
8. Solutions can be evaluated in terms of their concentrations. The solution with the higher solute concentration is hypertonic, and the solution with the lower solute concentration is hypotonic. If two solutions have the same solute concentration, they are isotonic.
9. By osmosis, when a selectively permeable membrane separates two solutions that have different concentrations, water flows in the direction that tends to equalize the solute concentration on each side of the membrane.

10. Plant cells are contained within rigid cell walls. Plasmolysis occurs when water moves out of plant cells, causing the plasma membrane to separate from the cell wall.
11. The rigid cell wall limits movement of water into a plant cell because it restricts cell expansion. The pressure of the contents of a plant cell against the cell wall is called turgor pressure. Turgor pressure supports nonwoody plants. Decreased turgor pressure, when the plant cells lose water, causes these plants to droop.

What determines the movement of molecules through a selectively permeable membrane?

1. Passive transport occurs spontaneously and does not require energy.
2. Facilitated diffusion, which is an increased rate of passive transport, depends on specific molecules in the membrane that aid the process. These transport molecules bind to the molecule to be moved and carry it across the membrane.
3. Active transport moves molecules against their concentration gradient and requires energy.
4. The sodium-potassium pump is a common example of active transport and it is coupled to ATP, which provides the energy.
5. Another type of active transport—cotransport—couples the transport of two substances and depends on special transmembrane proteins called cotransporters that transport the two molecules or ions. One moved substance or ion provides energy by moving along its concentration gradient, and that energy is used to move the other substance against its concentration gradient.

HOW DO MEMBRANES INTERACT WITH THE EXTERNAL ENVIRONMENT?

1. Cell walls, which are rigid external structures made of carbohydrate, surround the plasma membranes of most nonanimal cells.
2. The cell wall can continue to function after the cell inside it dies.
3. Animal cells may be surrounded by a network of various molecules, especially carbohydrates and proteins, collectively referred to as the extracellular matrix.

Membrane fusion allows the import and export of particles and bits of extracellular fluid

1. Biological membranes can quickly change shape.
2. Plasma membranes can extend outward and surround materials, then move them into the cell, forming vesicles around the material. Phagocytosis is such a process, by which relatively large particles are engulfed.
3. Endocytosis is basically the same process, but the cell takes in tiny amounts of materials by inward foldings of the membrane that form the vesicles. Pinocytosis is "cell drinking" in which bits of liquid and dissolved substances are taken in. Receptor-mediated endocytosis takes in only specific substances that bind to special receptor proteins on the cell surface.
4. Exocytosis is how the cell exports molecules and is essentially pinocytosis in reverse.
5. Regulation of exocytosis is an especially important cell function. For example, release of neurotransmitter from nerve cells is done by exocytosis.

How do cells communicate in multicellular organisms?

1. Cells can communicate with each other by using chemical signals such as hormones. Such chemical signals bind to specific receptor proteins on the cell surface or within the cell.
2. All living cells of flowering and higher plants are in intimate contact with each other through fine intercellular channels called plasmodesmata, which allow molecules to pass directly between cells.
3. Some animal cells communicate via gap junctions, where the plasma membrane of adjacent cells is very close and a protein called connexin allows exchange of chemical signals.
4. Adhering junctions usually consist of desmosomes, which are buttonlike structures physically holding two plasma membranes together. These provide strength.

5. Impermeable junctions weld cells together and prevent molecules from leaking out of the cells. A common type of impermeable junction is the tight junction, in which case the plasma membranes of adjacent cells actually fuse.
6. Most intercellular communication is done by chemical signals, such as hormones, which bind specifically to receptors on or in the cells with which they are communicating.

VOCABULARY BUILDING

In your own words, first write a brief definition, then a full explanation for each of the following terms. Include examples where appropriate. Complete this section from your memory—you will not learn the terms by simply copying definitions from the textbook. Once you have finished, check your responses against the information in the chapter and make any necessary corrections.

protoplasm —

cell theory —

plasma membrane —

nucleus —

nucleoid —

organelles —

cytoplasm —

cytosol —

cytoskeleton —

surface-to-volume ratio —

vesicles —

vacuoles —

chromatin —

chromosomes —

nuclear envelope —

nuclear pores —

ribosomes —

proteasomes —

endoplasmic reticulum (ER) —

lumen —

rough ER —

smooth ER —

Golgi complex —

glycoproteins —

lysosomes —

phagocytosis —

peroxisomes —

mitochondria —

plastids —

chloroplast —

photosynthesis —

thylakoids —

granum —

chromoplasts —

amyloplasts —

microtubules —

microtubule organizing centers (MTOCs) —

centrosome —

centriole —

tubulins —

microtubule-associated programs (MAPs) —

motors —

actin filaments —

microfilaments —

intermediate filaments —

flagellum —

cilia —

fluid —

fluid mosaic model —

selectively permeable —

concentration —

simple diffusion —

concentration gradient —

osmosis —

osmotic pressure —

hypotonic —

isotonic —

hypertonic —

cell wall —

plasmolysis —

turgor pressure —

passive transport —

active transport —

sodium-potassium pump —

facilitated diffusion —

cotransport —

extracellular matrix —

endocytosis —

pinocytosis —

receptor-mediated endocytosis —

exocytosis —

plasmodesmata —

gap junctions —

adhering junctions —

desmosomes —

epithelial cells —

epithelium —

tight junctions —

receptors —

CHAPTER TEST

The following test has four parts. Complete as much of the exam as you can from memory. If you cannot answer a question, skip it. Once you complete all that you can, try to answer any questions you skipped. If you still cannot answer them, consult your textbook for the answers. Once you have completed all sections of the test, check your answers for Parts 1 - 3 against those in the back of this book. Highlight any incorrect answers then review that material in your textbook. Correct your answers for future reference.

Part 1: Multiple Choice

For each of the following, select all correct responses—more than one may be correct.

1. Which of the following organelles is primarily involved in protein synthesis?
   a) mitochondrion
   b) ribosome
   c) lysosome
   d) peroxisome

2. Which of the organelles is considered the "powerhouse" of the cell, producing ATP?
   a) mitochondrion
   b) ribosome
   c) endoplasmic reticulum
   d) Golgi complex

3. In eukaryotic cells, DNA is located within the:
   a) cytoplasm
   b) cytosol
   c) nucleus
   d) vacuole

4. The plasma membrane contains which of the following?
   a) lipids
   b) carbohydrates
   c) nucleic acids
   d) proteins

5. Passive movement of substances from an area of high concentration to an area of low concentration is called:
   a) active transport
   b) simple diffusion
   c) osmosis
   d) exocytosis

6. Movement of water through a selectively permeable membrane and along its concentration gradient is called:
   a) active transport
   b) simple diffusion
   c) osmosis
   d) exocytosis

7. Facilitated diffusion requires:
   a) a concentration gradient
   b) energy
   c) transport proteins
   d) ATP

8. When comparing two solutions, the one with the highest solute concentration is:
   a) hypotonic
   b) isotonic
   c) hypertonic
   d) none of these

9. Which of the following is true of active transport?
   a) substances move from an area of low to high concentration
   b) energy is required
   c) special molecules in the membrane
   d) water is moved through the membrane

10. What is a common energy source used *directly* to drive cellular processes?
    a) lipids
    b) ATP
    c) sunlight
    d) starch

11. During which of the following processes do cells take in small bits of fluid?
    a) exocytosis
    b) phagocytosis
    c) pinocytosis
    d) receptor-mediated endocytosis

12. The process by which substances are exported from the cell is called:
    a) exocytosis
    b) phagocytosis
    c) pinocytosis
    d) receptor-mediated endocytosis

13. Most receptors in cell membranes are:
    a) proteins
    b) carbohydrates
    c) lipids
    d) ATP

14. An intercellular junction in which the plasma membranes of two adjacent cells actually fuse is the:
    a) gap junction
    b) adhering junction
    c) tight junction
    d) plasmodesmata

15. Which of the following provide the force for cell movements and shape changes?
    a) actin filaments (microfilaments)
    b) microtubules
    c) intermediate filaments, such as keratin
    d) plasma membrane

Part 2: Matching

For each of the following, match the correct term with its definition or example. More than one answer may be appropriate.

ribosome **Golgi complex** **rough ER** **smooth ER** **mitochondrion**

1. ____________________ Makes ATP.
2. ____________________ Site of protein synthesis.
3. ____________________ Attached to rough ER.
4. ____________________ Involved in synthesis and metabolism of lipids.
5. ____________________ Packages and labels glycoproteins.
6. ____________________ Modifies newly synthesized proteins.
7. ____________________ Detoxifies substances such as alcohol.

phagocytosis **pinocytosis** **exocytosis**

8. ____________________ This is "cell drinking."
9. ____________________ Plasma membrane extends to take in large molecules.
10. ____________________ Products made within the cell are wrapped in membranes then expelled from the cell.
11. ____________________ Plasma membrane folds inward, forming a vesicle around tiny droplets of liquid.

Part 3: Short Answer

Write your answers in the space provided or on a separate piece of paper.

1. Why is a plasma membrane arranged in a bilayer?

2. List three concepts included in the cell theory.

3. Why does rough endoplasmic reticulum have a rough appearance?

4. Which organelle contains strong digestive enzymes?

5. In general terms, what happens in a chloroplast?

6. List three functions of the cytoskeleton.

7. Why is the plasma membrane said to be selectively permeable?

8. What does the Golgi complex do?

9. List and briefly explain three processes by which substances are taken into a cell.

10. In dividing cells, complexes of DNA and protein form visible structures called what?

11. What specializations of the nuclear envelope allow rapid communication between the nucleoplasm and the cytoplasm?

12. What is the difference between a nucleus and a nucleoid?

13. List three common characteristics of all cells.

14. List three components of the cytoskeleton.

15. Where is ATP produced in the cell?

***Part 4:* Critical Thinking: Using Your Knowledge**

Answer each of these in essay form, using complete sentences and paragraphs. Provide as much information as you can. (For extra essay practice, write out answers to the Review and Thought Questions in your textbook.)

1. Explain numerous ways in which all cells are similar.

2. How do eukaryotic and prokaryotic cells differ? What about plant and animal cells?

3. In the last chapter, we saw that in all organisms, very complex organization and function is built from the same universally simple organization—just a few building molecules. In what ways is this same theme apparent in cell structure and function?

4. Discuss some of the advantages of cellular organization for multicellular organisms.

5. Describe the structural and functional organization of a eukaryotic cell. Next, discuss ways in which individual cells function as if they are independent organisms. Finally, consider parallels in the functioning of a human body and an individual cell (for example, what acts as the brain, what acts as the digestive system, etc.?).

6. Discuss the role of the cell membrane in maintaining homeostasis in the cell.

7. What do you predict would happen if a hospital patient was administered a hypertonic solution through an IV for an extended period, and why?

8. Dialysis machines work by passing the patient's blood through a system of dialysis tubing, which is a selectively permeable membrane, that is surrounded by dialysis fluid. Based on your knowledge of movements through membranes, how are substances moved out of the blood into the dialysis fluid? Is this fluid isotonic, hypotonic, or hypertonic to blood?

9. Discuss ways in which cells communicate with adjacent cells and with distant cells. How does the type of junction between adjacent cells affect their communication abilities?

10. Discuss all organelles and processes involved in the production, refinement, packaging, and exporting of a protein molecule that is destined for use in another cell.

Chapter 5
Directions and Rates of Biochemical Processes

KEY CONCEPTS

1. The laws of thermodynamics predict the direction of an energy transformation (such as a chemical reaction).
2. The laws of kinetics predict the rate of an energy transformation (such as a chemical reaction).
3. Work is the movement of an object against a force.
4. Potential energy is stored energy, whereas kinetic energy is the energy of moving objects.
5. ATP has potential energy.
6. The First Law of Thermodynamics states that the total amount of energy in any process remains constant—it can change form but can neither be created nor be destroyed.
7. The Second Law of Thermodynamics states that in any process, the amount of energy available to do work decreases, such as when heat is lost.
8. Exergonic processes release energy and occur spontaneously, whereas endergonic reactions consume energy and do not occur spontaneously. Rearrangement of chemical bonds between atoms may consume or release energy.
9. With the help of enzymes, an exergonic reaction, such as the splitting of ATP, can be coupled with an endergonic reaction, providing the energy needed to drive this reaction.
10. The equilibrium point of a reaction depends on the concentration of the reactants and products, and also on the free energy in the bonds between the atoms of each molecule.
11. Molecules move randomly and with varying energies. The random bouncing of molecules helps explain chemical equilibrium.
12. Enzymes hasten chemical reactions by lowering the activation energy of the reaction.
13. Enzymes lower the activation energy for a reaction by:
    - orienting substrate molecules to increase the chances of their interaction,
    - straining the covalent bonds within the substrate molecules, or
    - participating directly in the chemical reactions.
14. The rates of enzymatic reactions increase with temperature, but only to a certain point. Most enzymes also work within a narrow pH range.
15. Enzyme effectors may increase or inhibit enzyme activity.
16. Enzyme inhibitors, examples of effectors, prevent an enzyme from catalyzing a reaction by:
    - competing with the substrate for the active site, or
    - altering the shape of the enzyme.

EXTENDED CHAPTER OUTLINE

Ludwig Boltzmann: A man left behind or a man ahead of his time?

1. Nineteenth-century positivists asserted that only observable facts provided true knowledge.
2. Thermodynamics is the study of the relationships between different forms of energy.
3. Every event, reaction, or process is accompanied by a transformation of energy.
4. Kinetics is the study of the rates of reactions.
5. Thermodynamics tells us which way a reaction will go; kinetics tells us how fast it will go.

WHAT DETERMINES WHICH WAY A REACTION PROCEEDS?

1. In any process, energy changes form.
2. Thermodynamics has established rules that govern all energy changes.

How may work be converted to kinetic and potential energy?

1. Work is the movement of an object against a force.
2. Work can be stored as potential energy.
3. Kinetic energy is the energy of moving objects.
4. Moving molecules have kinetic energy in the form of heat.
5. Rearranging atoms in a chemical reaction is a form of work.
6. ATP stores potential energy that can be released later to do work.

How does thermodynamics predict the direction of a reaction?

1. In any process, energy is neither created nor destroyed—it only changes form.
2. The First Law of Thermodynamics states that the total amount of energy in any process stays constant.
3. Energy may switch from potential energy to kinetic energy and back.
4. Any heat generated during a process is a form of kinetic energy. Heat is the rapid and random movement of atoms and molecules.
5. The Second Law of Thermodynamics states that, in the course of any process, the energy available for work decreases.
6. An ATP molecule gives up part of its stored energy by splitting off one of its three phosphate groups. The resultant ADP still has potential energy, but less than the ATP had.
7. Heat, which results from the random motion of molecules, is largely wasted energy.
8. The calorie is the unit usually used to measure energy. One calorie equals the amount of energy needed to raise the temperature of one gram of water by 1°C.
9. Food energy is typically measured in Calories (with a capital C). One Calorie is equal to 1000 calories, which equals 1 kilocalorie.
10. Each time energy changes form, a little more energy turns to heat, which usually cannot be used to do more work.
11. In general, formation of a chemical bond between any two atoms releases energy. Breaking a chemical bond usually consumes energy. The amount of energy released varies.

How do changes in free energy predict the direction of a reaction?

1. Free energy is the energy in a system available for doing work. It is abbreviated G.
2. Each chemical has a certain amount of free energy, and each chemical reaction involves a "change in free energy."
3. The change in free energy is abbreviated as ΔG. It is calculated by the following equation:

$$\Delta G = G_{\text{products}} - G_{\text{reactants}}$$

4. Exergonic processes release energy, so the free energy decreases ($-\Delta G$).
5. Endergonic processes consume energy, so their free energy increases and they have a $+\Delta G$.

6. Exergonic processes occur spontaneously; endergonic processes do not—they require energy from another source.
7. In any process, the change in free energy depends only on the initial and final states.

What is the source of the free energy released or consumed during a reaction?

1. Energy is released when two isolated atoms form a chemical bond. The tighter the bond, the more energy released.
2. High-energy bonds, such as in ATP, are unstable bonds that give up energy to form stable, low-energy bonds. High-energy bonds are often represented by a "squiggle" (~).
3. The second and third phosphates of ATP are held by high-energy bonds.
4. The ΔG for conversion of ATP to ADP and phosphate is -7.3kcal/mole. (One mole = 6×10^{23} molecules.)
5. Bonds between atoms in a molecule have characteristic energies. Rearrangement of these bonds may consume or release energy.

How can one process provide the energy for another?

1. When one process (exergonic) provides the energy to drive a second process (endergonic), the two processes are said to be coupled.
2. Coupled chemical reactions together must be exergonic.
3. For almost all biological processes, the driving exergonic reaction is the splitting of ATP.
4. Biochemical reactions are usually coupled by enzymes, which are large protein molecules.
5. Enzymes do not effect the ΔG of a reaction, but they speed up the reaction.

How do concentration and entropy affect equilibrium?

1. The amount of disorder in the universe tends to increase.
2. Entropy is a formal measure of disorder. Entropy has a high value when objects are disordered (randomly arranged) and a low value when they are organized.
3. The tendency of entropy to increase can be used to do work.
4. Any process that converts an orderly arrangement to a less orderly one can perform work, but any process that converts a disorderly arrangement into an orderly one consumes energy.
5. Free energy—energy available to do work—depends on the concentration of the molecules.
6. The greater the concentration difference between two sides of a barrier, such as a cell membrane, the more work that can be done.
7. Whether a reaction occurs depends not only on the energy in the individual bonds, but also on the concentration of both the reactants and the products.
8. When the free energy of the reactants equals the free energy of the products, equilibrium is reached, and no further net change will occur.
9. The equilibrium of a reaction depends on:
   - the free energy in the bonds between the atoms of each molecule, and
   - on the concentration of the reactant and products.

WHAT DETERMINES THE RATE OF A CHEMICAL REACTION?

How does molecular motion help explain reaction rates?

1. Molecules and atoms are constantly in motion.
2. When a substance absorbs heat, its temperature increases and the atoms and molecules move around more rapidly.
3. Kinetics—based on the random movements of atoms—allows us to predict the rates of chemical reactions, and to understand how temperature and other environmental factors can alter these rates.

What stops a chemical reaction?

1. Temperature is a measure of the average kinetic energy of moving molecules. It is only an average value, so the molecules may have a range of kinetic energies.
2. Molecules move in the direction of lower free energy.
3. The random bouncing of molecules helps explain chemical equilibrium—molecules keep moving about until the free energy of the reactants and products becomes equal.

What starts a chemical reaction?

1. Every process must be started in the right direction.
2. Energy for reactions between molecules comes from random movements of molecules.
3. The minimum energy needed for a process to occur is called the activation energy.
4. The number of molecules in a space that can react depends on both the temperature and the amount of activation energy needed.
5. A catalyst is a substance that lowers the activation energy of a reaction but is not used up in the reaction.
6. A catalyst, by definition, is not changed by the reaction that it speeds up.
7. An enzyme is a biological catalyst, and all enzymes are proteins.
8. Speeding up reactions is essential to life.
9. Cells control their chemistry by regulating the production and activity of individual enzymes.

HOW DO ENZYMES WORK?

How does an enzyme bind to a reactant?

1. Enzymes speed up specific chemical reactions by binding to the reacting molecules, called substrates.
2. Enzymes have enormous catalytic power—the appropriate enzyme can make a chemical reaction proceed a million times faster than it could without an enzyme.
3. Enzymes are highly specific—each usually catalyzes only a single reaction.
4. Every enzyme has an active site—a groove or cleft on its surface by which it binds to the substrate.
5. The binding of the substrate and enzyme lowers the activation energy for a particular chemical reaction.
6. Although different enzymes may be able to bind to the same substrate, each enzyme lowers the activation energy for a different and specific reaction.
7. The substrate and enzyme bind temporarily by weak noncovalent bonds, such as hydrogen bonds, ionic bonds, hydrophobic interactions, and van der Waals attractions.
8. When the shape of the enzyme changes during interaction with the substrate, it is called an induced fit.
9. The binding of the enzyme and substrate is reversible, and the enzymes are used repeatedly.
10. Interactions between enzymes and substrates occur one molecule at a time, so the rate of chemical reactions also depends on how many enzyme molecules are available.

How does an enzyme lower the activation energy of a chemical reaction?

1. Formation of noncovalent bonds between the enzyme and substrate provide the free energy needed to form a more ordered state.
2. The enzyme itself always remains unchanged after the reaction is completed.
3. Enzymes bring substrates together in a position that favors a reaction.
4. When an enzyme binds to its substrate, it may strain and distort the covalent bonds of the substrate. When this occurs, the substrate is in a transition state—intermediate to the reactant and product—but stays in this state very briefly, perhaps a billionth of a second.

8. The conversion of ATP to ADP is what kind of a reaction?
   a) exergonic
   b) endergonic
   c) both
   d) neither

9. Entropy is a measurement of:
   a) free energy
   b) disorder
   c) heat
   d) kinetic energy

10. The minimum amount of energy needed for a process to occur is called:
    a) activation energy
    b) free energy
    c) potential energy
    d) kinetic energy

11. Which of the following may affect an enzyme's activity?
    a) temperature
    b) pH
    c) presence of inhibitor molecules
    d) all of these

12. Reactions in which small molecules are joined to form larger molecules are referred to as:
    a) catabolism
    b) molabolism
    c) anabolism
    d) inhibition

13. Catabolic reactions are generally:
    a) endergonic
    b) exergonic
    c) either of these, depending on the enzymes involved
    d) neither endergonic nor exergonic

14. Which of the following is not true of steric inhibition?
    a) The inhibitor binds to the enzyme's active site.
    b) It can be overcome by increasing the substrate concentration.
    c) It involves noncompetitive inhibitors.
    d) The inhibitor is similar in size and shape to the substrate.

15. Which of the following is not true of a catalyst?
    a) It lowers the activation energy.
    b) It is not changed by the reaction.
    c) It slows the rate of the reaction to allow better control of the process.
    d) It does not participate directly in the reaction.

***Part 2:* Matching**

For each of the following, match the correct term with its definition or example. More than one answer may be appropriate.

kinetic energy **potential energy** **free energy**

1. ______________ Energy of moving objects.
2. ______________ Energy available to do work.
3. ______________ Stored energy.
4. ______________ Decreases in exergonic processes.
5. ______________ Heat is an example.
6. ______________ ATP has a lot of this.

exergonic reactions **endergonic reactions**

7. ______________ Energy-consuming.
8. ______________ Occurs spontaneously
9. ______________ Often coupled to reactions that can provide the needed energy.
10. ______________ $+\Delta G$
11. ______________ Conversion of ADP to ATP
12. ______________ Require energy from another source

***Part 3:* Short Answer**

Write your answers in the space provided or on a separate piece of paper.

1. What is the most common unit of measurement for energy?

2. How does heat loss during a process support the Second Law of Thermodynamics?

3. Heat is an example of what type of energy?

4. What is the equation for the change in free energy (ΔG)?

5. What aspects of a reaction do the laws of thermodynamics allow us to predict?

6. Why are many biochemical reactions coupled?

7. What is the main function of a catalyst, such as an enzyme?

8. In terms of entropy, what kind of processes can perform work?

9. Briefly explain chemical equilibrium.

10. How does an enzyme bind to its substrate?

11. List two environmental factors that affect the rate of a chemical reaction.

12. Explain how enzyme and substrate concentration affect a chemical reaction's rate.

13. List two ways in which enzymes can lower the activation energy of a reaction.

14. Most human enzymes work best within what pH range?

15. What is meant by "competitive inhibitor"?

Part 4: **Critical Thinking: Using Your Knowledge**

Answer each of these in essay form, using complete sentences and paragraphs. Provide as much information as you can. (For extra essay practice, write out answers to the Review and Thought Questions in your textbook.)

1. Differentiate between kinetic and potential energy. Explain how they are related and, using an example, how they can be interconverted.

2. Differentiate between exergonic and endergonic reactions. How does this relate to any shift in the equilibrium of the reaction? How does this relate to how spontaneous the reaction is? How does this relate to whether the reaction is anabolic or catabolic?

3. Why would you get hot if you ran a race? If you raced against a well-trained athlete, which of you would probably generate the most heat, and why?

4. What are the advantages to having biochemical processes occur in numerous small steps, instead of a single step? Based on what you've learned in earlier chapters, would you expect different organisms to use many of the same reactions, or would each organism have unique reactions?

5. Many biochemical reactions are reversible, meaning they can go in either direction. How is this possible?

6. Heat is a catalyst and is often used in chemical experiments to speed up reactions. Heat is also released from many biological processes. Yet, in many organisms, such as humans, heat is not used as a catalyst. Why not?

7. Why is homeostasis essential for proper enzyme function?

8. Describe various ways by which nonsubstrate molecules could enhance or impair enzyme function.

9. The amount of disorder (entropy) in the universe tends to increase, and yet living organisms remain organized, in some cases in a very complex way. How is this possible?

10. In what way is DNA related to enzyme function? In how many ways could DNA affect biochemical processes occurring in your body?

ELECTRON TRANSPORT: HOW DOES THE ENERGY IN GLUCOSE REACH ATP?

How does the flow of electrons from electron donors to oxygen release energy to the phosphate bonds in ATP?

1. NAD^+ and FAD are the most important electron acceptors in energy metabolism. They transfer electrons from glucose to oxygen in a series of steps.
2. Each NAD^+ can accept two electrons and a hydrogen, reducing it to NADH. Each FAD can accept two electrons and two hydrogens, reducing it to $FADH_2$.
3. Overall, the process of cellular respiration transfers 24 electrons from glucose to oxygen. About 90% of the energy stored in the bonds within glucose is harnessed into NADH and $FADH_2$.
4. The two electrons received by each NADH or $FADH_2$ will ultimately reduce a single oxygen, which then combines with two hydrogens to form one water molecule.

Which molecules serve as electron carriers?

1. The flow of electrons occurs through a series of molecules called electron carriers. The pathway the electrons follow is called the electron transport chain. The electrons flow through a series of oxidation and reduction reactions.
2. Cellular respiration depends on a set of specific proteins, called cytochromes. They change color as they accept or donate electrons because they contain heme, which is the same pigment that colors red blood cells.

How do cells harvest the energy of electron transport?

1. The difference in hydrogen ion concentration across a membrane is called a proton gradient.
2. According to the chemiosmotic hypothesis, the energy from the electron transport chain drives ATP production through chemiosmosis—the harnessing of energy stored in a proton gradient created by the electron transport chain.

How do mitochondria generate a proton gradient?

1. Recall that mitochondria have two membranes. As a result, a mitochondrion provides four distinct locations in which chemical reactions can occur:
   - the outer membrane,
   - the intermembrane space (between the two membranes),
   - the inner membrane, and
   - the mitochondrial matrix, which is inside the inner membrane.
2. Most of the NADH and $FADH_2$ is produced within the mitochondrial matrix, and the electron transport chain is within the inner membrane.
3. As electron carriers pass electrons through the chain, hydrogen ions move from the matrix out to the intermembrane space, crossing the inner membrane. This movement establishes a proton gradient across the inner membrane as protons are pumped out of the matrix.

What pumps protons out of the mitochondrial matrix?

1. Electron transport complexes in the inner membrane use the energy derived from transporting electrons down the electron transport chain to pump protons across the membrane in one direction only.
2. Proton pumping generates a charge gradient as well as a pH gradient—recall that the protons are H^+, which determines pH. As the protons are pumped out, they leave behind an excess of negative ions.
3. This dual action results in an electrochemical gradient, which is a double gradient—a chemical gradient (pH difference) and an electrical gradient (charge difference).

How does the flow of protons back into the matrix cause the synthesis of ATP?

1. After being pumped out of the matrix, the only way the protons can return is through special channels in the inner membrane.
2. As the protons flow back, a protein complex called ATP synthase uses this flow to drive ATP synthesis.
3. Coupling factors are part of the inner membrane and look like miniature lollipops, each having a stick portion called F_0 and a top portion called F_1.
4. F_0 channels protons back into the matrix; F_1 is the coupling factor that is responsible for making ATP. How this works is not yet understood.

Are proton pumping and ATP synthesis really separate processes?

1. The chemiosmotic hypothesis explains that:
   - the electron transport chain creates an electrochemical gradient across the inner mitochondrial membrane, and
   - the F_0-F_1 complex couples the flow of protons down the gradient to the process of ATP synthesis.
2. Proton pumping and ATP synthesis are separate processes, but they can be coupled.

How much usable energy can a cell harvest from a molecule of glucose?

1. Cellular respiration produces 30 or 32 molecules of ATP for each molecule of glucose, depending on the tissue in which the process occurs. Until recently, the estimates were 36 or 38 molecules because the net ATP yield from each NADH and $FADH_2$ had been overestimated.
2. All but four of the ATP generated from one glucose are produced from oxidative phosphorylation.
3. In the absence of oxygen, glycolysis alone can produce only a net gain of two ATP.
4. Cellular respiration is an amazingly efficient process—about 50% of the energy in the glucose molecule is successfully extracted.

VOCABULARY BUILDING

In your own words, first write a brief definition, then a full explanation for each of the following terms. Include examples where appropriate. Complete this section from your memory—you will not learn the terms by simply copying definitions from the textbook. Once you have finished, check your responses against the information in the chapter and make any necessary corrections.

vitalism —

autotrophs —

heterotrophs —

metabolism —

cellular respiration —

aerobic —

anaerobic —

facultative anaerobe —

obligate anaerobe —

digestion —

glycolysis —

citric acid (Krebs) cycle —

oxidative phosphorylation —

pyruvate —

oxidizing agent —

reducing agent —

reduction-oxidation reactions (redox reactions) —

NAD^+ and NADH —

FAD and FAD_2 —

fermentation —

mitochondria —

coenzyme A —

catabolism —

anabolism —

electron carriers —

electron transport chain —

cytochromes —

heme —

proton gradient —

chemiosmosis —

chemiosmotic hypothesis —

intermembrane space —

mitochondrial matrix —

electrochemical gradient —

ATP synthase —

F_0 —

F_1 —

coupling factor —

cirrhosis of the liver —

CHAPTER TEST

The following test has four parts. Complete as much of the exam as you can from memory. If you cannot answer a question, skip it. Once you complete all that you can, try to answer any questions you skipped. If you still cannot answer them, consult your textbook for the answers. Once you have completed all sections of the test, check your answers for Parts 1 - 3 against those in the back of this book. Highlight any incorrect answers then review that material in your textbook. Correct your answers for future reference.

Part 1: Multiple Choice

For each of the following, select all correct responses—more than one may be correct.

1. Which of the following processes can occur without oxygen?
   a) citric acid cycle
   b) glycolysis
   c) oxidative phosphorylation
   d) fermentation

2. Where does digestion usually occur?
   a) outside the cytosol
   b) in the mitochondria
   c) in the cytosol
   d) always outside the cell

3. Where does glycolysis occur?
   a) outside the cytosol
   b) in the mitochondria
   c) in the cytosol
   d) always outside the cell

4. During which process is energy from broken bonds transferred to bonds in ATP?
   a) glycolysis
   b) oxidative phosphorylation
   c) citric acid cycle
   d) all of these

Chapter 7
Photosynthesis:
How do Organisms Get Energy From the Sun?

KEY CONCEPTS

1. Photosynthesis is the process that changes light energy into chemical energy.
2. In light, plants take in carbon dioxide and release oxygen as a waste product; but in the dark, respiration, which takes in oxygen and releases carbon dioxide, is their only energy source.
3. The overall reaction of photosynthesis is as follows:

$$6CO_2 + 6H_2O \rightarrow C_6H_{12}O_6 + 6O_2$$

carbon dioxide + water → glucose + oxygen

4. The light reactions are photophosphorylation—they use photons to convert ADP to ATP.
5. The light-independent reactions use this energy to make glucose from carbon dioxide.
6. Photons with shorter wavelengths have more energy than photons with longer wavelengths. In the visible range, different wavelengths appear as different colors.
7. An atom or a molecule can absorb a photon of light if the photon's energy matches the energy needed to boost an electron to a higher energy level. A molecule's absorption spectrum is the particular set of wavelengths the molecule can absorb.
8. Relatively few chlorophyll molecules, located in the photochemical reaction center, transfer electrons to $NADP^+$. Most chlorophyll molecules work with carotenoids, in the antenna complex, to trap and transfer light energy to the reaction center.
9. Photosystem I absorbs light with wavelengths of about 700 nm. Photosystem II absorbs light with wavelengths of about 680 nm.
10. Photosystem particles consist of chlorophyll, carotenoids, specific proteins, and other molecules and ions, and are inside chloroplasts and associated with the thylakoid membranes.
11. In cyclic photophosphorylation, excited electrons from P_{700} pass through an electron transport chain then back to P_{700}, creating a proton gradient that can generate ATP.
12. In the noncyclic route of photosystem I, the excited electrons from P_{700} help form NADPH. After this occurs, electrons are needed to replace those that were liberated from P_{700}.
13. In noncyclic photophosphorylation, photosystem II's P_{680} supplies electrons to P_{700} via an electron transport chain. The electron flow generates a proton gradient that is capable of generating ATP. P_{680} gets new electrons from water.
14. In photosystem II, oxygen is released to the atmosphere as electrons are stripped from water.
15. The light-independent reactions use the Calvin cycle, in which carbon dioxide is converted to glucose. The enzyme Rubisco is used to synthesize three-carbon sugars which can then be converted to glucose or other molecules, or regenerate ribulose biphosphate.
16. For each molecule of glucose synthesized, photosynthesis consumes 18 molecules of ATP and 12 molecules of NADPH. This represents the energy from about 72 photons of light.
17. C_4 plants minimize water loss and photorespiration while maximizing photosynthesis by maintaining high concentrations of carbon dioxide in special cells. C_4 plants are more efficient than C_3 plants but need more light than C_3 plants.

EXTENDED CHAPTER OUTLINE

The chemical evangelist

1. Photosynthesis changes light energy into the energy of chemical bonds in sugars.

HOW DO WE KNOW HOW PLANTS OBTAIN CARBON AND OXYGEN?

1. Leaves of plants have stomata—minute pores that allow air to pass to the leaves' interior.

What do plants and air do for each other?

1. In the dark, plants respire, taking in oxygen and releasing carbon dioxide.
2. In light, plants take carbon dioxide from the air and release oxygen as a waste product.
3. In photosynthesis, plants take in water from the soil and carbon dioxide from the air and use them to make glucose. In the process, the plants release oxygen back into the air.

Where do the atoms go in photosynthesis?

1. The overall chemical reaction of photosynthesis is as follows:

$$6CO_2 + 6H_2O \rightarrow C_6H_{12}O_6 + 6O_2$$
$$\text{carbon dioxide} + \text{water} \rightarrow \text{glucose} + \text{oxygen}$$

2. The overall reaction of photosynthesis is the exact opposite of cellular respiration.
3. Another way to write the chemical transformation of photosynthesis is:

$$CO_2 + H_2O \rightarrow C(H_2O) + O_2$$
$$\text{carbon dioxide} + \text{water} \rightarrow \text{carbohydrate} + \text{oxygen}$$

4. The oxygen released by plants comes from water, not from carbon dioxide.
5. Photosynthesis consists of two processes—light-dependent reactions and light-independent reactions.
6. The light-dependent reactions use energy from sunlight to generate energy-transfer molecules such as ATP. These reactions are also referred to as photophosphorylation because they use the energy in the photons to add phosphorus to ADP, thus forming ATP.
7. The light-independent (dark) reactions use the energy captured during the light reactions (in ATP and NADPH) to synthesize glucose from carbon dioxide.

HOW DO PLANTS COLLECT ENERGY FROM THE SUN?

What is light?

1. Light is the energy of photons—particles that travel on straight lines at the speed of light (186,000 miles per second).
2. Light behaves like both a particle and a wave.
3. Photons travel at the same speed but have different amounts of energy, depending on the wavelength—the distance between successive crests.
4. Most light is a mixture of wavelengths.
5. Visible light has wavelengths of about 390 nm to 760 nm, and different wavelengths appear as different colors.
6. A photon is a light particle, and it is also a package of pure energy—it has no mass.
7. Shorter wavelengths (more bluish) have more energy than longer wavelengths (more reddish). Energy, wavelength, and color of light all measure the same thing.
8. Brightness, or intensity, of the light relates to the number of photons, not the energy of the photons or wavelength of the light.
9. In photosynthesis, light excites electrons in particular molecules, thus increasing their free energy. Some of the free energy is captured and converted to high-energy chemical bonds.

How can we tell which molecules help with photosynthesis?

1. Different molecules absorb different wavelengths of light.
2. The green pigment in plants absorbs every color of light except green.
3. When a molecule absorbs a photon, the photon's energy kicks an electron into a higher-energy orbital. If the electron returns to the lower-energy orbital, it loses energy as another photon or as heat. Or, the electron can move to another molecule, transferring energy.
4. The particular set of wavelengths that a molecule absorbs is its absorption spectrum.
5. Pigments absorb visible light and can change their electronic arrangements with relatively little energy.
6. Any light-dependent process has an action spectrum, which reveals the relative effectiveness of different wavelengths in promoting the process.
7. The action spectrum of photosynthesis roughly follows the absorption spectrum of the green pigment chlorophyll. It does not exactly match this spectrum, though, indicating that other pigments, chlorophyl b and the carotenoids, are also involved.
8. Most of the chlorophyll molecules in a leaf associate with a few molecules of carotenoids to form a weblike antenna complex. These molecules trap light and transfer energy to a photochemical reaction center.
9. The photochemical reaction center contains a small number of chlorophyll molecules that accept electrons, transferring them to $NADP^+$.

Why do plants use two types of reaction centers?

1. In plants, photosynthesis involves two distinct but interacting sets of reactions, called photosystems.
2. Each photosystem relies on different wavelengths of light. Photosystem I contains chlorophyll called P_{700} that absorbs primarily light with wavelengths near 700 nm (deep red). Photosystem II contains chlorophyll called P_{680} that absorbs primarily light with wavelengths of about 680 nm (red).
3. Photosynthetic bacteria have only one photosystem.
4. The two photosystems in plants are distinct particles, but both are located within chloroplasts, which are associated with the thylakoid membranes that enclose stacked grana.
5. Each photosystem particle consists of chlorophyll and carotenoid molecules, proteins, and other molecules and ions.
6. Excited electrons from photosystem I have two path options:
   - Cyclic photophosphorylation, which produces ATP, and
   - Noncyclic photophosphorylation, which produces NADPH
7. When P_{700} absorbs light from the antenna complex, its free energy increases and it becomes a strong reducing agent (electron donor). In this state, P_{700} readily releases an electron to one of two electron transport chains, becoming oxidized.
8. In cyclic photophosphorylation, excited electrons from P_{700} pass through an electron transport chain and back to P_{700}, restoring it to its original state so the P_{700} is ready to absorb another photon.
9. Cyclic passage of electrons through the cytochrome complex creates a proton gradient across a membrane in the chloroplast which is capable of generating ATP. During this process, protons move from the stroma (space outside the thylakoids) into the thylakoid space.
10. Like the inner mitochondrial membrane, the thylakoid membrane contains ATP synthetase molecules which produce ATP as the protons flow back down their gradient.
11. Photosystem I alternatively employs a noncyclic electron path that does not produce ATP. Here, the energy of the electron from P_{700} is used to make NADPH—an electron carrier. In this route, P_{700} must receive another electron before it can again be an electron donor.

12. The noncyclic path of photosystem I supplies most of the reducing power for glucose synthesis.
13. In photosystem II, electrons move from excited P_{680} through an electron transport chain and finally reach the oxidized P_{700} that was produced by photosystem I. Through this mechanism, by regenerating reduced P_{700}, photosystem II supplies electrons to photosystem I even when there is no cyclic flow of electrons.
14. The electron transport chain in photosystem II also generates a proton gradient that drives ATP synthesis. This process is called noncyclic photophosphorylation.
15. The P_{680} needs to be reduced before it can function again. To accomplish this, photosystem II strips electrons from water, and the remaining oxygen is released to the atmosphere.
16. The evolution of photosystem II allowed the accumulation of oxygen in our atmosphere. This accumulation, in turn, triggered the evolution of animals and other heterotrophs.

How do the light reactions generate NADPH and ATP?

1. Together, photosystems I and II make up the light-dependent reactions of photosynthesis.
2. Absorption of light by the two photosystems causes:
   - the splitting of water,
   - production of ATP, and
   - production of NADPH.
3. Oxygen is a waste product released into the atmosphere.
4. ATP and NADPH contribute to the light-independent reactions of photosynthesis, which synthesize glucose.
5. The energy of two protons of light is transformed into the chemical energy of one molecule of NADPH and two molecules of ATP.
6. Plants use this stored energy to make glucose, which is then stored as starch, sucrose, or some other carbohydrate.
7. Plants transport sugar from cell to cell as sucrose in the same way that animals transport sugar from cell to cell as glucose.
8. The synthesis of glucose is the second part of photosynthesis.

HOW DO PLANTS MAKE GLUCOSE?

1. Glucose provides plants with their most important source of energy, as well as building blocks containing carbon, hydrogen, and oxygen.
2. The overall process of photosynthesis is the reverse of respiration (although the individual steps are not): $6\ CO_2 + 6\ H_2O \rightarrow C_6H_{12}O_6 + 6\ O_2$

What is the Calvin cycle?

1. The Calvin cycle is a set of reactions that converts carbon (from carbon dioxide) to glucose.

What happens to carbon dioxide in photosynthesis?

1. The Calvin cycle has three parts:
   - production of phosphoglycerate from combining carbon dioxide with an acceptor molecule;
   - conversion of phosphoglycerate into glyceraldehyde phosphate in two reactions that use ATP and NADPH produced in the light reactions; and
   - regeneration of carbon dioxide.
2. In part one of the Calvin cycle, carbon dioxide combines with the five-carbon ribulose biphosphate, forming a six-carbon compound that quickly separates into two molecules of phosphoglycerate.
3. Rubisco (ribulose biphosphate carboxylase), the enzyme catalyzing this reaction, is almost solely responsible for recycling carbon dioxide back into the biosphere and supplying carbon dioxide for all living things.

4. This enzyme performs its task slowly, so plants produce a lot of Rubisco to compensate for the sluggishness. Rubisco is probably the most abundant protein on the planet.
5. In part two of the Calvin cycle, the two molecules of phosphoglycerate produced in part one are each converted to glyceraldehyde phosphate—a carbohydrate. This is then converted to glucose or other molecules, or enters part three.
6. In part three of the Calvin cycle, three-carbon sugars made in part two are combined to regenerate ribulose biphosphate (which captures carbon dioxide for part one).
7. The Calvin cycle is the heart of the light-independent reactions of photosynthesis, using the enzyme Rubisco to synthesize three-carbon sugars from ordinary carbon dioxide.

How many photons does a chloroplast need to make one glucose?

1. For each molecule of glucose produced, six molecules of carbon dioxide must enter the Calvin cycle. Through six turns, the cycle produces twelve molecules of glyceraldehyde phosphate. Of these, two form glucose and the remaining ten regenerate six molecules of ribulose biphosphate.
2. The six turns of the Calvin cycle needed to produce a single molecule of glucose require twelve molecules of ATP and twelve molecules of NADPH. Production of the six molecules of ribulose biphosphate requires an additional six ATP.
3. The energy driving the Calvin cycle comes from the high-energy bonds in ATP and NADPH formed during the light-dependent reactions.
4. To summarize, for each molecule of glucose synthesized, photosynthesis consumes 18 molecules of ATP and 12 molecules of NADPH—the energy from about 72 photons of light. This is the equivalent of 3 molecules of ATP and 2 molecules of NADPH per carbon atom.

HOW DO PLANTS COPE WITH TOO LITTLE WATER OR CARBON DIOXIDE?

1. Plants require carbon dioxide, water, and photons to make the molecules needed for survival.
2. Plants have many adaptations allowing them to maximize their productivity—the efficiency with which they collect and use the raw materials they need.

What factors limit productivity?

1. The major factors governing the rate of photosynthesis are:
   - amount, wavelength, and intensity of incoming photons,
   - amount of water, and
   - amount of carbon dioxide.
2. The amount of light (intensity—the number of photons striking the leaves) is affected by shade, length of the day and the growing season, amount of cloud cover, and pollution.
3. Land plants take water from the soil. Some is used for photosynthesis, but most of the water evaporates, leaving through stomata in the leaves.
4. Stomata also allow oxygen out of and carbon dioxide into the leaves.
5. In hot and dry weather, the stomata close to save water. As they remain shut, oxygen levels rise and carbon dioxide levels drop.

How do plants prevent photorespiration?

1. Decreasing carbon dioxide levels slow or halt photosynthesis, and start a dangerous chemical pathway called photorespiration, in which plants waste much of their carbohydrate.
2. Rubisco requires a relatively high level of carbon dioxide to work properly—converting ribulose biphosphate to two molecules of phosphoglycerate. When carbon dioxide is limited, Rubisco instead converts ribulose biphosphate to one phosphoglycerate and one almost useless two-carbon sugar. Through this process, plants waste up to half the carbohydrates they produce.
3. Photorespiration not only wastes carbohydrate, but it also produces no ATP or NADPH.

4. Because carbon dioxide inhibits photorespiration, some plants have evolved a special biochemical path that allows them to maintain high carbon dioxide levels in their cells, even when their stomata remain closed much of the time.
5. In these plants, the first reaction of carbon dioxide produces a four-carbon compound—oxaloacetate (OAA) instead of the usual three-carbon phosphoglycerate. OAA, in turn, breaks down to release carbon dioxide, which immediately enters the Calvin cycle.
6. Plants that use the three-carbon phosphoglycerate (most plants) are C_3 plants. Plants that use the four-carbon oxaloacetate pathway are C_4 plants, and include corn and sugarcane.
7. C_4 plants thrive in hot, dry climates because they can minimize water loss by keeping their stomata closed while still providing enough carbon dioxide to maximize photosynthesis and minimize photorespiration.
8. C_4 plants waste far less carbohydrate than C_3 plants, but they require more ATP and light.
9. C_4 plants predominate in hot, dry climates that have a lot of light. C_3 plants predominate in temperate zones with less light and more water.
10. C_4 plants are especially efficient at capturing solar energy, converting as much as 8% of the solar energy falling on them into chemical energy in the bonds of carbohydrate molecules.
11. Desert-adapted plants, such as cacti, use a modified C_4 pathway called crassulacean acid metabolism (CAM). This pathway conserves water—the plants only open their stomata to collect carbon dioxide at night, when the air is cool and water loss is minimal.
12. The carbon dioxide that CAM plants take in at night is stored in a four-carbon molecule until the next day, when photosynthesis can proceed.
13. CAM plants' nighttime respiration has a high energy cost, so their growth rates are extremely low.

VOCABULARY BUILDING

In your own words, first write a brief definition, then a full explanation for each of the following terms. Include examples where appropriate. Complete this section from your memory—you will not learn the terms by simply copying definitions from the textbook. Once you have finished, check your responses against the information in the chapter and make any necessary corrections.

photosynthesis —

stomata —

photophosphorylation —

light-dependent reactions —

light-independent reactions —

photon —

wavelength —

visible light —

absorption spectrum —

pigments —

chlorophyll —

action spectrum —

carotenoids —

antenna complex —

photochemical reaction center —

photosystem I —

photosystem II —

P_{700} —

P_{680} —

stroma —

cyclic photophosphorylation —

noncyclic photophosphorylation —

Calvin cycle —

Rubisco —

photorespiration —

C_3 plants —

C_4 plants —

crassulacean acid metabolism (CAM) —

herbicide —

CHAPTER TEST

The following test has four parts. Complete as much of the exam as you can from memory. If you cannot answer a question, skip it. Once you complete all that you can, try to answer any questions you skipped. If you still cannot answer them, consult your textbook for the answers. Once you have completed all sections of the test, check your answers for Parts 1 - 3 against those in the back of this book. Highlight any incorrect answers then review that material in your textbook. Correct your answers for future reference.

Part 1: Multiple Choice

For each of the following, select all correct responses—more than one may be correct.

1. Photons of which wavelength of light have more energy?
   a) longer wavelength
   b) shorter wavelength
   c) all have the same energy
   d) wavelength and energy are unrelated

2. Chlorophyll in photosystem I best absorbs light of what wavelength?
   a) 680 nm
   b) 700 nm
   c) 720 nm
   d) none of these

3. Which of the following corresponds to the mitochondrial matrix?
   a) thylakoids
   b) stomata
   c) cytoplasm
   d) stroma

4. Which of the following is a solid storage form of glucose in plants?
   a) sucrose
   b) glucose
   c) Rubisco
   d) starch

5. Carbohydrate is produced during which part of the Calvin cycle?
   a) part 1
   b) part 2
   c) part 3
   d) all of them

6. Synthesis of one molecule of glucose requires how much light energy?
   a) 12 photons
   b) 18 photons
   c) 72 photons
   d) 680 photons

7. Which of the following is not a limiting factor for productivity?
   a) amount of incoming photons
   b) temperature
   c) amount of water
   d) amount of carbon dioxide

8. Which of the following movements does not occur at the stomata?
   a) oxygen moves out
   b) water moves out
   c) carbon dioxide moves in
   d) all of these do occur at the stomata
9. Which of the following is true of CAM plants?
   a) they do not have stomata
   b) their stomata are only open at night
   c) they require little light
   d) they rely on photorespiration for energy production
10. The electron needed to regenerate reduced P_{700} comes from:
    a) P_{680}
    b) P_{700}
    c) Calvin cycle
    d) carbon dioxide
11. The reactions that produce glucose from carbon dioxide are known as:
    a) photosystem I
    b) photosystem II
    c) Calvin cycle
    d) photorespiration
12. Where does photosynthesis occur?
    a) chloroplast
    b) mitochondrion
    c) cytoplasm
    d) stomata
13. How do C_4 plants avoid photorespiration?
    a) maintain high levels of carbon dioxide while stomata are closed
    b) maintain high levels of oxygen while stomata are closed
    c) maintain low levels of carbon dioxide while stomata are open
    d) maintain low levels of oxygen while stomata are open
14. The "cycle" of cyclic photophosphorylation restores:
    a) high carbon dioxide levels
    b) water
    c) glucose
    d) reduced P_{700}
15. Which of the following is not accomplished directly by the light-dependent reactions of photosynthesis?
    a) splitting of water
    b) production of ATP
    c) production of glucose
    d) production of NADPH

Part 2: Matching

For each of the following, match the correct term with its definition or example. More than one answer may be appropriate.

photosystem I **photosystem II**

1. ____________________ Uses P_{680}.
2. ____________________ Uses P_{700}.
3. ____________________ Performs cyclic photophosphorylation.
4. ____________________ Releases oxygen to the atmosphere.

oxygen **carbon dioxide** **water**

5. ____________________ Required for photosynthesis to occur.
6. ____________________ Original source of oxygen that is released from the plant.
7. ____________________ Used to make glucose during the Calvin cycle.
8. ____________________ Leaves the plant via the stomata.

C_3 plants **C_4 plants**

9. ____________________ Waste less carbohydrate.
10. ____________________ Require less light energy.
11. ____________________ Minimize water loss and photorespiration.
12. ____________________ Thrive in dry, hot climates.

Part 3: Short Answer

Write your answers in the space provided or on a separate piece of paper.

1. What must plants take in so they can make glucose?

2. What is the overall equation for photosynthesis?

3. How does light behave as both a particle and a wave?

4. Why does chlorophyll appear green?

5. Which wavelength of light is best absorbed by the chlorophyll in photosystem II?

6. Which part of photosynthesis generates the oxygen that we breathe?

7. How is the energy stored in the chemical bonds of ATP and NADPH used?

8. What type of sugar in a plant can move from cell to cell?

9. In what way is the Calvin cycle a "cycle"?

10. What is the total cost of making one molecule of glucose?

11. List three environmental factors that can reduce the productivity of photosynthesis.

12. How is photosystem I dependent on photosystem II?

13. List two ways in which humans are dependent on photosynthesis in plants.

14. Rubisco performs its task slowly, so how do plants compensate for this slowness?

15. What type of plants are most common in temperate zones?

Part 4: **Critical Thinking: Using Your Knowledge**

Answer each of these in essay form, using complete sentences and paragraphs. Provide as much information as you can. (For extra essay practice, write out answers to the Review and Thought Questions in your textbook.)

1. Explain how the processes of photosynthesis and cellular respiration are related.

2. Explain the process of photosynthesis. Thoroughly discuss events that occur during the light-dependent and light-independent reactions, and how these sets of reactions are related.

3. Explain the process of photorespiration. When and why does it occur, and why is it a problem for plants?

4. Differentiate between C_3 and C_4 plants. What environmental conditions is each adapted for, and what are the adaptations? How do their energy needs and productivities compare?

5. Epiphytes are a type of plant that is often found living high in the branches of trees in heavily forested areas, with no contact with the soil. What advantages might this location provide? With this adaptation, how can these plants receive their requirements for photosynthesis?

6. Explain the characteristics of light energy; discuss how it acts as both a particle and a wave.

7. How do you think the growth and general health of plants would differ if some were grown under green lights, some under red lights, some under yellow lights, and some under blue lights?

8. It is believed that if there were a major nuclear war, dust from the explosions would obscure the sunlight. Starting with plants, explain how this situation alone might destroy most life on our planet.

9. For years, people from northern climates have flocked to the desert Southwest, which has traditionally been a safe haven for allergy sufferers because the air has been so pure. Preferring the landscaping they left behind to the dry, barren rockiness they found, thesenew residents also brought a bit of home with them,. As the plants arrived, so did the allergens. However, the plants that thrive "back home" do not always do so well in their new locations. What reasons can you give for this? Also, many western rivers no longer run their full course, drying out long before they reach their destination. How do non-native plants, increased agriculture, and population booms in the West relate to this problem?

10. The greenhouse effect refers to the buildup of carbon dioxide in the atmosphere, which then decreases the amount of heat that can leave. This is leading to global warming which can have potentially devastating consequences. A major contributor to the greenhouse effect is the burning of fossil fuels (petroleum products). In the past, how have plants contributed to the current problem? Now, how are human actions involving plants contributing to this? How might plants be used to improve the situation?

3. Because DNA has a negative charge, the positively charged histones bind to it tightly.
4. DNA and histones associate into a form that looks rather like beads on a string. Each bead, along with its DNA strand, is called a nucleosome.
5. Each nucleosome contains a 146-nucleotide long stretch of DNA and 8 histone molecules.
6. Histones from all eukaryotes resemble one another.

How do DNA, histones, and other proteins form such compact structures?

1. DNA in chromosomes is more condensed than that in the diffuse chromatin.
2. Loops of DNA form and cluster in characteristic ways in each chromosome, which gives rise to chromosome banding.
3. Chromosome banding patterns can be used to detect abnormal chromosomes.

HOW DOES A CELL DIVIDE ITS CYTOPLASM?

1. Cytokinesis is the process by which a dividing cell partitions its cytoplasm, including the two nuclei that are formed during mitosis.
2. Cytokinesis and mitosis are separate processes, each using different molecular machinery.
3. Cytokinesis almost always accompanies mitosis, usually beginning during anaphase and ending after the end of telophase.
4. The plane of cell division is always perpendicular to the axis of the mitotic spindle.
5. In animal cells, a bundle of actin filaments forms a temporary structure—the contractile ring—that surrounds the dividing cell.
6. As the beltlike contractile ring tightens, the cell membrane is pulled into a deepening groove called the cleavage furrow.
7. At the end of telophase, the daughter cells are completely separated.
8. As the daughter cells separate, they must increase their cell membranes; this extra membrane material comes from extra membrane made by the parent cell during interphase.
9. Cytokinesis in a plant cell also requires building a new cell wall between the daughter cells.
10. The new cell wall begins in telophase as a small, flattened disc, called the early cell plate, between the two daughter nuclei.
11. The disc grows to become a cell plate that seals off the two daughter cells.

HOW DOES A CELL REGULATE PASSAGE THROUGH THE CELL CYCLE?

1. Cells divide rapidly during growth periods and more slowly, if at all, in mature organisms.
2. The only exception to this rule is the uncontrolled growth and division of cancer cells, which multiply at the expense of all other cells in the body.

How do normal cells determine when to stop dividing?

1. Cell senescence—cell aging—limits the number of times a cell can divide.
2. Regulatory mechanisms involved in growth control prevent uncontrolled division by allowing the cell cycle to proceed under some conditions but to stop under others.
3. Due to contact inhibition:
   - cells stop dividing during G when they run out of free space in which to spread (when neighboring cells all are in contact); and
   - cells that have moved beyond G_1 begin to divide when contact with adjacent cells ends.
4. Cancer cells do not exhibit contact inhibition.
5. Growth control depends on several types of proteins:
   - Growth factors,
   - Proteins that bind to growth factors,
   - Proteins that regulate expression of genes into RNA and protein, and
   - Tumor-suppressor proteins.

How do normal cells determine when it is time to divide?

1. Most cells divide only after they have first doubled their mass.
2. Once a cell proceeds beyond a certain point in G_1, the cell proceeds through the rest of the cell cycle. This point of no return is called Start.
3. Arrival at Start depends heavily on the environment of the cell—including nutrient availability and signals from other cells.
4. Egg cells are an exception to the rule that cells double their mass before dividing. Egg cells grow to enormous sizes without dividing.
5. Early embryos of many animals are also an exception—after fertilization, those derived from large eggs typically undergo a series of rapid cleavage divisions through which the embryo divides into many cells without increasing its mass.

What triggers the main events of mitosis?

1. Cell reproduction requires the coordination of three cycles:
   - duplication and packaging of DNA,
   - duplication of the centrosomes and the operation of the mitotic spindle apparatus, and
   - cytokinesis, which depends on the contractile ring.
2. All eukaryotes use almost identical mechanisms to coordinate and initiate these cycles.
3. Synchronous cell populations are made by:
   - Using rapidly reproducing cells from early embryos in marine invertebrates;
   - Blocking DNA synthesis with a chemical inhibitor; or
4. Fluctuating amounts of various proteins, called cyclins, seem to regulate the cell's passage through mitosis.
5. Each type of cyclin stimulates a specific enzyme, called a cyclin-dependent kinase, which alters cell proteins.
6. The periodic increase and decrease of cyclin is both necessary and sufficient for cell cycling.

VOCABULARY BUILDING

In your own words, first write a brief definition, then a full explanation for each of the following terms. Include examples where appropriate. Complete this section from your memory--you will not learn the terms by simply copying definitions from the textbook. Once you have finished, check your responses against the information in the chapter and make any necessary corrections.

chromosomes —

cell division —

daughter cells —

chromatin —

mitosis —

aster —

fertilization —

replicate —

binary fission —

homologous chromosomes —

homolog —

cell cycle —

cytokinesis —

interphase —

M —

G_1 —

S —

G_2 —

sister chromatids —

centromere —

prophase —

mitotic spindle —

microtubules —

centrioles —

kinetochore —

metaphase —

metaphase plate —

anaphase —

telophase —

histones —

nucleosome —

chromosome banding —

contractile ring —

cleavage furrow —

cell plate —

cell senescence —

growth control —

contact inhibition —

Start —

synchronous cell populations —

cyclin —

CHAPTER TEST

The following test has four parts. Complete as much of the exam as you can from memory. If you cannot answer a question, skip it. Once you complete all that you can, try to answer any questions you skipped. If you still cannot answer them, consult your textbook for the answers. Once you have completed all sections of the test, check your answers for Parts 1 - 3 against those in the back of this book. Highlight any incorrect answers then review that material in your textbook. Correct your answers for future reference.

Part 1: Multiple Choice

For each of the following, select all correct responses—more than one may be correct.

1. Division of the DNA in a cell is called:
   a) replication
   b) mitosis
   c) cell cycle
   d) cytokinesis

2. Which of the following processes occurs first?
   a) cytokinesis
   b) mitosis
   c) DNA replication
   d) all occur together

3. The materials and machinery used during cell reproduction come from:
   a) the father
   b) the mother
   c) both parents
   d) neither parent—they are made by the reproducing cell

4. Which of the following is not true of binary fission?
   a) It is the method by which prokaryotes divide.
   b) It typically involves only a single strand of DNA.
   c) The daughter DNA molecules attach to the plasma membrane.
   d) It is more complicated than mitosis.

5. How many pairs of chromosomes do humans have?
   a) 2
   b) 23
   c) 46
   d) 92

6. Each member of a pair of chromosomes is called a:
   a) chromatid
   b) chromatin
   c) homolog
   d) centromere

7. Which of the following lists, in correct order, the phases of the eukaryotic cell cycle?
   a) M, G_1, G_2, S
   b) G_1, G_2, M, S
   c) S, G_1, G_2, M
   d) M, G_1, S, G_2

8. During which stage does DNA replication (doubling) occur?
   a) metaphase
   b) interphase
   c) telophase
   d) prophase

9. If a cell runs out of nutrients, during which part of the cell cycle will the process stop?
   a) M
   b) G_1
   c) G_2
   d) S

10. Which of the following lists, in the correct order, the phases of mitosis?
    a) prophase, metaphase, anaphase, telophase
    b) prophase, anaphase, metaphase, telophase
    c) prophase, telophase, anaphase, metaphase
    d) anaphase, metaphase, telophase, prophase

11. Which of the following is the most distinct event that occurs during metaphase?
    a) nuclear membrane disappears
    b) chromosomes align along the cell's equator
    c) chromosomes separate and move towards opposite poles
    d) chromosomes disappear

12. Which of the following is the most distinct event that occurs during anaphase?
    a) nuclear membrane disappears
    b) chromosomes align along the cell's equator
    c) chromosomes separate and move towards opposite poles
    d) chromosomes disappear

13. Which of the following is not part of interphase?
    a) M
    b) G_1
    c) G_2
    d) S

14. The number of times that a cell can divide is limited by:
    a) growth control mechanisms
    b) contact inhibition
    c) histones
    d) cell senescence

15. What triggers mitosis?
    a) contact inhibition
    b) growth factors
    c) histones
    d) all of these

Part 2: **Matching**

For each of the following, match the correct term with its definition or example. More than one answer may be appropriate.

M **G_1** **S** **G_2**

1. ________________ Interphase
2. ________________ Nuclear division
3. ________________ Cytokinesis occurs.
4. ________________ The cell cycle will be arrested in this phase if nutrients are insufficient.
5. ________________ DNA replication occurs.
6. ________________ Chromosomes are visible.
7. ________________ Most of the cell's materials and machinery are doubled.
8. ________________ Preparations are made for mitosis and cytokinesis.

interphase **telophase** **prophase**

metaphase **anaphase**

9. ________________ This is not part of mitosis.
10. ________________ This is the last stage of mitosis.
11. ________________ Chromosomes separate.
12. ________________ Chromosomes are located in opposite poles.
13. ________________ Cytokinesis is occurring.
14. ________________ Chromosomes align at the middle of the cell.
15. ________________ Nucleoli disappear.

Part 3: Short Answer

Write your answers in the space provided or on a separate piece of paper.

1. Why do single-celled organisms undergo cell division?

2. What is the main difference between mitosis and cytokinesis?

3. What are the materials, machinery, and memory needed for cell division?

4. From where do the materials, machinery, and memory come?

5. What is binary fission?

6. List, in order, the three major stages of the cell cycle.

7. During what phase of the cell cycle is most of the work of reproduction done?

8. List, in order, the four phases of mitosis.

9. When does cytokinesis occur?

10. When does the mitotic apparatus disappear?

11. What are histones?

12. Of what is the contractile ring composed?

13. Explain cell senescence.

14. Explain how contact inhibition can work to promote or prevent cell division.

15. Explain the significance of Start to cell division.

Part 4: **Critical Thinking: Using Your Knowledge**

Answer each of these in essay form, using complete sentences and paragraphs. Provide as much information as you can. (For extra essay practice, write out answers to the Review and Thought Questions in your textbook.)

1. Thoroughly explain each stage of the cell cycle.

2. Make a chart listing the major events that occur in each stage of mitosis. Add an illustration depicting the chromosomes for each stage.

3. Explain the mitotic apparatus. When and how does it form? Where is it? From what is it made? How does it participate in mitosis?

4. Discuss mechanisms by which the cell cycle is regulated.

5. List some ways mentioned in the text in which HeLa cells have benefited other people. Using the knowledge you have gained in this chapter, explain why cancer cells, such as those of Henrietta Lacks, can be so valuable to researchers.

6. Explain mechanisms that are employed by a cell to regulate the process of cell division. What happens if these mechanisms fail? What are some reasons that they might fail?

7. In cancer, cells divide more rapidly than usual and without normal growth constraints. How can this situation lead to death?

8. If you compared cancer cells to normal cells in a mature adult, how would the cells' rates of respiration compare? Why? Clinically, what consequence does that have?

9. Different cells in your body experience different rates of cellular division. Which cells do you think might divide more rapidly, and why? Which cells might divide more slowly, and why? What happens with red blood cells?

10. Why does a cell divide into two smaller cells, instead of just continuing to grow into a much larger cell?

Chapter 9
From Meiosis to Mendel

KEY CONCEPTS

1. Phenotype results from the interaction of genotype and environment.
2. The germ cells of sexually reproducing organisms undergo meiosis to produce haploid gametes, which combine to form a diploid individual.
3. For meiosis, a cell replicates its DNA once but then divides twice, resulting in four daughter cells that are each haploid.
4. During anaphase I of meiosis, sister chromatids go to the same pole in the cell, but homologs, carrying different information, go to opposite poles.
5. Meiosis II resembles mitosis, but each resulting cell is haploid, rather than diploid.
6. Nondisjunction—failure of homologs to separate—produces gametes that have too many or too few chromosomes.
7. Sexual reproduction increases genetic diversity, but has a tremendous energy cost.
8. Mammals have a pair of sex chromosomes—XX in females and XY in males—and all other chromosomes are called autosomes. Humans have 22 pairs of autosomes and 1 pair of sex chromosomes.
9. Genes are located on the chromosomes.
10. Each sexually reproducing organism has 2 copies of each gene—one copy on each of the appropriate homologs.
11. By the Principle of Segregation, the 2 copies of each gene segregate during gamete production.
12. The Principle of Independent Assortment states that each gene in a pair is distributed independently during gamete production.
13. Crossing over, which is the physical exchange of material between homologous chromosomes, recombines alleles.

EXTENDED CHAPTER OUTLINE

WHY WAS THE CHROMOSOME THEORY OF INHERITANCE SO HARD TO ACCEPT?

1. Before the 17th century, a major theory held that miniature versions of future organisms were preformed in sperm or eggs. Most philosophers believed the egg had the materials for life and the sperm contributed some vital life force.

Blending inheritance: A wrong turn

1. Most biologists thought offspring differences resulted from "blending" of the parents' genes.
2. The blending model seemed to fit traits that vary continuously, but did not account for the reappearance of unblended characteristics in later generations.
3. Blending inheritance and Darwin's theory of natural selection were incompatible theories.
4. Most 19th-century biologists accepted blending inheritance and rejected natural selection.

Why did biologists doubt that chromosomes could carry information?

1. August Weismann argued that:
   - organisms inherit their traits by means of some information-carrying chemical located in the nucleus, and
   - since sexual reproduction involves fusion of hereditary material from each parent, the egg and sperm must each contain only half of the normal amount of hereditary material.
2. Meiosis is the process by which cells with the normal number of chromosomes divide to produce cells with only half the normal number of chromosomes.
3. In 1900, three scientists independently rediscovered Mendel's work and its significance. This gave birth to the field of genetics.
4. Walter Sutton was the first to explicitly state that genes are on the chromosome.

HOW DO ORGANISMS PASS GENETIC INFORMATION TO THEIR OFFSPRING?

1. Genetics is the study of inheritance.
2. Transmission genetics is the study of how variation is passed from one generation to the next; molecular genetics is the study of how DNA carries genetic instructions and how cells carry out these instructions.

How is phenotype related to genotype?

1. Everything about an organism constitutes its phenotype. Phenotype encompasses both physical and behavioral characteristics.
2. Genotype is the genetic constitution of a cell or organism—it is a collection of genes.
3. Genotype may be used to refer to:
   - all the genes in an organism, which is also called the genome, or
   - a subset of genes, or even one gene, that influences a certain trait.
4. A gene is a region of DNA. Most are structural genes, which specify a particular protein; others are regulatory genes, which regulate the expression of structural genes.
5. An organism's phenotype reflects both its genotype and its environment.
6. Identical twins have the same genotype, so differences between them are due to environmental differences.
7. Genes and environment affect the phenotype separately, but also together.
8. Phenotype is subject to natural selection. Crucial biological questions—how well does an individual compete with others and how much will it contribute to the next generation—depend on phenotype.

The same laws of inheritance apply to all sexually reproducing organisms

1. All sexually reproducing eukaryotes follow the same basic genetic rules.
2. The reasons for this universality are:
   - All organisms use DNA as the genetic material.
   - DNA in all eukaryotes is organized into chromosomes.
   - Almost all chromosomes exist in pairs at some time during the sexual life cycle.
   - These pairs of chromosomes behave in the same way during meiosis and fertilization in all eukaryotes.

HOW DO SEXUALLY REPRODUCING ORGANISMS KEEP THE SAME NUMBER OF CHROMOSOMES FROM GENERATION TO GENERATION?

1. Reproduction through mitosis (asexual reproduction) produces offspring with genes from just one parent. The offspring constitute a clone—a set of genetically identical individuals.
2. Sexual reproduction produces offspring that inherit genetic information from two parents.
3. Sexual reproduction combines already unique combinations of genes in new ways so each generation has unique individuals.
4. Sexual reproduction creates enormous diversity.
5. Pairs of matching chromosomes are called homologous chromosomes—each member of a pair is a homolog. The 46 human chromosomes are 23 pairs of homologous chromosomes.
6. Cells that contain two sets of chromosomes (homologous pairs) are called diploid.
7. Sexually reproducing organisms maintain the diploid state when each parent contributes only one homolog from each homologous pair of chromosomes—each parent contributes one copy of each of the 23 chromosomes, combining together to provide the offspring with a full diploid set of 46 chromosomes.
8. Chromosomes are passed from parent to offspring in special reproductive cells—gametes.
9. In humans, each gamete (ovum/egg in females, sperm/spermatozoon in males) is haploid.
10. Haploid gametes arise from diploid cells by meiosis—a process that distributes one haploid set of chromosomes to each of four germ cells.
11. In adults animals, germ cells are found in special gamete-producing organs called gonads (ovaries in females, testes in males).
12. All cells other than the germ cells are called somatic cells, and they undergo mitosis only—they cannot perform meiosis.

The first cell of the new generation has two sets of chromosomes

1. Fertilization is the union of two haploid gametes to form a diploid cell called a zygote. Syngamy is another name for this process.
2. All new generations begin with fertilization.
3. Gametes have unique adaptations making them suitable for fertilization.

How do the egg and the sperm find one another?

1. Every sexually-reproducing species has behavioral and structural adaptations that bring eggs and sperm together.
2. In animals, fertilization may be internal or external.

HOW DOES MEIOSIS DISTRIBUTE CHROMOSOMES TO THE GAMETES?

1. Meiosis consists of two cell divisions, meiosis I and meiosis II, which occur only in germ cells when they are making gametes.
2. DNA replication occurs before the first division (meiosis I) but does not occur before the second division (meiosis II).

3. The stages of the meiotic divisions are the same as those in mitosis:
   - During each prophase, chromosomes condense and the nuclear membrane breaks down.
   - During each metaphase, the chromosomes move to the equator.
   - During each anaphase, the chromosomes move toward the poles.
   - During each telophase, the nuclear membrane reforms.
4. The two divisions of meiosis produce four haploid daughter cells; in contrast, the single mitotic division produces two identical diploid cells.

During prophase I, homologous chromosomes form pairs

1. The same basic events occur as in mitotic prophase, except that as the chromosomes condense, the homologous chromosome pairs come together (synapsis). Each chromosome is made of two sister chromatids, so the pair consists of four chromatids, called a tetrad.
2. Synapsis—the alignment of the homologous pairs—is the central event of meiosis.
3. Crossing over, in which homologous chromosomes break and exchange equivalent pieces forming new combinations of genes, occurs during synapsis when the homologous chromosomes are in close proximity to each other.
4. Crossing over is visible by the crosslike chiasma. It is this configuration that allows crossing over to occur.
5. Prophase I occupies more than 90% of the total time of meiosis.
6. In humans, meiosis begins before birth, pauses at prophase I, and resumes after puberty.

During the rest of meiosis I, one chromosome from each homologous pair goes to each daughter cell

1. During metaphase I, each tetrad migrates to the equator.
2. During anaphase I, sister chromatids go to the same pole; homologs go to opposite poles.
3. Telophase I and cytokinesis occur quickly, then the cell immediately enters meiosis II. There is no DNA replication at this point.
4. The assortment of chromosomes in meiosis I is random—each pair of homologs aligns independently.
5. Meiosis is a major source of genetic diversity—a single human germ cell can combine the 23 chromosome pairs into any of 2^{23} (about 8 million) unique gametes.

Meiosis II distributes sister chromatids to daughter cells

1. Meiosis II resembles mitosis in haploid cells, except each reulting cell is haploid, rather than diploid.
2. Anaphase II separates the sister chromatids so that each chromosome of the daughter cell now consists of a single chromatid.
3. At the end of meiosis, from the original single diploid cell, four haploid daughter cells have been formed.

Disjunction and nondisjunction

1. Disjunction refers to when the homologous chromosomes (anaphase 1) or sister chromatids (anaphase II) separate.
2. Nondisjunction is when separation does not occur correctly during either anaphase I or anaphase II, resulting in a gamete with too many or too few chromosomes.
3. A cell or individual with the normal number of chromosomes is said to be euploid; one with an abnormal number is said to be aneuploid.
4. Trisomy 21, which produces Down syndrome, is one of the most common examples of aneuploidy. These individuals have three copies of chromosome 21 instead of the normal two because one of the gametes they received had two copies, instead of the normal one. Trisomy 21 usually occurs from nondisjunction during meiosis I.

WHY SEX?

1. Many organisms can reproduce either sexually or asexually.
2. Asexual reproduction is the quickest and most energy-efficient way to reproduce.
3. Sexual reproduction increases genetic (and thus phenotypic) diversity.

What good is genetic variation?

1. A genetically diverse population is far more likely to survive a changing environment than a genetically uniform population.
2. Genetic diversity is the raw material for natural selection and evolution.
3. Ultimately, all genetic variation arises from mutations—random, spontaneous changes in the genetic information—but sexual reproduction increases genetic diversity even more.

What are sex chromosomes?

1. In mammals, birds, and some other organisms, either sex may have an unmatched pair of chromosomes.
2. The X and Y chromosomes are called sex chromosomes. All other chromosomes are called autosomes.
3. A human cell has 22 pairs of autosomes and one pair of sex chromosomes.
4. In mammals, the Y chromosome carries far fewer genes than the X chromosome.
5. The two sex chromosomes behave the same way as any homologous pair of chromosomes.
6. The genotype XX determines female, whereas XY determines male.

HOW DO THE NUMBER AND MOVEMENTS OF CHROMOSOMES EXPLAIN THE INHERITANCE OF GENES?

1. A cell that has two copies of each chromosome also has two copies of each gene.
2. Alternative versions of the same gene are called alleles.
3. Most genes are simply instructions for making a certain protein, so alleles are variations in those instructions.
4. An organism is said to be homozygous for an allele when it has two copies of the same allele. An organism with two different alleles for a single gene is said to be heterozygous.
5. Alleles are designated by letters or short abbreviations.
6. True-breeding organisms are those that always produce offspring of the same type. True-breeders are homozygous.
7. Cross breeding is breeding two genetically distinct organisms, and the progeny are called hybrids.
8. The original parents in a cross are called the parental (P) generation, and the progeny are called the first filial (F1) generation. The second filial (F2) generation are the progeny of the F1 generation.

What are the genotypes and phenotypes of the F2 generation?

1. A Punnett square is a tool used to predict the genetic outcomes of a certain cross. Gametes from one parent are written along the top of the square and those of the other parent are written down the left side. Then the boxes are filled in with the genotypes that the particular gamete combination would produce.
2. A cross of two hybrids would produce a genotypic ratio of 1:2:1.

HOW DID GREGOR MENDEL DEMONSTRATE THE PRINCIPLES OF GENETICS?

1. Gregor Mendel was the first person to realize the significance of the ratios of different phenotypes in genetic crosses.
2. An allele is dominant when it alone determines the phenotype of a heterozygote; an allele is recessive when it contributes nothing to the phenotype of a heterozygote.

3. Now we know that a dominant allele usually specifies a functional protein, but a recessive allele usually specifies a protein that doesn't function.
4. Lethal recessive genes, such as Tay-Sachs, are rare because people who are homozygous for these alleles usually die before they reproduce.
5. When two alleles each contribute to the phenotype, the alleles lack dominance and are said to have partial dominance or codominance.
6. By convention, names of dominant and codominant alleles start with a capital letter and those of recessive alleles start with a small letter.

What did Mendel's F1 crosses show?

1. When individuals are crossed who have both a dominant and a recessive allele, the phenotypic ratio is 3:1.

The Principle of Segregation

1. According to the Principle of Segregation, each sexually reproducing organism has two "determinants" for each characteristic, and these two determinants segregate during the production of gametes. We now know these determinants are genes.
2. In a backcross, F1 heterozygotes are crossed with the homozygous parental stock that contained only the recessive allele. If it is done to reveal the genotype of the other parent, the cross is referred to as a testcross.

The Principle of Independent Assortment

1. Homozygous parents can produce only a single kind of gamete.
2. When individuals are crossed who are heterozygous for two traits, having a dominant allele for each trait, the phenotypic ratio that results is 9:3:3:1.
3. According to the Principle of Independent Assortment, each pair of genes is distributed independently during gamete formation.

WHAT WAS THE EVIDENCE FOR THE CHROMOSOMAL THEORY OF INHERITANCE?

1. Mendel's discoveries made no impact until 16 years after he died.
2. In 1902, Walter Sutton and Theodor Boveri showed that the behavior of chromosomes during meiosis accounted for Mendel's principles of segregation and independent assortment.

What were Sutton's arguments?

1. According to the Sutton-Boveri chromosomal theory of inheritance:
    - Chromosomes are actually pairs of chromosomes, one from each parent.
    - Synapsis consists of pairing the homologous maternal and paternal chromosomes.
    - The chromosomes retain their individual forms throughout the cell cycle.
    - Chromosomes contain genes.
    - Meiosis creates new combinations of genes in each generation.
    - Each chromosome carries a different set of genes, and all genes on one chromosome are inherited together.

How did Thomas Hunt Morgan bolster the theory of chromosomal inheritance?

1. As Morgan discovered, fruit flies have many advantages for genetic research.
2. In flies and other organisms with unmatched sex chromosomes, sex-linked genes have different inheritance patterns in males and females because they lie on chromosomes that also determine the gender of the offspring.
3. In mammals and flies, the X chromosome contains many more genes than the Y chromosome, and sex linkage therefore almost always refers to genes on the X chromosome.

Do genes on the same chromosome assort independently?

1. The principle of independent assortment depends on the independent assortment of chromosomes, but it cannot hold for genes that are physically attached to one another on the same chromosome.
2. Genes assort independently only when they are on separate chromosomes.
3. The tendency of genes on the same chromosome to stay together is called genetic linkage.
4. Linkage groups are sets of genes that do not assort independently.

Why did some flies have normal wings but purple eyes?

1. Homologous chromosomes frequently exchange material with each other, during meiosis, through crossing over.
2. This allows for unexpected results because the chromosomes in the gametes are different than the parents' chromosomes.
3. Homologous chromosomes frequently exchange material with each other during crossing over in meiosis, when single chromatids break and exchange material.
4. Crossing over results in a new chromosome that contains some genes from each of the homologs.

Can all the genes on a chromosome cross over and recombine?

1. The farther apart two genes are on the chromosome, the greater the likelihood that they will cross over.
2. Sturtevant realized that he could express the genetic distance between two genes as the chance that recombination would occur between them.
3. A map unit is the distance between two genes that would produce 1% recombinant gametes.
4. A genetic map, or linkage map, summarizes the distances between genes.
5. The position of a gene on a chromosome is its locus.

VOCABULARY BUILDING

In your own words, first write a brief definition, then a full explanation for each of the following terms. Include examples where appropriate. Complete this section from your memory—you will not learn the terms by simply copying definitions from the textbook. Once you have finished, check your responses against the information in the chapter and make any necessary corrections.

blending —

meiosis —

genetics —

transmission genetics —

molecular genetics —

phenotype —

genotype —

genome —

gene —

asexual reproduction —

clone —

sexual reproduction —

diploid —

homologous chromosomes —

homolog —

diploid —

gametes —

haploid —

egg/ovum —

sperm/spermatozoon —

germ cells/germ line —

gonads —

ovaries —

testes —

somatic cells —

syngamy/fertilization —

zygote —

meiosis I —

meiosis II —

prophase —

metaphase —

anaphase —

telophase —

synapsis —

chromatids —

tetrad —

crossing over —

chiasma —

disjunction —

nondisjunction —

euploid —

aneuploid —

trisomy 21 (Down syndrome) —

sex chromosomes —

autosomes —

alleles —

homozygous —

heterozygous —

true-breeding —

cross breeding —

hybrids —

parental (P) generation —

first filial (F1) generation —

second filial (F2) generation —

Punnett square —

dominant —

recessive —

partial dominance (codominance) —

Principle of Segregation —

backcross —

testcross —

Principle of Independent Assortment —

Sutton-Boveri chromosomal theory of inheritance —

hemizygous —

sex-linked genes —

genetic linkage —

linkage groups —

map unit —

genetic map/linkage map —

locus/loci —

spontaneous abortions —

CHAPTER TEST

The following test has four parts. Complete as much of the exam as you can from memory. If you cannot answer a question, skip it. Once you complete all that you can, try to answer any questions you skipped. If you still cannot answer them, consult your textbook for the answers. Once you have completed all sections of the test, check your answers for Parts 1 - 3 against those in the back of this book. Highlight any incorrect answers then review that material in your textbook. Correct your answers for future reference.

Part 1: Multiple Choice

For each of the following, select all correct responses—more than one may be correct.

1. From a single cell, at most, how many cells are formed at the end of meiosis?
   a) 1
   b) 2
   c) 4
   d) 8

2. The term genotype has which of the following meanings?
   a) the genetic constitution of an organism or cell
   b) all physical and behavioral characteristics of an individual
   c) a subset of genes that influence a particular trait
   d) all of these

3. Identical twins share the same:
   a) genotype
   b) phenotype
   c) genotype and phenotype

4. Which of the following is not a benefit of sexual reproduction?
   a) increased genetic diversity
   b) energy conservation
   c) greater likelihood that some offspring will survive
   d) greater chance the species can survive in a changing environment

5. Which of the following are haploid?
   a) somatic cells
   b) gametes
   c) zygote
   d) all of these

6. How many times does DNA replication occur during meiosis?
   a) 1
   b) 2
   c) 4
   d) 0

7. Which phase of meiosis lasts the longest?
   a) prophase I
   b) prophase II
   c) anaphase I
   d) anaphase II

8. What happens during anaphase I?
   a) homologs go to different poles
   b) homologs go to the same pole
   c) sister chromatids go to opposite poles
   d) sister chromatids go to the same pole

9. Nondisjunction would occur during which phase of meiosis?
   a) metaphase I or II
   b) telophase I
   c) anaphase I or II
   d) prophase II

10. An individual who is homozygous has:
    a) two different alleles for that gene
    b) two copies of the same allele for that gene
    c) only one allele for that gene
    d) no alleles for that gene

11. Hybrids are of what genotype?
    a) homozygous recessive
    b) homozygous dominant
    c) heterozygous
    d) any of these

12. When examining a single gene, what is the predicted genotypic ratio in the F2 generation if two F1 hybrids are crossed?
    a) all heterozygous
    b) 1:2:1
    c) 3:1
    d) all homozygous

13. In a testcross, an individual of unknown genotype is crossed with an individual of which genotype?
    a) homozygous recessive
    b) homozygous dominant
    c) heterozygous
    d) any of these

14. How many autosomes do humans have?
    a) 46
    b) 2
    c) 22 pairs
    d) 23

15. Gametes are produced in the:
    a) ovum
    b) somatic cells
    c) ovary
    d) testis

Part 2: Matching

For each of the following, match the correct term with its definition or example. More than one answer may be appropriate.

meiosis I **meiosis II** **mitosis**

1. ________________ Occurs in somatic cells.
2. ________________ Diploid cell produces four haploid cells.
3. ________________ Tetrads form.
4. ________________ Diploid cell produces two diploid cells.
5. ________________ Essential in sexual reproduction.
6. ________________ Process resembles mitosis but begins with a haploid cell.
7. ________________ Sister chromatids separate during anaphase.
8. ________________ DNA replication occurs immediately before this begins.

sexual reproduction **asexual reproduction**

9. ________________ Increases genetic diversity.
10. ________________ Produces clones.
11. ________________ Most energy-efficient.
12. ________________ Best suited for a changing environment.
13. ________________ Involves haploid gametes.
14. ________________ Produces unique individuals.

Part 4: Short Answer

Write your answers in the space provided or on a separate piece of paper.

1. Why are the blending model of inheritance and natural selection not compatible?

2. How do genotype and phenotype differ?

3. What is a gene?

4. In humans, what cells are haploid?

5. How does meiosis ensure that offspring have the normal number of chromosomes?

6. What is synapsis?

7. How does anaphase I of meiosis differ from anaphase II or mitotic anaphase?

8. How does nondisjunction lead to aneuploidy?

9. How many chromosomes of each type do humans have?

10. What genotype must an organism have if a recessive autosomal trait is expressed?

11. Why are disorders involving recessive sex-linked genes located on the X chromosome, such as hemophilia, more common in males?

12. What are some explanations for increased incidence of trisomy 21 with maternal age?

For questions 13-15, use this information:

In flamingos, imagine that a single dominant gene determines their ability to fly. This is the F gene. Another allele (f) causes them to turn to plastic and be fated to a life as a lawn ornament. Assume two heterozygous parents breed.

13. What is the genotype of the parents?

14. Design a Punnett square to determine the potential genotypes of their children.

15. How many of their children do you predict will be plastic?

***Part 4:* Critical Thinking: Using Your Knowledge**

Answer each of these in essay form, using complete sentences and paragraphs. Provide as much information as you can. (For extra essay practice, write out answers to the Review and Thought Questions in your textbook.)

1. Explain various ways in which sexual reproduction contributes to genetic diversity. Why is genetic diversity important?

2. Make a chart listing the major events that occur during each stage of meiosis I and II. Also, add illustrations of the chromosomes at each stage.

3. Thoroughly explain ways in which mitosis and meiosis are similar and different.

4. List and explain the six main points of the Sutton-Boveri chromosomal theory of inheritance.

5. Thoroughly explain how inheritance patterns will be different for traits that are controlled by genes located on autosomes (such as the ability to curl your tongue) and traits that are controlled by genes located on the X chromosome. Consider inheritance patterns for both dominant and recessive alleles, for males and females.

6. Some genes, such as the one that causes hairy ears, are located exclusively on the Y chromosome. How will this affect the inheritance pattern of these traits?

7. Identical twins have exactly the same genotype, and yet they are two distinct and unique individuals with differences. What are some specific things that can account for such differences?

8. People who inherit either a single X chromosome (XO) or an extra chromosome (XXX) may have some ill-effects, but usually live. But if a single Y chromosome and no X (YO) is inherited, the offspring will die before birth. Why is this situation lethal, but an X aneuploidy is not?

9. What events would have to occur to get an aneuploidy resulting in a zygote of YY? DO you think this offspring would be viable?

10. For trisomy 21, nondisjunction usually occurs in the female. Trace the steps of meiosis that produce an ovum that could cause trisomy 21. Trace the steps with the nondisjunction occurring in meiosis I, then do it a second time, with the nondisjunction occurring during meiosis II. What is the difference between the two paths?

Chapter 10
The Structure, Replication, and Repair of DNA

KEY CONCEPTS

1. DNA is composed of two polynucleotide strands wound together into a double helix.
2. Each nucleotide in DNA consists of a sugar, a phosphate, and a base. Sugar and phosphate groups form the "backbone," like the upright parts of a ladder, and the bases form the steps of the ladder.
3. The DNA bases always align so that A binds to T, and C binds to G.
4. Genes carry the information for polypeptide synthesis.
5. Viruses are parasitic assemblies of protein and either DNA or RNA that are capable of forcing host cells to manufacture more viruses.
6. DNA replication is semi-conservative. When DNA replicates, each half of the double helix acquires a new mate.
7. RNA polymerase creates an RNA primer that enables DNA polymerase to begin making a copy of each strand.
8. DNA replication occurs in replication forks where the two DNA strands have separated.
9. DNA replication occurs in only one direction. In the leading strand of a replication fork, nucleotides are added directly to the 3' end, but the other strand is produced in short fragments that are then joined together.
10. A mutation is any change in the sequence or number of nucleotides.
11. Mutations occur spontaneously, but they occur more frequently when DNA is exposed to mutagens.
12. Cells have elaborate mechanisms for repairing damage and for correcting mistakes made during DNA replication.

EXTENDED CHAPTER OUTLINE

WHAT IS THE STRUCTURE OF DNA?

1. When Rosalind Franklin started her work in 1951, the chemical makeup of DNA had been known for about 30 years.
2. DNA consists of two long polynucleotide strands.
3. Each nucleotide consists of a sugar, a phosphate, and a base. The sugar and phosphate units are linked together alternately into a sugar-phosphate "backbone." The bases are suspended from this backbone.
4. The four bases include two pyrimidines (cytosine and thymine) and two purines (adenine and guanine).
5. **A** always bonds with **T**; **C** always bonds with **G**.
6. The two polynucleotide chains run in opposite directions. Each strand has the third carbon (3') of one sugar attached by the phosphate group to the fifth (5') carbon of the next sugar. The chains are aligned so that the 3' end of one strand is at the 5' end of the other strand.
7. Overall, DNA resembles a ladder. The sugar and phosphate units form the upright parts, and
8. the bases form the rungs. The "ladder" is also twisted around its long axis.

How much did Franklin discover about the structure of DNA?

1. Franklin believed that DNA was probably a big helix with multiple parallel chains and the phosphates around the outside of the helix.
2. She also accurately measured the density of DNA, the number of water molecules associated with each nucleotide, and discovered that DNA structure differs when it is either wet or dry.
3. Her months of work yielded the main dimensions of DNA molecules and she suspected it was a double helix.

What were Chargaff's rules?

1. Erwin Chargaff found that:
   - although the proportions of the bases can vary widely from species to species, the amount if A is always equal to the amount of T, and the amount of C is always equal to the amount of G; and
   - the proportion of A plus T to C plus G is constant within a species.
2. This information became known as Chargaff's rules.
3. An AT pair has the exact same shape and size of a CG pair.

Franklin comes within two steps of the correct model

1. Franklin correctly concluded that DNA is a double helix, with a diameter of 2nm, and that the sugar-phosphate backbone is on the outside.
2. When Watson and Crick finalized their model, Franklin was two steps from solving it herself.
3. When Watson and Crick published their findings in *Nature*, Franklin was the one person whose work was not formally cited; instead, they explicitly asserted that they had been unaware of her work.
4. Watson and Crick's structure rested heavily on the unpublished data of Rosalind Franklin and her colleague, Raymond Gosling.
5. Rosalind Franklin died at age 37, never aware of her crucial contribution to science.

Modern science is a social endeavor

1. Franklin's failure to discover the structure of DNA was a direct result of the social isolation.
2. At least a half dozen scientists, through intellectual discourse, provided essential clues that guided Watson and Crick to their ultimate realization.

WHAT IS A GENE?

1. Genes specify the sequences of amino acids in polypeptides.
2. They are made of DNA.
3. The structure of DNA is especially favorable for their replication and repair.

How do genes affect biochemical processes?

1. The first clue that genes influence phenotype through biochemistry came from Archibald Garrod's work in 1902 on infants with the genetic disease alkaptonuria.
2. Garrod's work suggested that genes affect biochemical processes by specifying enzymes.

One gene—one enzyme

1. George Beadle and Edward Tatum induced mutations in *Neurospora* with x-rays, then looked for biochemical defects.
2. *Neurospora* is haploid, having only a single copy of each gene, so changes in the genes immediately show up in the phenotype.
3. Beadle and Tatum's work helped unravel previously unknown biochemical pathways because the mutations blocked the pathways at different steps.
4. Their work suggested that every gene somehow affects the function of a single enzyme, which was summarized as "One gene—one enzyme."

One gene—one polypeptide

1. Sickle-cell disease results from a recessive allele. The disease afflicts 1 out of every 625 African-Americans and is carried by 1 in 13.
2. Linus Pauling and his colleagues showed that the oxygen-bonding protein, hemoglobin, is different in sickle-cell patients.
3. A simple explanation for the disease emerged:
   - A recessive allele codes for an abnormal protein, HbS.
   - Heterozygotes have genes for both normal and abnormal proteins.
   - In the presence of HbA (normal), HbS does not cause the disease (heterozygotes are healthy).
   - Homozygotes for the recessive allele make only HbS and are sick.
   - In the absence of the normal protein, HbS causes sickle-cell disease.
4. Pauling extended "one gene—one enzyme" to "one gene—one protein."
5. In fact, hemoglobin consists of 4 polypeptide chains, and HbS differs from HbA in only a single amino acid (out of 146) in the β chains. So the gene did not affect the whole protein—merely one of its polypeptides.
6. "One gene—one polypeptide" means that genes generally influence phenotype by carrying the information needed for synthesis of polypeptides.

HOW DID BIOLOGISTS LEARN WHAT GENES ARE MADE OF?

1. To find what genes are made of, biologists turned to less complex, smaller organisms: bacteria and their viruses.

How could dead bacteria change the genetic properties of other bacteria?

1. In 1928, Frederick Griffith discovered that dead, virulent bacteria could transform harmless bacteria into virulent ones.

What was the transforming factor?

1. Transformation is thought to be the transfer of one or more genes from one organism to another.
2. In 1944, Oswald Avery and his associates concluded that the transforming factor is DNA.

What is a bacteriophage?

1. In 1915, Felix d'Herelle discovered that bacteria can be killed by a tiny infectious agent called a bacteriophage (or just phage).
2. A bactreriophage is a virus. It is not alive.
3. A virus is a molecular assembly of protein and either DNA or RNA—it is not a cell.
4. A virus parasitizes a cell, subverting the cell's machinery so it will produce more viruses.

What convinced biologists that genes are made of DNA?

1. Bacteriophages reproduce very rapidly, making them ideal for genetic studies.
2. Bacteriophages have genes, and, since the viruses consist of only DNA (or RNA) and protein, the genes had to either be DNA or protein.
3. Radioactively labeled phages showed that when a phage infects a bacterium, it is the DNA that enters the bacterium, not the protein.
4. Avery put forth a convincing argument that DNA was the genetic material, yet scientists remained skeptical.
5. Most scientists incorrectly believed that:
   - DNA would have to vary from one species to the next to account for all the diversity, but it is the same in all species; and
   - DNA was a "tetranucleotide" in which the same sequence of 4 bases was merely repeated over and over.
6. Chargaff's rules showed that species differences did exist, at the level of the bases.
7. It took from 1903 to 1952 to persuade biologists that DNA is the genetic material.

HOW DOES ONE STRAND OF DNA DIRECT THE SYNTHESIS OF ANOTHER?

1. Each strand of DNA has the capacity to direct the synthesis of the other, because a base on one strand can match with only one other base.
2. The structure of DNA explains both how cells copy their DNA and how genes specify polypeptides.

How do dividing cells copy the DNA?

1. The two strands of DNA are complementary, and the sequence of the bases in one strand can determine the sequence of bases in the other.
2. DNA replication is semiconservative—half of each parent DNA molecule will be present in each daughter molecule.
3. Each new DNA molecule is half old and half new—each half of the original double helix receives a new mate.

How does a cell begin assembling a new strand of DNA?

1. DNA is made through polymerization, a process by which many small molecules (nucleotides) are linked to form a larger molecule (DNA).
2. DNA polymerase is the enzyme that joins the nucleotides together.
3. Each strand of the DNA acts as a template for the assembly of a complementary strand.
4. DNA polymerase can only add nucleotides to an already existing polynucleotide called a primer.
5. DNA polymerase always adds nucleotides to the 3' end of the primer, so the polynucleotide grows in a single direction—from the 5' end to the 3' end.
6. The particular nucleotide added at each point depends on the base sequence of the template strand.
7. DNA replication does not begin with a DNA primer, but rather with a temporary RNA primer.

8. The RNA primer arises from the action of primase—an RNA polymerase that copies short stretches of DNA into RNA without requiring a primer.
9. DNA polymerase then uses the RNA primer's 3' end to start synthesizing the DNA primer.

How does a cell copy both strands of DNA simultaneously?

1. DNA in dividing tissue looks like long, thin molecules with bubbles; and as DNA replicates, these bubbles expand.
2. Each bubble consists of two replication forks—Y-shaped regions of DNA where the two strands of the helix are unzipped.
3. As replication proceeds, the forks move away from each other.

How does DNA replicate in the 3' to 5' direction?

1. As the replication fork moves along the DNA, it is moving in the 5' to 3' direction along one strand, but in the 3' to 5' direction along the other strand.
2. Nucleotides must be added to the 3' end.
3. The second (lagging) stand of DNA is replicated discontinuously.
4. After the fork moves about 1,000 nucleotides, primase produces a short piece of complementary RNA, which serves as a primer for DNA polymerase.
5. DNA polymerase then adds nucleotides to the 3' end of the RNA primer, which produces Okazaki fragments—short stretches of RNA connected to about 1000 DNA nucleotides of a DNA strand.
6. DNA polymerase then removes the RNA primer and an enzyme called DNA ligase joins the Okazaki fragments together.

How does DNA synthesis begin?

1. DNA replication always begins at a special DNA sequence called the origin of replication, with the help of various enzymes that open, untwist, and separate the two strands.
2. In eukaryotes, each cell contains thousands of origins of replication, and groups of 20 to 50 origins, called replication units, form replication forks at the same time.

WHAT IS THE ULTIMATE SOURCE OF GENETIC DIVERSITY?

1. Every living organism shares the same common chemistry.
2. Each species has its own characteristic DNA, but scientists believe life began on Earth just once.
3. Mutations are permanent changes in the DNA sequence.
4. Different alleles of a gene are created by mutations in the original DNA sequence.
5. All genetic differences ultimately depend on the accumulation of mutations in the genome.
6. The vast majority of mutations are either harmful or meaningless.
7. Occasionally, a new mutation is useful under certain circumstances, and in these cases, natural selection may increase its representation in the population.

What causes mutations?

1. Mutations occur spontaneously as random events in cells.
2. Occasionally, mutations occur because DNA polymerase places the wrong nucleotide in a sequence during the S (synthesis) phase of the cell cycle.
3. Certain agents, called mutagens, increase the rate of mutation.
4. Examples of mutagens are UV light, radiation, and many chemicals.

How often do mutations occur?

1. Changes in the DNA sequence occur spontaneously all the time.
2. The mutation rate is how often mutations occur; specifically, it is the number of changes per nucleotide per generation.

3. The mutation rate depends on:
   - how often a sequence mutates, and
   - how efficiently the cell repairs these mutations.
4. Mutation hot spots are DNA sequences that have a higher mutation rate than others.
5. The average mutation rate in bacteria during ordinary DNA replication is only one mutation in 10 million.

What kinds of mutations are there?

1. Point mutations change rather small sections of DNA—one or several nucleotide pairs.
2. Point mutations are of the following types:
   - base substitution—one base is replaced by a different one;
   - insertion—one or more nucleotides are incorrectly added; or
   - deletion—one or more nucleotides are removed.
3. Chromosomal mutations change rather large regions of chromosomes.
4. Chromosomal mutations are of the following types:
   - deficiencies—deletions that are larger than a few nucleotides;
   - translocations—exchange of material between two chromosomes;
   - inversions—a segment of DNA is flipped 180° from its normal position;
   - duplications—part of a chromosome is incorrectly repeated; or
   - aneuploidy—abnormal chromosome number, typically trisomy or monosomy.

HOW DO CELLS HANDLE MISTAKES IN THE NUCLEOTIDE SEQUENCE OF DNA?

1. Cells must be able to repair damaged DNA to maintain the continuity of life.
2. Both prokaryotic and eukaryotic cells have sophisticated mechanisms to repair damaged DNA or errors that occur during replication.
3. Repair mechanisms depend on
   - the redundancy of information in the structure of DNA, and
   - enzymes that recognize commonly occurring errors and remove them.
4. After correction enzymes have removed the error, DNA polymerase creates a new, correct stretch of DNA.
5. DNA ligase joins the new DNA with the rest of the strand.
6. Errors do still slip through this elaborate system.

VOCABULARY BUILDING

In your own words, first write a brief definition, then a full explanation for each of the following terms. Include examples where appropriate. Complete this section from your memory—you will not learn the terms by simply copying definitions from the textbook. Once you have finished, check your responses against the information in the chapter and make any necessary corrections.

pyrimidines (cytosine and thymine) —

purines (adenine and guanine) —

5' end —

3' end —

Chargaff's rules —

Part 3: Short Answer

Write your answers in the space provided or on a separate piece of paper.

1. What is meant by semiconservative replication?

2. What is meant by the two strands of DNA being complementary?

3. Why is a virus not a cell?

4. List the four bases found in DNA.

5. How did Chargaff's rules show that, although DNA is the same in all species, it still allows for genetic differences?

6. Purines and pyrimidines are of different sizes, so how is the diameter of DNA constant?

7. Why is the lack of recognition of Rosalind Franklin's work an example of how science is a social endeavor?

8. To what end of a polynucleotide are nucleotides added?

9. Ultimately, all genetic differences depend on what?

10. List some examples of mutagens.

11. What are mutation hot spots?

12. What are the three components of a nucleotide?

13. In what way are nucleotides linked together to give DNA its characteristic structure?

14. List and explain three examples of point mutations.

15. In general, how do cells correct errors?

Part 4: Critical Thinking: Using Your Knowledge

Answer each of these in essay form, using complete sentences and paragraphs. Provide as much information as you can. (For extra essay practice, write out answers to the Review and Thought Questions in your textbook.)

1. Give highlights of the historical quest for the true structure of DNA. What were Franklin's contributions? What were Watson and Crick's contributions? Why were Franklin's efforts not rewarded?

2. Explain the overall process of DNA replication. How does the process differ for the two strands? Use a drawing to illustrate the events.

3. Explain ways in which the structure of DNA is ideally suited for the replication and repair processes.

4. Thoroughly discuss mutations. What are they, how do they occur, what types are there, and how can these effect the cell, the organism, and the species?

5. A botanist discovers a new plant that has a very aromatic flower. When doing a molecular analysis, he discovers that its DNA is 22 percent adenine. Based on this, how much thymine does it have? How much guanine and cytosine?

6. Skin cancer rates are skyrocketing, and it is believed that the thinning of the ozone layer, which protects us from ultraviolet radiation, may be the culprit. Explain a possible mechanism by which thinning of the ozone layer may lead to skin cancer.

7. What would be the outcome of a mutation that altered DNA polymerase so that it was nonfunctional?

8. In what ways does the structure of DNA protect it from mutations?

9. Some genes are regulatory genes, rather than structural genes. How can these genes tie in with the idea of "one gene—one polypeptide."

10. During DNA replication, one strand is replicated continuously, but the other is replicated discontinuously. Which strand do you think is more susceptible to mutation? Why aren't the two strands replicated from opposite ends but continuously?

Chapter 11
How Are Genes Expressed?

KEY CONCEPTS

1. The genetic code translates the language of DNA and RNA into the language of polypeptides. Three nucleotides—a codon—specify one amino acid.
2. The genetic code is nearly universal and redundant.
3. Genetic information flows from DNA to RNA to polypeptides.
4. Transcription is the process by which the information contained in the DNA's nucleotides is stored in mRNA molecules. In eukaryotes, this process occurs in the nucleus.
5. Translation is the process by which the information stored in the mRNA is used to build a polypeptide. In eukaryotes, this process occurs in the cytoplasm.
6. During transcription, RNA polymerase, with help from transcription factors, produces molecules of mRNA by copying the antisense strand of DNA, starting at a promoter sequence.
7. Pre-mRNA contains long stretches of noncoding sequences, called introns, that are removed. Then the remaining sequences (exons) are spliced together to form mature mRNA.
8. A single gene can encode several different polypeptides.
9. During translation, ribosomes link amino acids together. The small subunit of a ribosome reads the mRNA, the larger subunit makes the peptide bonds between the amino acids.
10. Transfer RNA (tRNA) participates in translation by first bonding to a certain amino acid, then bonding to the correct mRNA codon.
11. The ribosome continues its translation until it reaches a stop codon.
12. Cells "address" the polypeptides with signal peptides that direct the polypeptides to the correct destination.
13. Cells have many ways in which to regulate gene expression, but the most efficient and common way is through transcriptional control.
14. Regulatory genes, which include both the regulatory site and genes for transcription factors, regulate the amount of polypeptide made by a cell.
15. Eukaryotes can regulate gene expression in more ways than prokaryotes because of:
    - the presence of a nuclear membrane,
    - how the DNA is folded and packaged, and
    - the many compartments inside the cell.

EXTENDED CHAPTER OUTLINE

HOW WAS THE GENETIC CODE DISCOVERED?

1. Once Watson and Crick discovered DNA's structure, scientists everywhere wanted to solve the next mystery—the genetic code.
2. In 1955, Marianne Grunberg-Manago discovered an enzyme that could link nucleotides to form long polynucleotide chains, then she began making artificial RNA polymers.
3. Marshall Nirenberg set out to synthesize the polypeptides that make proteins.
4. Nirenberg and Johann Heinrich Matthei together asked "What kinds of RNA stimulate the synthesis of polypeptides?" They believed RNA must carry a message from DNA (in the nucleus) to the ribosome (in the cytoplasm), where protein synthesis usually occurs.
5. Matthei worked with the artificial RNA polymers Grunberg-Manago had created and got spectacular results.
6. Nirenberg and Philip Leder showed that a three-nucleotide length of RNA, now called a codon, could specify an amino acid.
7. Nirenberg, Gobind Khorana, and others completed a dictionary for the genetic code in 1966. Nirenberg and Khorana won a Nobel Prize for this work in 1968.

What is the genetic code like?

1. The DNA/RNA language has an alphabet of just four nucleotides, and the polypeptide alphabet has 20 amino acids.
2. Each set of three nucleotides, called a codon, specifies one amino acid.
3. The genetic code is redundant—several codons may have the same meaning.
4. Of the 64 triplets, only 61 specify amino acids. The remaining three (UAA, UAG, and UGA) are nonsense, or stop, codons, which signal the end of a polypeptide, rather than coding for an amino acid.
5. The genetic code is said to be universal because the same code is used by almost all species. (Mitochondria use a slightly different code.)

GENETIC INFORMATION FLOWS FROM DNA TO RNA TO POLYPEPTIDES

1. RNA is always an intermediate in protein production.
2. Crick's central dogma of molecular biology states that "DNA specifies RNA, which specifies proteins."
3. The process by which DNA transmits its information to RNA is called transcription.
4. The information in the RNA, which was made from the DNA, is what is used to make polypeptides.
5. The RNA that is transcribed from the DNA is messenger RNA (mRNA), and these molecules contain the same information that is stored in the DNA.
6. RNA differs from the DNA in three ways:
   - the backbone sugar is ribose, not deoxyribose;
   - uracil (U) replaces thymine (T) as one of the pyrimidine bases; and
   - RNA is usually single-stranded.
7. Translation is the process by which the mRNA message is converted into strings of amino acids.
8. The overall process of polypeptide synthesis can be summarized as:

$$\text{DNA} \xrightarrow{\text{transcription}} \text{RNA} \xrightarrow{\text{translation}} \text{polypeptide}$$

9. In eukaryotes, transcription occurs in the nucleus and translation occurs in the cytoplasm, at the ribosome, where amino acids are joined together.

RNA POLYMERASE TRANSCRIBES DNA INTO RNA

1. A gene is expressed when the cell makes that gene's polypeptide.
2. Gene expression starts when the enzyme RNA polymerase transcribes DNA into mRNA.

How does RNA polymerase begin transcription?

1. A special sequence of DNA, called a promoter, acts as a start signal for RNA polymerase. When the enzyme recognizes the promoter, it unwinds and separates the two DNA strands and attaches to the promoter.
2. In bacteria, RNA polymerase binds to the promoter and starts making an RNA strand that is complementary to the DNA template strand.
3. Eukaryotic RNA polymerases do not recognize the promoter. They are directed to the right promoter by special proteins called activators or transcription factors.
4. Both eukaryotic and prokaryotic RNA polymerases stop transcription at special sequences called terminators.

How does the cell alter the mRNA transcribed from the DNA?

1. In prokaryotes, as soon as the mRNA is made, it binds to ribosomes and begins polypeptide synthesis.
2. In eukaryotes, the mRNA made during transcription is called pre-mRNA and undergoes modifications before it is mature mRNA that can be translated.
3. The modifications to mRNA include:
   - capping the 5' end with GTP;
   - adding a poly-A "tail" to the 3' end of most mRNA molecules; and
   - removing large pieces of RNA and splicing the remainders together.
4. Genes are interrupted by nucleotide sequences that do not code for amino acids. In the mRNA, these sections are called introns (intervening sequences) and they are cut out. The parts left, called exons (expressed sequences) are then spliced together.
5. Most of the DNA in a gene consists of introns. No one knows what introns do.

One gene—Heaven only knows how many polypeptides

1. Cells can sometimes transcribe many different mRNAs from the same gene, so a single gene may encode several polypeptides and the cell may express all of these various polypeptides at once.
2. Most molecular biologists now say that a gene is a DNA sequence that is transcribed as a unit and encodes either a single polypeptide or a set of related polypeptides.

HOW DOES A CELL TRANSLATE mRNA INTO A POLYPEPTIDE?

1. Types of RNA include:
   - messenger RNA, which carries the genetic information from the DNA;
   - transfer RNA, which carries each amino acid to the corresponding mRNA codon at the ribosome; and
   - ribosomal RNA, which is part of the ribosome where polypeptides are made.

What do ribosomes do?

1. Ribosomes, found in all cells, link amino acids together into polypeptides.
2. Ribosomes are made of proteins and ribosomal RNA, and have a small and a large subunit.
3. Ribosomes' main tasks are to read the information in the mRNA and translate it into a sequence of amino acids.
4. The grouping of nucleotide triplets is called the reading frame from translation.
5. Molecules of mRNA contain start and stop signals.

6. Each ribosome contains two grooves—a groove on the smaller subunit reads the mRNA, recognizing start and stop signals; a groove on the larger subunit holds the growing polypeptide.
7. The grouping of nucleotide triplets is called the reading frame for translation.

How do tRNAs serve as adapters between mRNAs and amino acids?

1. A transfer RNA (tRNA) molecule is folded into a cloverleaf shape. One loop attaches to the correct amino acid. The opposite side has a sequence of three nucleotides, called the anticodon, which binds to the appropriate mRNA codon.
2. A given tRNA has only one kind of anticodon and can pick up only one kind of amino acid.

How do tRNAs recognize the correct amino acid?

1. Linkage of tRNAs to their corresponding amino acids to form "charged" tRNAs—tRNAs joined to their amino acids by high-energy bonds—depends on a set of enzymes.
2. Each charging enzyme has two binding sites—one holds a certain type of tRNA, the other holds the corresponding amino acid.

How do ribosomes begin translation?

1. Translation begins with initiation—when the ribosome's small subunit attaches to the mRNA.
2. The initiation (start) site in mRNA is always the codon AUG, which specifies methionine.
3. The initiation process aligns the mRNA for proper reading and moves the first amino acid into place.

How do ribosomes make peptide bonds?

1. Elongation of the polypeptide has three steps:
   - putting the next amino acid in position;
   - forming a peptide bond between the growing chain and the next amino acid; and
   - moving the ribosome to the next codon.
2. The large ribosomal subunit has three pockets. The P (peptide) site and the A (amino acid) site bind two tRNAs to adjacent mRNA codons, and the E site binds each tRNA as it leaves the ribosome.
3. When the P and A sites are both filled, an enzyme within the ribosome links the amino acids with a peptide bond.
4. The tRNA in the P site falls away, via the E site, and the ribosome moves forward, so that the tRNA in the A site is now in the P site, attached to the polypeptide chain, and this leaves the A site ready to accept the next tRNA.
5. The cycle continues until all the codons are read.

Several ribosomes can read a single strand of mRNA simultaneously

1. Several ribosomes may be attached to a single mRNA at one time, each overseeing the production of identical polypeptides. This complex is called a polysome.
2. A cell can make many copies of a polypeptide at once.

How does polypeptide synthesis stop?

1. A stop codon (UAA, UAG, or UGA) in the mRNA signals the end of the polypeptide chain. This process is called termination.
2. When a stop codon is reached, there is no corresponding tRNA, so, instead, a release factor (an enzyme) binds to the stop codon, and the enzyme that formed the peptide bonds cuts the polypeptide from the last tRNA, releasing it into the cytoplasm.
3. Once a polypeptide chain is made, it must be folded into a functioning protein.

How do cells direct newly made proteins to different destinations?

1. Many newly made polypeptides have "address labels," which are blocks of 4 to 12 extra amino acids called signal peptides, that direct the polypeptides to the proper location either inside or outside the cell.

HOW DO CELLS REGULATE GENE EXPRESSION?

2. Cells contain thousands of genes, some of which are expressed as polypeptides all the time, some occasionally, and many never.
3. Prokaryotes transcribe almost all of their genes, but eukaryotes regulate gene expression more closely.
4. A eukaryotic cell produces different proteins according to its needs.
5. A cell can make thousands of copies of a polypeptide, or only two or three copies.
6. Cell specialization in most multicellular organisms requires differential expression of genes in cells that have identical genomes.
7. A typical eukaryotic cell transcribes only about 20% of its DNA into RNA.
8. Cells can regulate gene expression at many levels.
9. Transcriptional control, in which cells increase or decrease the amount of mRNA transcribed, is the most common and efficient means of regulation.
10. In posttranscriptional control, the cell transcribes the mRNA but modifies the pre-mRNA before it reaches the cytoplasm.
11. In translational control, the cell finishes the mRNA but regulates the rate at which it is translated.
12. In posttranslational control, the cell makes the polypeptide, but then modifies its structure and behavior.

How do prokaryotes regulate gene expression?

1. Bacteria usually regulate gene expression by controlling transcription.
2. Bacteria must respond to their environment.
3. When a bacterium's environment contains high concentrations of amino acids, it devotes most of its energy to rapid division; when amino acids are not available, the bacterium rapidly synthesizes amino acids.
4. To accomplish this, the bacterium must make the right enzymes at the right time.
5. An operon is a stretch of DNA that includes several genes under coordinated control.
6. Regulatory genes are those that help govern the amount of a polypeptide made by the cell. Some regulatory genes direct the synthesis of transcription factors, which are proteins that influence the expression of other genes.
7. The lactose (*lac*) operon in *E. coli* is the most studied operon in prokaryotes. This system allows the bacterium to produce enzymes needed to metabolize lactose only when the enzymes are needed.
8. Jacob and Monod discovered two genes that regulate expression of the *lac* operon. One is a structural gene that codes for the *lac* repressor protein, which prevents expression of the *lac* operon. The other gene is the operator—a regulatory site to which the *lac* repressor can bind.
9. The operator overlaps the promoter, so when the *lac* repressor is attached to the operator, RNA polymerase cannot attach to the promoter to initiate transcription. The enzymes are not made.
10. When lactose is present, it binds to another site on the *lac* repressor, and this prevents the *lac* repressor from binding to the operator. Transcription can occur and the enzymes needed to use the lactose are synthesized.
11. The *lac* repressor is an example of a negative regulator—a molecule that inhibits transcription.

Positive regulation in prokaryotes

1. Bacteria also use positive regulators (or activators)—molecules that increase the rate of transcription.
2. Catabolite activator protein (CAP) is a positive regulator. CAP has two binding sites—one for a particular DNA sequence and one for cyclic AMP.
3. Cyclic AMP is a signaling molecule that conveys signals from one cell to another. High cyclic AMP concentrations mean that glucose is unavailable so another energy source, such as lactose, must be used.
4. In the presence of lactose, CAP can bind to the promoter to stimulate expression of the *lac* operon. This increases the rate at which RNA polymerase binds and starts transcription.
5. The combination of negative and positive control allow bacteria to quickly respond to environmental changes.

How do eukaryotes regulate gene expression?

1. Most control in eukaryotes also occurs at the transcription level.
2. Presence of a nuclear membrane has important consequences, such as:
   - Only proteins in the nucleus can contribute to regulation;
   - The mRNA can be modified before it is released to the cytoplasm for translation (this is posttranscriptional regulation).
3. Eukaryotic transcription requires a much more elaborate set of activators than in prokaryotes.
4. Eukaryotes also have many intracellular compartments and elaborate mechanisms for the distribution of newly made proteins. Varying the rate at which regulatory proteins are sent into the nucleus is another means of gene regulation.

How complex is the problem of gene regulation in eukaryotes?

1. A transcription factor can affect transcription in two ways:
   - it can change the rate of transcription, or
   - it can produce a permanent change in chromatin structure that makes a gene more or less accessible for transcription.
2. Some DNA sequences that regulate gene expression identified so far are:
   - promoters, which specify the starting point and direction of transcription;
   - response elements (enhancers), which usually stimulate transcription of adjacent DNA without serving as promoters;
   - silencers, which inhibit transcription; and
   - termination signals.
3. Biologists estimate that mammals have about 5000 transcriptional activators and a corresponding number of response elements. Each transcriptional activator may interact with multiple response elements, and each response element can respond to more than one transcription factor.
4. In eukaryotes, the huge number of both regulatory proteins and their corresponding binding sites to DNA allow nearly unlimited interactions, so gene expression is an incredibly complex issue.

A single transcription factor may affect many genes

1. The absence of a single transcription factor (for example the androgen receptor) can significantly change gene expression and thus the resulting phenotype.

VOCABULARY BUILDING

In your own words, first write a brief definition, then a full explanation for each of the following terms. Include examples where appropriate. Complete this section from your memory—you will not learn the terms by simply copying definitions from the textbook. Once you have finished, check your responses against the information in the chapter and make any necessary corrections.

codon —

genetic code —

nonsense (stop) codons —

universal —

central dogma —

transcription —

messenger RNA (mRNA) —

translation —

RNA polymerase —

promoter —

activators (transcription factors) —

terminators —

introns —

exons —

ribosomal RNA (rRNA) —

reading frame —

transfer RNA (tRNA) —

anticodon —

initiation —

start site —

elongation —

polysome —

termination —

release factor —

signal peptide —

transcriptional control —

posttranscriptional control —

translational control —

posttranslational control —

operon —

lac operon —

induces —

lac repressor —

operator —

negative regulator —

positive regulator —

catabolite activator protein —

androgen receptor —

androgen insensitivity syndrome —

acridine dyes —

template (antisense) strand —

sense strand —

missense mutation —

conditional mutations —

nonsense mutation —

silent mutation —

frameshift mutation —

reversion —

reversion (suppressor) mutation —

antibiotic —

CHAPTER TEST

The following test has four parts. Complete as much of the exam as you can from memory. If you cannot answer a question, skip it. Once you complete all that you can, try to answer any questions you skipped. If you still cannot answer them, consult your textbook for the answers. Once you have completed all sections of the test, check your answers for Parts 1 - 3 against those in the back of this book. Highlight any incorrect answers then review that material in your textbook. Correct your answers for future reference.

Part 1: Multiple Choice

For each of the following, select all correct responses—more than one may be correct.

1. How many letters are in the DNA alphabet?
   a) 2
   b) 4
   c) 20
   d) 23

2. How many codons are possible?
   a) 4
   b) 20
   c) 46
   d) 64

3. Which of the following does not use the same universal genetic code as humans?
   a) plants
   b) most bacteria
   c) mitochondria
   d) all of these

4. What does a codon specify?
   a) a nucleotide
   b) an amino acid
   c) a polypeptide
   d) a protein

5. Which of the following is not a way that mRNA differs from DNA?
   a) mRNA contains the sugar ribose
   b) mRNA is composed of nucleotides
   c) mRNA contains uracil instead of thymine
   d) mRNA is single-stranded

6. Pre-mRNA contains noncoding sections that are removed before the mRNA leaves the nucleus. These noncoding sequences are called:
   a) introns
   b) exons
   c) point mutations
   d) codons

7. Which type of RNA carries an amino acid to the ribosome to be linked with others?
   a) mRNA
   b) tRNA
   c) rRNA
   d) all of these

8. What determines the order in which amino acids are joined at the ribosome to form a polypeptide?
   a) tRNA
   b) rRNA
   c) mRNA
   d) the ribosome

9. How do polypeptides get directed to their correct destinations?
   a) Signal peptides, which are extra blocks of amino acids, act as "address labels."
   b) Messenger RNA molecules carry the polypeptides to their destinations.
   c) Transfer RNA molecules carry the polypeptides to their destinations.
   d) Spindle fibers form and direct the polypeptides' movements.

10. Modification of the pre-mRNA is an example of what type of control?
   a) transcriptional
   b) posttranscriptional
   c) translational
   d) posttranslational

11. A stretch of DNA that contains several genes under coordinated control is called a(n):
   a) polysome
   b) codon
   c) transcriptional factor
   d) operon

12. Most regulation of gene expression occurs at what level?
   a) transcriptional
   b) posttranscriptional
   c) translational
   d) posttranslational

13. How long is a codon?
   a) 1 nucleotide
   b) 2 nucleotides
   c) 3 nucleotides
   d) Codon length varies with the species.

14. A point mutation that changes the codon for an amino acid to one that specifies a stop codon is called a:
    a) conditional mutation
    b) nonsense mutation
    c) frameshift mutation
    d) suppressor mutation

15. Insertions or deletions cause what type of mutation?
    a) conditional mutation
    b) nonsense mutation
    c) frameshift mutation
    d) suppressor mutation

***Part 2:* Matching**

For each of the following, match the correct term with its definition or example. More than one answer may be appropriate.

translation	**transcription**
1. ________________	Occurs in the nucleus in eukaryotes.
2. ________________	Involves DNA and mRNA.
3. ________________	Information in mRNA is used to build a polypeptide.
4. ________________	Occurs at the ribosome.
5. ________________	Involves three types of RNA.
6. ________________	Codons are matched to anticodons.

mRNA **tRNA** **rRNA** **DNA**

7. ________________	Always located in the nucleus in eukaryotes.
8. ________________	Carries information from the DNA to the ribosome.
9. ________________	Carries amino acids to the ribosome.
10. ________________	Is a component of the ribosome.
11. ________________	Moves from the nucleus to the cytoplasm.
12. ________________	Attaches to the ribosome.

***Part 3:* Short Answer**

Write your answers in the space provided or on a separate piece of paper.

1. What is meant by saying that the genetic code is redundant?

2. What are the nonsense (stop) codons?

3. What is stated by Crick's central dogma?

4. What happens during transcription?

5. Why are sections of pre-mRNA removed before the mRNA is translated?

6. What enzyme is involved in transcription?

7. What is the reading frame for translation?

8. What does the large subunit of the ribosome do?

9. What starts initiation of translation?

10. What is the start site for translation, and what amino acid does it specify?

11. What are the three steps in polypeptide elongation?

12. What are the A and P sites on the ribosome for?

13. Why do bacteria regulate expression of their genes?

14. Why can the sense strand of DNA not be transcribed?

15. What is a conditional mutation?

Part 4: Critical Thinking: Using Your Knowledge

Answer each of these in essay form, using complete sentences and paragraphs. Provide as much information as you can. (For extra essay practice, write out answers to the Review and Thought Questions in your textbook.)

1. Thoroughly discuss the steps involved in synthesis of a polypeptide. Include all details about transcription and translation.

2. Describe the components and functioning of the *lac* operon. Why is it needed? How does CAP affect functioning of this operon?

3. Discuss ways in which eukaryotic gene expression is regulated. Include discussion of regulation at each stage of polypeptide synthesis.

4. Explain various reasons that eukaryotic regulation of gene expression is more complex than regulation in prokaryotic cells. What benefits are there for eukaryotes by having this increased complexity of control?

5. How is the four-letter genetic alphabet translated into the twenty-amino acid polypeptide alphabet?

6. What is meant by saying there is redundancy in the genetic code? In what ways is this advantageous? What is the significance of this when a mutation occurs?

7. Which type of mutation do you think would be more disruptive—a base insertion, a base deletion, or a base substitution, and why?

8. What explanations can you give for the universality of the genetic code? How might this relate to the idea of evolution of species from common ancestors?

9. What are some possible reasons for why most of our DNA is composed of introns which do not code for polypeptides? Why might these intervening sequences be there? Where might they have come from? Do you think they serve any purpose?

10. What harmful effects might occur from overuse of antibiotics and other antibacterial compounds?

Chapter 12
Jumping Genes and Other Unconventional Genetic Systems

KEY CONCEPTS

1. Unconventional genetic systems include all genes that are not part of a cell's nuclear genome, such as mitochondrial and chloroplast DNA and all mobile genes.
2. A virus is a large complex that contains a core of nucleic acid and protein and is enclosed in a protein capsid or membrane envelope.
3. Viruses can infect a cell, and the infection may be helpful, harmful, or harmless to the cell.
4. Viruses enter cells through various mechanisms, then force the cell to make viral proteins.
5. A plasmid is a circular piece of DNA or RNA, capable of autonomous replication, which carries genes from one bacterial cell to another.
6. A transposon is a mobile DNA sequence that can only replicate when it is part of the host cell's DNA. Transposons include a gene for the enzyme transposase, which catalyzes insertion into new sites.
7. Both eukaryotes and prokaryotes contain mobile genes, which can move within the genome of the host cell and affect expression of other genes.
8. Tumor viruses integrate their DNA into that of a eukaryotic host, then force the host to replicate viral and genomic DNA. The viral proteins disrupt the host cell's ability to control cell division, causing a tumor to form.
9. Episomes are viruses or plasmids that switch back and forth between integrated and autonomous replication as needed.
10. Retroviruses use reverse transcriptase to transfer information from RNA to DNA, which violates the central dogma.
11. Molecular biologists believe that, through evolution, transposons acquired the ability to replicate autonomously, making them plasmids. Plasmids then later acquired the genes for making a protein coat.
12. Mitochondria and chloroplasts rely on both their own genetic information and that in the nucleus.
13. Energy organelles—mitochondria and chloroplasts—contain independent genetic systems and probably evolved from ancient associations with prokaryotes.

EXTENDED CHAPTER OUTLINE

JUMPING GENES AND INDIAN CORN

1. Multicolored Indian corn is "normal." Pale corn like we buy results from mutation, but this mutation can completely or partly reverse itself.
2. Barbara McClintock identified a gene that could move into a pigment gene causing changes in expression of the gene.
3. She coined the term "transposon" to describe such genes that jump around in the genome, but they are also called "jumping genes."
4. McClintock also realized that regulatory genes can finely regulate expression of structural genes.
5. Due to its complexity, for 30 years geneticists did not even try to understand McClintock's work.
6. In 1983, McClintock was awarded the Nobel Prize for her tremendous contributions to biology.
7. Unconventional genes are either not in the regular genome of the cell nucleus, or they move about the genome.
8. DNA in mitochondria and in chloroplasts does not move, but it is considered unconventional because it is passed down through the organelles without sexual recombination.
9. Jumping genes, or transposons, are the simplest of the mobile genes (genes that can move around). Other examples are bacterial plasmids and viruses.

WHAT IS A VIRUS?

1. A virus is not alive—it is an assemblage of mostly nucleic acids, arranged in a core, and a protein or membrane coat.
2. Most viruses do not cause disease.
3. Bacteriophages are bacteria-infecting viruses.
4. Some of a virus's genes direct synthesis of proteins and nucleic acids that make up the virus, but other genes specify proteins that regulate the host cell's DNA.

What is the form of a virus?

1. Viruses have many shapes, but most are either a long helix or spherical. Some have very complex forms.
2. A virus's outside surface is either a protein coat, called a capsid, or a membrane envelope similar to a cell membrane.
3. The nucleic acid may be either DNA or RNA.

How do viruses make more viruses?

1. Viruses can make more viruses in many ways, but their histories always include:
   - the virus attaches to the host cell;
   - the viral nucleic acid enters the cell;
   - the cell synthesizes proteins specified by the virus's genes;
   - the cell replicates the viral DNA or RNA;
   - the new viral protein and nucleic acids made by the host cell are assembled into new viruses (viral particles); and
   - the new viruses are released from the cell.

How does a virus attach to a cell?

1. A given kind of virus can only infect certain cell types.
2. Specific proteins on the surface of the virus attach to specific receptors on the surface of the host cells.

How does the viral DNA or RNA enter the host cell?

1. How the viral nucleic acid enters the host cell depends on the kind of virus and the kind of cell. Some viruses have an enzyme that digests a hole in the cell wall. Other viruses enter cells by endocytosis, and still others are enclosed by bits of the host cell's membrane, making it easy for them to enter the host cell.

How are virus particles manufactured and released?

1. Once inside the cell, the virus's genes direct synthesis of viral proteins in many ways. In most cases, the host cell is forced to make viral proteins instead of its own.
2. Lytic viruses contain a viral enzyme that causes the cell to rupture (lysis) once new viruses are made, releasing the new viruses. Not all viruses destroy the host cell.

HOW DO VIRUSES AND OTHER MOBILE GENES REPLICATE THEIR GENES?

1. A virus is the most elaborate of the mobile genes—many others consist of only DNA or RNA, but they also include sequences that force the host cell to make copies of their genes.
2. Mobile genes have two reproductive strategies:
   - In autonomous replication, the genes are separate from the host's DNA and can be replicated at any time.
   - In integrated replication, the genes become integrated into the host's DNA and are replicated along with those of the host (they can only be replicated when the host's genes are replicated).
3. Some mobile genes switch between the two reproductive strategies as circumstances change.

How do plasmids replicate autonomously?

1. A plasmid is a small piece of circular DNA or RNA in a bacterial or yeast cell.
2. Most plasmids exist as independent entities, instead of integrating into the host cell's DNA.
3. Many plasmids can replicate autonomously because they contain origins of replication.
4. A cell may contain up to 1000 copies of each plasmid, and many different kinds of plasmids.
5. Bacteria can engage in conjugation—a sexlike process during which two bacteria temporarily join together and exchange genetic material—which allows plasmids to move between bacteria.
6. Plasmids can offer benefits for their hosts. Some contain resistance (R) factors that make the cell resistant to antibiotics.

Many mobile genes can replicate only when integrated into a host cell's DNA

1. A transposon is mobile DNA that can only replicate when it is integrated into the host cell's DNA.
2. Transposons, or jumping genes, either move to new locations or duplicate themselves so they can be repeatedly inserted elsewhere.
3. Each transposon contains a gene for the enzyme transposase, which catalyzes insertion into new sites.
4. Transposons are common in all cells.
5. Eukaryotic transposons are surprisingly mobile—they can jump around in the genome, duplicating themselves, disrupting other genes, deleting and repeatedly duplicating whole stretches of DNA, breaking chromosomes, and even attaching part of one chromosome to another.
6. By rearranging the genome and disrupting genes, transposons increase genetic variability.
7. A simple transposon specifies only transposase.

8. Two simple transposons located close to each other can jump together and carry with them the host DNA that lies between them. These arrangements are called complex transposons.
9. Due to the mobility of complex transposons, antibiotic resistance can quickly and easily spread from one bacteria to many others, including different types of bacteria.

Tumor viruses also insert themselves into chromosomal DNA

1. Viruses that cause tumors multiply only when their eukaryotic host cells do.
2. Genes of most tumor viruses become integrated into their host's DNA and stay there, using the host's enzymes to express their viral genes.
3. Some viral genes, called oncogenes, cause production of viral proteins that disrupt the cell's normal control of cellular division, leading to tumor formation.

Episomes can replicate both autonomously and through integration

1. Episomes are viruses or plasmids that switch between autonomous and integrated replication.
2. The integrated form of a virus is called a provirus (prophage in a bacteriophage).
3. During the lytic cycle, a virus's DNA is separate from the host's and the virus lyses the host.
4. The lysogenic cycle consists of the provirus/prophage cycle and the lytic cycle.
5. Viruses in the provirus cycle cannot infect other cells because they do not lyse their hosts—they stay inside the host.
6. Because the viral DNA is integrated into the host DNA, viral DNA is replicated each time host DNA is, and all daughter cells will be infected with the provirus and the organism cannot get rid of the virus.
7. Certain environmental stresses, such as UV light or various chemicals, can trigger the lytic cycle and activate the virus.

How do RNA viruses reproduce and express their genes?

1. Many mobile genes do not use double-stranded DNA as their genetic material.
2. Some viruses and plasmids have RNA genes.
3. RNA viruses all rely on standard base pairing to make a complementary strand.
4. Plus-strand viral RNA works like mRNA, actually coding for viral proteins. Minus-strand viral RNA is antisense RNA, and it is transcribed into RNA that works like mRNA.

RNA tumor viruses present special problems

1. Howard Temin and David Baltimore independently discovered that RNA tumor viruses contain reverse transcriptase (as well as the gene for it), which copies RNA into DNA.
2. Viruses that use reverse transcriptase are called retroviruses, and the information flows from RNA to DNA (reverse of the normal), violating the central dogma.
3. In most viruses, only the nucleic acids enter the host cell, but in retroviruses reverse transcriptase enters along with the RNA.
4. Once DNA is made from the RNA via reverse transcriptase, the DNA is integrated into the host cell's DNA.

HOW DO MOBILE GENES EVOLVE?

Although viruses are not alive, they are still subject to natural selection

1. Hosts that evolve ways to resist viral infection are more likely to survive and reproduce.
2. Some bacteria produce restriction enzymes that help guard against viral infection.
3. Restriction enzymes cut all foreign DNA into little pieces by cutting at specific target sites.
4. As hosts evolve ways for resisting viral infection, viruses evolve mechanisms to get past these defenses.
5. Viral multiplication cycles have evolved to become highly efficient.

6. Viral regulatory genes are so powerful that the host cell's machinery recognizes them over its own genes.
7. Every gene in a virus is important, unlike the eukaryotic genome that is filled with junk DNA.

How may viruses have evolved?

1. Some scientists think viruses originated as single-celled organisms that became parasites within other cells, eventually losing most cellular components as they let the host's machinery do all the work.
2. Others theorize that, over time, transposons acquired the ability to replicate autonomously, becoming plasmids.
3. Plasmids then may have acquired genes for making protein coats.

MITOCHONDRIA AND CHLOROPLASTS CONTAIN THEIR OWN DNA

1. Because chloroplasts and mitochondria contain their own DNA, almost every eukaryotic cell carries at least one genetic system independent of a nucleus.
2. The reproduction of mitochondria and chloroplasts depends on cellular machinery that is specified both by the DNA in the nucleus and by the DNA in the organelle itself.
3. Mitochondria and chloroplasts contain complete systems for transcription and translation.

Studies of the DNA and RNA of energy organelles support the theory that they derive from ancient prokaryotes

1. Mitochondrial and chloroplast DNA is circular like prokaryotic DNA.
2. Ribosomes in these two organelles are more like bacterial ribosomes than they are like ribosomes in the cytoplasm of eukaryotic cells.
3. The endosymbiotic theory holds that present-day mitochondria and chloroplasts are descended from prokaryotes that began residing inside ancient eukaryotic cells.
4. The genetic code used by mitochondria and chloroplasts differs slightly from that in prokaryotes and eukaryotic cytoplasm, suggesting a very ancient origin that predates establishment of a universal genetic code.

VOCABULARY BUILDING

In your own words, first write a brief definition, then a full explanation for each of the following terms. Include examples where appropriate. Complete this section from your memory—you will not learn the terms by simply copying definitions from the textbook. Once you have finished, check your responses against the information in the chapter and make any necessary corrections.

transposon (jumping gene) —

mobile gene —

unconventional genetic system —

plasmid —

virus —

bacteriophage —

icosahedron —

capsid (membrane envelope) —

lysis —

lytic virus —

autonomous replication —

integrated replication —

episome —

conjugation —

resistance factor (R factor) —

recombinant plasmid —

transposase —

simple transposon —

complex transposon —

oncogene —

proto-oncogene (cellular oncogene) —

viral oncogene —

lytic cycle —

provirus (prophage) —

lysogenic cycle —

reverse transcriptase —

retrovirus —

restriction enzymes —

endosymbiotic theory —

CHAPTER TEST

The following test has four parts. Complete as much of the exam as you can from memory. If you cannot answer a question, skip it. Once you complete all that you can, try to answer any questions you skipped. If you still cannot answer them, consult your textbook for the answers. Once you have completed all sections of the test, check your answers for Parts 1 - 3 against those in the back of this book. Highlight any incorrect answers then review that material in your textbook. Correct your answers for future reference.

Part 1: Multiple Choice

For each of the following, select all correct responses—more than one may be correct.

1. Which of the following is not true about viruses?
   a) they are not alive
   b) they do not replicate
   c) most of them do not cause disease
   d) some do not contain DNA

2. A bacteriophage is a:
   a) virus
   b) bacteria
   c) mitochondrion
   d) disease-causing gene

3. Some mobile genes that are separate from the host DNA can replicate even when the host DNA does not by:
   a) integrated replication
   b) translation
   c) autonomous replication
   d) all of these

4. Plasmids are found in:
   a) bacteria
   b) viruses
   c) bacteriophages
   d) yeast

5. Conjugation involves exchange of genetic material between:
   a) two virus particles
   b) two bacteria
   c) a virus and a bacteria
   d) a virus and a eukaryotic cell

6. Which of the following catalyzes insertion of DNA into new locations?
   a) plasmid
   b) reverse transcriptase
   c) transposase
   d) oncogene

7. Which of the following is a mobile sequence of DNA that can replicate only when it is integrated into the host's DNA?
   a) plasmid
   b) transposon
   c) R factor
   d) mitochondrion

8. Viral genes whose expressed proteins interfere with the host cell's control of cell division are called:
   a) oncogenes
   b) transposons
   c) proviruses
   d) R factors

9. Episomes are capable of which of the following?
   a) autonomous replication
   b) integrated replication
   c) They are capable of both of these.
   d) They are not capable of either of these.

10. The integrated form of a virus is a(n):
    a) oncogene
    b) provirus
    c) episome
    d) phage

11. Which of the following is true of retroviruses?
    a) They use reverse transcriptase.
    b) Their genetic material is RNA.
    c) The direction of flow of the genetic information violates the central dogma.
    d) All of these are true.

12. What does a restriction enzyme do?
    a) It restricts replication of the host cell's DNA.
    b) It restricts host cell division.
    c) It cuts out foreign DNA.
    d) It gives bacteria resistance to some antibiotics.

13. According to the endosymbiotic theory, mitochondria and chloroplasts descended from:
    a) plasmids
    b) prokaryotes
    c) viruses
    d) nuclei of very ancient cells

14. What does a plasmid have that allows it to replicate autonomously?
    a) reverse transcriptase
    b) restriction enzymes
    c) R factors
    d) origin of replication

15. A simple transposon:
    a) only specifies transposase
    b) replicates only once
    c) is a bacteriophage
    d) uses reverse transcriptase

Part 2: Matching

For each of the following, match the correct term with its definition or example. More than one answer may be appropriate.

plasmid **virus** **transposon**

1. ________________ A bacteriophage is one.
2. ________________ Found in bacteria or yeast cells.
3. ________________ Consists of a nucleic acid core and a protein coat or membrane envelope.
4. ________________ These mobile genes can only replicate with the host's DNA.
5. ________________ These genes are so numerous that they likely account for much of a eukaryotic cell's "junk" DNA.
6. ________________ May contain R factors that confer resistance, such as to antibiotics, to their host.

lytic cycle **provirus**

7. ________________ Cannot infect other cells in this state.
8. ________________ DNA is integrated with that of the host.
9. ________________ Host cell is caused to rupture.
10. ________________ Viral activation triggers this.

Part 3: Short Answer

Write your answers in the space provided or on a separate piece of paper.

1. What is meant by an unconventional genetic system?

2. How can plasmids and viruses initiate replication when their host's DNA is not replicating?

3. Differentiate between autonomous and integrated replication.

4. List three examples of mobile genes.

5. What is a lytic virus?

6. What is an episome?

7. What is a complex transposon?

8. How do tumor viruses differ from lytic viruses?

9. What are the two components of the lysogenic cycle?

10. What mechanism do biologists suspect is at work when viral oncogenes cause tumor growth?

11. How can a virus stay in a cell for a long time without causing any ill effects?

12. What is a retrovirus?

13. How do restriction enzymes protect bacteria from viral infection?

14. List two ways in which viruses have evolved to become highly efficient.

15. List two advantages to using phages to fight infections.

Part 4: **Critical Thinking: Using Your Knowledge**

Answer each of these in essay form, using complete sentences and paragraphs. Provide as much information as you can. (For extra essay practice, write out answers to the Review and Thought Questions in your textbook.)

1. Thoroughly explain, using illustrative examples, the six stages common to all viruses.

2. Explain the relationship between the evolution of a host's protective mechanisms and a virus's infective mechanisms.

3. Most of our antibiotics are becoming useless as bacteria acquire resistance to these drugs. Explain how this resistance arises, including the role of plasmids. How has overprescription of antibiotics affected this situation? Why is this increase in drug-resistant bacteria a concern?

4. Explain how viruses may offer another disease treatment option. What are the limitations and potential problems? What are the advantages?

5. Discuss the endosymbiotic theory. What is it? What evidence supports this theory?

6. Discuss the means by which viruses replicate their genes.

7. Summarize the various treatment approaches now being used to fight HIV infection. By what mechanisms does each work? Which ones would not be effective if HIV were not a retrovirus? Why are these drugs so expensive?

8. Explain how oncogenes function. Do you think presence of these genes always causes cancer? How is environment involved?

9. In what ways might mobile genes be subject to natural selection? What advantages do they have over their nonmobile counterparts?

10. Viruses lack most characteristics of life and are not considered alive. Yet they do have certain characteristics shared by living organisms. Thoroughly discuss both sides of this issue. What does this say about their evolution? Why do they reproduce?

Chapter 13
Genetic Engineering and Recombinant DNA

KEY CONCEPTS

1. Genetic manipulation depends on choosing organisms with desirable genetic properties from a population that contains natural or induced variation.
2. Recombinant DNA techniques join together pieces of DNA in combinations that do not occur in nature, and allow production of relatively large amounts of pure genes and gene products.
3. Recombinant DNAs can reprogram prokaryotes and eukaryotes to have new properties.
4. Molecular biologists can design useful organisms by inserting or destroying genes that code from proteins involved in specific biochemical pathways.
5. Restriction enzymes allow preparation of DNA fragments of defined length and sequence.
6. DNA ligase can link together DNA fragments from different sources to produce recombinant DNA.
7. Vectors are DNA sequences from plasmids, viruses, or other mobile genes that can carry recombinant DNA into cells.
8. Eukaryotic genes cannot be expressed in bacteria, so molecular biologists use reverse transcriptase to first make a DNA copy (cDNA) of a eukaryote's mRNA (which has undergone posttranscriptional modification, including removal of introns). Then the cDNA can be cloned by usual methods.
9. Proteins that need to be altered cannot be expressed in bacteria because the bacteria lack the capacity to do this modification.
10. Cloning and PCR technologies allow researchers to make thousands or millions of copies of individual genes.
11. Two kinds of molecular probes allow biologists to locate specific genes—hybridization probes are used to detect genes, and antibodies can detect synthesis of specific proteins in bacterial colonies that contain recombinant DNA.
12. Gene therapy as a treatment for genetic diseases is still highly experimental.
13. Creating transgenic mammals poses many difficult problems.
14. Advances in genetic engineering have raised many moral and ethical questions that need to be addressed by society as a whole, not just the scientific community.

EXTENDED CHAPTER OUTLINE

THE MAVERICK

1. Kary Mullis developed the polymerase chain reaction (PCR), which specifically and repetitively copies any segment of DNA between any two defined nucleotide sequences.
2. Mullis received a Nobel Prize for this work in 1993.
3. Today, using PCR, a single copy of a DNA sequence can be multiplied into millions or billions of copies in less than a day.
4. Uses of PCR seem endless and include:
   - DNA fingerprinting for solving crimes;
   - locating viral DNA in blood and tissue samples;
   - early detection of cancer and genetic defects in human eggs and embryos; and
   - studies of DNA contained in ancient plant and animal specimens.
5. PCR technology today is a $1-billion-a-year business.

WHAT IS GENETIC ENGINEERING?

1. Thousands of years ago, farmers began practicing artificial selection, breeding both plants and animals. Brewers and cheese makers also learned to select the best strains of microorganisms to use as a "starter."
2. Biotechnology is the use of living organisms for practical purposes.
3. Modern genetic engineering transgresses the natural boundaries between species.

Genetic engineering makes use of both natural and induced variation

1. To make penicillin in amounts needed for human use, scientists had to find better ways to grow the mold and choose strains of mold that could produce more penicillin.
2. The best soil sample for penicillin—one from Peoria, Illinois—was subjected to mutagens that caused various mutations, most of which were useless. But a few of these induced mutations increased penicillin production.
3. Biotechnology selects useful traits from a range of variation, which may be natural or induced by mutagens.

Knowing biochemical pathways helps molecular biologists design useful organisms

1. Today, molecular biologists know so much about biochemical pathways in some organisms that they can predict what type of mutation will produce the desired result.
2. By understanding what causes tomatoes to soften, using genetic engineering can damage a gene that controls production of ethylene (which triggers softening of the tomatoes) so they can be vine-ripened and remain firm for shipping.

WHAT IS RECOMBINANT DNA AND HOW IS IT USEFUL?

1. By the early 1970s, molecular biologists could cut and paste pieces of DNA from different organisms, and by the early 1980s they could alter individual nucleotides in a gene.

How do molecular biologists use recombinant DNA?

1. Recombinant DNA is a DNA molecule that consists of two or more DNA segments that are not found together in nature.
2. Molecular biologists can now make recombinant DNA that fulfills particular requirements, the resulting recombinant DNA molecules are sometimes referred to as "designer genes."

3. Development of recombinant DNA technology depended on many different areas of research, including:
   - bacterial restriction enzymes,
   - DNA replication and repair,
   - replication of viruses and plasmids, and
   - chemical synthesis of specific nucleotide sequences.

How do restriction enzymes cut up a genome?

1. The enzymes DNAase and RNAase cut the links between nucleotides in DNA or RNA.
2. To isolate a specific piece of DNA, a researcher must use restriction enzymes, each of which is a special DNAase that cuts DNA only at a specific sequence.
3. Bacteria use restriction enzymes to protect themselves from viruses and other mobile genes.
4. The sequence recognized by a particular restriction enzyme is called a restriction site.
5. Some restriction enzymes cut both of the DNA strands at the same point, producing "blunt" ends, while others make staggered cuts, producing bare sections of each single strand called a "sticky" end, which can then join to other fragments through complementary base pairing.
6. A restriction enzyme cuts multiple copies of DNA any place a particular sequence occurs, producing a matched set of restriction fragments—DNA pieces that begin and end with a restriction site.
7. Researchers can compare the sizes of restriction fragments to establish a restriction map, which shows the positions of the restriction sites within the piece of DNA.
8. Researchers can join the restriction fragments together with others to make unique combinations—recombinant DNA.

How do molecular biologists join restriction fragments together?

1. DNA ligase links separate pieces of DNA into a continuous strand, so it is used to link restriction fragments together.

How do molecular biologists express recombinant DNA in bacteria and other hosts?

1. Plasmids are capable of forcing a bacterial cell to make several copies of a single gene.
2. Using restriction enzymes and DNA ligase, researchers made streamlined plasmids that contained only two parts:
   - an origin of replication (needed to start DNA synthesis), and
   - one or more genes for antibiotic resistance, to act as a marker.
3. With DNA ligase, researchers can link any gene to such a plasmid, then the recombinant plasmid ensures that the gene is copied and transcribed in the bacterial cell.
4. Researchers then grow the bacteria in an antibiotic. Only bacteria with the recombinant plasmid have the antibiotic resistance, so only the genetically engineered cells survive.
5. The term vector refers to any organism that spreads disease from one organism to another, so it is now also used to refer to anything that spreads genes from organism to organism.
6. Genetic vectors include plasmids, bacteriophages, viruses, and transposable elements.
7. Actual genes from eukaryotes cannot be expressed in bacteria because bacteria lack the machinery to recognize and remove introns from eukaryotic pre-mRNA (posttranscriptional modification), and if they cannot remove the introns, they cannot translate the gene.
8. Researchers get around this problem by using retroviruses, which use RNA as their nucleic acid, with the aid of the enzyme reverse transcriptase.
9. Reverse transcriptase, found in retroviruses, copies almost any RNA into DNA.
10. Researchers start with mature RNA for the desired gene (its introns have been removed). Reverse transcriptase then copies that RNA into a molecule of DNA, called complementary DNA (cDNA), which is complementary to the mRNA.
11. The cDNA contains no introns and can be inserted into bacteria for cloning.

12. Genes for polypeptides that need modification cannot be expressed in bacteria because the bacteria lack the capacity to perform this modification. These genes must be expressed in eukaryotic cells.
13. Through these techniques, bacteria and various eukaryotic cells can mass-produce large quantities of proteins that would otherwise be available only in small amounts.

How do researchers make multiple copies of recombinant DNA?

1. Through the 1970s and 1980s, copies of recombinant DNA were made through cloning, in which a single recombinant DNA molecule was put into a bacterial host cell. The plasmid induced the cell to make several copies of the gene, and researchers induced the cell to divide rapidly.
2. This process is called cloning of recombinant DNA because it produces a clone, or exact copy, of the recombinant DNA.
3. Molecular biologists also refer to genetically identical cells or organisms as clones.
4. Kary Mullis' invention of the polymerase chain reaction (PCR) allows researchers to make millions or billions of copies of a DNA sequence in a test tube in less than a day.
5. PCR depends on the ability to make two specific polynucleotides that correspond to sequences at each end of the sequence to be copied. These sequences are examples of oligonucleotides.
6. Each oligonucleotide serves as a primer for DNA polymerase.
7. The steps in the polymerase chain reaction are as follows:
   - DNA is heated, which causes the strands to unwind and separate.
   - The DNA is cooled, and the two primers bind to the complementary sequences—one on each strand.
   - DNA polymerase copies the sequence on each strand.
   - The temperature is raised again, separating the template strand and newly made complementary copy.
   - The temperature is lowered and DNA polymerase again copies the sequence.
   - The cycle is repeated—raising and lowering the temperature—resulting in a rapid and exclusive multiplication of the specific DNA sequence.
8. Cloning and PCR technologies can make millions of copies of individual genes.

How do biologists find the right DNA sequence in a recombinant DNA library?

1. The genome of any organism can be cut into pieces with restriction enzymes.
2. A gene library is a collection of thousands or millions of restriction fragments from a single genome.
3. A complete genomic library contains at least one representative of each sequence in a genome.
4. A recombinant DNA library can be constructed from mRNA alone, using reverse transcriptase, and is called a cDNA library.
5. Geneticists have developed techniques for finding and cataloging the tens of thousands of sequences in a gene library.
6. A probe is a short sequence of single-stranded DNA whose sequence is complementary to a small part of the sequence in the gene the researchers are looking for.
7. Antibodies are immune system proteins that recognize and bind to other proteins, so they can be used as probes to detect the protein product made by that DNA.
8. A probe can be used as a hybridization probe which, by base-pairing with DNA in the library, forms hybrid DNA—DNA whose complementary strands come from different sources.

9. The probes can easily locate the DNA sequence for which it is complementary. Once it binds to the DNA sequence, the probe can easily be located if it has been chemically or radioactively labeled.
10. The technique of hybridizing a labeled DNA probe to find the right piece of DNA in a gene library is called molecular hybridization.
11. Antibodies are proteins with surfaces that recognize the shapes of foreign molecules.
12. Labeled antibodies can be used to identify bacterial colonies that are synthesizing the protein for which the desired gene codes. Once the bacteria that contain the gene are located, cloning can be done.
13. Hybridization probes detect genes in recombinant DNA clones, cells extracts, and cells; antibodies detect synthesis of specific proteins in colonies of bacteria containing recombinant DNA.

HOW CAN BIOLOGISTS STUDY THE EXPRESSION AND STRUCTURE OF MANY INDIVIDUAL GENES?

1. Researchers can use a cDNA or an oligonucleotide probe to measure the quantity of a specific mRNA present in a tissue.
2. New methods allow biologists to measure the levels of thousands of mRNA types at the same time.
3. To do these measurements, researchers use DNA arrays—thin, dime-sized wafers housing thousands of tiny spots of DNA in a fixed arrangement, each corresponding to a single gene.

RECOMBINANT DNA CAN REPROGRAM CELLS TO MAKE NEW PRODUCTS

1. Hundreds of biotechnology, chemical, and pharmaceutical companies have begun to use recombinant DNA to make proteins useful to medicine, agriculture, and the chemical industry.

Genetically engineered bacteria and eukaryotic cells can make useful proteins

1. Recombinant DNA technology allows the reprogramming of cells to make an extraordinary number of products.
2. Recombinant DNA techniques allow preparation of safer and uncontaminated vaccines because the immunizing molecules are made of viral DNA rather than from active viruses.
3. Virologists isolate a single piece of viral DNA that encodes a single viral protein, because a single viral protein cannot cause disease, but can stimulate the immune system to provide immunity.
4. Growth factors are rare proteins that keep specific kinds of animal cells alive, stimulate cell division, or trigger particular types of cell specialization. Recombinant DNA techniques have produced relatively large amounts of pure growth factors, allowing researchers to study their structure and action.
5. Genetically engineered bacteria and eukaryotic cells can make useful proteins, including enzymes, vaccines, drugs, and human proteins for treating genetic disorders.

RECOMBINANT DNA CAN GENETICALLY ALTER ANIMALS AND PLANTS

1. Transgenic organisms carry recombinant DNA in their genomes, and the added DNA is called a transgene.

How do researchers produce a transgenic mammal?

1. For a gene to be expressed in all the appropriate cells of an animal, researchers must put the transgene into the zygote before the beginning of embryonic development—then all the cells will contain the engineered DNA.

2. Researchers must surgically remove eggs from the female's abdomen. The eggs are fertilized *in vitro* with sperm from a male, and the resulting zygote is then injected with the engineered gene.
3. Transgenic mammals must be made one at a time.
4. So far, only one transgenic organism is born for every 100 injected zygotes.
5. Engineering of transgenic organisms faces serious obstacles. For example, even seemingly incorporated genes may be lost in subsequent cell divisions, and the transgenes are inserted into the host's DNA randomly, which may disrupt important genes.
6. Genetic knockouts—animals in which a specific gene has been inactivated—allow study of how specific genes affect phenotype, and of preparing lab specimens with specific disorders. However, the results are not always predictable.
7. Recently, researchers have created "knockin" mice, in which the allele has been replaced, rather than merely added.
8. In a knockin, the allele goes to exactly the right place and the old allele is eliminated.
9. Interpreting the phenotypic results of knockout organisms is also problematic.

The genetic engineering of plants is easier than that of animals

1. The most rapidly expanding area of genetic engineering is agriculture.
2. Plant cells are generally easier to clone than animal cells.
3. Genetic engineering of plants also has the potential to be enormously lucrative.
4. Examples of genetic engineering in agriculture include inserting bacterial or fungal resistance into crops, and inserting herbicide resistance, such as with Roundup-Ready® seed, sold by Monsanto Corporation, which grows plants resistant to the herbicide Roundup®, also sold by Monsanto.
5. Researchers can engineer the Ti (tumor-inducing) plasmid in a species of soil bacteria, so that the plasmid will carry a certain gene into the genome of a plant.
6. In the gene gun technique, tiny particles of gold or tungsten are coated with recombinant DNA fragments (with antibiotic resistance), then fired into a young plant. The plant is then chopped, its pieces cultured with antibiotic, and the cells that survive contain the recombinant DNA.
7. More and more applications of genetic research are being tried in agriculture.

What are the risks of recombinant DNA in agriculture?

1. The long-term consequences of moving genes from one organism to another are unknown.
2. Genetically engineered plants can, for example, exchange their new genes with wild cousins, either conventionally or through plasmids or other vectors.
3. Milkweed was intentionally dusted with pollen from insect-resistant corn, and the milkweed, which is required for the monarch butterfly's life cycle, killed monarchs who fed on it.
4. Genetically engineered seed are patented, so farmers cannot save seed from year to year, as they previously did.
5. The European Union has placed a moratorium on importing genetically modified food.

The application of recombinant DNA technology poses moral questions for society

1. Moral and ethical issues are raised by recombinant DNA's precision. For example, the diagnosis of many genetic diseases is possible now, but treatment for many is not.
2. Potential employers and insurance companies want to know what people's genetic flaws are so they can avoid hiring or insuring people likely to get sick.
3. Many people worry about applying genetic engineering to humans.
4. What constitutes a genetic disease may become controversial.
5. As in the case of any new advance, the use of technology is a concern for society as a whole, not just for scientists.

VOCABULARY BUILDING

In your own words, first write a brief definition, then a full explanation for each of the following terms. Include examples where appropriate. Complete this section from your memory—you will not learn the terms by simply copying definitions from the textbook. Once you have finished, check your responses against the information in the chapter and make any necessary corrections.

polymerase chain reaction (PCR) —

biotechnology —

recombinant DNA —

restriction enzymes —

restriction site —

restriction fragments —

restriction map —

DNA ligase —

vector —

reverse transcriptase —

complementary DNA (cDNA) —

clone —

oligonucleotides —

gene library —

probe —

antibodies —

hybridization probe —

hybrid DNA —

DNA array —

transgenic organisms —

transgene —

Ti plasmid —

gene gun —

CHAPTER TEST

The following test has four parts. Complete as much of the exam as you can from memory. If you cannot answer a question, skip it. Once you complete all that you can, try to answer any questions you skipped. If you still cannot answer them, consult your textbook for the answers. Once you have completed all sections of the test, check your answers for Parts 1 - 3 against those in the back of this book. Highlight any incorrect answers then review that material in your textbook. Correct your answers for future reference.

Part 1: Multiple Choice

For each of the following, select all correct responses—more than one may be correct.

1. For which of the following can the polymerase chain reaction be used?
   a) DNA fingerprinting in criminal investigations
   b) detection of trace amounts of viral DNA, such as from HIV, in blood
   c) screening for genetic defects in eggs and embryos
   d) all of these

2. What job does a DNAase perform?
   a) It joins RNA fragments to DNA fragments.
   b) It separates DNA nucleotides.
   c) It joins DNA fragments together.
   d) none of these

3. Which of the following is the sequence that is recognized by a restriction enzyme?
   a) DNA primer
   b) restriction site
   c) restriction map
   d) vector

4. A restriction enzyme is a special type of:
   a) DNAase
   b) DNA ligase
   c) RNA
   d) reverse transcriptase

5. What role does DNA ligase play in making recombinant DNA?
   a) It initiates DNA synthesis.
   b) It separates the plasmid DNA.
   c) It locates the selected gene.
   d) It joins the DNA fragments together.

6. What is the main role of reverse transcriptase in recombinant DNA technology?
   a) Reverse transcriptase allows researchers to use retroviruses to clone genes.
   b) Researchers can change bacterial genes into eukaryotic genes.
   c) Researchers can start with mature eukaryotic mRNA, which has had its introns removed.
   d) Reverse transcriptase allows genetic engineers to remove genes from the genome.

7. Which of the following can be correctly considered a clone?
   a) human identical twins
   b) plants grown from cuttings
   c) multiple copies of recombinant DNA manufactured in a bacterium
   d) All of these are clones.

8. Why are bacteria commonly used in recombinant DNA technology?
   a) They are antibiotic resistant.
   b) Their plasmids can force them to make several copies of the recombinant DNA.
   c) They are a ready source—bacteria are everywhere.
   d) The bacteria cannot infect humans when they contain recombinant DNA.

9. Which of the following would a researcher use to cut eukaryotic DNA at a particular sequence?
   a) restriction enzyme
   b) DNA ligase
   c) DNA polymerase
   d) hybridization probe

10. When genetic engineers prepare plasmids for use in recombinant DNA, what do the streamlined plasmids contain (before the eukaryotic gene is spliced in)?
    a) an origin of replication
    b) all of the bacterial genes
    c) restriction fragments
    d) a gene or genes for antibiotic resistance

11. Researchers can start with mature eukaryotic mRNA, then use reverse transcriptase to make:
    a) pre-mRNA
    b) tRNA
    c) cDNA
    d) plasmid DNA

12. PCR directly involves which of the following?
    a) DNA polymerase
    b) two DNA primers—one that marks each end of the desired DNA sequence
    c) alternating heating and cooling cycles
    d) bacteria

13. For a gene to be expressed in all cells in an animal, the transgene is put into:
    a) the egg
    b) the sperm
    c) the zygote
    d) the infant shortly after birth

14. Which of the following is a limitation to the success of transgenic organisms?
    a) The recombinant DNA is often not incorporated into the zygote.
    b) When incorporated, the recombinant gene is inserted into the chromosomes randomly.
    c) The transgenic organism already has two alleles of its own for that gene.
    d) all of these

15. Which of the following can be used as vectors?
    a) plasmids
    b) bacteriophages and other viruses
    c) transposable elements
    d) any of these

Part 2: **Matching**

For each of the following, match the correct term with its definition or example. More than one answer may be appropriate.

polymerase chain reaction **recombinant DNA technology**

1. ________________ Uses a vector.
2. ________________ Relies on temperature changes to cause repeated cycles of DNA production.
3. ________________ Very rapidly mass-produces DNA sequences.
4. ________________ Allows reprogramming of cells so they make new products.
5. ________________ Used to create transgenic organisms.
6. ________________ Used to clone DNA for study, such as in DNA fingerprinting.

restriction enzyme **reverse transcriptase** **DNA ligase**

7. ________________ Used by bacteria to defend themselves against viral infection.
8. ________________ Used during DNA replication.
9. ________________ Used by retroviruses to produce DNA from their RNA.
10. ________________ Can hook almost any gene into DNA from a vector, such as a plasmid.
11. ________________ Used to prepare DNA fragments for recombination.
12. ________________ Used to prepare cDNA from mature eukaryotic mRNA.

Part 3: **Short Answer**

Write your answers in the space provided or on a separate piece of paper.

1. Who were some of the first people to use biotechnology?

2. Explain the genetic solution to the problem of needing to pick tomatoes before they have developed full flavor in order to avoid damage during transport.

3. What marks the ends of a restriction fragment?

4. What is used to link together restriction fragments?

5. Why is antibiotic resistance usually included in recombinant DNA?

6. Why is a vector used in recombinant DNA technology?

7. Why must cDNA be made?

8. List two ways in which recombinant DNA can be mass-produced.

9. List two tools that researchers can use to locate certain genes in a gene library.

10. To what does the term "designer genes" refer?

11. What is a genetic knockout?

12. What is a genetic knockin?

13. What is an oligonucleotide used for in PCR?

14. What is recombinant DNA?

15. List two ways in which bacteria can be forced to create a large clone of a certain gene.

Part 4: Critical Thinking: Using Your Knowledge

Answer each of these in essay form, using complete sentences and paragraphs. Provide as much information as you can. (For extra essay practice, write out answers to the Review and Thought Questions in your textbook.)

1. Thoroughly explain the polymerase chain reaction. What is the procedure? What are its limitations and advantages? How can it be used in criminal cases?

2. Thoroughly explain how recombinant DNA is made. How did various lines of research and study lead to this technology?

3. Genetic engineering has many applications in the field of agriculture. For example, plants are being engineered to have a longer shelflife in the stores. How would these crops benefit farmers? How would they benefit you? What potential risks might there be?

4. When Monsanto Corporation introduced its Round-up Ready® seed, environmentalists raised many concerns, among them that this type of product would increase the use of herbicides. What are the risks of such increased usage?

5. Monsanto countered the complaints of environmentalists by claiming that these genetically-engineered plants would actually decrease the amount of herbicide used. Which side do you think is correct? Using the Internet or other resources, find out what has actually happened to the sales of the herbicide Roundup® (which is also manufactured by Monsanto).

6. World hunger and starvation continue to be serious problems. Crop yields increase while prices paid to farmers plummet. Many American farmers have recently seen their worst years and are being forced out of the profession because they cannot earn a living at it. The European Union now refuses to import genetically engineered crops, which has forced many grain elevators in the United States to either refuse these crops or have separate storage for them, which increases the cost to farmers. Farmers cannot save seed from one year to the next, and are paying additional "technology" fees when they purchase many products, including seed and chemicals. At this point, do you think genetic technology is helping or hurting farmers? How is it impacting the global economy and problems of hunger and starvation? What changes would you recommend to make this technology more beneficial?

7. Read your local newspaper and watch the national news on television or listen to it on the radio for one week. Keep a log of how many reports are about genetic engineering and genetic research. For each, list the potential benefits and potential risks.

8. In the O. J. Simpson murder trial, DNA evidence seemed to back up the prosecutions' claims that Mr. Simpson was guilty. The defense pointed out some significant flaws in the collection and testing of the DNA evidence. A pivotal point in the trial occurred when one of the individuals who collected the samples was shown on videotape handling a sample without wearing gloves. Why was this so crucial?

9. You work in a crime lab and are brought a piece of cloth from the crime scene that contains a few drops of blood. You decide to use PCR to increase the amount of DNA you are working with. You estimate that your sample contains 20,000 copies of someone's DNA. If you performed 20 cycles of PCR, how many copies would you have in the end?

10. Media coverage of transgenic organisms, Roundup-Ready® crops, and whole-animal cloning, to name a few, have brought genetic engineering into everyone's home, and you have just read about many potential applications of genetic research. The possibilities seem almost endless. Yet there is great controversy surrounding whether the technology will be used for good or for evil purposes, and what possible long-term effects might be. The European Union has taken a stance and will no longer import genetically-altered crops. Should there be limitations on what research and experiments can be done? If so, who should decide what those limits are? Where would you draw the line, and why?

Chapter 14
Human Genetics

KEY CONCEPTS

1. Even though humans are more than 99% genetically identical, all humans, except identical twins, are genetically unique.
2. DNA fingerprinting identifies variations in DNA sequences taken from blood or other tissues and provides an estimate of how common that particular pattern of variation is in the general population.
3. Human races blur into one another, so defining consistent races is not possible. Only about 9% of all human genetic variation is among so-called races.
4. Alleles vary in their expressivity and penetrance, so genotype does not define phenotype.
5. Phenotype is always a product of the interaction between genotype and environment.
6. Pleiotropy is the capacity of one gene to have diverse effects on phenotype, and all genes are pleiotropic.
7. Expression of each trait is the result of many genes acting together.
8. A single gene may have many functioning alleles.
9. Recessive alleles are not necessarily rare, nor are dominant alleles necessarily common.
10. The Human Genome Project promises to map and sequence the entire human genome by about 2002.
11. Pedigree analysis allows researchers to find a genetic marker for a particular trait that shows approximately where on the genome a gene is located.
12. Positional cloning allows researchers to locate a particular gene precisely on a chromosome.
13. Genetic counselors can tell couples whether they carry certain alleles and the likelihood that their offspring will express certain traits.
14. Amniocentesis allows a geneticist to look for chromosomal or other genetic abnormalities in a four-month-old fetus.
15. Chorionic villus sampling allows geneticists to test for the same genetic defects as amniocentesis, but earlier in the pregnancy.
16. Medical researchers can select human eggs that are free of certain genetic defects, but the technology is cumbersome and expensive.
17. Gene therapy techniques are still fraught with problems.

EXTENDED CHAPTER OUTLINE

TO MAP THE GENOME OR HUNT THE HUNTINGTON'S GENE?

1. Huntington's disease is a classic genetic disease. An affected person passes the disease to half of his or her offspring, indicating the disease is caused by a single dominant gene.
2. Individuals have numerous small differences at the DNA level, and these "silent" differences are in DNA sequences that are not transcribed and translated (such as introns).
3. Choosing a single disease on which to focus their efforts gave researchers a sense of mission and a single, clear-cut goal that was emotionally appealing.
4. The genetics of humans is no different than the genetics of other organisms, but human genetic *research* is—humans cannot be crossed at the whim of researchers.

GENETIC DIVERSITY IN HUMAN BEINGS

1. Every individual has a unique set of gene fragments called a DNA fingerprint.
2. DNA for a DNA fingerprint can be extracted from any cell with a nucleus.
3. With the use of PCR, researchers can now analyze DNA from bare traces of cells.

How reliable is DNA fingerprinting?

1. The 95% of the DNA that is noncoding has no central role in phenotype, so it can mutate without significant effect on the individual.
2. Noncoding DNA accumulates mutations and contains far more variation than coding DNA.
3. Except for identical twins, every individual is genetically unique and DNA sequencing can reliably distinguish every individual in a population.
4. To create a DNA fingerprint, researchers use restriction enzymes to cut the DNA into restriction fragments.
5. The length of several restriction fragments varies from one individual to another. Such fragments are called restriction fragment length polymorphisms (RFLPs).
6. Gel electrophoresis is used to separate the RFLPs by size—shorter fragments travel farther through the gel.
7. When hybridized to DNA probes in a Southern blot, the fragments produce a banded pattern—each genome yields a unique pattern of dark bands.
8. Fingerprinting labs use a sample of noncoding DNA that is known to contain many variants, and calculate the probability that someone else in the population has the exact same variations.
9. The probability that someone else has the same variants depends on:
   - the number of genetic markers examined (the greater the number, the more specific the DNA fingerprint will be); and
   - how the "general population" is defined.
10. Geneticists have estimates for the probabilities in large populations that have been studied.
11. DNA is extremely durable, and, even if it were damaged, the changes would be random and extremely unlikely to match any other person's fingerprint.
12. In criminal cases, sample contamination is of little concern because the contaminating DNA would be distinct from the suspect's.
13. DNA fingerprinting has led to not only criminal convictions, but also release of innocent prisoners, and identification and return of kidnapped children.

What are human "races"?

1. Attempts to group people into human "races" have been problematic because there have always been tribes or nations that would not fit into any known group.
2. Most groups do not stand out from those around them—they blend and inevitably mix—so human races are never pure.
3. Studies of different alleles have convinced researchers that human races, in the biological sense, do not exist.
4. Genetic variation in humans is greater within populations than it is among individuals.
5. Of all human genetic variation, 85% is among individuals within a country or a continent, another 6% is among populations from the same continent, and only 9% comes from differences between peoples ("races") from different continents.

THE RELATIONSHIP BETWEEN GENOTYPE AND PHENOTYPE IS COMPLEX

1. Every individual, as expressed in phenotype, results from complex interactions between gene and environment.

Genes are expressed to different degrees

1. Expressivity is the range of phenotypes associated with a given genotype—two individuals with the same allele may express it very differently.
2. While expressivity is a measure of how strongly a trait is expressed, penetrance describes the likelihood that an individual with a dominant allele will show the phenotype usually associated with that genotype.
3. Complete penetrance means that the dominant allele will always be expressed; incomplete penetrance means that a dominant allele will not be expressed in certain conditions.
4. Both the environment and interactions with other genes play crucial roles in determining the expression of individual genes.

A single gene can affect many traits

1. About 70% of all cases of cystic fibrosis result from deletion of just three base pairs in a single gene, causing deletion of a single amino acid in the CFTR polypeptide, and this disruption causes a multitude of physiological complications.
2. One mutation can have wide-ranging effects on phenotype.
3. Pleiotropy is the capacity of one gene to have diverse effects. Every gene that has been studied is pleiotropic to some degree.

A single trait can be influenced by many genes

1. Phenotype depends on interactions between many genes and the effects of the environment.
2. Most traits are polygenic—they are influenced by more than one gene.
3. Many genes are now believed to be backup systems for other genes, so the genome has several ways to accomplish the same goal.
4. In humans, the vast majority of traits are polygenic and probably redundant.
5. Studying polygenic traits is extremely difficult.

A single gene may have multiple alleles

1. A gene may exist in many alternate forms, each of which contains a distinct sequence of nucleotides that specifies a distinct polypeptide.
2. The ABO blood group has three alleles—I^A, I^B, and i. I^A and I^B are codominant, meaning both are expressed, and they code for different surface markers on red blood cells. The i allele is recessive and does not code for a surface marker.

THE HUMAN GENOME PROJECT

1. Each of the 6 billion human beings on Earth harbor enormous genetic diversity, yet we are all more than 99% genetically identical.

What is the Human Genome Project?

1. The Human Genome Project is a 15-year, $3 billion project jointly funded by the Department of Energy and by NIH's National Center for Human Genome Research.
2. Benefits of the Human Genome Project include the following:
   - Linkage maps have already been made for more than 50,000 genes, restriction enzyme cut sites, or other markers.
   - A physical map has been produced for each human chromosome by first cutting the chromosomes into restriction fragments, identifying sequences that can act as landmarks, then sequencing to the next landmark.
   - The project plans to sequence all 3 billion base pairs on one set of chromosomes (23 pairs) for several anonymous volunteers, then load the information into a computer to create an electric version of the genome.
   - By reducing costs needed for sequencing, the project will allow scientists to study variations in the human genome and it will also determine the DNA sequence of several other species.
   - The project will develop new methods for studying newly identified gene products.
3. Work on the Human Genome Project began in laboratories around the world in 1990, under the direction of James Watson.
4. To simplify the work, Craig Venter extracted mRNA from human tissues and used it to make cDNA of only the parts of the genome that actually code for proteins, leaving out all the introns and junk DNA.
5. Thousands of researchers are now involved.
6. By the end of 1998, researchers had completed both the genetic and physical maps of human DNA.
7. By the end of 1999, researchers had complete sequences of several organisms and had determined the complete sequence of one human chromosome (number 22, the smallest).
8. By the end of 2003, geneticists expect a complete, "polished" sequence for human DNA with no gaps and no mistakes.

How do researchers locate genes for specific diseases?

1. The Human Genome Project has provided a gene library of actual gene fragments stored in test tubes, and a computer database listing the nucleotide sequence of each human chromosome.
2. Family pedigrees show inheritance patterns and are useful for studying inheritance of obvious traits that are passed in a simple manner.
3. In pedigree analysis, researchers collect blood samples from family members that carry a given disease, then use DNA fingerprinting to look for RFLP alleles that are consistently associated with the disease. By this technique, the approximate location of the Huntington's gene was determined in 1983.
4. Once the approximate location of a gene is found between two RFLPs, positional cloning is used to pinpoint the gene's specific location.
5. In positional cloning, researchers first assemble a library of clones that cover the region between the two RFLP markers, then string them together from one marker to the next, forming a "contig."
6. Researchers next hybridize human DNA with labeled DNA from another organism. Only sequences that have stayed almost the same through time hybridize—sequences that do not hybridize are very different from each other and likely represent noncoding DNA, so they are ignored.

7. Next, researchers look for mRNA from tissue in which the gene is expressed, then compare conserved sequences from diseased tissues to similar sequences from healthy individuals, looking for sequences that are always associated with the disease.
8. Positional cloning is time-consuming, but is getting easier.
9. Geneticists now know that few familial diseases result from changes in single genes, and they hope to gain more understanding of polygenic inheritance.
10. Pharmacogenomics is a new field of study, which examines how genetic differences result in different responses to medications.

Of what use is a sequenced gene?

1. Once a gene has been sequenced, geneticists can test individuals to see if they have a normal allele or a disease-causing allele.
2. Knowledge of genes does not always make a difference—although some alleles of the BRCA1 or BRCA2 genes give a woman an 85% chance of developing breast cancer, most women who develop breast cancer have neither of these alleles, and knowledge of these genes has not enabled researchers to improve either prevention or treatment of breast cancer.
3. Knowledge of a gene's sequence may help researchers understand how the gene causes diseases.

PREVENTING GENETIC DISEASE

1. Only about 3% of all human diseases are caused by defects in a single gene.

Genetic counseling: How can parents decide?

1. One type of genetic testing simply tests couples who are considering having a baby to see if they carry alleles for certain genetic diseases—tests are now available for more than 100 different genetic defects.
2. Once the parents' genotypes are known, genetic counselors can calculate the probability that a child will have a certain defect.

Prenatal testing: Can parents tell before?

1. Once a woman becomes pregnant, physicians can check fetal cells for genetic defects.
2. In amniocentesis, amniotic fluid, which contains fetal cells, is withdrawn and the fetal chromosomes are checked.
3. Amniocentesis can detect about 100 genetic abnormalities.
4. Amniocentesis is performed at about 15 or 16 weeks into the pregnancy, and it carries about a 0.5% risk of inducing a spontaneous abortion.
5. About 10 to 15% of known pregnancies abort spontaneously before the 20th week due to abnormalities in the fetus.
6. Chorionic villus sampling involves removing fetal cells from part of the placenta, and it can be performed much earlier than amniocentesis—as early as 6 to 12 weeks after conception.
7. The risk to the fetus is about the same in amniocentesis and chorionic villus sampling.
8. For now, prenatal testing implies only one decision—whether or not to abort the pregnancy.
9. One method allows parents to avoid the abortion decision—test tube baby technology allows a doctor to harvest eggs from a mother who carries one disease-causing allele and select eggs that do not carry that allele. The eggs are then fertilized *in vitro* and implanted into the mother.
10. The Human Genome Project is producing potential new genetic tests at the rate of about one per week.

Gene therapy: Can we fix it later?

1. Gene therapy is a technology that allows researchers to deliver normal genes into tissues of people who have defective genes.
2. To date, this technology has been used to try to treat 13 different genetic diseases. As of this writing, not one type of gene therapy has completely cured anybody. [*Author's note: as this book was going into print, the first apparently true genetic cure was just being announced, but the information was too new to be included in this book.*]
3. In most cases, gene therapy can produce a partial and temporary cure that must be renewed every few months.
4. All human gene therapy is focused on somatic cells, rather than germ cells, but only genetically engineered germ cells can pass altered genes on to future generations.
5. Researchers would need to replace nonfunctional genes in either the gametes or the zygote in order to effect a permanent cure, and although knockin technology makes this a possibility, the technique is too new for use in humans.
6. For the near future, gene therapy in humans will be directed only at somatic cells.
7. Currently, researchers cannot insert genes into chromosomes in human cells without risk of damaging other genes, and the genes are rarely incorporated into the chromosomes, so they fail to replicate.
8. Many researchers are now exploring ways to get genes into stem cells, which are dividing progenitor cells that can develop into many alternative cell types. Engineered stem cells can now be inserted into tissues most affected by a certain disease.
9. Most stem cell populations come from embryos, and many people opposed to abortion fear researchers and physicians may directly or indirectly promote abortion to increase stem cell availability. However, some stem cells can be harvested from adults.
10. In 1999, after much debate and discussion, the National Institutes of Health established strict procedures and guidelines for obtaining approval for any stem cell research.
11. Gene therapy has had expenditures of $400 million per year and attempted over 100 clinical trials in humans, but some of these trials were poorly designed and none has demonstrated the ability to cure a genetic disease. However, new ideas and lines of research do look hopeful.

VOCABULARY BUILDING

In your own words, first write a brief definition, then a full explanation for each of the following terms. Include examples where appropriate. Complete this section from your memory—you will not learn the terms by simply copying definitions from the textbook. Once you have finished, check your responses against the information in the chapter and make any necessary corrections.

restriction fragment length polymorphisms (RFLPs) —

DNA fingerprint —

expressivity —

penetrance —

pleiotropy —

polygenic —

multiple alleles —

Human Genome Project —

pedigree analysis —

positional cloning —

amniotic fluid —

amniocentesis —

chorionic villus sampling —

gene therapy —

CHAPTER TEST

The following test has four parts. Complete as much of the exam as you can from memory. If you cannot answer a question, skip it. Once you complete all that you can, try to answer any questions you skipped. If you still cannot answer them, consult your textbook for the answers. Once you have completed all sections of the test, check your answers for Parts 1 - 3 against those in the back of this book. Highlight any incorrect answers then review that material in your textbook. Correct your answers for future reference.

Part 1: Multiple Choice

For each of the following, select all correct responses—more than one may be correct.

1. In humans, the greatest variation between two people is found in:
   a) the number of chromosomes
   b) where individual genes are located on their chromosomes
   c) the DNA sequences that code for proteins
   d) noncoding DNA sequences

2. Which of the following are used to create DNA fingerprints?
   a) DNA polymerase
   b) restriction enzymes
   c) DNA ligase
   d) all of these

3. About how much of our DNA does not code for proteins?
   a) 5%
   b) 30%
   c) 50%
   d) 95%

4. Which of the following is likely to cause errors in DNA fingerprinting and could lead to conviction of an innocent person?
   a) DNA is fragile, so old samples will deteriorate and give false results.
   b) Samples are easily contaminated and foreign DNA could match an innocent suspect's.
   c) Very few DNA markers are examined.
   d) all of these

5. How much genetic variation can be contributed to differences between races?
   a) 6%
   b) 9%
   c) 91%
   d) 85%
6. Which of the following has a majort effect on phenotype?
   a) amount of noncoding DNA
   b) environment
   c) genotype
   d) length of RFLPs
7. How strongly a trait is expressed is referred to as a gene's:
   a) expressivity
   b) dominance
   c) penetrance
   d) allele
8. Which of the following is true about a dominant allele that has incomplete penetrance?
   a) It is always expressed fully.
   b) The degree to which it is expressed varies.
   c) It is expressed in some situations but not in others.
   d) It is never expressed.
9. Pleiotropy refers to:
   a) multiple genes controlling a single trait
   b) one gene having multiple effects
   c) having multiple alleles for a single gene
   d) all of these
10. When two alleles are both expressed, they are said to be:
   a) pleiotropic
   b) codominant
   c) polygenic
   d) polymorphic
11. Pedigree analysis is useful for:
   a) analyzing patterns of simple inheritance
   b) identifying the general location of a gene
   c) specifically locating a gene
   d) all of these
12. "Zoo blots" are used in:
   a) pedigree analysis
   b) DNA fingerprinting
   c) positional cloning
   d) prenatal testing
13. Contigs have been employed for which technology?
   a) amniocentesis
   b) chorionic villus sampling
   c) gene therapy
   d) positional cloning

14. Which of the following is a limitation of amniocentesis?
    a) Results are not available until near the time that the fetus becomes viable outside the womb, making the abortion decision more difficult for many parents.
    b) There is a risk of spontaneous abortion as a result of the procedure.
    c) Not all genetic defects are testable with this procedure.
    d) All of these are limitations.

15. Which of the following is a limitation of current gene therapy technology?
    a) Research is limited to treating only somatic cells, which cannot pass on the engineered gene.
    b) Engineered genes are usually placed into the nucleus but not incorporated into the chromosomes, so they do not replicate.
    c) Researchers cannot yet insert genes into chromosomes in human cells without risking damage to other genes.
    d) All of these are current limitations of gene therapy.

Part 2: Matching

For each of the following, match the correct term with its definition or example. More than one answer may be appropriate.

coding sequences of DNA **noncoding sequences of DNA**

1. ________________ Consists of exons.
2. ________________ Accounts for about 95% of human DNA.
3. ________________ Has no central role in phenotype.
4. ________________ Used for DNA fingerprinting.
5. ________________ Carries instructions for making proteins.

amniocentesis **chorionic villus sampling**

6. ________________ Done at 15 or 16 weeks after conception.
7. ________________ Fetal cells withdrawn from placenta.
8. ________________ Carries a 0.5% risk of spontaneous abortion.
9. ________________ Results are available earlier in the pregnancy.
10. ________________ Cannot correct defects if they are found.

Part 3: Short Answer

Write your answers in the space provided or on a separate piece of paper.

1. Explain restriction length polymorphisms.

2. Humans are genetically different from each other by what percentage?

3. From where does DNA used for DNA fingerprinting come?

4. What two factors affect the probability that another person will have the same DNA fingerprint results?

5. What two arguments have defense attorneys used to try to prevent DNA fingerprinting from being used against their clients?

6. What are two explanations for why it is difficult to group people into human races?

7. What does complete penetrance of a dominant allele tell you about the phenotype?

8. What is meant by pleiotropy?

9. What is meant by traits being polygenic and redundant?

10. Of what blood type are individuals with the following genotypes?
    a) I^Ai b) I^AI^B c) ii d) I^AI^A

11. The Human Genome Project began in what year and under whose direction?

12. Explain the relationships between the alleles that determine the ABO blood group.

13. When doing pedigree analysis, what is DNA fingerprinting used for?

14. What are two controversies surrounding testing for the BRCA1 and BRCA2 genes?

15. At this point in time, what are the options available to parents who discover, through prenatal testing, that their unborn child has a serious genetic defect?

Part 4: Critical Thinking: Using Your Knowledge

Answer each of these in essay form, using complete sentences and paragraphs. Provide as much information as you can. (For extra essay practice, write out answers to the Review and Thought Questions in your textbook.)

1. Thoroughly explain the process used for DNA fingerprinting. What are arguments against its use and how do its proponents counter those arguments? What are some of its applications?

2. Explain the major areas of focus in the Human Genome Project. What are some of the controversies surrounding this project? What are ways in which knowledge gained from this project is being put to use?

3. Health insurance is under fire in the United States. Many managed care programs are being criticized for providing too little care at too high a cost. The costs of insuring people who require expensive medical care gets passed on to all people covered by that insurance—as the company's costs increase, your premiums increase. Should insurance companies be required to pay for genetic therapies that offer a temporary treatment at best, but not a cure?

4. Insurance companies are for-profit businesses. Should insurance companies have access to the results of genetic screening, such as tests for determining one's risk of cancer? Should the companies have the right to refuse to insure people who are at a high genetic risk if it will save money for all the other people who they cover? What risks should or should not be covered? Who should make these decisions?

5. A test is available that determines if a woman carries certain alleles for two breast cancer genes (BRCA1 and BRCA2). Presence of abnormal alleles for these genes greatly increases a woman's risk of breast cancer, but most women who develop breast cancer do not have these alleles. Currently, the only "preventive" treatments for breast cancer are early detection or preventive mastectomy, in which the breasts are removed before they can become diseased. As a female, would you want to be tested for the gene? (Men, assume for this question that you are making the decision for your mother, sister, or wife). How would a family history of breast cancer affect your decision? If you discovered that you do carry the gene, how would your life be affected? What if you learned that you do not carry the gene? Would you want your mother, sister, or daughter to be tested? How would knowledge that any of them carried the gene affect you?

6. Assume, whether you are female or male (about 1% of all breast cancer occurs in males), that you have a family history of breast cancer, but your genetic test shows that you do not carry the alleles that have a high correlation to breast cancer. Would you consider preventive mastectomy? If your test was positive, would you consider this treatment option? What factors, other than strictly health concerns, were involved in your decision?

7. Should insurance companies pay for preventive mastectomy in women who have a family history of breast cancer and who test positive for the breast cancer alleles?

8. Amniocentesis and chorionic villus sampling can often allow physicians to inform parents that the child they have conceived carries genes for a potentially devastating disease, leaving them with the decision to continue the pregnancy or abort it. If you were faced with that decision, how would you decide? What factors would impact your decision? Would your decision be based on the specific genetic defect? How would you decide if you knew the particular disease would likely lead to an early and agonizing death for your child?

9. Most human traits are polygenic and pleiotropic. First, explain what that means. Next, explain why it makes genetic research so challenging. Finally, explain what advantages, if any, there are to this situation.

10. When it began, the Human Genome project met with some resistance because DNA was to be sequenced from only two individuals—one male and one female, and both were of European descent. Some people were offended by the lack of inclusion of minorities. How valid were their concerns? What, if any, advantages are there to using more volunteers from different ethnic backgrounds?

Chapter 15
What is the Evidence For Evolution?

KEY CONCEPTS

1. Evolution is an idea with a long, varied, and controversial history.
2. Darwin's theory of evolution is a synthesis of several ideas which can be roughly divided into two parts:
   - a group of ideas that describe evolution, and
   - a group of ideas that describe how and why evolution works.
3. Few biologists since Darwin have doubted that evolution has occurred because the evidence he offered was unassailable.
4. Darwin asserted that:
   - the world is ever-changing and very old;
   - species are made of individuals;
   - species are plastic; and
   - all organisms are related by descent from a common ancestor.
5. The order in which organisms appear in the fossil record is consistent with the theory of evolution.
6. Species all over the world are most closely related to those that live nearby.
7. Although it was not intended, Linnaeus' system of classification accurately reflects actual relatedness of organisms.
8. Homologous and vestigial structures suggest an evolutionary process by which ancient structures are redesigned or abandoned.
9. During embryological development, organisms frequently pass through stages that resemble the organisms from which they evolved.
10. At the molecular level, molecules of closely related organisms are more similar than those of more distant relatives.
11. Natural selection, the driving force behind evolution, is the process by which individuals who have traits that make them better suited for life in their particular circumstances are more likely to survive and reproduce, passing on the genes for the successful traits to the next generation.

EXTENDED CHAPTER OUTLINE

THE SCOPES TRIAL

1. The Tennessee law prohibiting the teaching of evolution was never intended to be enforced, but the American Civil Liberties Union (ACLU) wanted to test the law in a federal court so they advertised for a teacher who would volunteer to be arrested under that law.
2. John Scopes had never taught evolution, but he once substituted for a biology teacher and assigned the textbook pages that covered evolution.
3. The 1925 Scopes trial was rather dull, partly because both sides hoped for a conviction (the ACLU wanted to appeal to a higher court).
4. Scopes was convicted, but then acquitted on a technicality, so the ACLU was not able to appeal and the law remained in effect until 1968.
5. In 1973, antievolutionists in Arkansas, Tennessee, and Louisiana passed identical bills calling for equal time for teaching evolution and creationism; the law was later declared unconstitutional (it violated separation of church and state). Other states have made similar attempts.
6. Creationism is the belief that the universe was created by the Judeo-Christian God in seven days.
7. In 1982, Federal Judge William Overton overturned Arkansas' law, stating that it was clearly intended to promote religion in public schools. As to the debate about creationism being "science," he correctly stated that "While anybody is free to approach a scientific inquiry in any fashion they choose, they cannot properly describe the method used as scientific if they start with a conclusion and refuse to change it regardless of the evidence.
8. In 1987, the United States Supreme Court upheld a judge's decision to strike down a similar Louisiana law, effectively ending the dispute over teaching evolution that began 128 years earlier.
9. In the 1990s, a different antievolution maneuver began. Instead of asking for creationism to also be taught, antievolutionists requested that evolution not be taught.
10. In December 1999, the Kansas State Board of Education approved a curriculum that disallowed testing on evolution. Students will be required to learn about natural selection—Darwin's mechanism for evolution—but not that species can evolve through it, or even that all organisms on Earth are related.
11. Darwin's book, *On the Origin of Species*, began the evolution debate. His three major ideas that many found offensive were:
    - The Earth is very old.
    - One species can change into another.
    - Humans and apes are related.

WHAT IS EVOLUTION?

1. The first part of evolution theory is the idea that species change over time, which Darwin referred to as "descent with modification."
2. Evolution can be defined as a heritable change in organisms over time, and it also states that all life on Earth is descended from a few types of cells.
3. The second part of Darwin's theory is:
    - his specific mechanism for change, called natural selection, and
    - a description of how he thought evolution occurs, called gradualism.
4. Gradualism is the idea that species evolve gradually and continuously through a steady accumulation of changes.

5. Natural selection is the differential reproductive success of individuals due to genetically inherited traits.
6. A change is adaptive if it increases an organism's chances of passing its genes on to future generations. Whether a trait is adaptive depends on the organism's environment.
7. Artificial selection is the process by which new lines are created by selecting the most desirable individuals for breeding.
8. Darwin's theory was based on four ideas:
   - The world is ever-changing and very old.
   - Species change.
   - Species are composed of populations of individuals.
   - Every species and group of species is descended from a common ancestral species.
9. Essentialism, a belief put forth by Plato, maintains that individuals are distorted versions of an ideal, or essential, form.
10. Darwin recognized that species are not ideal forms, but groups of individuals.

EVOLUTION BEFORE DARWIN

1. Darwin was not the first person to believe that organisms evolve.

How old is the idea of evolution?

1. In the 6th century, B.C., Anaximander argued that
   - the world was not created abruptly,
   - animals evolved, and
   - humans and all vertebrate animals descended from fish.
2. Essentialism, along with creationism, formed a nearly unbreachable intellectual barrier to the idea of evolution.
3. In the 17th and 18th centuries, almost everyone in Europe believed the Earth was created in 4004 B.C.
4. By the end of the 18th century, several species had been discovered that are not mentioned in the Bible.
5. By the middle of the 19th century, geologists were certain the Earth was far older than most religions believed.

How did Linnaeus's ideas lay the groundwork for evolution?

1. Carolus Linnaeus (1707–1778) invented the science of taxonomy—the naming and grouping of organisms.
2. Using humans as an example, and moving from the largest to the most specific grouping, the hierarchical system of classification is as follows:
   - class—Mammalia
   - order—Primates
   - family—Hominidae
   - genus—*Homo*
   - species—*sapiens*
3. Each grouping in this system is a taxon (plural is taxa).
4. Linnaeus failed to ask why organisms fall so naturally into families, firmly believing each species was created by God.

Why was Count Buffon afraid to discuss evolution?

1. George-Louis Leclerc Buffon noted similarities in closely related species and stated that "...we should not be wrong in supposing that with sufficient time she [nature] could have evolved all other organized forms from one primordial type," but he quickly backed off to placate the Catholic Church.

What made Hutton think that the Earth was older than 6000 years?

1. James Hutton (1726–1797) promoted the idea that the Earth is eternal, with "no vestige of a beginning—no prospect of an end." This, too, was considered heresy.

Catastrophism: How did Cuvier try to reconcile Genesis and geology?

1. Baron Georges Cuvier invented both comparative anatomy (the detailed study of anatomies of different species) and paleontology (the study of fossils).
2. Fossils are the remains or imprints of past life. They confused people because they looked like plants or animals (not familiar ones), but they were made of stone.
3. Cuvier argued for catastrophism—the belief that organisms found in the fossil record at one point but not later perished in some catastrophe, then new species were created, so the fossil record merely showed numerous cycles of catastrophe and new creation.
4. Catastrophism allowed for the age of the Earth and the fossil record without implying that Genesis was wrong.

Uniformitarianism: Lyell rejects catastrophism

1. Charles Lyell (1797–1875) rejected catastrophism. He put forth the theory of uniformitarianism, which states that the processes that now mold the Earth's surface—erosion, sedimentation, and upheaval—are the ones that have always molded it.

Why was Lamarck ignored?

1. Jean Baptiste Lamarck (1744–1829) recognized a clear line of descent from older fossils to more recent fossils to modern species: the oldest fossils looked the least like modern organisms; later fossil groups included increasing numbers of familiar species.
2. Lamarck is best known for his theory of inheritance of acquired traits, in which he stated that acquired traits, such as lengthening of a giraffe's neck as it stretches to reach leaves, are passed on to offspring.
3. We now know that only traits with a genetic basis can be inherited, not acquired traits.

How did Darwin convince others that species evolve?

1. By the mid-19th century, natural science was quite popular with the lay public.

Darwin's education

1. Charles Robert Darwin (1809–1882) was entirely caught up in the natural science craze.
2. He lacked ambition, so his father chose a career for him—medicine.
3. His lack of ambition again forced his father to make a decision—he sent him to Cambridge University to study for the clergy. But the clergy was filled with naturalists, and Charles rediscovered his boyhood passion for nature.
4. He took a trip as a ship's naturalist on the H.M.S. Beagle in 1831.
5. Darwin read up on Lyell's work, about the planet being very old and change occurring very slowly.
6. Darwin's work required that he collect numerous specimens and he also took detailed notes.
7. It was six more years before Darwin started outlining his theory, and by then he had also discovered Malthus' work.

Malthus: Too many children

1. Thomas Robert Malthus predicted that, without some check on population growth, human populations would face a continuing, ferocious struggle for existence in the face of limited resources.
2. From Malthus' ideas, Darwin reasoned "that any being, if it vary however slightly in any manner profitable to itself, under the complex and sometimes varying conditions of life, will have a better chance of surviving, and thus be naturally selected."

Wallace scoops Darwin

1. In 1858, Alfred Russel Wallace (1823–1913) independently hit upon natural selection as a mechanism for evolution. Although he communicated with Darwin, Darwin did not finish his writings until after Wallace had drafted his.
2. Darwin's book, *On the Origin of Species by Means of Natural Selection, or the Preservation of Favoured Races in the Struggle for Life*, is considered the single most influential scientific book ever written—when it appeared in 1859, all 1250 copies of the first printing sold in a single day.

WHAT IS THE EVIDENCE FOR EVOLUTION?

1. Evolution is heritable change in populations of organisms, which Darwin called "descent with modification," and it is also "speciation," which is the division of one species into two or more.
2. The book's strength was Darwin's overwhelming evidence—from the fossil record, the distribution of plants and animals (biogeography), taxonomy, comparative anatomy, comparative embryology, and domestic breeding.

The fossil record tells a story of evolution

1. Fossils are distributed consistently. Rocks of the same age contain about the same organisms.
2. The order in which organisms are deposited in the fossil record suggests a path of evolution that is supported by other biological fields.
3. Recent fossils look most like modern organisms.

How do fossils form?

1. Fossils are the preserved remains or imprints of past life.
2. An animal or plant may become a fossil only if it dies under the right conditions for preservation—most organisms are instead consumed.
3. When an organism is buried in a bog or at the bottom of a sea or lake, it becomes covered by layers of sediment, and as these layers build and increase the pressure, the layers become sedimentary rock.
4. Most fossils are marine or fresh water organisms (because they live in the best environment for fossilization), and only hard, inedible parts of animals and woody parts of plants are likely to last long enough to become fossils.
5. Because the process of fossilization is chancy, the fossil record is incomplete.

What does the fossil record show?

1. The fossil history is divided into chapters as follows:
   - A period spans 30 to 75 million years and contains distinctive life forms in the fossil record.
   - Periods are grouped into four long eras, each of 65 million to several billion years.
   - The most recent "chapters" are subdivided into relatively short epochs.
2. The Precambrian Era comprises most of Earth's history, from the beginning about 4.6 billion years ago and is subdivided into six eras.
3. The first fossils (3.8 billion years ago) resemble modern prokaryotes.
4. The first eukaryotes appeared about 2 billion years ago. New evidence may age them to 2.7 billion years ago.

The Cambrian explosion

1. The first multicelled organisms appear in the fossil record at the end f the Precambrian Era, about 800 million years ago. The first multicelled animals don't appear until about 640 million years ago.
2. By the end of the Paleozoic Era, all major groups of animals had appeared. This era consists of six periods:
   - The Cambrian—this marks the beginning of the abundant fossil record and is referred to as the "Cambrian explosion." Fungi, algae, and trilobites predominate; vertebrates appeared. By the end of the Cambrian, about 500 million years ago, some mass extinction had wiped out whole classes and families.
   - The Ordovician and the Silurian—more fish with bony skeletons, starfish, and nautiluslike animals appeared, as did the first land plants.
3. The Devonian—the oceans receded dramatically, increasing dry land mass; fish continued to diversify tremendously and this is also known as the Age of Fish.
   - The Carboniferous—this is marked by diversification and proliferation of the Amphibia (the first vertebrates to become firmly established on land), the first reptiles, flying insects, ferns, horsetails, and cone–bearing trees. These fossilized swamps today are our main sources of fossil fuels—coal, oil, and natural gas.
   - The Permian—The Carboniferous and Permian together are often called the Age of Amphibia. During the Permian, the continents came together as a single supercontinent—Pangea—and the climate became arid, causing mass extinction of about 90% of all marine species.
4. The Mesozoic Era, also called the Age of Reptiles, started about 248 million years ago and ended 65 million years ago. It has three periods: the Triassic, the Jurassic, and the Cretaceous, which hosted another mass extinction.
5. During the Triassic period, reptiles diversified enormously and ancestors of the first mammals appeared. The Mesozoic also saw many new marine invertebrates. Dinosaurs appeared in the late Triassic and disappeared at the mass extinction at the end of the Cretaceous period. Organisms that survived the extinction included many reptiles, insects, flowering plants, birds, and mammals.
6. The Cenozoic Era is the briefest, lasting from 65 million years ago to the present. Insects, flowering plants, modern birds, and modern mammals have flourished and diversified.

Taxonomy

1. Linnaeus' hierarchical classification system looks exactly like a family tree.
2. The careful anatomical comparisons that Linnaeus and his successors made reflect actual relatedness.

Comparative anatomy

1. Species descended from a common ancestor may evolve in very different directions yet retain many of the same characteristics.
2. Homologous structures are similar structures seen in two or more species.
3. Homologous structures may perform the same or vastly different functions.
4. Convergence is the opposite of homology, and is when unrelated structures have a similar shape and perform similar functions. This type of trend is called convergent evolution.
5. Evolution is opportunistic.
6. Intermediate forms are related organisms appearing in the fossil record that differ from both their ancient ancestors and from their modern relatives, with traits between the two.
7. A vestigial structure is one that has little or no function—it is a remnant. But vestigial structures are indicators of biological history and support the theory of evolution.

Comparative embryology

1. The early embryos of all vertebrates are alike.
2. Developing organisms frequently pass through stages that resemble the organisms from which they evolved.

Comparative molecular biology

1. An organism's genes and gene products (proteins) are a clear record of its heredity.
2. All cells rely on the same molecular machinery:
   - DNA to carry genetic information;
   - RNA, ribosomes, and about the same genetic code to translate that information into proteins;
   - the same 20 amino acids to build proteins; and
   - ATP to carry energy.
3. This molecular universality implies a common heritage to all life.
4. By comparing proteins produced by different organisms, biologists can construct hierarchies of relatedness—phylogenetic trees—much like those based on other comparative studies.
5. Phylogenetic trees can also be constructed by sequencing individual genes from different species.
6. Phylogenetic trees reflect the relatedness between species revealed by the fossil record.
7. In any two species, the number of molecular differences in a shared molecule is proportional to the amount of time that has elapsed since the two species diverged. These changes are used to develop molecular clocks for different species, and these clocks can predict when two species diverged from one another, even of no common ancestor is found in the fossil record.
8. Comparative molecular biology confirms and elaborates on the story told by comparative anatomy and the fossil record.

Biogeography

1. Biogeography is the study of the past and present distribution of plant and animal species.
2. Species all over the world are most closely related to those that live nearby, which can be attributed to Darwin's theory of common descent.

NATURAL SELECTION: DARWIN'S MECHANISM FOR EVOLUTION

1. Darwin believed that natural selection, gradually and over immense amounts of time, gave rise not just to new species, but to new genera, families, and all higher taxa.

Artificial selection

1. Breeders obtain varied forms from a single species by carefully selecting and breeding only individuals that show certain desired traits most strongly.
2. Artificial selection can transform the characteristics of a breed if:
   - individual organisms vary in their characteristics,
   - these variations are heritable, and
   - the breeder consistently selects certain traits in each generation.
3. Artificial selection does not result in production of new species, but it does demonstrate how much genetic variation exists in populations and how much change can occur in a short time.

Darwin's argument for natural selection

1. Natural selection must occur because nature cannot sustain unlimited exponential growth.
2. In artificial selection, the decisions of the breeder determine which individuals produce the next generation; in natural selection, the individuals that manage to survive and reproduce will produce the next generation.

3. The rate at which changes accumulate depends on:
   - the intensity of selection,
   - the extent of inherited variation, and
   - the extent of variation within the population.

Natural selection analyzed

1. Natural selection has been broken down to a series of facts and conclusions:
   - Superfecundity—all species would overpopulate if all their offspring survived.
   - Not all offspring reproduce—disease, predation, limited access to resources, etc. prevent many in the population from reproducing.
   - Individual variation—reproductive individuals in a population differ.
   - Heredity—many individual differences are heritable.
   - Adaptive traits are, by definition, those that are passed on differentially.
   - Adaptive traits accumulate within a population.
2. Over the years, biologists have proposed many alternatives to natural selection.

VOCABULARY BUILDING

In your own words, first write a brief definition, then a full explanation for each of the following terms. Include examples where appropriate. Complete this section from your memory—you will not learn the terms by simply copying definitions from the textbook. Once you have finished, check your responses against the information in the chapter and make any necessary corrections.

creationism —

"descent with modification" —

evolution —

gradualism —

natural selection —

artificial selection —

essentialism —

species —

genus —

family —

order —

class —

taxon —

taxonomy —

comparative anatomy —

paleontology —

fossils —

catastrophism —

uniformitarianism —

theory of inheritance of acquired characteristics —

period —

era —

epoch —

Precambrian Era —

Paleozoic Era —

vertebrates —

Mesozoic Era —

Cenozoic Era —

homologous structures —

convergent evolution —

intermediate forms —

vestigial structures —

branchial arches —

phylogenetic tree —

molecular clock —

biogeography —

absolute age —

decay —

isotope —

half-life —

plates —

CHAPTER TEST

The following test has four parts. Complete as much of the exam as you can from memory. If you cannot answer a question, skip it. Once you complete all that you can, try to answer any questions you skipped. If you still cannot answer them, consult your textbook for the answers. Once you have completed all sections of the test, check your answers for Parts 1 - 3 against those in the back of this book. Highlight any incorrect answers then review that material in your textbook. Correct your answers for future reference.

Part 1: Multiple Choice

For each of the following, select all correct responses—more than one may be correct.

1. Whether or not a particular trait is adaptive depends mostly on:
   a) the organism's genotype
   b) the organism's age
   c) the organism's environment
   d) all of these

2. Which of the following summarizes Darwin's idea of gradualism?
   a) The Earth was created gradually over a very long period.
   b) Species evolve gradually, through steady accumulation of changes.
   c) Life originated gradually from increasing complex organization of molecules.
   d) All of these are included in Darwin's idea of gradualism.

3. Which of the following is not a factor in natural selection?
   a) disease
   b) predation
   c) selected breeding
   d) speed

4. Plato is associated with which of the following beliefs?
   a) evolution
   b) creationism
   c) essentialism
   d) uniformitarianism

5. Which of the following is listed in correct taxonomic order?
   a) species, genus, family, order, class
   b) genus, species, family, order, class
   c) species, genus, family, class, order
   d) family, genus, species, order, class

6. Uniformitarianism maintains that:
   a) Layers of the fossil record are evidence of repeated catastrophes.
   b) The same forces that molded the Earth's surface in the past continue to do so.
   c) Only one type of organism can be found in each layer of the fossil record.
   d) All individuals are distortions of one ideal form.

7. According to Lamarck, species change by inheritance of:
   a) genetic traits
   b) coding sequences of DNA
   c) acquired traits
   d) all of these

8. Which of the following spans the most time?
   a) epoch
   b) period
   c) era
   d) these are all words for the same amount of time

9. Most fossils are of what type of organisms?
   a) marine or fresh water organisms
   b) birds
   c) insects
   d) land-dwelling reptiles

10. Most of the Earth's history has occurred in which era?
    a) Paleozoic
    b) Precambrian
    c) Mesozoic
    d) Cenozoic

11. The first chordates appeared before the end of what period?
    a) Cambrian
    b) Devonian
    c) Carboniferous
    d) none of these

12. Dinosaurs appeared and disappeared during which era?
    a) Paleozoic
    b) Precambrian
    c) Mesozoic
    d) Cenozoic

13. Which is the current era of geological time?
    a) Paleozoic
    b) Precambrian
    c) Mesozoic
    d) Cenozoic

14. Which of the following is true about homologous structures?
    a) They arose from a common ancestor.
    b) They may have the same function.
    c) They may have very different functions.
    d) All of these are true about homologous structures.

15. Phylogenetic trees are derived by:
    a) comparative anatomists
    b) comparative embryologists
    c) paleontologists
    d) molecular biologists

Part 2: Matching

For each of the following, match the correct term with its definition or example. More than one answer may be appropriate.

evolution **creationism**

1. ________________ All life was created by God.
2. ________________ The Earth is relatively young.
3. ________________ Species change gradually.
4. ________________ Each species is created individually.
5. ________________ Humans and apes are related species.
6. ________________ One species can change into another.
7. ________________ "Survival of the fittest."
8. ________________ Extinction, as shown in the fossil record, indicates a catastrophe occurred at that time.

Precambrian Era **Paleozoic Era** **Mesozoic Era** **Cenozoic Era**

9. ________________ Dinosaurs appear and disappear.
10. ________________ The Age of the Amphibians.
11. ________________ Longest era.
12. ________________ Appearance of first prokaryotes, eukaryotes, and multicellular organisms.
13. ________________ The greatest diversity of evolution occurred.
14. ________________ The Age of Reptiles.
15. ________________ Ancestors of the first mammals appeared.
16. ________________ The shortest era.

Part 3: Short Answer

Write your answers in the space provided or on a separate piece of paper.

1. In the Scopes trial, why did both sides hope for a conviction?

2. According to Darwin's theory of evolution, how did life originate?

3. What is meant by an "adaptive" trait?

4. How do artificial selection and natural selection differ?

5. Starting with the class, identify the taxon to which modern humans belong.

6. What two sciences did Baron Georges Cuvier invent?

7. How did catastrophism explain the fossil record?

8. What theory was put forth to reject catastrophism?

9. Who coined the term *biology*, and what was it in reference to?

10. How did Lamarck think that species changed over time?

11. Although not all biologists agree that natural selection is the driving force behind evolution, after Darwin's book was printed, all biologists do agree that evolution occurs. Why was Darwin's work so convincing?

12. List, in order from oldest to most recent, the four eras of geologic history.

13. Why do most dead organisms not become part of the fossil record?

14. How do organisms get trapped in rock?

15. What are three factors that determine the rate at which a species evolves?

Part 4: **Critical Thinking: Using Your Knowledge**

Answer each of these in essay form, using complete sentences and paragraphs. Provide as much information as you can. (For extra essay practice, write out answers to the Review and Thought Questions in your textbook.)

1. Explain the issues at the heart of the debate over evolution vs. creationism. In 1996, Pope John Paul II gave a speech in which he stated that evolution as a theory seems quite valid and that creationism and evolution need not be viewed as contradictory. How might this be so? The debate rages on, often quite heatedly, with neither side willing to make any concessions to the other. Why is this such a controversial issue? What is your position?

2. Use examples to thoroughly explain how natural selection works and leads to evolution of a species.

3. Darwin was not the first person to believe in evolution. Review the work of Darwin's predecessors that helped lead Darwin to his conclusions. Critique each individual's work in terms of the science they used and the conclusions they drew.

4. Summarize the evidence in support of evolution from each of these areas: fossil record, and biogeography. What are the weaknesses and strengths of each discipline's approach?

5. Describe some examples of artificial selection that affect your life. Describe some ways in which natural selection may operate within our own species.

6. Explain how fossils develop and how the fossil record has been used by those who support and those who deny the theory of evolution. Why are there gaps in the fossil record? How does that affect its importance?

7. Many people believe humans are the superior species, and yet many of our own activities jeopardize our health, and possibly, ultimately, our species' survival. What are some of these activities and how do they possibly relate to our future evolutionary success?

8. Most biologists would agree that if there were an all out nuclear war on our planet, humans would not survive. Which organisms do you think would be the most successful and the most likely to survive such a catastrophe, and why?

9. Humans exert considerable control over our environments. We are rather far removed from most natural forces that would have affected the survival of our early ancestors. In what way has our ability to minimize the impact of natural selection on our survival benefited us? In what ways might our comfortable, sheltered lives be detrimental?

10. Is it likely all species currently alive will be alive in a million years? Based on the planet's history, how important is it that all species survive? What will likely happen to life on Earth if there is another major global catastrophe?

Chapter 16
Microevolution: How Does a Population Evolve?

KEY CONCEPTS

1. August Weismann proposed that the hereditary material is in the nucleus, is molecular, and is not susceptible to influence by developments in the rest of the organism; inheritance of acquired traits was proven to not be possible.
2. Sexual reproduction, by creating new combinations of traits, acts as a source of new variations for natural selection.
3. The original source of genetic variation is gene mutations or chromosomal mutations.
4. A population is polymorphic for a given gene if that gene has more than one allele.
5. Most traits are polygenic, meaning they are controlled by many genes.
6. Genetic variation spreads when sexual reproduction and recombination mix these changes into endless new combinations.
7. Most natural populations vary enormously.
8. Mutations that do not affect phenotype are not subject to selection.
9. In the absence of genetic variation, evolution would cease.
10. The Hardy-Weinberg Principle shows that sexual reproduction maintains a stable distribution of genotype frequencies through the generations, called the Hardy-Weinberg equilibrium. Evolution of populations can be measured against this equilibrium.
11. In small, isolated populations, random drift can cause significant changes in gene frequencies. In large populations, drift occurs so slowly that it is a less significant factor than the effects of mutation, gene flow, and selection.
12. Gene flow increases variation within a local population, but makes adjacent populations more alike.
13. Natural selection cannot, by itself, eliminate lethal recessive alleles.
14. A balanced polymorphism may result from heterozygote superiority or from different selective pressures in time and space without the same population.

EXTENDED CHAPTER OUTLINE

A FATAL DISAGREEMENT

1. In the 1920s, Nikolai Vavilov was the world's expert on biogeography of wheat.
2. Trofim Lysenko firmly believed in Lamarckism—inheritance of acquired traits.
3. Lysenko became quite popular with Stalin and was put in charge of all Soviet agriculture.
4. Lysenko ordered tens of thousands of farms to test a theory about cross-pollinating wheat.
5. Vavilov and fellow biologists correctly predicted that cross-pollinating most of the Soviet wheat crop would eliminate many of the country's best wheat varieties—and it did.
6. Vavilov continued to speak out in favor of genetics, which Lysenko disdained, and in 1940 Vavilov was arrested for "wrecking" Soviet agriculture, eventually receiving a sentence of life in prison, where he died of starvation.
7. Lysenko's influence died shortly after Stalin, when crop failures forced Khrushchev to buy crops from the United States and Canada. Genetic research was resumed, after being abandoned for thirty years.

HOW ARE VARIANTS CREATED AND MAINTAINED?

1. Even when biologists accepted that evolution occurs, they could not agree on how it worked.
2. We now know that Darwin's theory of natural selection was largely correct.

Why did Charles Darwin accept blending inheritance?

1. Darwin thoroughly understood selection—the differential survival and reproduction of genetic variants—but he didn't know where variation came from.
2. The gene was completely unknown in 1859.
3. Most scientists, including Darwin, thought inheritance was blended—offspring had a blend of both parents' traits.
4. In 1866, Gregor Mendel published his research showing the particulate nature of inheritance.
5. Mendel's research showed that the individual units of inheritance (now known as genes) remain intact from one generation to the next, instead of blending with others.

Why was the idea of blending inheritance a problem?

1. In 1867, Fleeming Jenkin, an engineer, announced that natural selection could never work in sexually reproducing organisms because interbreeding would quickly dilute any new traits. Jenkin was mistaken in almost everything he said.
2. Thanks to Mendel's work, we now know that:
   - inheritance is particulate, not blending, and
   - instead of diluting traits, sexual reproduction creates an assortment of different traits.
3. The idea of inheritance of acquired traits had been universally accepted for hundreds of years, and Darwin turned to it in his uncertainty.
4. The theory of inheritance of acquired characteristics states that:
   a) environmental changes create changes in the needs of organisms;
   b) changes in needs allow changes in the organisms' behaviors to satisfy the needs;
   c) behavioral changes result in increased use, and thus development, of certain body parts (old or new); and
   d) these changes are passed on to offspring.

HOW DID BIOLOGISTS COME TO REJECT THE THEORY OF ACQUIRED CHARACTERISTICS?

1. August Weismann was one of the first biologists to completely reject the inheritance of acquired characteristics and the first to suggest a coherent theory of molecular genetics.
2. Weismann tested the theory and, in 1883, published a paper rejecting Lamarckism entirely.
3. Wallace (Darwin's co-discoverer of evolution) and Weismann alone believed that natural selection, not inheritance of acquired traits, was the major mechanism for evolution.
4. Weismann correctly theorized that the genetic material was all in the nucleus.
5. Weismann's "continuity of life" theory states that:
   - the germ plasm is separate from the rest of the body from the beginning; nothing in the development of the environment of the body can influence the germ cells or the hereditary material; and
   - only the germ plasm is passed on to the offspring; thus
   - acquired characteristics cannot be inherited.

How did Weismann explain variation in individuals?

1. Weismann correctly concluded that sexual reproduction brings together old traits into new combinations, thus producing variation upon which natural selection could work.

How did the rediscovery of Mendel's work transform biology?

1. In 1900, three biologists (Hugo Marie de Vries, Karl Franz Joseph Correns, and Erich Tschermak) working independently, rediscovered Mendel's laws.
2. William Bateson, de Vries, and other Mendelians (geneticists) argued that evolution occurred in leaps and bounds, called saltations, by means of genetic mutation.
3. They believed that evolution occurred suddenly, from one generation to the next, and that mutations in an individual were the cause of new species.
4. Naturalists knew from observations that variation is continuous and heritable, but they overreacted to the Mendelians' rejection of natural selection, insisting that Mendelian genetics did not apply to the kind of variation that causes evolution.
5. Both Mendelians and Naturalists were wrong, and the battle raged for about 40 years.

How did the modern synthesis reconcile genetics and natural history?

1. The modern synthesis, a marriage of genetics and evolutionary theory, occurred in the late 1930s, and helped end the internal feud between geneticists and naturalists.
2. Population genetics combines Mendelian genetics with recognition that evolution can occur in populations.
3. Population genetics explains, in precise mathematical terms, the processes by which variation is generated and passed on within populations of organisms.
4. These processes are called microevolution which, in the narrow sense, refers to changes in the frequencies of alleles of genes in a population.
5. Macroevolution refers to processes by which species and higher groupings of organisms originate, change, and go extinct.
6. Population genetics explains variation at the population level while also providing a basis for natural selection and other evolutionary forces.

WHAT IS THE SOURCE OF THE VARIATION THAT FUELS EVOLUTION?

1. The two chromosomes in a pair are homologous—each possesses the same genes in the same positions.
2. Egg and sperm each have only one chromosome from each pair, so each parent contributes one chromosome for each pair to the offspring.
3. Genetic variation arises when a gene mutates into different forms, called alleles.
4. If each chromosome in the pair has the same allele, the individual is homozygous for that gene.
5. If each chromosome has a different allele for the same gene, the individual is heterozygous.

What is genetic variation?

1. Most genetic variability comes from gene mutations.
2. Chromosome mutations, which affect the number or arrangement of genes on a chromosome, can change several proteins at once.
3. Most chromosomal mutations are lethal or very damaging in animals (less so in plants), but some are actually beneficial.
4. If a population has one or more alleles of a given gene, it is said to be polymorphic.
5. If members of a population come in two or more forms, the population is also said to be polymorphic.

Are traits determined by single genes?

1. The Human Genome Project has focused public attention on single genes which, when mutated, can cause diseases.
2. Most traits result from interactions between numerous genes and the environment.
3. Most traits are polygenic—governed by more than one gene—and they vary continuously within a population.
4. The curves describing such traits typically have a bell shape. The breadth of the curve is a measure of the variability of the trait, and the average of a trait may vary considerably between populations.

What determines the genetic variability of a population?

1. Populations that contain many alleles vary more than populations that have few alleles per gene.
2. Three factors determine the genetic variability of a population:
   - the rate at which mutations accumulate in the DNA;
   - the rate at which changes spread through a population (by sexual reproduction); and
   - the rate at which deleterious mutations are eliminated from a population by natural selection.

How often do mutations occur?

1. Mutations cause random changes in protein structure, and most are damaging.
2. On average, each human baby has one or two mutations.
3. Mutations arise constantly as a result of random processes.

How do mutations spread through populations?

1. The rate at which a population evolves depends on how rapidly new mutations spread through the population.
2. The principal way genetic variation spreads is through sexual reproduction and recombination.
3. Recombination mixes mutations that have arisen independently into endless new combinations.

How much do populations vary genetically?

1. Electrophoresis has been widely used to estimate the degree of polymorphism in many populations.
2. Another way of expressing the level of genetic variation in a population is to estimate the likelihood that any individual will be heterozygous for a given gene.
3. In humans, about 7% of the genes that code for proteins are heterozygous.

Is all genetic variation subject to natural selection?

1. The only variants important for natural selection are those that contribute to changes in the whole individual in ways that affect the individual's ability to survive and reproduce.
2. Phenotype is a result of both the genotype and effects of the environment.
3. Selection acts on phenotype.
4. Most lethal mutations are not subject to selection unless they are homozygous.
5. Most mutations occur in noncoding parts of the DNA so they do not affect phenotype and are, therefore, not subject to natural selection.
6. Mutations that do not affect phenotype are not subject to natural selection.

Why is genetic variation essential to evolution?

1. The guiding force of evolution is natural selection.
2. Mutations that best equip their possessors to survive and reproduce will increase in frequency in each generation.
3. Evolution by natural selection depends on the continued existence of variation.

How do populations maintain genetic equilibrium?

1. According to the Hardy-Weinberg Principle, sexual reproduction by itself does not change the frequencies of alleles within a population.
2. All the genes of all the individuals in a population constitute the gene pool.
3. The allele frequency is the number of times the allele is present in the population divided by the total number of chromosomes on which the gene appears.
4. To calculate allele frequency: if there are two alleles for a certain gene (*A* and *a*),
   - the frequencies of one allele (*A*) is designated as p,
   - the frequency of the other (*a*) is q, and
   - $p + q = 1$.
5. To calculate genotype frequency: again assuming two alleles, *A* and *a*,
   - the frequency of the *AA* homozygote is p^2,
   - the frequency of the heterozygote (*Aa*) is 2pq,
   - the frequency of the *aa* homozygote is q^2, and
   - $p^2 + 2pq + q^2 = 1$
6. The frequencies of the individual alleles in a population do not change as a result of sexual reproduction.
7. Hardy-Weinberg equilibrium is a stable distribution of genotype frequencies maintained in a population from generation to generation.
8. Attainment of Hardy-Weinberg equilibrium requires that five conditions be met:
   - random mating,
   - large population size,
   - no mutations,
   - no breeding with other populations, and
   - no selection.
9. In reality, these conditions are rarely met, so gene frequencies change and evolution occurs.
10. The Hardy-Weinberg equilibrium is a baseline against which the evolution of populations can be measured.

WHAT CAUSES ALLELE FREQUENCIES TO CHANGE?

1. Of all the causes of changing allele and genotype frequencies, only natural selection consistently leads to adaptive changes in allele frequencies.

How can new mutations cause changes in allele frequencies?

1. Mutations provide the raw material for evolution.
2. Random drift can eliminate mutations or their spread through a population.

How do chance events change gene frequencies?

1. Sampling errors are chance fluctuations that affect small samples but not large ones.
2. In small populations, sampling errors may cause allele frequencies to change randomly from generation to generation.
3. Random drift refers to changes in gene frequency that are due to random events, not mutation, selection, or immigration. Such random events can lead to changes in gene frequency within small populations.
4. In the founder effect, an entire population arises from a small group of individuals.
5. When a large population is reduced to a few surviving individuals by some sort of random disaster or harsh selection pressure, it is said to have passed through a bottleneck. The individuals then found a new population.
6. With both the founder effect and after a bottleneck, the gene pool of the new population is a tiny sample of the original gene pool. Since the allele frequencies have changed, the population is said to have evolved in the absence of natural selection.
7. In humans, the founder effect, along with inbreeding and increased homozygosity, are believed responsible for high incidences of certain genetic disorders in certain populations.
8. Random drift can play an important role in the spread of new mutations.
9. Over time, a new mutation will either disappear or spread through the population, even in the absence of selection. The smaller the population, the less time this takes.
10. Even in large populations, random drift will tend to cause alleles to disappear.
11. In the absence of natural selection, new mutations, recombination, and migration, drift would cause the frequencies of all alleles in even large populations to go to 0% or 100%, and evolution would cease.

How does nonrandom mating cause changes in allele frequencies?

1. Assortative mating occurs when individuals choose mates on the basis of the mate's genotype (reflected in their phenotype).
2. Positive assortative mating occurs when individuals choose mates on the basis of similarities in their genotypes.
3. Positive assortative mating tends to increase the number of homozygotes.
4. Inbreeding—mating among close relatives—is an extreme form of positive assortative mating that greatly increases the number of homozygotes.
5. In sexual species at least, inbred individuals who are homozygous for many genes are, on the average, less healthy than individuals with greater heterozygosity.
6. Most species are outbreeders—they cross-breed with usually not-too-related individuals, and often have complex physical or behavioral adaptations to promote this.
7. Outbreeders use negative assortative mating—they preferentially choose mates somewhat unlike themselves.

How does breeding between populations change allele frequencies?

1. Hardy-Weinberg equilibrium requires that a population not breed with other populations, but such isolation rarely occurs.
2. Gene flow between populations causes changes in allele frequencies.
3. Gene flow between populations is common; gene flow from outside populations increases genetic variation, but the same gene flow also makes adjacent populations more alike.

How does selection decrease the frequency of certain traits?

1. Natural selection reduces the frequency of deleterious alleles and is a powerful way to change gene frequencies.
2. While selection can dramatically change allele frequencies, even the most lethal recessive alleles can still persist indefinitely as heterozygotes.

How does selection increase the frequency of certain traits?

1. In the same manner in which selection can reduce the incidence of a certain trait, it can also increase the incidence of others. This is seen, for example, when bacteria, parasites, and pests develop resistance to chemicals meant to destroy them.
2. The darkening of moths and butterflies to give better protective coloration in sooty environments is now called industrial melanism. The darker forms have a selective advantage over the light forms in the dark, sooty environment.

In what directions does natural selection push populations?

1. Selection may push populations in an adaptive direction, or it may have other effects.
2. Directional selection shifts the frequency of one or more traits in a particular direction. Industrial melanism is an example of this type of selection, as is development of resistance.
3. Directional selection is common in changing environments and can lead to evolution of new species.
4. Stabilizing selection tends to act against extreme phenotypes so that the average is favored.
5. Stabilizing selection is characteristic of stable environments because such environments consistently favor the average and rarely or never favor extreme variants.
6. Disruptive selection is the opposite of stabilizing selection: it increases the frequency of extreme types in a population at the expense of intermediate forms.
7. Sexual dimorphism refers to having differences between the sexes in a population. Such dimorphism may result from sexual selection—the differential ability of individuals with different genotypes to acquire mates.
8. Two types of sexual selection are recognized—male competition and female choice.
9. In male competition, a male competes with other males for territory or access to females.
10. Female choice often involves courtship behavior, which allows the female to assess the male.
11. Females typically invest more time and energy into each of their offspring, so the female must be more selective in choosing a mate.
12. Most species have evolved quite complex rituals that enable the female to decide if her prospective mate is genetically worthy of her.

HOW CAN NATURAL SELECTION PROMOTE GENETIC VARIATION?

1. Genetic heterogeneity in a population is expected if different alleles of the same gene are equally useful.
2. Such a situation creates a balance of different alleles in a population, called a balanced polymorphism.
3. Populations may maintain high frequencies of alleles that are deleterious or even lethal when homozygous; in such cases, polymorphism results from selective superiority of the heterozygote.

VOCABULARY BUILDING

In your own words, first write a brief definition, then a full explanation for each of the following terms. Include examples where appropriate. Complete this section from your memory—you will not learn the terms by simply copying definitions from the textbook. Once you have finished, check your responses against the information in the chapter and make any necessary corrections.

blending inheritance —

inheritance of acquired characteristics —

germ plasm —

saltations —

modern synthesis —

population genetics —

microevolution —

macroevolution —

homologous —

alleles —

homozygous —

heterozygous —

polymorphic —

polygenic —

mutations —

phenotype —

genotype —

gene pool —

allele frequency —

Hardy-Weinberg equilibrium —

sampling errors —

random drift —

founder effect —

bottleneck effect —

assortative mating —

positive assortative mating —

inbreeding —

outbreeders —

negative assortative mating —

gene flow —

natural selection —

directional selection —

stabilizing selection —

disruptive selection —

sexual dimorphism —

sexual selection —

female choice —

male competition —

balanced polymorphism —

CHAPTER TEST

The following test has four parts. Complete as much of the exam as you can from memory. If you cannot answer a question, skip it. Once you complete all that you can, try to answer any questions you skipped. If you still cannot answer them, consult your textbook for the answers. Once you have completed all sections of the test, check your answers for Parts 1 - 3 against those in the back of this book. Highlight any incorrect answers then review that material in your textbook. Correct your answers for future reference.

Part 1: Multiple Choice

For each of the following, select all correct responses—more than one may be correct.

1. Changes in the frequencies of alleles in a population are referred to as:
   a) microevolution
   b) macroevolution
   c) natural selection
   d) independent assortment

2. An individual who carries two copies of the same allele for a certain gene is:
   a) heterozygous
   b) homologous
   c) homozygous
   d) analogous

3. Polymorphic means that:
   a) The population has more than one allele of a given gene.
   b) Each trait is controlled by multiple genes.
   c) The population size varies often.
   d) Members of the population exist in two or more forms.

4. Genetic variation depends on:
   a) evolution
   b) natural selection
   c) mutation
   d) sexual selection

5. Alleles are?
   a) different forms of a single gene
   b) acquired traits
   c) homologous structures
   d) minor fluctuations in gene frequency

6. Selection acts on:
   a) phenotype
   b) genotype
   c) acquired traits
   d) all of these

7. The equation $p^2 + 2pq + q^2 = 1$ is used to calculate:
   a) frequency of alleles
   b) frequency of genes
   c) frequency of genotypes
   d) amount of variation in a population

8. Which of the following is not a condition that must be met to attain Hardy-Weinberg equilibrium?
   a) random mating
   b) small population
   c) no mutations
   d) no selection

9. Only which of the following leads to adaptive changes in allele frequencies?
   a) nonrandom mating
   b) mutations
   c) small population size
   d) natural selection

10. When individuals choose mates on the basis of similarities to the individual's own genotype, it is called:
    a) positive assortative mating
    b) outbreeding
    c) crossbreeding
    d) negative assortative mating

11. Inbreeding is an extreme example of:
    a) negative assortative mating
    b) crossbreeding
    c) sampling error
    d) positive assortative mating

12. Gene flow from outside populations has which of the following effects?
    a) increases genetic variation
    b) increases similarity between adjacent populations
    c) establishes Hardy-Weinberg equilibrium
    d) all of these

13. Industrial melanism is an example of:
    a) stabilizing selection
    b) sexual selection
    c) directional selection
    d) disruptive selection

14. With which type of selection are intermediate forms selected against?
    a) stabilizing selection
    b) sexual selection
    c) directional selection
    d) disruptive selection

15. When a small subset of a larger population establishes a new population, the change in gene frequencies that results is referred to as the:
    a) bottleneck effect
    b) founder effect
    c) positive assortative mating
    d) outbreeding

Part 2: Matching

For each of the following, match the correct term with its definition or example. More than one answer may be appropriate.

blending inheritance **Lamarckism** **natural selection**

1. ________________ Weismann disproved this.
2. ________________ In time, this would decrease genetic variation.
3. ________________ This is now believed to be the major force driving evolution.
4. ________________ Inheritance of acquired traits.
5. ________________ Acts on phenotype.
6. ________________ Belief that offspring's phenotype would be intermediate to both parents' phenotypes.

directional selection **stabilizing selection** **disruptive selection**

7. ________________ Common in a relatively constant environment.
8. ________________ Favors extreme phenotypes.
9. ________________ Increasing antibody resistance in bacteria is an example.
10. ________________ A likely cause of the evolution of new species.

Part 3: Short Answer

Write your answers in the space provided or on a separate piece of paper.

1. List the four basic points of theory of inheritance of acquired traits.

2. How did Weismann's theory of the continuity of the germ plasm rule out Lamarckism?

3. What was involved in the modern synthesis?

4. What discipline focuses on microevolution?

5. What is the source of most genetic variation?

6. The rate at which a population evolves depends on what?

7. How does genetic variation spread through a population?

8. In the equation "$p + q = 1$," what do p and q represent?

9. What five conditions must be met to attain Hardy-Weinberg equilibrium?

10. Explain assortative mating.

11. List two examples by which populations that derive from other populations can evolve in the absence of natural selection.

12. What would happen to the rate of evolution if random drift ultimately led to a completely homozygous population?

13. Why does natural selection not eliminate all deleterious alleles?

14. Why does stabilizing selection favor the average phenotype?

15. What are two types of sexual selection?

Part 4: **Critical Thinking: Using Your Knowledge**

Answer each of these in essay form, using complete sentences and paragraphs. Provide as much information as you can. (For extra essay practice, write out answers to the Review and Thought Questions in your textbook.)

1. Explain how Darwin made mistakes which took him in directions away from natural selection.

2. Explain Hardy-Weinberg equilibrium—what conditions must be met, what equations are used, what does the equilibrium, or lack of this equilibrium, tell us about a population?

3. Knowing the five conditions that must be met for Hardy-Weinberg equilibrium, give specific examples of how the human population does not meet these conditions.

4. Explain the impact the environment can have on the type of selection that occurs in a population. Also discuss how changes in the environment can change the type of selection at work. Give specific examples.

5. Explain why inbreeding can have negative effects on a population. In humans, why was inbreeding relatively frequent in the past, but less common now?

6. Vavilov died in prison even though he had been correct. His main failure was that he did not disprove Lamarckian inheritance. If you were a scientist at that time, keeping in mind that genes were not yet known to exist, how might you have scientifically tested and disproved Lamarck's inheritance of acquired characteristics?

7. Many plant and animal species are currently endangered, and tremendous efforts have been taken to restore breeding populations for some of these. For example, in the Pacific Northwest, lumbering has been halted in many areas to protect the spotted owl. How would you weigh the pros and cons of such a situation, knowing that your decision could mean survival or extinction of a species?

8. Many critics of policies that protect endangered species at the expense of economical interests argue that species can be saved in captivity, such as in zoos, and breeding programs can be used to rebuild the population. Yet many of these breeding programs are unsuccessful. Based on your knowledge of population genetics, what might be some reasons for this?

9. If natural selection favors individuals with adaptive traits, explain why natural selection does not rid a population of lethal or disease-causing alleles. Might there be any advantage to keeping all these alleles in a population?

10. Individuals inherit genes from their parents, yet selection acts on phenotype. Using specific examples, explain this. How might selection favor different genes for a single trait? Next, again using examples, explain how environment determines which phenotype has the selective advantage.

Chapter 17
Macroevolution: How Do Species Evolve?

KEY CONCEPTS

1. Species are groups of actually or potentially interbreeding populations, which are reproductively isolated from other such groups.
2. A population may form a separate species when it becomes reproductively isolated. Populations may become reproductively isolated in many ways.
3. Many kinds of adaptations contribute to the reproductive isolation of new species.
4. Allopatric speciation is speciation proceeding from geographical isolation. Once two populations are geographically isolated, independent selective pressures and genetic drift can further differentiate the two populations.
5. Sympatric speciation—formation of species without geographical isolation—is controversial.
6. Parapatric speciation may occur if gene flow between adjacent but nonoverlapping populations is sufficiently limited.
7. Adaptive radiation—generation of diverse new species from a single ancestral species—is the key to biological diversity.
8. Evolution at the species level occurs as rapidly as ever, but the evolution of new phyla seems to have ceased. The Cambrian explosion may have been a one-time phenomenon.
9. All species go extinct eventually. The rate of extinctions since the Cambrian has been constant except for five mass extinctions, the cause of which is yet to be determined.
10. Researchers assume that the first hominids had to walk long distances and that this new lifestyle created selective pressures for an upright stance, but we may never know exactly where our ancestors were walking or why.
11. Anthropologists do not agree whether *Homo sapiens* evolved in Africa. In the last 2 million years, several species of *Homo* appear to have been living on Earth at the same time.

EXTENDED CHAPTER OUTLINE

WHAT DARWIN SAW AT THE GALÁPAGOS ISLANDS

1. The island-bound plants and animals were related to those on the mainland.
2. Each endemic species had evolved on one of the islands, descended from migrants from other islands or the mainland.
3. According to the 19th-century theological concept of special creation, similar habitats should support identical species and, since each species was thought to have been specially created by the Judeo-Christian God, species were not supposed to be related to one another.
4. Darwin realized that five million years ago, the volcanic Galápagos Islands had burst from the sea, devoid of life. Gradually, plants and animals began to colonize the islands.
5. Some organisms survived and reproduced, others did not. Some organisms changed islands, interbreeding with other populations; others stayed in one place, slowly evolving away from their cousins on other islands and the mainland parent population. One species became many.
6. Macroevolution is the origin and multiplication of species.

WHAT IS A SPECIES?

1. To date, biologists have identified about 1.4 million species of living organisms—give or take a hundred thousand. This number is probably less than one tenth of the total number of species currently living.
2. Of all known species, more than half are insects, and nearly half of the insects are beetles—290,000 species, representing 20% of all species of organisms.

Why is it so hard to define a species?

1. Appearance alone fails to define a species. Similar looking species are common, and two species may even look identical.
2. A sibling species is one that is extremely similar but reproductively isolated from another.
3. Every individual in a species is unique—ideal types do not exist.
4. We now think that species are, for the most part, discrete and very real entities.
5. The Biological Species Concept, a modern definition of a species, was stated in 1942 by Ernst Mayr as follows: "Species are groups of actually or potentially interbreeding populations, which are reproductively isolated from other such groups." A biological species, then, is the largest unit of a population in which gene flow is possible.
6. The Biological Species Concept is a tricky definition because:
   - it does not apply to asexual organisms;
   - it does not apply easily to extinct organisms (did they interbreed?);
   - paleontologists must give taxonomic species names to fossil organisms on the basis of morphological differences in the hard parts alone; and
   - paleontological species are subjective designations.
7. Geographically isolated organisms—whether very different or very similar in appearance—may or may not be capable of interbreeding.

What about other taxonomic categories?

1. Many species have races or subspecies (morphologically distinct subpopulations) that can interbreed.
2. Almost any subpopulation can be distinguished from others on the basis of some cluster of characteristics, so many biologists consider the concept of race or subspecies to be a normal manifestation of natural variation without any real biological meaning.
3. The differences among human "races" constitute only about 15% of human genetic diversity.

4. Subspecies that do not interbreed may be in the process of separating into species and are sometimes called incipient species.
5. A higher taxon is usually defined by a few characters that consistently distinguish its members from the species in other taxa. Division of a group of a species into separate genera, families, or higher taxa can be very subjective.

HOW DO SPECIES FORM?

1. Species must become reproductively isolated to become separate species—gene flow between them must cease. Such isolation often originates as geographical isolation.

What barriers reproductively isolate populations and species?

1. Some populations merely become reproductively isolated and then evolve apart; others first evolve separate adaptations then develop barriers to reproduction to retain their adaptations.
2. Populations become reproductively isolated through one of two mechanisms:
   - Prezygotic barriers prevent the fusion of the sperm and egg to form a zygote, and
   - Postzygotic barriers make a zygote either inviable or sterile.
3. Prezygotic barriers also include differences in:
   - how the species live (ecological isolation),
   - when they reproduce (temporal isolation),
   - mating behaviors (behavioral isolation),
   - the complementarity of male and female reproductive organs (mechanical isolation), and
   - the compatibility of their gametes (gametic isolation).
4. Postzygotic barriers are at work when fertilization leads to a zygote that either dies or fails to reproduce.
5. A hybrid is an individual that results from cross-fertilization between two different species or between two distinct populations. Hybrids are usually not viable (hybrid inviability) or they produce offspring that are weak or sterile (hybrid breakdown).
6. In plants, changes in chromosome number can simultaneously create new variations and cause their immediate reproductive isolation. Polyploid plants contain two or more complete sets of chromosomes.
7. Plants formed from gametes of different species are ordinarily sterile (if not inviable) and rarely produce gametes because the parents' chromosomes are not homologous, so they cannot pair during meiosis. Pairing can occur if the hybrid doubles its chromosomes (allopolyploid).
8. Gametes of the plant hybrid can only form viable zygotes with each other, so they become reproductively isolated from both paternal populations.

How does geographic isolation initiate speciation?

1. Populations usually must be geographically isolated from the parent population before other barriers to reproduction arise.
2. Isolating mechanisms preserve the isolated population's differences and allow it to speciate—form a new species.
3. Species formation by geological isolation is called allopatric speciation.

How can species form without geographical isolation?

1. Formation of separate species without geographical isolation is called sympatric speciation.
2. Some barriers to gene flow may instantaneously isolate subpopulations from their parent populations. Polyploidy in plants is an example of this.

Can species form with limited gene flow?

1. Parapatric speciation occurs when a species splits into two populations that are geographically separated but still have some contact, so parapatric speciation is intermediate to allopatric and sympatric speciation.
2. Reproductive barriers can isolate parapatric populations if:
   - the environments of the two populations cause different selective pressures on each, and
   - if gene flow between the two populations is small.

HOW DO SPECIES MULTIPLY?

1. The separation of one species of organisms into two (or more) species is also called divergent evolution.
2. Convergent evolution—the independent development of similar features in separate groups of organisms—is not the opposite of divergent evolution, but a separate, unrelated idea.
3. The best understood example of divergent evolution is adaptive radiation, the generation of diversely adapted new species from a single ancestral species.

How does adaptive radiation occur on chains of islands?

1. Adaptive radiation is the key to biological diversity and distinctly characterizes the evolution of life.
2. New species generally require reproductive isolation and one or both of two other conditions:
   - distinct selective pressures (leading to different adaptations), or
   - small population size (so genetic drift can operate).
3. The distribution of species in chains of islands provides a record of the arrival of immigrants from the mainland. A similar pattern is seen in isolated lakes, mountaintops, and forests.
4. In the Galápagos, each island differs slightly in geology, terrain, vegetation, and insect life, so the finch population on each island experienced different selective pressures. Through allopatric speciation, separate species formed.
5. Plants also undergo adaptive radiation.

How do species diversify on continents?

1. Two ways that species diversity on continents increases are:
   - dispersal—the spread of a taxon through a large area, and
   - vicariance—the fragmentation of an already dispersed species or a group of species.
2. In vicariance, a widely dispersed population can be fragmented either through the extinction of populations in between or through creation of geographic barriers.
3. The fossil record makes sense in terms of the movements of the continents.
4. Australia became separated from all the other continents near the end of the Cretaceous Period, some 65 million years ago, after which Australian species evolved in isolation.
5. With the exception of bats, which flew there, and rats, which must have rafted from island to island, all the native mammals of Australia are marsupials. The marsupials of Australia all appear to have evolved from a primitive, opossum-like marsupial of the Cretaceous.
6. Evolution of the marsupials in Australia parallels that of the placental mammals elsewhere, in what is an equally extraordinary display of large-scale convergent evolution.

THE CAMBRIAN EXPLOSION

1. Until the Cambrian Period almost 600 million years ago, life on Earth was limited to a few soft, single-celled organisms such as algae and bacteria.
2. At the beginning of the Cambrian, life bloomed—all of the modern animal phyla that have fossilizable skeletons appeared.
3. Taxonomists have based the classification of animal phyla on basic body plans.

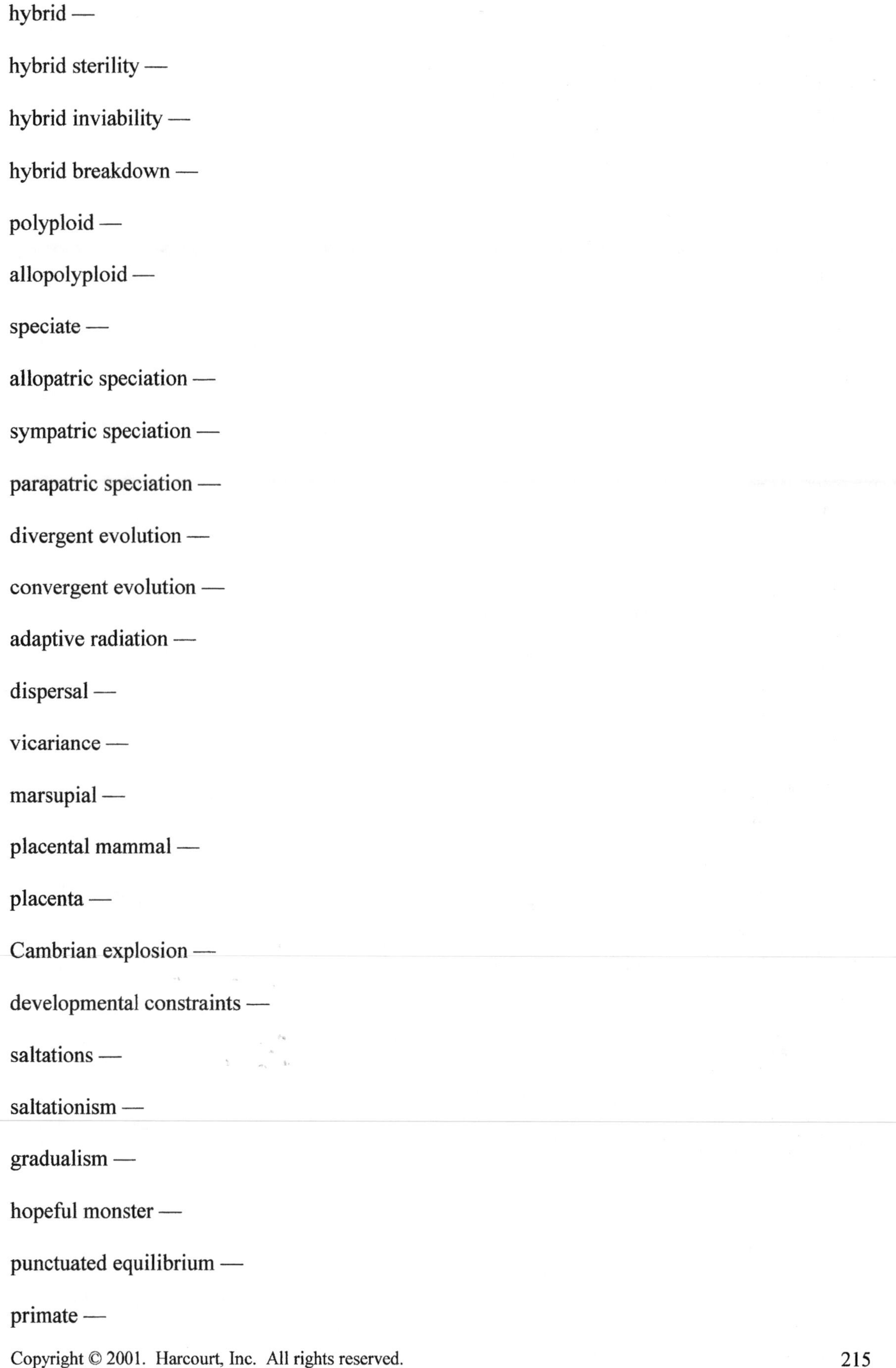

hybrid —

hybrid sterility —

hybrid inviability —

hybrid breakdown —

polyploid —

allopolyploid —

speciate —

allopatric speciation —

sympatric speciation —

parapatric speciation —

divergent evolution —

convergent evolution —

adaptive radiation —

dispersal —

vicariance —

marsupial —

placental mammal —

placenta —

Cambrian explosion —

developmental constraints —

saltations —

saltationism —

gradualism —

hopeful monster —

punctuated equilibrium —

primate —

platyrrhines —

catarrhines —

hominoid —

physical anthropologist —

Pongidae —

Hominidae —

bipedal —

australopithecine —

Cro-Magnon —

Neanderthal —

CHAPTER TEST

The following test has four parts. Complete as much of the exam as you can from memory. If you cannot answer a question, skip it. Once you complete all that you can, try to answer any questions you skipped. If you still cannot answer them, consult your textbook for the answers. Once you have completed all sections of the test, check your answers for Parts 1 - 3 against those in the back of this book. Highlight any incorrect answers then review that material in your textbook. Correct your answers for future reference.

Part 1: Multiple Choice

For each of the following, select all correct responses—more than one may be correct.

1. Which of the following characteristics is the primary determinant of a species?
   a) live in same area
   b) can reproduce with each other
   c) similar appearance
   d) same ancestors through evolutionary time
2. Which of the following cannot be classified as species by the Biological Species Concept?
   a) sibling species
   b) asexual organisms
   c) extinct organisms
   d) None of these can be classified using this definition.
3. Which type of isolation is required for speciation to occur?
   a) reproductive
   b) geographical
   c) dietary
   d) none of these

4. Allopatric speciation occurs when a population separates:
   a) into different geographic locations
   b) within the same geographical location
   c) into different locations but the two groups still have contact
   d) under any of these conditions

5. The evolutionary divergence of a single line of organisms is called:
   a) allopatric speciation
   b) adaptive radiation
   c) sympatric speciation
   d) convergent evolution

6. Taxonomic classifications of animal phyla are based on what?
   a) geographic location
   b) coloration
   c) behavior
   d) body plan

7. According to biologists' estimates, what percent of all species that ever lived are extinct?
   a) 10%
   b) 42%
   c) 50%
   d) 99%

8. Which of the following views is the most recent explanation of how evolution occurs?
   a) punctuated equilibrium
   b) saltationism
   c) gradualism
   d) These are all from the same time.

9. To what order of mammals do apes, monkeys, and lemurs belong?
   a) hominoids
   b) primates
   c) sapiens
   d) Hominidae

10. Which of the following distinguish all hominids from all other apes?
    a) hominids have rectangular jaws
    b) hominids have an opposable toe
    c) hominids are bipedal
    d) All of these are traits of hominids but not of other apes.

11. Modern humans are of what genus and species?
    a) *Australopithecus afarensis*
    b) *Homo erectus*
    c) *Homo sapiens*
    d) *Australopithecus anamensis*

12. Which of the following includes humans?
    a) catarrhines
    b) hominoids
    c) Hominidae
    d) all of these

13. Incipient species are defined as:
    a) species that live in different locations
    b) species that are native to their location
    c) subspecies that do not interbreed
    d) different species that do interbreed, creating nonviable offspring

14. Convergent evolution is defined as:
    a) multiple species evolving together and becoming a single species
    b) a single species evolving into two or more species
    c) the opposite of divergent evolution
    d) none of these

15. Separate species are formed through:
    a) allopatry
    b) sympatry
    c) parapatry
    d) all of these

Part 2: Matching

For each of the following, match the correct term with its definition or example. More than one answer may be appropriate.

allopatric speciation **sympatric speciation** **parapatric speciation**

1. ________________ Separate species form without geographical isolation.
2. ________________ Separate species form by geographical isolation.
3. ________________ Separate species form by geographic isolation but the two groups also maintain contact with each other.
4. ________________ This is intermediate between the other two types of speciation.
5. ________________ Darwin's Galápagos Islands finch species arose this way.

saltationism **gradualism** **punctuated equilibrium**

6. ________________ This view was held by Darwin.
7. ________________ This view was held by Mendelians.
8. ________________ Species arise abruptly from one generation to the next.
9. ________________ This view was promoted by Niles Eldredge and Stephen Gould.
10. ________________ Evolution occurs slowly and continuously.
11. ________________ Species change very little most of the time, but when they do, the change occurs over a relatively brief period.
12. ________________ This is the most recent view, and it is a compromise view.

Part 3: Short Answer

Write your answers in the space provided or on a separate piece of paper.

1. Of the known species, how many are insects?

2. Why do paleontologists have trouble with the Biological Species Concept?

3. What is meant by reproductive isolation?

4. What is hybrid sterility?

5. Differentiate between divergent and convergent evolution.

6. List two ways in which vicariance can occur.

7. Most mammals in the United States are of what type?

8. What are developmental constraints, and what is it argued that they explain?

9. Summarize the fluctuations in the rate of extinction through history.

10. What effect did the mass extinctions have on diversity?

11. List two possible explanations for the gaps in the fossil record.

12. What circumstances can lead to rapid speciation in small, isolated populations?

13. What are the two main ideas of punctuated equilibrium?

14. What do physical anthropologists study?

15. Where are the first hominids believed to have evolved?

Part 4: **Critical Thinking: Using Your Knowledge**

Answer each of these in essay form, using complete sentences and paragraphs. Provide as much information as you can. (For extra essay practice, write out answers to the Review and Thought Questions in your textbook.)

1. Differentiate between allopatric, sympatric, and parapatric speciation. Describe specific situations in which each type might occur.

2. Describe various mechanisms that can lead to reproductive isolation, and possibly speciation. Discuss specific prezygotic and postzygotic barriers and provide examples for each.

3. Describe the evolution of humans, including our ancient ancestors. What changes have been noted at various stages in our evolution?

4. Tropical rain forests contain Earth's greatest species diversity, but human actions threaten those species' survival. What do you predict about the extinction of current tropical species, future speciation, and global effects of losing so much diversity?

5. Explain other ways in which human actions are contributing to the current mass extinction. To what extent do you think we control our own fate? What do you think that fate will be?

6. Humans generally believe that we are far removed from natural selection because we control our environment more than our environment controls us. What are some ways in which the environment is still controlling us? Do these environmental influences currently exert any selection pressures on humans?

7. Humans generally believe that we are the superior organisms, yet many humans in science argue that bacteria or insects are far superior to humans in terms of nature. What arguments could you put forth supporting superiority of either of these groups of organisms over humans? (Hint: get on the Internet and look up the Biosphere 2 project. Why did it fail?)

8. Genetically, chimpanzees and humans differ in their DNA by less than 1%. Humans and chimpanzees cannot reproduce, yet horses and donkeys can. Why, as similar as the DNA is, are humans and chimps distinctly different species and reproductively isolated?

9. Many people wrongly believe that Darwin put forth that humans are directly descended from apes. That was not his position, nor is that what modern science believes. Correct the fallacy by explaining what the true relationship is between humans and apes.

10. You have clearly seen how geographical isolation can lead to speciation. Using Darwin's finches as an example, if the island finches were brought to the mainland, do you think they would eventually evolve back into the mainland species from which they originated? Why or why not?

Chapter 18
How Did the First Organisms Evolve?

KEY CONCEPTS

1. Organisms do not arise spontaneously from nonliving matter.
2. The conditions available for the origin of life differ from those for spontaneous generation in three dramatic ways:
   - the long period of time available,
   - the absence of life, and
   - the absence of oxygen.
3. Life appeared just a few hundred million years after the Earth's crust hardened.
4. Most biologists believe that life originated on Earth—no one knows enough about conditions in the rest of the universe to guess how extraterrestrial life might have evolved or safely arrived on Earth.
5. The first organic molecules could have come from asteroids, meteors, or comets.
6. Prebiotic synthesis is plausible, but deriving the likely biochemical pathways depends on knowing the composition of the early atmosphere.
7. Scientists can construct a reasonable scenario for the evolution of the first biological molecules and for the generation of abiotic cell-like structures.
8. Life required the evolution of cells, reproduction, and metabolism.
9. Evolution of the first cells required a way to couple protein synthesis to genetic instructions.
10. Eukaryotic cells probably arose through a symbiotic association of prokaryotic cells.
11. Mitochondria are probably older than chloroplasts.
12. Mitochondria and chloroplasts, which resemble prokaryotes, probably began their association with eukaryote cells as endosymbionts.

EXTENDED CHAPTER OUTLINE

CAN ORGANISMS ARISE SPONTANEOUSLY FROM NONLIVING MATTER?

1. Until the last few hundred years, most people believed that new life arose spontaneously and daily from nonliving materials.
2. Francesco Redi, an early modern experimental biologist, performed controlled experiments challenging the idea of spontaneous generation.
3. Redi's experiments convinced most scientists that spontaneous generation was incorrect. But, within 10 years, van Leeuwenhoek used his new microscope to find microorganisms everywhere—most people thought these microorganisms arose by spontaneous generation.
4. In 1864, Louis Pasteur, the founder of microbiology, designed a simple but irrefutable experiment that resolved the debate. Some of Pasteur's open flasks—still sterile after more than 130 years—are on display in Paris. Pasteur concluded,"All life from life."

WHERE DOES LIFE COME FROM?

If spontaneous generation is impossible, how did life originate?

1. Darwin and others have shown that all modern species have evolved from previous ones.
2. Scientists cannot say with certainty how life arose on Earth—they can speculate, based on an understanding of astronomy, planetary science, geophysics, and biology.
3. Because the origin of life is not an ongoing, testable process as evolution is, biologists can only surmise how life might have arisen.

When and where did life first appear?

1. Cambrian and later fossils are usually hard bones and shells preserved in sedimentary rock, but most Precambrian organisms had few hard parts—their soft remains usually dissolved or decayed, and persisted only when caught in an extremely fine-grained rock called chert.
2. The oldest and most primitive fossils resemble modern prokaryotes and are about 3.5 billion years old, but researchers have dated fossil organisms even older—some 3.8 billion years.
3. Paleontologists assume that fossil stromatolites were photosynthetic, so photosynthesis probably evolved at least 3 billion years ago.
4. Two general explanations are offered for the relatively sudden appearance of prokaryotic organisms soon after the Earth had cooled:
   - Life arose spontaneously from nonliving organic molecules in the primitive environment.
   - Life arrived from elsewhere in the universe.
5. Until recently, no evidence existed for life on the other planets in the solar system. In August 1996, however, researchers discovered potential signs of life in a meteorite composed of 3.6 billion-year-old martian rock—it contains what look like tiny fossils and complex organic molecules that are produced when organisms decay or burn.
6. Most biologists now believe that life must have developed from nonliving molecules on the early Earth's surface, a process called prebiotic evolution.

PREBIOTIC EVOLUTION: HOW COULD COMPLEX MOLECULES EVOLVE?

1. By the 1930s, biochemists realized that all organisms are made from the same building blocks—sugars, lipids, amino acids, and nucleotides.
2. Alexander Ivanovich Oparin and J.B.S. Haldane proposed that these building blocks formed in the primitive environment. The original synthesis of these materials, they argued, was prebiotic and therefore abiotic—before life and therefore without life.
3. Spontaneous generation had to occur instantly, but origin of life from nonliving matter may have taken as long as 300 million years.

What was the Earth like when it was young?

1. The Earth formed about 4.6 billion years ago, when a spinning cloud of rocks, dust, and gas formed the sun and its planets.
2. The material at the center condensed and became the sun. The rest of the cloud formed planets, comets, meteors, and other objects of the solar system.
3. Gravity compacted the Earth, and, for a billion years, huge meteors bombarded the planet.
4. The compacting and the bombardment kept the molten Earth from solidifying for another half a billion years.
5. After 500 to 600 million years, the surface cooled enough to form a thin crust and dense gases escaped from the interior through cracks and volcanoes to form the primitive atmosphere.
6. Among the gases that escaped was water vapor. Immense quantities of vapor condensed and torrential rains pummeled the Earth for millions of years, creating the oceans and eroding the surface.
7. About 3.8 billion years ago the first sediments and the first life appeared.

The evolution of the atmosphere

1. The modern atmosphere is 78% nitrogen, 21% oxygen, about 1% inert argon gas, varying amounts of water vapor, and about 0.1% is traces of carbon monoxide, carbon dioxide, sulfur and nitric oxides, and other gases.
2. The early atmosphere had almost no oxygen—oxygen came much later, probably as a by-product of photosynthesis.
3. Absence of oxygen meant absence of an ozone layer, which shields the Earth from intense ultraviolet (UV) light from the sun. Modern organisms need that protection.
4. Synthetic reactions are almost all thermodynamically uphill, but energy was provided by:
   - UV radiation from the sun,
   - heat from the Earth's interior, and
   - lightening accompanying the storms that filled the oceans.

Is prebiotic synthesis plausible?

1. In 1953, Stanley Miller simulated the early Earth inside a simple, glass apparatus. He added a mixture of hydrogen, methane, and ammonia gases, let them circulate past flashing electrodes, and then analyzed the contents. Organic molecules had formed, including amino acids, precursors of amino acids, and the bases of nucleic acids.
2. As long as free oxygen was absent, the same building blocks appeared, and other sources of energy also worked—most notably ultraviolet light, the principal energy source available before the Earth developed its protective ozone layer.
3. It seemed likely that the primitive oceans could have become loaded with organic molecules capable of becoming the building blocks of life—this is the primordial soup. The amino acids most easily produced in these abiotic experiments are the ones most common in proteins today.
4. Recent research by geoscientists suggests that the major components of the atmosphere were more likely carbon dioxide and nitrogen, with water, traces of carbon monoxide, and free hydrogen. Such an atmosphere would not have allowed the amino acid synthesis suggested by Miller's experiment. However, other lines of thought still think prebiotic synthesis of biological building blocks may have been possible.

Was prebiotic synthesis necessary?

1. Asteroids, meteors, and comets are high in complex organic compounds created during the formation of the solar system—as much as 10 million kg of complex organic molecules might have showered the Earth each year.

How can biological building blocks assemble outside of cells?

1. In the Miller atmosphere, adenine easily forms from five molecules of hydrogen cyanide but the other nucleic acid bases require more complex reactions, so adenine could have been the first nitrogenous base to form. If so, it would not be surprising for its activated product, ATP, to have become the universal energy currency of life.
2. ATP itself may have formed rather easily—mixtures of adenine, ribose, and phosphate can form ATP when exposed to ultraviolet light.
3. Nucleotides can combine in the test tube to form short RNAs. These RNAs then form double-stranded molecules that follow the Watson-Crick rules for base pairing.
4. ATP in the primordial soup may have contributed to synthesis of activated amino acids, and solutions of such activated amino acids can spontaneously form polypeptides. Sidney Fox suggested that such polypeptides could have formed on volcanic cinder cones and washed into the sea.
5. The materials of life are not the same as life. Two important steps would have to occur:
   - the evolution of reproduction, and
   - the evolution of metabolism.
6. Evolution of reproductive ability depends on development both of the genetic code and of a mechanism of translation.
7. Primitive cells—protobionts—could concentrate organic molecules and maintain an internal environment different from that of their surroundings. They could develop the first suggestion of metabolism.

How can prebiotic "cells" form?

1. Polypeptides, nucleic acids, and polysaccharides in solution spontaneously concentrate into discrete, tiny balloonlike droplets called coacervates, which have a discrete inside and outside separated by a membrane-like boundary. If enzymes are in the solution, these will be bound up in the coacervates, which then act like little cells.
2. Other mixtures can also give rise to isolated, cell-like structures, such as proteinoid microspheres and liposomes (some of which have two-layered membranes).
3. None of these self-organizing structures are cells. Protobionts show some of the characteristics of life, and existence of protobionts shows that cell-like structures could have formed spontaneously in the primordial soup.
4. Primitive metabolic pathways could have developed from need: as chemicals were used up, coacervates that could convert a compound into a needed one would continue.

EARLY BIOTIC EVOLUTION: HOW DID THE FIRST GENETIC SYSTEM EVOLVE?

Could RNA have served as the original genetic material?

1. A modern cell stores its genetic information as DNA, transcribes the information into RNA, and then translates the message into specific polypeptides.
2. In modern cells, RNA plays a central role in translation, supporting the idea that RNA could have been the primitive genetic material (DNA cannot form without enzymes, which are proteins, but proteins cannot be made without the instructions in DNA).
3. Some RNA can spontaneously create itself and replicate in a test tube. In the presence of an RNA template, ordered sequences of five to ten nucleotides polymerize according to the standard base-pairing rules.
4. In 1989, Jennifer A. Doudna and Jack Szostak created an artificial RNA enzyme, or ribozyme, that could make new RNA molecules.
5. Many researchers think the first enzymes weren't RNA or protein, but some other molecule.
6. RNA-based cells could allow evolution of cells that used proteins as enzymes and stored genetic information in the more stable DNA.

How did translation evolve?

1. All organisms use the same triplet genetic code, which suggests that the system originated with life itself.
2. Because the code is universal among both prokaryotes and eukaryotes, it must have completed its refinement before the time of the common origin of all modern organisms.

How did ribosomes evolve?

1. The primitive translational machinery likely depended primarily on RNA. The anatomy of rRNAs also supports the idea that rRNAs directly participated in protein synthesis.

How did tRNAs evolve?

1. All modern cells attach amino acids to transfer RNAs (tRNAs) before assembling a polypeptide.
2. The tRNAs probably diverged from a single tRNA early in the history of life.

How did the first cells use genetic information?

1. The first true cells would have needed the following properties:
   - a boundary,
   - enzymes to extract energy from chemicals in the environment,
   - energy storage in ATP,
   - RNA that specified the structure of specific enzymes,
   - RNA replication, using the Watson-Crick rules of base pairing,
   - specification of each amino acid by a triplet codon, with most of the information conveyed by one or two bases, and
   - primitive tRNAs to connect amino acids to nucleotides.
2. The first enzymes probably arose by random assembly of amino acids into short polypeptides.
3. The first genetically useful RNAs probably had only short stretches of information.
4. Reproduction, protein synthesis, and metabolism must have evolved together.

HOW DID MODERN CELLS EVOLVE?

1. The first clearly eukaryotic fossils are about 1.7 billion years old.

How did early life affect the environment?

1. The first cells were heterotrophs—organisms that obtain energy only by degrading existing organic molecules.
2. The earliest organisms likely obtained both energy and building blocks from preexisting organic molecules in the primordial soup. But, with no oxygen available, they could not do true respiration.
3. Plants and other modern autotrophs obtain energy from the sun by photosynthesis or from inorganic molecules.
4. Of the three ways to extract energy from molecules—photosynthesis, respiration, and fermentation—only fermentation is common to all bacteria and eukaryotes, so this metabolic pathway is likely the oldest and had already evolved in the earliest cells.

The evolution of photosynthesis

1. The most abundant energy source was not inorganic molecules, it was sunlight.
2. The first step in harnessing solar energy would have been development of a proton pump.
3. The largest potential source of electrons, however, is water, which is the electron source for nearly all modern photosynthetic organisms.

The first global toxic waste problem

1. The steps required to separate the hydrogen atoms from water also released huge quantities of molecular oxygen.
2. Development of photosynthesis drastically altered the Earth's early environment.
3. Photosynthetic cyanobacteria flourished nearly unchecked for millions and millions of years.
4. The availability of oxygen allowed the evolution of a new class of heterotrophic organisms that could use oxygen to break down organic molecules.
5. The advantage of exploiting oxygen was so great that today anaerobic organisms persist in only a few environments.
6. Some new organisms that developed the ability to breathe oxygen then developed the ability to eat the organisms that created this caustic waste. The first autotrophs thus provided the basis of the food chain.

How did mitochondria and chloroplasts arise?

1. Almost all eukaryotes—protists, fungi, plants, and animals—have mitochondria, the organelles that produce most of a cell's energy. Plants and some protists, however, have chloroplasts. The first eukaryotes probably had mitochondria, but no chloroplasts.
2. Because mitochondria depend on oxygen to generate useful energy, they could only evolve after photosynthetic prokaryotes had filled the world with oxygen.
3. Because they use a genetic code slightly different from the universal code, mitochondria are believed to be quite ancient, and probably older than chloroplasts.

Are mitochondria and chloroplasts endosymbionts?

1. A close association of two organisms, one of which lives inside the other, is called endosymbiosis. Endosymbiosis is mutually beneficial—it provides an internal source of food for the host and mobility and protection for the inner organisms.
2. Chloroplasts resemble cyanobacteria in size and internal organization.
3. Most biologists now favor the view that both chloroplasts and mitochondria are the endosymbiotic descendants of free-living prokaryotes, but some biologists still consider the endosymbiosis theory too speculative.
4. Most biologists think that there is as yet no convincing argument for the origin of eukaryotic nuclei, though a strong candidate has recently emerged.

Did eukaryotes arise through the proliferation of internal membranes?

1. Opponents of the endosymbiotic theory argue that eukaryotes arose by the proliferation of membranes to form new cellular compartments.
2. For the moment, at least, most biologists seem to find endosymbiosis the most convincing general explanation for the origin of eukaryotic cells.

VOCABULARY BUILDING

In your own words, first write a brief definition, then a full explanation for each of the following terms. Include examples where appropriate. Complete this section from your memory—you will not learn the terms by simply copying definitions from the textbook. Once you have finished, check your responses against the information in the chapter and make any necessary corrections.

prebiotic evolution —

abiotic —

proteinoids —

protobionts —

coacervates —

proteinoid microspheres —

liposome —

heterotrophs —

autotrophs —

endosymbiosis —

CHAPTER TEST

The following test has four parts. Complete as much of the exam as you can from memory. If you cannot answer a question, skip it. Once you complete all that you can, try to answer any questions you skipped. If you still cannot answer them, consult your textbook for the answers. Once you have completed all sections of the test, check your answers for Parts 1 - 3 against those in the back of this book. Highlight any incorrect answers then review that material in your textbook. Correct your answers for future reference.

Part 1: Multiple Choice

For each of the following, select all correct responses—more than one may be correct.

1. How old are the oldest fossils that have been dated?
   a) 124,000 years
   b) 3 billion years
   c) 3.8 billion years
   d) 15 billion years

2. About how long ago did the Earth form?
   a) 4.6 billion years ago
   b) 6.4 million years ago
   c) 300 million years ago
   d) 10.2 billion years ago

3. What percentage of the current atmosphere is oxygen?
   a) 78%
   b) 21%
   c) 45%
   d) 97%

4. In the primitive atmosphere, lack of an ozone layer meant the Earth's surface was exposed to tremendous amounts of ultraviolet light from the sun. Why was there no ozone layer?
   a) Ozone is produced by photosynthesis, which had not yet evolved.
   b) Ozone is a waste product from decaying organic matter, which had not yet evolved.
   c) Ozone is made of oxygen, which was almost absent in the primitive environment.
   d) Ozone escaped the atmosphere because Earth's young gravity could not hold it.

5. Recent research suggests that the primitive atmosphere contained no:
   a) methane
   b) carbon dioxide
   c) nitrogen
   d) ammonia

6. Abiotic synthesis of which of the following does not seem plausible?
   a) RNA
   b) polypeptides
   c) ATP
   d) synthesis of all of these seems plausible

7. Evolution of what processes would have been required before organic molecules gave rise to life?
   a) reproduction
   b) breathing
   c) metabolism
   d) none of these

8. The primitive genetic material was most likely:
   a) amino acids
   b) DNA
   c) a coacervate
   d) RNA

9. Which of the following must have evolved at the same time as reproduction?
   a) protein synthesis
   b) metabolism
   c) nuclei
   d) all of these

10. The first cells are believed to have been:
   a) autotrophs
   b) eukaryotes
   c) heterotrophs
   d) both b and c

11. Which method of extracting energy from molecules is common to all bacteria and eukaryotes?
    a) photosynthesis
    b) fermentation
    c) respiration
    d) all of these
12. What was the most abundant energy source for the earliest organisms?
    a) inorganic molecules
    b) organic molecules
    c) ATP
    d) sunlight
13. Cyanobacteria were the major contributors of what to the atmosphere?
    a) carbon dioxide
    b) water
    c) oxygen
    d) methane
14. Most biologists believe that the eukaryotic nucleus developed from:
    a) endosymbiosis with ancient prokaryotes
    b) chloroplasts
    c) mitochondria
    d) None of these—no convincing argument has been put forth yet to explain this development.
15. Which of the following probably first became associated with eukaryotes through endosymbiosis?
    a) chloroplasts
    b) RNA
    c) mitochondria
    d) all of these

Part 2: Matching

For each of the following, match the correct term with its definition or example. More than one answer may be appropriate.

spontaneous generation **prebiotic evolution** **colonization from space**

1. ________________ Life arose from nonliving matter.
2. ________________ Francesco Redi disproved this theory.
3. ________________ Living organisms arrived on Earth from other planets.
4. ________________ Organic molecules were synthesized naturally in the primitive environment, and spontaneously became arranged.

primitive atmosphere	**current atmosphere**
5. ________________	Has 78% nitrogen.
6. ________________	No ozone layer.
7. ________________	Mostly composed of carbon dioxide and nitrogen.
8. ________________	Relatively large amount of oxygen as a result of photosynthesis.

***Part 3:* Short Answer**

Write your answers in the space provided or on a separate piece of paper.

1. Why can scientists not know with certainty how life arose on Earth?

2. Using an example (not from the book), explain why people of the past thought that life resulted from spontaneous generation. What information were they missing?

3. The oldest and most primitive fossils resemble which modern organisms?

4. What two explanations have been offered for the sudden appearance of prokaryotes?

5. Is there life on Mars?

6. How did the primitive atmosphere form?

7. The modern atmosphere contains what proportions of gases?

8. List three energy sources that could have driven chemical reactions after the Earth's crust formed.

9. How did protobionts show signs of metabolism?

10. What properties would the first true cells have needed?

11. What are the three processes used for extracting energy from molecules, and in what order did they evolve?

12. What is the largest source of electrons?

13. Why was the evolution of photosynthesis a problem for some organisms?

14. Did mitochondria evolve before or after photosynthesis, and why?

15. In what way is endosymbiosis mutually beneficial?

Part 4: Critical Thinking: Using Your Knowledge

Answer each of these in essay form, using complete sentences and paragraphs. Provide as much information as you can. (For extra essay practice, write out answers to the Review and Thought Questions in your textbook.)

1. Explain how spontaneous generation was disproven, and why prebiotic evolution and colonization from space are so difficult to prove or disprove.

2. Do you think life in the universe is restricted to our planet? Why can we not say with certainty that there is no other life out there? What criteria do we use to define life so that we will know it if we find it? Are these valid criteria?

3. Recap the origin of Earth, and explain how it is believed that life originated from the primitive environment.

4. Compare and contrast the primitive environment and our current environment. What are some of the features that make life for us possible now but not in the primitive environment? Can all of the earliest organisms live in today's environment?

5. What are some examples that indicate that the current environment is still changing? The ozone layer, absent in the primitive atmosphere, is thinning and has holes in it. If this continues, how might it affect humans?

6. You find yourself in a debate with a classmate who insists that, due to the simplicity of their structure, viruses surely must have been the first form of life on our planet. Do you agree or disagree? What are your arguments in support of your position?

7. Most scientists do not believe that attempts to explain the origin of life by special creation (Creationism) are necessarily wrong. Instead, they believe that this idea is not scientific, and thus outside of the realm of their professions. Explain this position.

8. A college biology textbook from 1978 states that life originated on Earth 2 billion years ago. Yet your current text says the oldest fossils of living organisms are dated at 3.8 billion years. How do you explain the difference?

9. By most hypotheses, life came into existence very slowly a very long time ago. Yet the fossil record shows a tremendous proliferation of life forms in a relatively slow period. Why did life not show a linear progression, but rather a very sluggish start followed much later by a boom in diversification?

10. How many reasons can you offer to explain why the fossil record is limited in providing us with information about the very first organisms?

Chapter 19
Classification: What's In a Name?

KEY CONCEPTS

1. Today, we traditionally recognize the following hierarchy of classification: species are grouped into genera, genera into families, families into orders, orders into classes, classes into phyla, and phyla into kingdoms.
2. The official name of a species is a binomial—a two-part name with both genus and species.
3. Linnaeus standardized both the naming of organisms (with a Latin binomial) and the classification of organisms (into a strict hierarchy).
4. Biological species are groups of actually or potentially interbreeding populations that are reproductively isolated from other such groups.
5. A species is a reproductive unit, a genetic unit, and an ecological unit.
6. Homologous characteristics indicate relatedness, while analogous characteristics indicate similar adaptations in unrelated organisms.
7. Phenetics is a classification method that tries to avoid classical taxonomy's subjectivity.
8. Pheneticists classify species and higher groupings according to a "numerical taxonomy," in which they assign equal value to every characteristic that they list.
9. Cladists group organisms according to a system of logic designed to clarify evolutionary relationships between groups.
10. The rules of cladistics force cladists to ignore similarities in physiology and lifestyle when grouping species into higher taxa.
11. Molecular approaches to classification are extremely useful, but they are not necessarily more objective than traditional methods.
12. Classification is made more difficult by the fact that the forces of natural selection shape organisms (and their genes) in ways that may obscure patterns of relatedness.
13. The six-kingdom classification reflects differences and similarities in cellular organization and lifestyle, rather than strict evolutionary relationships.

EXTENDED CHAPTER OUTLINE

IS THE RED WOLF A SPECIES?

1. In 1967, the federal government designated the red wolf an endangered species.
2. Government biologists realized that to save the red wolf from extinction, they must capture all remaining wolves and breed them in captivity.
3. Only 14 of the 400 captured animals produced descendants that looked like wolves. This small group has produced hundreds of offspring, of which many have been released.
4. The red wolf reintroduction program is the most successful captive-breeding program.
5. In 1791, naturalists designated the red wolf a subspecies of the gray wolf, but later biologists decided the red wolf was a separate species. In 1979, Ronald Nowak, after careful comparisons, concluded that the red wolf was the ancient ancestor of both the gray wolf and the coyote, appearing before either of them.
6. In 1970, David Mech proposed that the red wolf was neither a unique species nor subspecies, but a hybrid produced by interbreeding between the gray wolf and the coyote.
7. With government funding, Robert Wayne and Susan Jenks searched the red wolf genome for some unique stretch of DNA that neither the coyote nor the gray wolf had—something that would clearly identify the red wolf as a separate species.
8. To Wayne's and Jenk's surprise, the mitochondrial DNA of the red wolf included no sequences unique to red wolves. They concluded that the red wolf is a hybrid of the gray wolf and the coyote.
9. Wayne's results threatened to discredit the reintroduction program, strip the red wolf of its endangered status, and further undermine the increasingly battered public image of the federal Endangered Species Act itself.
10. For now, the red wolf hybrid has not lost protection under the Endangered Species Act, but its hybrid status certainly weakened its claim to protection.
11. To protect the red wolf, the U.S. Fish and Wildlife Service pressured Wayne to avoid the word "hybrid" in his papers and to substitute terms like "intergrade" species.
12. Scientists continue to disagree about the red wolf's status, at least in part due to politics.

WHY IS IT HARD TO NAME AND CLASSIFY ORGANISMS?

1. A taxonomist is a biologist who classifies organisms.

Why do we name things?

1. Naming things also allows us to:
   - communicate what we know about something;
   - name distinctions that we might not otherwise have noticed; and
   - classify them, or assign them to groups.

What did Linnaeus's system of classification accomplish?

1. In a hierarchy, larger groups contain smaller groups, which contain still smaller groups.
2. Biologists use a hierarchical taxonomy established by the 18th-century Swedish physician and biologist Carolus Linnaeus.
3. A genus is a group of species that share many morphological characteristics.
4. Today, we traditionally recognize the following hierarchy of classification:
   - species are grouped into genera,
   - genera are grouped into families,
   - families are grouped into orders,
   - orders are grouped into classes,
   - classes are grouped into phyla, and
   - phyla are grouped into kingdoms.

5. From largest group to smallest, you can remember the list by the following mnemonic device: "**K**ings **P**lay **C**hess **O**n **F**ine-**G**rained **S**and." (Or, make up your own!)
6. Humans are members of the domain Eukarya, kingdom Animalia, phylum Chordata, class Mammalia, order Primates, family Hominidae, genus *Homo*, and species *sapiens*.
7. A taxon is a group of organisms at any level of the classification hierarchy.
8. The official name of a species is a binomial, a two-part name with both the genus and species (such as *Homo sapiens*), usually written in italics. Binomial names are Latin or Greek.

Who names a species?

1. The first person to describe a new species assigns it its name.

How do we recognize species?

1. Plato argued that the individual objects of the world are all variations of universal "types"—forms that share an ideal underlying plan.
2. Aristotle extended Plato's concept of types to organisms. In this view—the typological species concept—each species consists of individuals that are variants of a fixed underlying plan. A typological species cannot evolve and become something different.
3. The typological species concept has two important limitations:
   - all species change over time; and
   - the individual members of species may vary.
4. Twentieth-century biologists mostly define biological species as groups of actually or potentially interbreeding populations that are reproductively isolated from other such groups. In this view—the biological species concept—individuals are members of a species only in relation to other species.
5. Two objections to the biological species concept arise:
   - we often lack information about whether two groups can interbreed, and
   - many recognized species can and do interbreed.
6. A species is a reproductive unit in the sense that the species is the group of potential mates.
7. A species—sexually reproducing or not—is also a genetic unit because it is a gene pool.
8. Finally, a species is an ecological unit; within a given habitat, a species interacts in a predictable fashion with other species in the environment.
9. Bacteria are not generally reproductive units or even genetic units but only ecological ones, and the way that bacteria interact with their environment can change radically with the sudden acquisition of a single gene.
10. In reality, biologists usually classify species by how they look, relying on morphology—the form or visual appearance of organisms—for the provisional definition of a species.
11. Morphologically defined species typically correspond to biologically defined species because reproductive isolation promotes evolution of divergent morphological characters.

HOW DO BIOLOGISTS GROUP SPECIES?

1. Although biologists rarely disagree on boundaries separating individual species, they bicker constantly about how to subdivide species into subspecies and how to place species into larger groups, such as genera, families, orders, phyla, kingdoms, and now domains.
2. Systematics is the study of the kinds and diversity of organisms and all relationships among them. In contrast, taxonomy is the theoretical study of classification, including the study of the principles, procedures, and rules of classification.
3. Similarity by itself does not always indicate relatedness, and which similarities are important is often a matter of much dispute.
4. Taxonomy invites dissent. As a result, biologists now use a least four methods for classifying organisms.

Classical evolutionary taxonomy

1. Classification by classical taxonomy relies explicitly on the judgment and experience of the taxonomist. For living organisms, morphological traits are most often used—external form, internal anatomy, tissue types, and chromosomal. Embryological evidence is also used.
2. Similarities inherited from a common ancestor are called homologous. Similarities that perform the same function but evolved separately are analogous.
3. When classifying organisms, a classical taxonomist must consider which characteristics are likely to be homologous and which only analogous, disregarding analogous ones.
4. Distinguishing analogous structures from homologous ones can be difficult.

How do advocates of phenetics classify organisms?

1. Phenetics—a classification method that tries to avoid the subjective choices of classical taxonomy—is based on all observable characteristics.
2. The phenetic taxonomist characterizes the phenotype of an organism by a series of yes or no questions. Each question defines a unit character—a characteristic of an organism that cannot be split into a set of finer descriptions.
3. Phenetics assigns equal weight to each unit character.
4. Pheneticists record answers to questions in a binary (1 or 0) code (a yes = 1 and a no = 0).
5. The pheneticist assembles a scorecard for each organism, then uses a computer to sort organisms according to their degree of similarity. Pheneticists (sometimes called "numerical taxonomists") argue that the resulting groupings are objective measures of similarities. This classification makes no distinction between homologous traits (indicating relatedness) and analogous traits (which may be from convergent evolution and not because the organisms are biologically related).
6. Phenetics is objective, but it is meaningless in terms of biological relatedness.

How do advocates of cladistics classify organisms?

1. Cladistics is another approach to classification. A cladistic taxonomist first describes the groupings of organisms to show their phylogeny or evolutionary history. Cladists, therefore, present their groupings of organisms in evolutionary trees called cladograms, based on shared, derived characteristics.
2. According to cladists, every taxon above the species level should be monophyletic—each taxon should include the ancestral species and all its descendants.
3. Cladists reject taxons that are polyphyletic (include descendants of more than one ancestor).
4. Refer to the cladogram in Figure 19.5 of your textbook. Note that these five primates are grouped together by shared characteristics, but that differences are also pointed out.
5. Constructing the cladogram requires that we note characters in which the species differ. The cladistic groupings correspond to the classical taxonomic groupings.
6. The cladists' argument that phylogeny must be the basis of classification leads them to reject the entire taxon of reptiles. Reptiles are paraphyletic—some of the descendants of reptiles have evolved into organisms not considered reptiles. Specifically, birds and mammals are descended from early reptiles, yet are not classified as reptiles. Similarly, fish would not be considered a class, because lungfish gave rise to amphibians, reptiles, birds, and mammals.
7. Bacteria defy classification by cladistics. They are usually classified phenetically.
8. Cladistics has contributed a great deal to current understanding of evolutionary relationships.

Do DNA and protein sequences provide "objective" criteria for classification?

1. The sequence of a protein evolves in much the same way as morphological features, so comparisons of amino acid sequences suggest relationships between species.

2. Molecular biologists may not always know whether similar sequences are homologous or analogous, so base sequences are not automatically more objective measures of relatedness than morphological characteristics.
3. Proteins also evolve at different rates according to how much selective pressure they experience. Divergence of sequences, then, reflects natural selection. It makes sense when classifying organisms, then, to give priority to the proteins that have evolved slowly.

Why do cladists, pheneticists, and classical taxonomists disagree?

1. Cladists insist that classification must reflect evolutionary history.
2. Pheneticists rely solely on observed differences in modern organisms.
3. If all genes and morphological features changed at constant rates, then cladists, pheneticists, and traditionalist taxonomists would nearly always agree on classification. Instead, characteristics of organisms change in response to the chance appearance of new mutations, changing geography, and the diverse forces of natural selection.

HOW MANY GROUPS OF ORGANISMS ARE THERE?

1. Linnaeus divided all organisms into two kingdoms—the plants and the animals.
2. In 1969, Robert H. Whittaker of Cornell University proposed a five-kingdom classification that most biologists accepted. Whittaker put all the prokaryotes into the kingdom Monera, then divided the eukaryotes into four kingdoms—protists, fungi, plants, and animals.
3. Archaea are anatomically and metabolically distinct from all other bacteria and so diverse that some taxonomists argue that the Archaea comprise three kingdoms of organisms.
4. Most researchers now agree that all kingdoms can be grouped into three domains: Archaea, Eubacteria, and Eukarya.
5. The five- and six-kingdom classifications (and your text) distinguish among organisms in two ways: cellular organization, and source of energy.
6. Eukarya, including protists, fungi, plants, and animals, all have nuclei and membrane-bound organelles. The prokaryotic Archaea and Eubacteria do not.
7. Eukarya can be grouped by energy source—plants use photosynthesis, animals eat other organisms, and fungi absorb chemicals from their environments. Protists are usually, although not always, single-celled eukaryotes that do not fit into any of the other three kingdoms of eukaryotes.

VOCABULARY BUILDING

In your own words, first write a brief definition, then a full explanation for each of the following terms. Include examples where appropriate. Complete this section from your memory—you will not learn the terms by simply copying definitions from the textbook. Once you have finished, check your responses against the information in the chapter and make any necessary corrections.

taxonomist —

hierarchy —

genus —

taxon —

binomial —

typological species concept —

biological species —

biological species concept —

reproductive unit —

genetic unit —

ecological unit —

systematics —

taxonomy —

homologous —

analogous —

phenetics —

cladistics —

phylogeny —

monophyletic —

polyphyletic —

paraphyletic —

CHAPTER TEST

The following test has four parts. Complete as much of the exam as you can from memory. If you cannot answer a question, skip it. Once you complete all that you can, try to answer any questions you skipped. If you still cannot answer them, consult your textbook for the answers. Once you have completed all sections of the test, check your answers for Parts 1 - 3 against those in the back of this book. Highlight any incorrect answers then review that material in your textbook. Correct your answers for future reference.

Part 1: Multiple Choice

For each of the following, select all correct responses—more than one may be correct.

1. Which of the following is listed in the correct order?
   a) species, genus, order, family, class, phylum, kingdom
   b) species, genus, family, order, class, phylum, kingdom
   c) species, genus, class, family, order, phylum, kingdom
   d) species, family, genus, class, order, phylum, kingdom

2. A species' official name is a binomial that includes what names?
   a) genus and species
   b) species and kingdom
   c) kingdom and phylum
   d) species and family

3. Which of the following is seen as a limitation of the biological species concept?
   a) All species change over time.
   b) Biologists do not always know if two groups can interbreed.
   c) Individual members of species vary.
   d) Many recognized species can and do interbreed.

4. A species can be recognized as which of the following?
   a) a reproductive unit
   b) an ecological unit
   c) a genetic unit
   d) all of these

5. According to the biological species concept, which of the following characteristics defines a species?
   a) It is geographically isolated from other species.
   b) Members of a species can potentially interbreed.
   c) Species are reproductively isolated from each other.
   d) Individuals in one species are variants of a fixed underlying plan.

6. Biologists most often classify species based on their:
   a) genetic makeup
   b) geographic location
   c) morphology
   d) evolutionary history

7. Similarities that are inherited from a common ancestor are said to be:
   a) homologous
   b) classical
   c) analogous
   d) none of these

8. Analogous characteristics indicate:
   a) relatedness
   b) similar adaptations in unrelated organisms
   c) members of the same species
   d) none of these

9. Phenetics bases classification on which of the following?
   a) evolutionary history
   b) genotype
   c) phenotype
   d) all of these

10. Which of the following classification systems uses a binary code to "score" organisms?
    a) classical taxonomy
    b) phenetics
    c) cladistics
    d) molecular classification

11. Which of the following classification systems is based on organisms' phylogeny?
    a) classical taxonomy
    b) phenetics
    c) cladistics
    d) molecular classification
12. With which of the following classification systems would reptiles not be a recognized taxon?
    a) classical taxonomy
    b) phenetics
    c) cladistics
    d) molecular classification
13. Which of the following kingdoms does not contain eukaryotes?
    a) Protists
    b) Monera
    c) Fungi
    d) Animals
14. How do the five- and six-kingdom classification systems distinguish between organisms?
    a) cellular organization
    b) size
    c) source of energy
    d) evolutionary relationships
15. Which of the following would also consider embryological evidence for classification?
    a) classical taxonomists
    b) cladists
    c) pheneticists
    d) molecular biologists

***Part 2:* Matching**

For each of the following, match the correct term with its definition or example. More than one answer may be appropriate.

typological species concept	**biological species concept**
1. ________________	A species cannot evolve and become something different.
2. ________________	Species are defined relative to other species.
3. ________________	Members of species are variations of a fixed type.
4. ________________	Species are reproductively isolated from each other.
5. ________________	Based on ideas of Plato and Aristotle.
6. ________________	Twentieth-century biologists mostly believe this concept.

classical taxonomy **cladistics** **phenetics**

7. ________________ Organisms are classed based on as many traits as possible.
8. ________________ Classification is based on evolutionary history.
9. ________________ Presence or absence of traits are "scored" as a 1 or 0.
10. ________________ A taxon includes one ancestral species and all its descendants.

Part 3: Short Answer

Write your answers in the space provided or on a separate piece of paper.

1. How did Wayne and Jenks conclude that the red wolf is not a separate species?

2. To what domain and kingdom do humans belong?

3. What are two limitations of the typological species concept?

4. In what ways are species reproductive, genetic, and ecological units?

5. Why are bacteria not generally reproductive or genetic units?

6. Why do morphologically defined species often correspond to biologically defined species?

7. What is the difference between homologous and analogous similarities?

8. Which classification systems rely on comparing as many morphological features as possible?

9. In what ways is phenetics not ideally objective?

10. Why do cladists reject the taxon of reptiles?

11. How do base sequence analyses lead to classifications that typically parallel those based on morphological features?

12. Cladistics classifies organisms primarily on the basis of what?

13. List three factors that can affect the rate at which organisms' characteristics change.

14. In what ways do Archaea differ from other bacteria?

15. What are the six kingdoms used for classification purposes in your textbook?

Part 4: Critical Thinking: Using Your Knowledge

Answer each of these in essay form, using complete sentences and paragraphs. Provide as much information as you can. (For extra essay practice, write out answers to the Review and Thought Questions in your textbook.)

1. Explain why scientists have struggled over how to define a species. Discuss the typological species concept and the biological species concept, and the limitations of each definition.

2. Evaluate the four classification systems discussed in this chapter. What are some problems faced by using each system?

3. Why is the classification of organisms so difficult?

4. How do you feel about the amount of resources used in the United States to prevent extinction of endangered species? Are these programs and expenditures necessary, and if so, in what ways? What are the costs (other than monetary) and benefits (other than direct benefit to the species) of these programs? Should the government expand or limit these programs, and on what is your answer based?

5. The question of which is the best way to classify forms of life often leads to heated debates among scientists. How important is this issue? Why is a classification system needed? How important is it that everyone agrees on one best system? What are the advantages and disadvantages to having multiple classification systems in use?

6. Suppose one of our space missions retrieves specimens that we believe to be living organisms. If you were a scientist, how would you go about classifying these new organisms?

7. Which classification system or systems would you use for your space "critters," and why?

8. If your space specimens appeared to be unlike anything on Earth, which of the classification systems might you be able to use? Which ones could you not easily use?

9. If the space specimens were unlike anything found on our planet, how would you be able to determine if they were truly alive, remains of living organisms, or nonliving matter?

10. Is it possible that we could find life on another planet or the moon and not recognize it as life because of flaws in our approach to classification? How do we define life?

Chapter 20
Prokaryotes: How Does the Other Half Live?

KEY CONCEPTS

1. Most pathogenic bacteria make us ill by releasing toxic chemicals that kill our own cells.
2. Koch's postulates are used to determine whether a microorganism causes a specific disease.
3. Resistance genes accumulate in bacterial populations through natural selection and by moving from strain to strain by means of plasmids and other mobile genes. Resistance may allow a bacterium to survive exposure to an antibiotic or to actively destroy any antibiotic in its environment.
4. Structural characteristics of prokaryotes distinguish them from eukaryotes.
5. The defining characteristic of prokaryotes is the absence of a nucleus or other membrane-enclosed organelles.
6. The DNA of prokaryotes forms a single circle of double-stranded DNA.
7. Presence of large amounts of peptidoglycan distinguishes the prokaryotic cell wall from the eukaryotic cell wall. Gram-positive bacteria have a thick, peptidoglycan-rich cell wall that leaves them susceptible to penicillin and other narrow-spectrum antibiotics.
8. Prokaryotes move in a directed fashion by gliding on slime or rotating propeller-like flagella.
9. A streamlined metabolism and a simple means of DNA replication allow bacteria to grow and divide more rapidly than eukaryotes—prokaryotes are successful because of their rapid growth rates and metabolic versatility.
10. Prokaryotes are metabolically far more diverse than eukaryotes, which contributes to their enormous success. Prokaryotes' metabolic abilities also provide a basis for classification.
11. The Archaea consist of three groups, the most widespread and ecologically important of which are the methanogens.
12. Biologists classify prokaryotes according to:
    - how they acquire energy and building blocks,
    - whether they photosynthesize, and
    - whether they use or tolerate oxygen.
13. Biologists divide the Eubacteria into three main groups:
    - the aerobic heterotrophs,
    - the anaerobic heterotrophs, and
    - the autotrophs.
14. The most ecologically important Eubacteria are:
    - the cyanobacteria (the main photosynthetic prokaryotes) which fix carbon; and
    - the nitrogen-fixing bacteria, which supply organisms with fixed nitrogen for making proteins.

EXTENDED CHAPTER OUTLINE

TRACKING DOWN A KILLER

1. In 1983, Scott Holmberg showed that feeding antibiotics to farm animals can sicken and even kill people. Antibiotic-fed farm animals can carry especially deadly strains of salmonella.
2. Epidemiologists study the incidence and transmission of diseases in populations.
3. In the United States, large numbers of farm animals are fed low doses of antibiotics to increase their growth rate. This practice kills bacteria that are sensitive to antibiotics, encouraging the proliferation of antibiotic-resistant bacteria.
4. Antibiotic-resistant bacteria are unusually dangerous—people are 21 times more likely to die from infection by antibiotic-resistant salmonella than from one by nonresistant salmonella.

PROKARYOTES AND PEOPLE

1. The number of bacteria in your mouth at this moment exceeds the total number of people that have ever lived. You carry many more bacteria on your skin and in your gut than you have cells in your body.
2. Prokaryotes are single-celled organisms with no nuclei and include Eubacteria and Archaea.
3. Most of the total mass of life forms on our planet is bacteria.
4. Bacteria are the most numerous, diverse, and ancient organisms.

How do bacteria make us ill?

1. Most bacteria are useful, but bacteria can also be pathogens—agents that cause disease.
2. To cause a disease a bacterium usually must:
   - invade the host, multiplying on or in the host's tissues and cells; and
   - produce a toxin that destroys or interferes with the host's normal processes.
3. Most bacterial diseases result from bacterial toxins. These toxins may be endotoxins (toxic substances that are components of the bacterial cells) or exotoxins (secreted by the bacteria).
4. In the late 19th century, Robert Koch, a German physician, established rigorous criteria for identifying the pathogen for a given disease. Koch's postulates are:
   - The same pathogen must be present in every person with the disease.
   - The pathogen must be capable of being grown in a pure culture—uncontaminated with other organisms.
   - The pure-cultured pathogen must cause the disease when introduced into an experimental animal.
   - The same pathogen must be present in the infected animal after the disease develops.
5. Koch's postulates are not always fulfilled (syphilis, Chlamydia, HIV, influenza cannot be grown in pure cultures).
6. Many pathogenic bacteria are normal residents of our bodies that are out of place or have multiplied when the body's defenses weaken.

How do bacteria acquire resistance to antibiotics?

1. Antibiotics are chemicals that kill bacterial cells without harming our own eukaryotic cells.
2. Pathogenic bacteria now exist that are resistant to all antibiotics.
3. Most antibiotics are naturally synthesized by soil bacteria or fungi—soil bacteria produce so many chemicals that are toxic to one another that pathogenic bacteria usually die when exposed to ordinary dirt. But these producers are also resistant to their own toxins, and this is where resistance originates.

4. Antibiotics may have numerous effects, including:
   - preventing the bacteria from synthesizing a cell wall (penicillin does this);
   - preventing bacteria from expressing their genes;
   - interfering with DNA folding;
   - inhibiting synthesis of folic acid, a precursor to the energy molecule ATP; or
   - blocking protein synthesis.
5. Antibiotic resistance occurs in two main varieties:
   - Some resistance genes allow bacteria to survive exposure to an antibiotic.
   - Other resistance genes allow bacteria to release enzymes that destroy the antibiotic.
6. In all populations of bacteria, a few individuals carry genes for resistance to antibiotics.
7. When humans or other animals are exposed to antibiotics, the resistant bacteria flourish in the absence of competing bacteria, multiplying to become the majority through natural selection.
8. Bacteria can pass fully functioning antibiotic-resistance genes to completely different kinds of bacteria, so resistance spreads rapidly.
9. Bacteria engage in a process called conjugation, in which one bacterium passes plasmids and other DNA to another bacterium.
10. Because plasmids themselves pass genes around to one another by means of jumping genes, plasmids accumulate genes for resistance. Plasmids can carry resistance to as many as eight classes of antibiotics.
11. Multiple antibiotic resistance is the rule, rather than the exception.

HOW DO WE KNOW THAT AN ORGANISM IS A PROKARYOTE?

1. The single characteristic that sets prokaryotes apart from all other organisms is the absence of a true nucleus—a prokaryote's DNA is in a concentrated mass called a nucleoid. Prokaryotes also lack any other membrane-bound organelles.
2. The DNA in the nucleoid consists of a single molecule of double-stranded DNA that forms a closed loop.
3. A prokaryote may also contain one or more smaller loops of DNA called plasmids, which can replicate continuously.
4. Prokaryotes exchange genetic information so freely that reproductive isolation is meaningless in bacteria.
5. Prokaryotic cells come in a variety of shapes—the most common are:
   - the rod-shaped bacilli,
   - the spiral-shaped spirilla, and
   - the spherical cocci.

Most bacteria have complex cell walls

1. Most bacterial cells have rigid cell walls outside their plasma membranes, made mostly of a mixture of polysaccharides and polypeptides called peptidoglycan.
2. The peptidoglycan cell wall is important to biologists for two reasons:
   - it helps distinguish two major groups of bacteria, and
   - it makes some bacteria susceptible to antibiotics.
3. In the 1880s, Danish bacteriologist Hans Christian Gram found that when he treated bacteria with a particular purple dye, some kinds of bacteria became permanently stained while others could be washed clean. Today bacteriologists call these two categories of bacteria gram-positive and gram-negative, after Gram's stain.
4. Gram-negative bacteria have a thin, three-layered cell wall, whose outermost layer does not bind the gram stain. Gram-positive bacteria have a thick, single-layered cell wall containing at least 20 times as much peptidoglycan as the gram-negative bacteria.

5. Penicillin and other narrow-spectrum antibiotics kill gram-positive bacteria only. Broad-spectrum antibiotics, such as tetracycline, kill both gram-negative and gram-positive bacteria.
6. The bacterium that causes pneumonia has a gelatinous layer of polysaccharides and proteins, called a capsule, outside its cell wall. Capsules protect disease-causing bacteria from attack by the white blood cells, so these bacteria are more deadly (virulent).
7. Other prokaryotes possess a slime layer—a tangled web of polysaccharides—which, like a capsule, lies outside the cell wall.

How do bacteria move?

1. Some prokaryotes move by secreting slime and gliding on it.
2. Other prokaryotes use a flagellum—the prokaryotic flagellum is a thin, rotating protein filament attached to the outside of the cell.
3. Many bacteria use their flagella for directed movement, or taxis.
4. Chemotaxis is taxis in response to varying concentrations of specific chemicals.

How do prokaryotes grow so rapidly?

1. Because bacteria keep their DNA in a nucleoid instead of a nucleus, they can replicate their DNA continuously, if necessary, to divide very rapidly.
2. The rapid proliferation of prokaryotes allows a favorable new mutation, such as antibiotic resistance, to spread quickly to a huge number of progeny.

Prokaryotes are metabolically diverse

1. The relentless sameness of all eukaryotes' metabolic pathways strongly supports the idea that we are all descended from a common ancestor.
2. In contrast, prokaryotes use a great variety of biochemical pathways.
3. Because prokaryotes are so diverse, they flourish where eukaryotes could never survive.
4. Over billions of years, prokaryotes have evolved an amazing variety of adaptations that contribute to their continuing success, including:
   - spore formation, which allows bacteria to survive, multiply, and disperse no matter how hostile the environment;
   - production of antibiotics that kill competitors;
   - use of resistance transfer factors and plasmids to spread antibiotic resistance from one bacterium to another; and
   - the ability to obtain energy from the waste products of other species.

HOW DO BIOLOGISTS CLASSIFY PROKARYOTES?

1. Because prokaryotes do not reproduce sexually, taxonomists cannot use the biological species concept to classify prokaryotes.
2. Reconstructing the phylogeny of bacteria is equally difficult, as the fossil record for bacteria is nearly nonexistent.
3. Instead, taxonomists have relied on morphological and metabolic features, but morphology can be quite similar while metabolism is quite different, so it is very difficult to classify most bacteria as species.

How do biologists classify the Archaea?

1. Prokaryotes fall into two large domains: Archaea and Eubacteria.
2. The Archaea consist of at least two kingdoms.
3. The Archaea differ from all other bacteria and from all eukaryotes in:
   - the composition of their ribosomes, and
   - the kinds of lipids in their cell membranes.
4. The cell walls of Archaea lack peptidoglycan.

5. The two kingdoms of Archaea differ in how they derive energy from hydrogen gas.
6. The first kingdom—Euryarchaeota—includes methanogens, which produce methane, and halophiles, which live in extremely salty conditions.
7. The second kingdom—Crenarchaeota—includes thermoacidophiles, which live in hot sulfur springs.
8. Methanogens are all obligate anaerobes and live principally in two environments:
   - swamps and sewage treatment plants, and
   - the guts of animals, especially cows, termites, and others that digest cellulose.
9. If methanogens did not recycle carbon from organic molecules back into the atmosphere, other organisms would gradually run out of carbon.

How do biologists classify the Eubacteria?

1. Microbiologists classify Eubacteria into at least 25 kingdoms, which are broad categories based on how they derive food:
   - autotrophs obtain carbon atoms directly from carbon dioxide;
   - heterotrophs (which include most prokaryotic species) derive carbon atoms from organic molecules.
2. Autotrophs may be divided into:
   - photoautotrophs, which use light energy to drive the synthesis of organic compounds from CO_2, and
   - chemoautotrophs, which derive energy by oxidizing inorganic substances such as hydrogen sulfide (H_2S) or ammonia (NH_3).
3. Heterotrophs fall into two groups:
   - photoheterotrophs use light energy but also require organic compounds;
   - chemoheterotrophs use no light, relying exclusively on organic molecules for both energy and carbon atoms.
4. Heterotrophic prokaryotes may be:
   - decomposers, which absorb nutrients from dead organisms, or
   - symbionts, which absorb nutrients from living organisms.
5. Symbionts that cause illness are called pathogens.
6. Decomposers recycle organic material back into the environment.
7. Heterotrophs that require oxygen for respiration are called obligate aerobes; those that can grow either with or without oxygen are called facultative aerobes.
8. Heterotrophs for which oxygen is a deadly poison are called obligate anaerobes.
9. The Eubacteria include aerobic heterotrophs, anaerobic heterotrophs, and autotrophs.

The aerobic heterotrophs

1. The aerobic heterotrophs have tremendous biochemical diversity.
2. Nitrogen-fixing aerobic bacteria may be the most important prokaryotes—they provide nitrogen in forms (such as ammonia and amino acids) from which other organisms can make proteins.
3. All organisms rely on nitrogen-fixing bacteria to provide nitrogen in a usable form. The bacteria process, called nitrogen fixation, is done at a great energy cost.
4. The pseudomonads, known as "the weeds of the bacterial world," are a large and aggressive group of bacteria that seem to live everywhere. They require oxygen and produce CO_2. The extraordinary success of pseudomonads comes from their metabolic diversity.
5. Omnibacteria are the most common organisms on Earth. They can all perform aerobic respiration and have the extraordinary ability to use nitrate as an electron acceptor instead of oxygen.

The anaerobic heterotrophs

1. The anaerobic heterotrophs are mostly obligate anaerobes. They include:
   - fermenting bacteria, which derive energy and carbon atoms from a variety of organic compounds in the absence of oxygen, and
   - the spirochetes, which have a distinctive corkscrew shape.

The autotrophs

1. The autotrophs consist of four phyla:
   - the chemoautotrophs, which photosynthesize without oxygen, and
   - three phyla of photoautotrophs, which photosynthesize with oxygen.
2. Photoautotrophs consist of anaerobic photosynthetic bacteria, cyanobacteria, and chloroxybacteria. Cyanobacteria are three main photosynthetic bacteria, so they fix carbon.

How has DNA sequencing changed the classification of prokaryotes?

1. The work of Carl Woese (University of Illinois, Urbana) led to recognition of Archaea as a separate domain of life, based on the genes found in their ribosomal RNA.
2. Complete DNA sequencing was completed on two Archaea species in 1997, which showed that in some important ways Archaea resemble eukaryotes more than Eubacteria. But sequencing of other Archaea show them more related to Eubacteria.
3. DNA sequencing is allowing further information to be gathered about individual genes.

VOCABULARY BUILDING

In your own words, first write a brief definition, then a full explanation for each of the following terms. Include examples where appropriate. Complete this section from your memory—you will not learn the terms by simply copying definitions from the textbook. Once you have finished, check your responses against the information in the chapter and make any necessary corrections.

epidemiologists —

pathogens —

endotoxins —

exotoxins —

Koch's postulates —

antibiotics —

nucleoid —

bacilli —

spirilla —

cocci —

peptidoglycan —

gram-positive —

gram-negative —

capsule —

slime layer —

flagellum —

taxis —

chemotaxis —

Archaea —

methanogens —

halophiles —

thermoacidophiles —

autotrophs —

heterotrophs —

photoautotrophs —

chemoautotrophs —

photoheterotrophs —

chemoheterotrophs —

decomposers —

symbionts —

obligate aerobes —

facultative aerobes —

obligate anaerobes —

nitrogen-fixing bacteria —

pseudomonads —

omnibacteria —

actinobacteria —

myxobacteria —

nitrogen fixation —

pseudomonads —

omnibacteria —

fermenting bacteria —

spirochetes —

anaerobic photosynthetic bacteria —

cyanobacteria —

chloroxybacteria —

CHAPTER TEST

The following test has four parts. Complete as much of the exam as you can from memory. If you cannot answer a question, skip it. Once you complete all that you can, try to answer any questions you skipped. If you still cannot answer them, consult your textbook for the answers. Once you have completed all sections of the test, check your answers for Parts 1 - 3 against those in the back of this book. Highlight any incorrect answers then review that material in your textbook. Correct your answers for future reference.

Part 1: Multiple Choice

For each of the following, select all correct responses—more than one may be correct.

1. An endotoxin is:
   a) a chemical that enters and then kills a bacterium
   b) a toxic substance that is part of the bacterium
   c) a toxic substance that is secreted by the bacterium
   d) a toxic substance produced by the human body to destroy bacteria

2. Which of the following violates Koch's postulates?
   a) All people with the disease have the same pathogen.
   b) The pathogen can be grown in a pure culture.
   c) The cultured pathogen does not cause disease when introduced into uninfected animals.
   d) The same pathogen is present in any experimental animal that has developed the disease.

3. Most bacteria make us ill by:
   a) rapidly outnumbering our normal healthy cells
   b) destroying our normal bacteria
   c) depleting our nutrients
   d) releasing toxic chemicals

4. Antibiotics are designed to kill:
   a) bacteria
   b) eukaryotic cells
   c) prokaryotes
   d) diseased animal cells

5. Which of the following is a mechanism by which antibiotics work?
   a) prevent synthesis of the bacterial cell wall
   b) prevent expression of bacterial genes
   c) block protein synthesis in the bacteria
   d) all of these

6. How can antibiotic-resistance genes benefit bacteria?
   a) They can allow the bacteria to manufacture their own antibiotics.
   b) They can allow bacteria to survive exposure to an antibiotic.
   c) They can allow bacteria to release enzymes that destroy the antibiotic.
   d) They can change the antibiotic to a harmless one.

7. Antibiotic-resistance genes are easily passed from bacterium to bacterium via:
   a) mitosis
   b) meiosis
   c) invasion of a host cell
   d) conjugation

8. DNA in bacteria is located in:
   a) the nucleus
   b) plasmids
   c) the nucleoid
   d) all of these

9. Spherical-shaped bacteria are called:
   a) cocci
   b) bacilli
   c) spirilli
   d) plasmidi

10. Which of the following is not useful for classifying bacteria?
    a) morphological characteristics
    b) biochemical similarities and differences
    c) evolutionary history
    d) molecular variations

11. Which of the following bacteria use light energy to synthesize organic compounds?
    a) photoautotrophs
    b) chemoautotrophs
    c) photoheterotrophs
    d) chemoheterotrophs

12. Oxygen is a deadly poison for:
    a) obligate anaerobes
    b) facultative aerobes
    c) obligate aerobes
    d) all of these

13. The most common organisms on Earth are the:
    a) nitrogen-fixing bacteria
    b) omnibacteria
    c) pseudomonads
    d) myxobacteria

14. Fermenting bacteria are:
    a) autotrophs
    b) anaerobic heterotrophs
    c) aerobic heterotrophs
    d) all of these

15. Which of the following is not a type of Archaea?
    a) methanogens
    b) halophiles
    c) pseudomonads
    d) thermoacidophiles

***Part 2:* Matching**

For each of the following, match the correct term with its definition or example. More than one answer may be appropriate.

gram-positive bacteria	**gram-negative bacteria**
1. ________________	These include *E. coli.*
2. ________________	These include staphylococci and streptococci.
3. ________________	These are killed by narrow-spectrum antibiotics.
4. ________________	These are killed by broad-spectrum antibiotics.
5. ________________	These have a thin, three-layered cell wall.

Archaea	**Eubacteria**
6. ________________	This includes two kingdoms.
7. ________________	These include methanogens.
8. ________________	These are divided into autotrophs and heterotrophs.
9. ________________	These have no peptidoglycan in their cell walls.
10. ________________	These include nitrogen-fixing bacteria.

***Part 3:* Short Answer**

Write your answers in the space provided or on a separate piece of paper.

1. What do epidemiologists study?

2. How does feeding antibiotics to farm animals increase antibiotic resistance in bacteria?

3. What two things must a pathogen usually accomplish to cause disease?

4. List Koch's postulates.

5. Why are human cells relatively safe from antibiotics?

6. Where are most antibiotics produced naturally?

7. List three ways in which antibiotics kill bacteria.

8. Why does antibiotic resistance spread so rapidly through bacterial populations?

9. What are the most common prokaryotic shapes?

10. What do the terms "gram-positive" and "gram-negative" mean?

11. How do capsules aid pathogenic bacteria?

12. How can bacterial populations grow so rapidly?

13. List four adaptations that make bacteria successful.

14. List three ways in which the Archaea differ from other bacteria and eukaryotes.

15. List and explain three types of heterotrophs that vary in how they use oxygen.

Part 4: **Critical Thinking: Using Your Knowledge**

Answer each of these in essay form, using complete sentences and paragraphs. Provide as much information as you can. (For extra essay practice, write out answers to the Review and Thought Questions in your textbook.)

1. Antibiotics are over-prescribed in the United States. Many patients insist on a pill when they are ill, so physicians may prescribe antibiotics even for illnesses that the drugs cannot treat (such as viral infections). Based on Holmberg's work, is this overprescription a good practice? What health risks might be associated with this widespread use of antibiotics? If you were a physician, how would you handle a patient that insists on medication when none is required?

2. It is believed that after a major global catastrophe, such as nuclear war, bacteria would be among the organisms surviving. Discuss the unique aspects and adaptations of prokaryotes that might enable them to be so successful under such circumstances.

3. Discuss, in detail, how resistance is passed between bacteria. Is this an intentional process? How does natural selection promote increased antibiotic resistance?

4. As more and more household cleaners and personal hygiene products contain antibacterial agents, what do you predict will happen to this resistance?

5. Describe the classification systems used for Archaea and Eubacteria, and why there are so many difficulties.

6. Farmers need to maximize the return on their investment. A handful of antibiotic in the feed is relatively cheap compared to cattle dying from infections. Explain how antibiotic use in livestock can cause an increased health risk. What measures would you suggest for farmers to take to minimize this risk while still being cost-effective?

7. It is believed that bacteria may exist elsewhere in the universe. If a meteor crashed into Earth, how would you go about testing it for presence of bacteria? If you found some, how would you go about classifying them?

8. What kinds of precautions would you take in testing space material for bacteria, and why? How many different potential threats to human health might bacteria from another planet pose?

9. Some bacteria are now resistant to all of our antibiotics. How, with as many advances as there have been in medicine, did we let this happen? Why haven't pharmaceutical companies stayed ahead of this situation, releasing new drugs? What might be some difficulties in trying to design new antibacterial drugs? Keeping in mind that organisms evolve together, and what you just learned about natural toxins and resistance, what might be some good sources for new antibiotics?

10. Global warming is believed to be, in large part, due to an excessive buildup of carbon dioxide in the atmosphere, trapping heat close to the planet. Odd as it sounds, cows and other farm animals are considered to be major contributors to this phenomenon. How are cows, bacteria, and global warming all associated?

Chapter 21
Classifying the Protists and Multicellular Fungi

KEY CONCEPTS

1. Protists are diverse, but all are eukaryotes and metabolically more uniform than prokaryotes.
2. Algae are protists that perform photosynthesis.
3. Protozoa are heterotrophic protists, most of which are motile.
4. Dinoflagellates are among the oldest and most primitive of the eukaryotes.
5. Euglenophytes can live as either autotrophs or heterotrophs.
6. Phaeophytes (brown algae) are the largest and most complex of the protists.
7. Rhodophytes (red algae) have distinctive pigments and complex sexual cycles.
8. Chlorophytes (green algae) are more like plants than other protists, and are also the most diverse of the plantlike protists.
9. Rhizopoda (amebas) move and ingest food with pseudopodia.
10. Foraminifera have tests studded with little pores, and complex sexual life cycles.
11. Sporozoans are nonmotile, spore-forming protozoa that live as parasites in animals.
12. Paramecia and other ciliophora (ciliates) are single-celled, ciliated, free-living heterotrophs.
13. Zoomastigina (flagellates) are single-celled and use flagella to move.
14. Acrasiomycota (cellular slime molds) live as amebas and as multicellular slugs.
15. Myxomycota (plasmodial slime molds) live as amebas or as multinucleated plasmodia.
16. Oomycota resemble fungi, but their spores have flagella.
17. Fungi are filamentous, lack flagella, have distinctive cell walls, absorb food, and reproduce by means of spores.
18. Bread molds and other zygomycetes are coenocytic and usually reproduce asexually by means of spores.
19. The truffles, yeasts, mildews, and other ascomycetes have septa dividing the nuclei of their hyphae, and no sporangia. They produce sexual spores in little sacs called asci.
20. Mushrooms, rusts, and other basidiomycetes are named for the club-shaped basidia that produce spores. They have septa and distinctive life cycles.
21. *Penicillium* and other deuteromycetes always reproduce asexually.
22. Lichens are mutually beneficial symbiotic associations of fungi and photosynthetic algae or cyanobacteria. In contrast to their photosynthetic partners, lichen fungi cannot live alone.
23. Mycorrhizae are fungi on the roots of plants that supply minerals to the plant.

EXTENDED CHAPTER OUTLINE

THE FUNGUS FIGHTERS

1. In 1948, two government researchers set out to find a drug that could safely cure deadly fungal infections in humans.
2. People with damaged immune systems are all susceptible to fungal infections. For such people, fungi are a potentially life-threatening problem.
3. In the 1940s, when antibiotics came into use, medical experts discovered that patients treated with antibiotics became highly susceptible to fungal infections—many of the bacteria killed by the antibiotics kept fungi at bay.
4. Elizabeth Hazen found two strains of bacteria that could kill fungi, but she needed to isolate the antifungal substances for further testing. She sent samples of bacteria to Rachel Brown, who then extracted compounds for Hazen to test.
5. One isolated substance was only mildly toxic and killed virtually any fungus it contacted. It was eventually named Nystatin. As the first treatment for fungal infections in humans, Nystatin was tremendously useful.
6. Between 1955 and 1979 alone, royalties from sales of the drug came to $13.4 million. Neither Brown nor Hazen ever accepted any of the enormous wealth that resulted from their discovery.
7. One important difference between fungi and the protists is their organization—most fungi are multicellular, but most protists are single-celled.
8. Some protists form colonies, but these are primitive collections of nearly identical cells.
9. Fungi resemble plants and animals in having complex internal structures and specialized cells.
10. Most fungi are remarkably alike and make an easily recognized kingdom of organisms. But the protists are so diverse that some biologists believe they should be divided into several kingdoms.
11. Protists, because they are mostly single-celled, are considered more primitive than the fungi.

WHAT ARE PROTISTS?

1. Protists include:
   - algae (which photosynthesize like plants),
   - protozoans (which act more or less like animals), and
   - some fungus-like forms.
2. Protists flourish wherever there is moisture.
3. Protists are eukaryotic. All can reproduce asexually by fission, and some can perform meiosis and reproduce sexually. All derive energy from glycolysis and respiration.
4. Protists lack the biochemical diversity of the prokaryotes, but no other kingdom includes so many different kinds of cells.

Why are protists so diverse?

1. When biologists regarded all organisms as either plants or animals, they classified all the multicelled protists as plants and all the single-celled protists as either algae (which, like plants, photosynthesize) or as protozoa (which, like animals, move under their own power).
2. Protists are eukaryotic organisms that are defined by not belonging to any of the other three eukaryotic kingdoms.
3. Biologists disagree on the number of phyla, or even if all protists should be in the same kingdom.
4. Your textbook follows the classification scheme of Margulis and Schwartz, who divide the protists into 27 phyla, mostly on the basis of their cellular structures and life cycles.

How do protists reproduce?

1. Every protist species goes through two phases in its life cycle: reproduction and dispersal.
2. Dispersal may take many forms—some swim or crawl; others scatter seedlike cysts (enclosed structures that contain reproductive cells); still others infect a mobile host.
3. Reproduction may be asexual or sexual.
4. Some protists undergo alternation of generations, passing through both haploid and diploid phases.

ALGAE ARE PHOTOSYNTHETIC (PLANTLIKE) PROTISTS

1. Algae are relatively simple plantlike organisms that live in moist environments.
2. Most algae are photoautotrophs—organisms that supply their energy by photosynthesis—but some can also live as heterotrophs.
3. Algae use chlorophyll *a* and other pigments to capture light for photosynthesis. Most protists do not have chlorophyll *b*.

Dinoflagellates

1. Most dinoflagellates are single-celled organisms that float freely as plankton in warm oceans. A few can form colonies, and some live symbiotically.
2. Photosynthetic dinoflagellates are the ultimate source of food in the coral reef and other communities.
3. The best known dinoflagellates are those that grow rapidly to cause red tides.
4. Many dinoflagellates are bioluminescent.
5. Almost all dinoflagellates have a rigid wall, or test, made of cellulose and coated with silica. The test has two grooves where the two flagella are embedded. Together, the two flagella create spinning that gives these organisms their name.

Chrysophytes (golden algae and diatoms)

1. Golden yellow pigments give the golden algae or chrysophytes their name.
2. Each chrysophyte family contains both single-celled and colonial forms and many have hard tests.
3. Most chrysophytes are plankton, common in temperate lakes and ponds. One group lives in the oceans.
4. Chrysophytes reproduce asexually in one of two ways. A single swarmer cell may swim away to found another colony, or the cells of a mature colony may split into two groups which form new colonies.
5. The diatoms are some of the most numerous and beautiful chrysophytes. Although diatoms can reproduce sexually, they usually reproduce asexually.

Euglenophytes

1. The euglenophytes derive their name from light-detecting eye spots which enable them to live a double life.
2. In the dark, they live heterotrophically, breaking down dissolved organic matter.
3. When light is available, euglenophytes use their eyespots and a single flagellum to swim into the sunshine.
4. Euglenophytes photosynthesize with both chlorophyll *a* and chlorophyll *b*.
5. Euglenophytes have no cell wall and store energy in a polysaccharide called paramylon, instead of in starch.
6. Euglenophytes all reproduce asexually.

Phaeophytes (brown algae)

1. The phaeophytes (brown algae), with the rhodophytes, make up most of the seaweeds. They are multicellular and include the biggest protists in the world—giant kelps.
2. Most phaeophytes reproduce sexually, but meiosis produces spores rather than gametes.
3. Phaeophyte spores germinate into new multicellular organisms in which all the cells are haploid. This haploid form is called a gametophyte—it produces gametes.
4. The diploid form is the sporophyte—it produces spores.
5. The individual cells of phaeophytes resemble those of other protists.
6. Phaeophytes never produce chlorophyll *b* and do not synthesize starch.

Rhodophytes (red algae)

1. Most of the world's seaweeds are rhodophytes, or red algae, and nearly all are marine.
2. The coralline algae deposit calcium carbonate in their cell walls and contribute to the building of coral reefs.
3. The cell walls of red algae occasionally contain cellulose but, more often, they contain another polysaccharide that makes them slippery. Two such polysaccharides—agar and carrageenan—are useful to humans.
4. All rhodophytes reproduce sexually and have complex life cycles. Meiosis can occur right after fertilization, with no sporophyte stage. Alternatively, the zygote can develop into a diploid organism, which later produces haploid spores that grow into male and female gametophytes.

Chlorophytes (green algae)

1. Chlorophytes, or green algae, are more like plants than any other protists—plants probably arose from a chlorophyte ancestor.
2. Chlorophytes have both chlorophylls *a* and *b* and make starch, which they store within their chloroplasts as do plants. Among the rest of the protists, only the euglenophytes have both chlorophylls *a* and *b*, and only the dinoflagellates make starch.
3. *Chlorella* is among the most widespread of the green algae.
4. *Chlorella*-like species live as photosynthetic symbionts within protozoa and many invertebrates. (A close association between two organisms is called symbiosis.)

PROTOZOA ARE HETEROTROPHIC (ANIMAL-LIKE) PROTISTS

1. The term protozoa describes the protists that most resemble little animals.
2. The protozoa include a wide variety of heterotrophs, most of which obtain their food by phagocytosis (ingesting particles or other cells).
3. Most protozoa are single-celled, motile, and grouped into eight phyla. Rhizopoda, ciliophora, and zoomastigina are distinguished by how they move; sporozoa have characteristic life cycles; and forminifera have distinctive shells.

Rhizopoda (amebas)

1. Rhizopoda, or amebas, are common throughout the world.
2. An ameba continuously changes its shape by means of pseudopodia—temporary extensions of membrane-enclosed cytoplasm.
3. They move in a directed manner toward food, then envelop the food in their pseudopodia.
4. All amebas reproduce through simple cell division.

Foraminifera

1. Foraminifera (forams) are marine organisms whose tests are full of tiny holes.
2. Two large groups are free-floating plankton.
3. Forams have complex sexual life cycles. Unlike most other protozoans, they have true alternation of generations, with extended haploid and diploid phases.
4. Although forams are unicellular, they are subdivided into multichambered shells connected by the cytoplasm.

Sporozoans

1. All sporozoans are parasites.
2. Sporozoans reproduce by a combination of asexual and sexual processes, including the alternation of generations.
3. Sporozoans have very complex life cycles, usually involving two or more different hosts.
4. Plasmodia cause malaria.
5. The complex like cycles of sporozoans ensure continued success.

Ciliophora (ciliates)

1. Ciliophora (ciliates) include about 8000 species of free-living single-celled heterotrophs that live in both fresh water and in salt water.
2. Their name reflects presence of thousands of cilia that help the ciliates move and feed.
3. The best-known ciliate is probably *Paramecium*.

Zoomastigina (flagellates)

1. The zoomastigina, or flagellates, are a diverse group of single-celled heterotrophs.
2. All zoomastigotes have a least one long flagellum; some have thousands.
3. Some zoomastigotes are free-living, but others are parasites.
4. Reproduction of zoomastigotes is mostly asexual, but some have a sexual cycle as well.

FUNGUSLIKE PROTISTS

Acrasiomycota (cellular slime molds)

1. Acrasiomycota, or cellular slime molds, are peculiar. They are animal-like at some stages of their lives, and plantlike at others.
2. When nutrients are plentiful, they flourish as ameba-like cells, consuming bacteria and other food particles by phagocytosis.
3. When food is scarce, the amebas aggregate to form a multicellular stage that looks like a miniature garden slug. The slug secretes a slime track and wanders in search of food and light, then settles down and develops a mushroomlike fruiting body that produces spores.
4. Each spore turns into an ameba.
5. Reproduction is asexual.

Myxomycota (plasmodial slime molds)

1. Myxomycota—the plasmodial, or acellular, slime molds—closely resemble the cellular slime molds in appearance and life cycle.
2. A plasmodial slime mold develops a plasmodium—a mass of cytoplasm with many nuclei but no boundaries between cells.
3. Plasmodial slime molds also have a sexual cycle. The amebas are haploid and two of them can form a diploid zygote.
4. Unlike the cellular slime mold's slug, the plasmodium can both feed and move about.
5. When food or water becomes scarce, the slug forms a fruiting body, undergoes meiosis, and releases haploid spores.

Oomycota

1. Oomycota most nearly resemble true fungi.
2. They are either:
   - parasitic, deriving nourishment from living organisms, or
   - saprophytic, deriving nourishment from dead organisms.
3. Oomycota extend threads into the tissues of their hosts, release digestive enzymes, and absorb the resulting organic molecules.
4. Oomycota have flagella on their motile spores.
5. They reproduce sexually, and their life cycles include both haploid and diploid phases.
6. Oomycota form water molds that decompose and recycle nutrients from dead organisms, and others cause potato blight.

What can we learn from DNA about the diversity and evolution of protists?

1. No complete protist genomes have been sequenced as of this writing.
2. DNA analysis has increased understanding that:
   - protists are more diverse than previously believed, and
   - endosymbiosis has been a major factor in protist evolution.
3. Protist mitochondria appear to have had a distinct history. Even protists that lack mitochondria have mitochondria-derived DNA.
4. Other evidence suggests a complex evolutionary history lacking any clear segregation of plantlike algae vs. animal-like protists.

HOW DO TRUE FUNGI DIFFER FROM OTHER ORGANISMS?

1. Fungi are prodigious decomposers, recycling the nutrients of dead plants and animals.
2. Fungi grow anywhere with sufficient warmth, moisture, and energy-rich organic molecules.
3. During the American Revolution, the British lost more ships to fungal dry rot than to the fledgling American Navy.
4. In World War II, fungal infections of the skin took more men out of action in the South Pacific than did battle wounds.
5. Like animals, fungi are heterotrophic, but fungi enclose their cells in rigid walls, like plants.
6. Fungi are distinct from both plants and animals:
   - they take in food differently than animals,
   - their cell walls are chemically different from those of plants, and
   - they lack the embryonic stages of plants and animals.

How can we characterize the fungi?

1. Most fungi are molds composed of masses of threadlike filaments, called hyphae, which contain many nuclei.
2. In coenocytic fungi, the nuclei all lie in a common cytoplasm.
3. In dikaryotic fungi (which have two nuclei per cell), walls called septa separate the nuclei within a filament.
4. Hyphae grow, branch, and intertwine to form a mass called a mycelium.
5. Fungi lack mobility. They disperse in two ways only:
   - the mycelium can move from place to place by growing, or
   - the mycelium can produce spores that the wind scatters.
6. Fungal cells walls, made of a tough polysaccharide called chitin, are distinctive.
7. Fungi obtain food by infiltrating the bodies of organisms and other food sources with long, thin hyphae. The fungi then secrete digestive enzymes that break down the food, and absorb, rather than ingest, the resulting nutrients.
8. Fungi may reproduce asexually or sexually. Most fungi reproduce vegetatively (asexually).

9. A single mycelium can produce separate individuals by:
   - simple cell division,
   - breaking up of existing hyphae, or
   - budding.
10. Fungi also reproduce by making and dispersing hardy spores.
11. Spores, which are always haploid, may be made in a variety of intricate structures, called sporophores.
12. In general, asexual spores form when food is plentiful and the environment is otherwise welcoming, and fungi reproduce sexually when food is scarce or the environment has become unfavorable.

How do mycologists classify fungi?

1. Fungi lack motility and cell walls differ from those in protists.
2. Fungi lack embryos and have unique acquisition of nutrients.
3. Fungi can be grouped by presence or absence of hyphae.
4. Most fungi are capable of reproducing both sexually and asexually:
   - Asexual reproduction may occur vegetatively or through the production of spores by mitosis.
   - Sexual reproduction always occurs by means of spores.

Zygomycetes (conjugation fungi)

1. The mycelium of a zygomycete, such as bread mold, is coenocytic: it consists of hyphae without dividing septa and all the nuclei share a common cytoplasm. The only exception is two spore-forming structures.
2. Most of the time, zygomycetes reproduce asexually by forming spores in spore-containing capsules called sporangia.
3. Zygomycetes always use spores whether reproducing sexually or asexually.

Ascomycetes (sac fungi)

1. Ascomycetes are a large and diverse group that includes most yeasts, the powdery mildews, the molds, the cup fungi, the bread mold *Neurospora*, and the truffles.
2. Ascomycetes, or sac fungi, are so named because their sexual spores are always produced in little sacs, called asci.

Basidiomycetes (club fungi)

1. Basidiomycetes, or club fungi, include many of the most familiar fungi, such as mushrooms, bracket fungi, and puffballs.
2. These fungi take their name from the tiny, club-shaped basidium they use to produce spores.
3. A basidiocarp is the equivalent of a fruit, and may develop overnight.
4. Rusts and smuts ruin billions of dollars worth of wheat and other grains in the United States and elsewhere every year.
5. Rusts and smuts do not form basidiocarps. Instead, they are parasites with complex life cycles, usually involving hosts of two species.

Deuteromycetes (imperfect fungi)

1. Deuteromycetes are considered an artificial category. As their modes of reproduction become known, they are recategorized into one of the other phyla.
2. The deuteromycetes include many species that are important to humans. Some are parasites on the skin causing ringworm, athlete's foot, and other skin diseases. Many compete with bacteria by secreting antibiotics. Some are predaceous.

WHAT IMPORTANT ECOLOGICAL ROLES DO FUNGI PLAY?

1. Fungi have two important ecological roles, both of which depend on the powerful enzymes that fungi secrete:
   - Fungi decompose organic matter and make nutrients available to other organisms.
   - Fungi also participate in close mutually beneficial symbiotic associations with other organisms.

What role do fungi play in lichens?

1. Lichens may be leafy, shrubby, or crusty.
2. Lichens are not plants, but symbiotic associations of fungi with photosynthetic partners.
3. In all but about 20 lichen species, the fungal partner is an ascomycete.
4. The photosynthetic partner in a lichen is either a green alga or a cyanobacterium.
5. Fungi extract nutrients from rock and other harsh microenvironments. By doing so, fungi allow the algae or cyanobacteria to grow where they otherwise could not, while the photosynthetic algae provide a continuous source of energy to their fungal partners.
6. Lichens grow where no plant could survive.
7. Lichens usually reproduce asexually.

What role do mycorrhizae play in the lives of plants?

1. Mycorrhizae are symbiotic associations between fungi and the roots of plants
2. Mycorrhizae are not parasites—they help the plant by supplying phosphate and metal ions.
3. Mycorrhizae help plants pull nutrients from the soil.

Where did the fungi come from?

1. Because fungi have no hard parts, the fossil record of fungi is poor. However, fossil fungi can be found at least back to the Ordovician Period, 430 to 500 million years ago.
2. Most mycologists think the first fungi were zygomycetes, which have the simplest life cycle.

VOCABULARY BUILDING

In your own words, first write a brief definition, then a full explanation for each of the following terms. Include examples where appropriate. Complete this section from your memory—you will not learn the terms by simply copying definitions from the textbook. Once you have finished, check your responses against the information in the chapter and make any necessary corrections.

protists —

plankton —

motile —

pseudopodia —

algae —

protozoa —

cysts —

alternation of generations —

dinoflagellates —

test —

chrysophytes —

euglenophytes —

trypanosomes —

phaeophytes —

spores —

gametophyte —

sporophyte —

rhodophytes —

chlorophytes —

symbiosis —

rhizopoda —

foraminifera (forams) —

sporozoans —

ciliophora (ciliates) —

zoomastigina (flagellates)—

acrasiomycota —

myxomycota —

plasmodium —

oomycota —

molds —

hyphae —

coenocytic —

dikaryotic —

septa —

mycelium —

chitin —

sporophores —

zygomycetes —

sporangia —

ascomycetes —

asci —

basidiomycetes —

basidium —

basidiocarp —

deuteromycetes —

lichens —

mycorrhizae —

CHAPTER TEST

The following test has four parts. Complete as much of the exam as you can from memory. If you cannot answer a question, skip it. Once you complete all that you can, try to answer any questions you skipped. If you still cannot answer them, consult your textbook for the answers. Once you have completed all sections of the test, check your answers for Parts 1 - 3 against those in the back of this book. Highlight any incorrect answers then review that material in your textbook. Correct your answers for future reference.

Part 1: Multiple Choice

For each of the following, select all correct responses—more than one may be correct.

1. Protists are considered to be more primitive than fungi primarily because protists are:
   a) smaller
   b) single-celled
   c) prokaryotic
   d) all of these

2. Which of the following are means by which protists may be motile?
   a) Some are plankton that float in water.
   b) Some use flagella.
   c) Some use pseudopodia.
   d) All of these give protists mobility.

3. Which organisms can reproduce very rapidly and cause red tides?
   a) dinoflagellates
   b) chrysophytes
   c) euglenophytes
   d) phaeophytes

4. Who are the largest and most complex protists?
   a) dinoflagellates
   b) chrysophytes
   c) euglenophytes
   d) phaeophytes

5. Which are the only photosynthetic protists that have both chlorophyll *a* and chlorophyll *b*, and also make and store starch?
   a) phaeophytes
   b) chlorophytes
   c) rhodophytes
   d) euglenophytes

6. Which of the following is true of most protozoa?
   a) They are single-celled.
   b) They are motile.
   c) They are heterotrophs.
   d) All of these are true of most protozoa.

7. Amebas use pseudopodia to:
   a) move
   b) reproduce
   c) surround their food
   d) none of these

8. Which of the following is a distinguishing characteristic of the foraminifera?
   a) They have numerous pseudopodia.
   b) They are photosynthetic.
   c) They have tests with numerous tiny holes.
   d) They usually live as parasites.

9. A Paramecium is a:
   a) sporozoan
   b) flagellate
   c) ciliophora
   d) none of these

10. Which of the following protists most closely resembles fungi?
    a) acrasiomycota
    b) ascomycetes
    c) myomycota
    d) oomycota

11. Molds are composed of masses of threadlike filaments called:
    a) hyphae
    b) pseudopodia
    c) septa
    d) mycelium

12. The cell walls of fungi are made of:
    a) cellulose, like in plants
    b) chitin, like in hard shells
    c) agar
    d) They have no cell walls, like in animals.

13. Mushrooms are examples of:
    a) zygomycetes
    b) ascomycetes
    c) basidiomycetes
    d) deuteromycetes

14. The imperfect fungi are considered:
    a) the oldest of the fungi
    b) funguslike protists
    c) an artificial category
    d) strictly sexual reproducers

15. Where are mycorrhizae found?
    a) in rocks
    b) in oceans
    c) on plant roots
    d) inside various animal hosts

Part 2: Matching

For each of the following, match the correct term with its definition or example. More than one answer may be appropriate.

algae **protozoa** **fungi**

1. ____________________ These absorb, rather than ingest, their food.
2. ____________________ These are photosynthetic.
3. ____________________ These are protists.
4. ____________________ These may be part of a lichen.
5. ____________________ Yeast is one of these.
6. ____________________ These include the foraminifera and ciliates.

euglenophytes	**myxomycota**	**ciliophora**	**ascomycetes**

7. ____________________ This is a true fungus.
8. ____________________ This is a funguslike protist.
9. ____________________ This is an algae.
10. ____________________ This is a protozoa.
11. ____________________ These have light-detecting eye spots.
12. ____________________ These develop a plasmodium.

Part 3: **Short Answer**

Write your answers in the space provided or on a separate piece of paper.

1. What three categories of phyla can the protists be grouped into?

2. What is meant by saying that most algae are photoautotrophs?

3. Differentiate between a gametophyte and a sporophyte.

4. Which of the algae deposit calcium carbonate that contributes to the coral reefs?

5. List at least three differences between algae and plants.

6. In what two ways can chrysophytes reproduce asexually?

7. What are pseudopodia?

8. List three characteristics of the sporozoans.

9. In what general ways do the funguslike protists resemble other protists and the true fungi?

10. What is meant by the oomycota being saprophytic?

11. Why is the fungi's role as decomposers ecologically important?

12. List three examples of harmful effects of fungi that affect humans.

13. In what way do true fungi resemble both plants and animals?

14. Explain how fungi obtain food.

15. Explain the composition of a lichen.

Part 4: **Critical Thinking: Using Your Knowledge**

Answer each of these in essay form, using complete sentences and paragraphs. Provide as much information as you can. (For extra essay practice, write out answers to the Review and Thought Questions in your textbook.)

1. Antibiotics are used routinely to kill bacteria. Frequently, people undergoing antibiotic therapy for a bacterial infection develop "thrush," which is a fungal infection. Why?

2. Make your own chart listing all phyla discussed in the chapter, and include distinguishing characteristics that will help you differentiate between them.

3. Explain the various ecological roles played by fungi.

4. List as many human health hazards as you can from each phyla discussed in this chapter.

5. List beneficial aspects from as many phyla as you can.

6. Why do you think protists exhibit such diversity?

7. Explain the symbiotic relationship of lichens. In what way does this benefit the environment and other organisms?

8. What is so important about the role fungi play by acting as decomposers? What might be a consequence if all fungi in a large area were eradicated?

9. Athlete's foot is a fungal disease, and is easily caught and very hard to cure. What are some explanations for this?

10. Explain, using examples, ways in which animals, plants, and fungi all rely on each other.

Chapter 22
How Did Plants Adapt to Dry Land?

KEY CONCEPTS

1. Plants are multicellular photosynthetic organisms that develop from embryos.
2. Evidence for plant evolution comes from fossils and from comparisons of living species. The first plants probably evolved from a common ancestor resembling a green alga, and vascular plants predate nonvascular plants.
3. Land offers plants more light, carbon dioxide, and nutrients than the sea. But land plants need adaptations for absorbing and retaining water, for fertilization with little or no free-standing water, and in the case of larger plants, for structural support.
4. Plants exhibit alternation of generations—a sexual life cycle in which alternating haploid and diploid phases are both multicellular.
5. The nonvascular plants, or bryophytes (the mosses, liverworts, and hornworts), lack specialized elements for transporting water and nutrients.
6. Although bryophytes lack a specialized transport system, all parts of their bodies are adapted to absorb water, which gives them a spongy feel.
7. In a bryophyte's life cycle, the haploid gametophyte phase dominates.
8. The development of specialized vascular elements enabled plants to transport water and nutrients, grow large, and diversify.
9. In vascular plants, the diploid phase dominates the life of the plants.
10. Ferns have megaphylls (leaves with networks of transport tubes) and independent gametophytes.
11. Psilotophytes, the simplest of the living vascular plants, lack true leaves.
12. Lycophytes have true roots, stems, and simple leaves.
13. Equisetophytes have true roots, stems, and complex leaves. Their stems are jointed, and their outer cell walls are reinforced with silica.
14. Seeds, which enable plants to colonize relatively dry environments, consist of a diploid zygote and a source of food, encased in a seed coat.
15. Conifers produce male and female gametophytes in cone-shaped strobili.
16. Cycads produce male and female gametophytes in strobili that lie on separate plants.
17. Ginkgos resemble cycads in their life cycle and conifers in their growth patterns.
18. Of the gymnosperms, the gnetae—a small, diverse group of cone-bearing desert plants—most resemble angiosperms.
19. Flowers attract animals that distribute pollen.
20. A fruit is a ripened ovary that encloses and protects seeds, and usually enhances dispersal.
21. Flowers reproduce by double fertilization—two sperm nuclei from the pollen grain fertilize two ova from the ovary, resulting in a diploid zygote and a triploid cell that forms the nutritious endosperm.
22. The male sporophyll is the stamen—including anther and filament—and the female sporophyll is the carpel (or pistil)—including style and ovary. The stamens and carpels are generally surrounded by a corolla of petals and a calyx of sepals.

EXTENDED CHAPTER OUTLINE

THE LOVES OF PLANTS

1. When 18th century biologists began to classify plants, they could not agree on what traits to use. Carolus Linnaeus's system ultimately came into use.
2. Linnaeus based his classification solely on the reproductive structures of flowers.
3. Although many of Linnaeus's species designations—genus and species—remain in use today, his classification of groups above the genus level have been abandoned for systems that more accurately reflect evolutionary relatedness.

HOW DID PLANTS ADAPT TO LIFE ON LAND?

1. Through photosynthesis, plants (together with bacteria and protists) provide all the energy we derive from food, whether that food is itself animal or plant.
2. Plants living a million years ago captured the energy that we now collect as coal, oil, or gas.
3. Plants provide the oxygen we breathe.
4. Plants support the entire web of terrestrial life—without plants, no other kinds of organisms could live on land.
5. Plants are multicellular organisms that perform photosynthesis and develop from embryos.
6. All plants are descended from protists.

What are the advantages and disadvantages of life on land?

1. The land offers three major advantages for photosynthetic organisms:
   - more light,
   - more carbon dioxide, and
   - a more reliable supply of minerals and other nutrients.
2. To colonize the land, plants had to be able to:
   - obtain, conserve, and distribute water;
   - accomplish fertilization with little or no water; and
   - support their own weight.
3. The evolution of plants depended principally on four adaptations:
   - a waxy cuticle, which coats the plant's exposed surfaces and reduces water loss through evaporation;
   - the ability to absorb water from dew, rainwater, or groundwater;
   - enclosed reproductive organs (gametangia) in which gametes form; and
   - enclosed sporangia in which spores form.

What is alternation of generations?

1. The life cycle of every sexually reproducing organism includes a diploid phase and a haploid phase.
2. In plants, the haploid stage is a multicellular gametophyte that generates ova and sperm through mitosis, rather than meiosis.
3. Plants exhibit alternation of generations—a sexual life cycle in which the alternating haploid and diploid phases are both multicellular.

How are vascular plants different from nonvascular plants?

1. Plants may be divided into two groups:
   - the tracheophytes, which are vascular plants, and
   - the bryophytes, which are nonvascular plants.
2. The tracheophytes include all the most familiar living plants.

3. Tracheophytes have pipelike tissues to conduct water and nutrients throughout the plant.
4. The tracheophytes' vascular system allows these plants to grow far larger than nonvascular plants, sometimes reaching enormous sizes.
5. Bryophytes are mosses and their relatives. Their lack of a vascular system limits their numbers.
6. Most botanists refer to the major groups of plants as divisions—equivalent to the phyla of animals.
7. Most botanists count 10 or 12 divisions of living plants, of which all but one (the bryophytes) are vascular plants.

How do we know about the evolution of plants?

1. The fossil record for plants is excellent.
2. The oldest plant fossils date from the beginning of the Silurian Period, about 430 million years ago. These fossils are not only of the oldest plant fossils, but of the oldest terrestrial organisms.
3. All plants:
   - show alternation of haploid and diploid generations,
   - contain both chlorophylls *a* and *b*,
   - have cell walls made of cellulose, and
   - are multicellular.
4. Almost all plants store energy as starch.
5. All plants develop from embryos surrounded by nonreproductive (sterile) tissue.
6. During cell division, all plants use a microtubular structure (the phragmoplast) to build a new cell wall between the daughter cells.
7. Most botanists believe both nonvascular and vascular plants descended from green algae.
8. Plants also have stomata—tiny mouthlike pores that open and close to control the flow of carbon dioxide and other gases in and out of the plant.

What does the fossil record tell us about the evolution of plants?

1. The earliest bryophyte fossils date from the Devonian Period, about 400 million years ago, while the earliest vascular plants appear at the beginning of the Silurian Period, 430 million years ago.
2. Botanists divide the vascular plants into those that produce seeds and those that do not.
3. Seedless plants buried deep beneath the layers of sediment turned into the fossil fuels. The large amounts of carbon in these fossils give the Carboniferous Period its name.
4. During the Carboniferous Period, plants grew taller and taller in an adaptive "light war." Low-growing ferns became so successful that the later Carboniferous Period is called the "Age of Ferns."
5. Seed plants, which first appeared in the late Devonian Period, were the progymnosperms, ancestors of the modern gymnosperms.
6. Gymnosperms are modern seed plants that do not have fruits or flowers, including pine trees and other conifers, and palmlike cycads.
7. At the end of the Paleozoic Era, about 225 million years ago, the Earth's climate changed dramatically, and for 100 million years the conifers and great palmlike cycads dominated.
8. About 130 million years ago flowering plants appeared.
9. Flowering plants are the most abundant on the planet, and fall into two groups: dicots (broadleafed) or monocots.

How did angiosperms evolve?

1. The angiosperms are flowering plants, and form the largest division of plants.
2. Angiosperms enclose their seeds in a vessel that we call fruit.
3. The first fossils of flowering plants are from the Cretaceous Period, about 125 million years ago.
4. In the 125 million years since the angiosperms first appeared, they have diversified both morphologically and biochemically. Among these adaptations are improved transport systems and deciduous leaves.

THE NONVASCULAR PLANTS, OR BRYOPHYTES, INCLUDE MOSSES, LIVERWORTS, AND HORNWORTS

1. The nonvascular plants—bryophytes—include the mosses, which give the division its name, and the less common liverworts and hornworts.
2. All bryophytes lack specialized vascular systems, so they depend on freestanding water for photosynthesis and fertilization by free-swimming sperm.
3. Bryophytes play important but poorly understood roles in the world's ecosystems, recycling carbon and absorbing nutrients and toxic metals. In some areas, bryophytes are the principal harvesters of solar energy, so all other organisms depend on them.
4. The most obvious part of the bryophyte is not the diploid phase, but rather the haploid phase.

How do bryophytes absorb water?

1. Bryophytes lack the specialized transporting tissues of the vascular plants.
2. They anchor to surfaces by specialized structures called rhizoids, which are not specialized for absorbing water and nutrients.
3. Instead, all parts of a bryophyte are absorbent.
4. In a moist environment, many plants may crowd together as a large mat.
5. Because direct sunlight dries them out, bryophytes usually grow in the shade.

How do bryophytes reproduce?

1. The most prominent phase of a bryophyte's life is the haploid gametophyte, which produces the gametes. Because the gametophyte is already haploid, it produces gametes through mitosis, not meiosis.
2. Gametes unite in the female gametophyte to form the diploid phase—the sporophyte.
3. The sporophyte produces haploid spores by means of meiosis, but these are not gametes.
4. When the haploid spore finds itself in a welcoming environment, its cells enter mitosis and the bryophyte grows directly into a leafy, green gametophyte.
5. The haploid gametophytes may be male or female.
6. Each sex develops gametangia—structures in which the male or female gametes develop. The male gametangium, called an antheridium, produces sperm; the female gametangium, called an archegonium, produces and houses the ova.
7. Fertilization occurs within the archegonium.
8. Immediately after fertilization, the diploid zygote begins to divide by mitosis to produce the sporophyte.
9. The sporophyte develops a fruitlike sporangium. Cells inside the sporangium produce the haploid spores, which can develop into a new mossy gametophyte.
10. Dispersal is essential to bryophyte reproduction.
11. Once in a hospitable environment, the haploid spores germinate, divide by mitosis, and develop into mature, leafy, male or female gametophytes.

THE VASCULAR PLANTS, OR TRACHEOPHYTES, INCLUDE SEED PLANTS AND SEEDLESS PLANTS

1. The tracheophytes (vascular plants) are more diverse and numerous than the bryophytes.

Of what advantage is a transport system to a plant?

1. The ability of vascular plants to transport water and nutrients allows them to grow to huge sizes and to develop specialized tissues.
2. Vascular plants also have more division of labor than the bryophytes.
3. Their transport systems contain two distinct kinds of conducting tubes:
   - xylem carries water and minerals from roots to the photosynthesizing leaves, and
   - phloem distributes the sugars and other organic molecules made in the leaves to the rest of the plant.
4. Vascular plants also produce a hard, supporting material called lignin which usually lines the walls of the xylem.

How do vascular plants reproduce and disperse?

1. In vascular plants, the diploid sporophytes are prominent.
2. In colonizing new habitats, seed plants do not rely on the chance that a dispersed seed will fall into a welcoming environment. Instead, the seed is encapsulated in a tough coat that protects the embryo inside.
3. A seed can withstand long periods of cold and drought.
4. Once the seed germinates, the embryo uses the stored food to finish development.

Four divisions of tracheophytes lack seeds

1. Four divisions of living vascular plants lack seeds:
   - pterophytes (ferns),
   - psilotophytes,
   - lycophytes, and
   - sphenophytes.
2. In most ferns, fronds grow from a rhizome—a horizontal stem that spreads on or below the ground.
3. Most seedless plants have microphylls—leaves with just one conducting vein. In contrast, fern fronds are megaphylls—large leaves with several veins that form complex networks of transport tubes.
4. Reproduction in ferns and other seedless plants resembles that in the bryophytes, so seedless plants require a wet environment to provide at least a film of water through which the sperm can travel.
5. As in other vascular plants, a fern's sporophyte is much more prominent than the gametophyte. It has leaves, called sporophylls, that are specialized for reproduction. Clustered on the underside of the sporophylls are groups of sporangia, called sori, which are diploid and produce haploid spores by meiosis.
6. Each spore germinates to form a tiny nonvascular gametophyte which is typically flat, thin, heart-shaped, and less than 1 cm wide.
7. The psilotophytes contain just two genera: *Psilotum* and *Tmesipteris*. Both have photosynthetic leaflike scales instead of true leaves.
8. Unlike other vascular plants, *Psilotum* lacks true roots and leaves. Underground, it has a rhizome, with tiny branches called rhizoids. Above ground, its vertical stem is forked and bears sporangia.
9. Lycophytes (also called lycopods) have true roots, stems, and leaves, and also a simple arrangement of xylem and phloem.

10. Lycopod leaves are microphylls, and some are sporophylls, with sporangia on their upper surfaces. The sporophylls may look like ordinary leaves or be distinct and grouped into spore-producing cones called strobili.
11. The living equisetophytes are members of the single genus *Equisetum* (horsetails).
12. Equisetophytes have true roots, stems, and complex leaves. The scalelike leaves grow from the joints and are thought to be megaphylls.

How do seed plants reproduce?

1. In seed plants, gametophytes have no independent lives—they develop within the tissues of the sporophytes.
2. Many seedless plants are homosporous—producing just one kind of spore and one kind of gametophyte.
3. All the seed plants are heterosporous—producing two kinds of gametophytes (male and female) from two kinds of spores on two kinds of sporangia.
4. The sporophytes produce haploid microspores (male) and megaspores (female) in separate sporangia.
5. Microspores and megaspores develop into distinct male and female gametophytes.
6. The female gametophyte stays within the megasporangium, safe inside the main plant body. The megasporangium (nucellus) is covered by one or two additional layers of tissue (integuments) which later develop into the protective seed coat. This combined structure—nucellus and integuments—is the ovule.
7. Small, male gametophytes are grains of pollen, which are released from the microsporangium and carried to the female gametophyte.
8. At fertilization, the sperm and ovum fuse to form a diploid zygote.
9. After fertilization, the ovule and its contents become a seed.
10. The seed includes:
    - the new sporophyte embryo;
    - the female tissues that nourish the developing embryo; and
    - the nucellus and integuments (now the seed coat) from the previous sporophyte generation.

Gymnosperms: Four divisions of seed plants without flowers

1. The four divisions of gymnosperms—conifers, cycads, ginkgos, and the gnetae—are vascular plants with seeds, but lacking flowers and fruits.
2. Conifers take their name from their cones, which are male and female strobili.
3. Many conifers have needlelike leaves that are well adapted for dry conditions. A needle's small surface area and heavy cuticle minimize evaporation.
4. Conifer stems are heavily reinforced with lignin, which makes the wood strong enough to support even the largest buildings.
5. Most conifers bear male and female cones on the same tree.
6. The cycads are large-leafed plants that look like palms.
7. True palm trees are flowering plants, while cycads have neither flowers nor fruits—they bear naked seeds on the scales of cones near the top of each plant. Individual plants bear either male or female cones, but not both.
8. Cycads contain toxins that can be quite damaging, even causing paralysis.
9. The ginkgo is the last living species in the division Ginkgophyta, the rest of whose members went extinct as long ago as 280 million years ago. The male trees have cones (strobili), while the female carries fleshy (and smelly) seeds at the ends of short stalks.
10. Ginkgos resemble cycads in life cycle but resemble conifers in growth patterns.

11. The Gnetophyta (gnetae) are a small but diverse group of cone-bearing desert plants.
12. Of all the gymnosperms, the gnetae most resemble angiosperms. Some of their strobili look like flowers, and the vessels of their leaves and stems resemble those of angiosperms. But, they have naked seeds and are certainly gymnosperms.

THE ANGIOSPERMS ARE SEED PLANTS WITH FLOWERS AND FRUITS

1. The diversity of the angiosperms dwarfs that of all other divisions of plants. Angiosperms, also called Anthophyta, are the most widespread and familiar of plants.
2. All angiosperms share two important adaptations:
   - flowers, which promote fertilization, usually by insects or other animals; and
   - fruits, which promote the survival and dispersal of seeds.

Of what use are flowers?

1. Flowers are reproductive structures that ensure the distribution of pollen.
2. Because wind pollination is so inefficient, wind-pollinated flowers release huge amounts of pollen.
3. Many flowers attract insects, birds, and even bats and other small mammals, whose visits are rewarded with a sugary fluid called nectar, a share of the high protein pollen, or nutritious petals or other flower parts.
4. Most flowers are specialized to ensure pollination by just one group of animals.
5. Insects and flowers provide a striking example of coevolution, the mutual adaptation of two separate evolutionary lines.
6. Coevolution occurs when two species are significant factors in the lives of one another.
7. Coevolution ensures a food supply for the insects and pollination for the plants.

Of what use are fruits?

1. A fruit is a mature ovary that encloses and protects seeds.
2. Many fruits promote seed dispersal by animals or wind.
3. An animal usually digests the fleshy part of such fruits, but the seeds pass unharmed through the digestive system. Seeds excreted by the animal can end up far from the plant that produced them, together with a nourishing supply of fertilizer.

How do angiosperms reproduce?

1. In angiosperms, as in all the other seed plants, the diploid sporophyte generation is more prominent than the haploid gametophyte.
2. The male and the female gametophytes of angiosperms are even tinier than those of the gymnosperms and live their whole lives inside the sporophyte's flowers.
3. The flowers produce haploid microspores (male) and megaspores (female) in separate sporangia, from which the gametophytes derive.
4. The carpel is the flower part that contains the ovule, within which the female gametophyte has developed.
5. From the male microspores, the pollen tube grows and two sperm nuclei develop in it.
6. In the female, meiosis yields four haploid megaspores, only one of which survives. That megaspore divides three times by mitosis to produce the embryo sac—the mature female gametophyte.

7. Fertilization in angiosperms is different from that in any organism—there are actually two separate fertilizations, together called a double fertilization:
   - one sperm nucleus fuses with the nucleus of the ovum to form the zygote, and
   - the other sperm nucleus fuses with the two nuclei of the central cell to form a triploid nucleus called the primary endosperm nucleus.
8. An endosperm provides food for the developing embryo, and in some cases for the germinating seedling.

What are the parts of a flower?

1. Angiosperms are heterosporous—separate sporangia produce male spores (microspores) and female spores (megaspores).
2. In angiosperms, the male and female sporophylls are flower parts instead of cones. The male sporophylls are stamens, and the female sporophylls are carpels.
3. Each carpel (or pistil) encloses one or more female sporangia (ovules) in a swollen base called the ovary.
4. Pollen first falls on the stigma, the carpel's receptive area.
5. After fertilization and development, the ovule becomes the seed, and the ovary the fruit.
6. Just outside the carpels lie the stamens—the male sporophylls. Each stamen consists of an anther, which is a thick pollen-bearing structure, and a filament, which is a thin stalk that connects the anther to the base of the flower.
7. Outside the stamens is a whorl, or corolla, of petals. The fourth (outermost) whorl is the calyx, a whorl of leaflike parts called sepals that protect the flower before it opens.
8. Flower structures vary widely—those that lack one or more of the four whorls are said to be incomplete; those that lack either stamens or carpels are said to be imperfect.
9. Fruits are as diverse as the flowers that produce them.

Why are there so many species of angiosperms?

1. More than 275,000 species of angiosperms are recognized, and they are found in almost every type of habitat.
2. Some plants double the chromosomes in the zygote, forming chromosome sets that have a diploid set from each parent, not the usual haploid set. Fertilizations with these doubled gametes produces plants that are polyploid, usually containing two or four sets of genes.
3. Polyploid plants are often more adaptive than diploid plants, and polyploidy has become a common source of new angiosperm species.

VOCABULARY BUILDING

In your own words, first write a brief definition, then a full explanation for each of the following terms. Include examples where appropriate. Complete this section from your memory—you will not learn the terms by simply copying definitions from the textbook. Once you have finished, check your responses against the information in the chapter and make any necessary corrections.

plant —

cuticle —

gametangia —

sporangia —

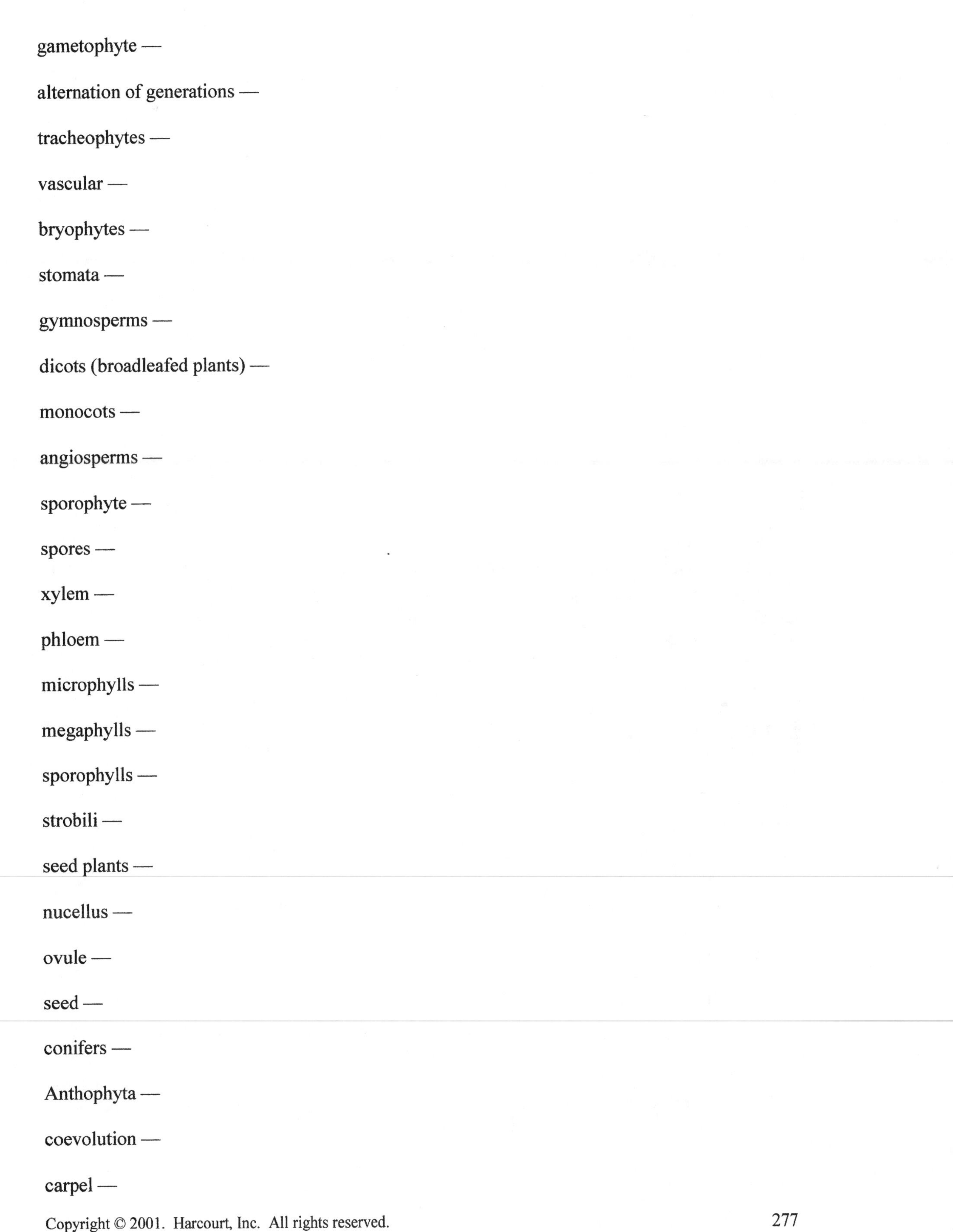

gametophyte —

alternation of generations —

tracheophytes —

vascular —

bryophytes —

stomata —

gymnosperms —

dicots (broadleafed plants) —

monocots —

angiosperms —

sporophyte —

spores —

xylem —

phloem —

microphylls —

megaphylls —

sporophylls —

strobili —

seed plants —

nucellus —

ovule —

seed —

conifers —

Anthophyta —

coevolution —

carpel —

double fertilization —

endosperm —

stamens —

ovary —

anther —

filament —

corolla —

petals —

calyx —

sepals —

CHAPTER TEST

The following test has four parts. Complete as much of the exam as you can from memory. If you cannot answer a question, skip it. Once you complete all that you can, try to answer any questions you skipped. If you still cannot answer them, consult your textbook for the answers. Once you have completed all sections of the test, check your answers for Parts 1 - 3 against those in the back of this book. Highlight any incorrect answers then review that material in your textbook. Correct your answers for future reference.

Part 1: Multiple Choice

For each of the following, select all correct responses—more than one may be correct.

1. Linnaeus's system of classifying plants was based solely on:
   a) plant size
   b) leaf structure
   c) flower structures
   d) reproductive structures

2. Which of the following is an ecological contribution made by plants?
   a) the oxygen we breathe
   b) fossil fuels
   c) all energy we derive from food
   d) all of these

3. Which of the following is not one of the three advantages for plants that live on land?
   a) more light for photosynthesis
   b) more oxygen for photosynthesis
   c) more reliable nutrient source
   d) These are all major advantages for land plants.

4. Which of the following is a nonvascular plant?
   a) conifer
   b) angiosperm
   c) moss
   d) fern

5. What is the function of a plant's stomata?
   a) They regulate the flow of carbon dioxide and other gases.
   b) They are the sites of photosynthesis.
   c) They absorb nutrients and water from the soil.
   d) They help the roots anchor the plant in the ground.

6. Gymnosperms are characterized by which of the following?
   a) have seeds
   b) are nonvascular
   c) have fruit
   d) lack flowers

7. Which is the largest division of plants?
   a) gymnosperms
   b) bryophytes
   c) angiosperms
   d) tracheophytes

8. What part of a bryophyte is absorbent, allowing water to enter?
   a) rhizoids
   b) leaves
   c) flowers
   d) all parts are absorbent

9. Which is the most prominent stage in vascular plants?
   a) haploid sporophyte
   b) haploid gametophyte
   c) diploid sporophyte
   d) diploid gametophyte

10. Which of the following has true roots, stems, and simple leaves?
    a) lycophytes
    b) psilotophytes
    c) equisetophytes
    d) pterophytes

11. Ginkgos are one division of:
    a) bryophytes
    b) gymnosperms
    c) angiosperms
    d) pterophytes

12. Fruits and flowers are found in the:
    a) bryophytes
    b) gymnosperms
    c) angiosperms
    d) pterophytes

13. Which of the following is the most important function of flowers?
    a) They promote dispersal of seeds.
    b) They promote fertilization.
    c) They attract animals that make nectar.
    d) none of these

14. Double fertilization results in which of the following?
    a) a haploid zygote
    b) a diploid zygote
    c) endosperm
    d) flowers

15. Which of the following is not a male part of the flower?
    a) anther
    b) carpel
    c) stamen
    d) filament

Part 2: Matching

For each of the following, match the correct term with its definition or example. More than one answer may be appropriate.

bryophytes **equisetophytes** **gymnosperms**

1. ____________________ These are vascular plants.
2. ____________________ These have true roots, stems, and leaves.
3. ____________________ These lack fruit.
4. ____________________ These include mosses.
5. ____________________ These have seeds, but no fruit or flowers.

fruit **flower**

6. ____________________ This houses the reproductive structures.
7. ____________________ This is a mature ovary.
8. ____________________ This protects the seeds.
9. ____________________ This is designed to promote fertilization.
10. ____________________ This is designed to enhance dispersal.

Part 3: Short Answer

Write your answers in the space provided or on a separate piece of paper.

1. List three benefits plants provide to other organisms.

2. List four adaptations that the evolution of land plants depended on.

3. The first plants likely evolved from a common ancestor that resembled what organism?

4. What period is known as the "Age of Ferns?"

5. Bryophytes depend on freestanding water for what two purposes?

6. What is the main role of the bryophyte's rhizoids?

7. What are the two components of a vascular plant's transport system?

8. Which are the simplest of the vascular plants?

9. What is meant by saying that all seed plants are heterosporous?

10. What are the contents of a typical seed?

11. What is the major difference between gymnosperms and angiosperms?

12. List the four divisions of gymnosperms.

13. In what ways are the gnetae similar to both gymnosperms and angiosperms?

14. What do animals that are attracted to flowers receive in return for performing pollination?

15. What part of a plant becomes the fruit?

***Part 4:* Critical Thinking: Using Your Knowledge**

Answer each of these in essay form, using complete sentences and paragraphs. Provide as much information as you can. (For extra essay practice, write out answers to the Review and Thought Questions in your textbook.)

1. Discuss the evolution of plants. What evidence is used to piece together this history?

2. Discuss the special needs that arose as plants began to live on land. What special problems did life on land pose? What adaptations were needed to ensure survival? Provide specific examples of these adaptations.

3. Describe the process of fertilization in angiosperms, including the function of the flowers and fruits. What advantages does this system offer the plants?

4. Explain how plants can be both the cause and the solution to global warming, which results from carbon dioxide accumulation in the atmosphere.

5. Explain the benefits of flowers and fruit. Why do you think these structures evolved?

6. Bryophytes lack a vascular system. In what ways was this beneficial? In what ways was it a limitation?

7. Explain coevolution and how it might occur between plants and animals. Create your own fictional plant and an animal that would coevolve. Discuss specific characteristics of each that would lead to special adaptations in the other.

8. What are the advantages of having a plant that attracts only a single kind of animal?

9. Under what circumstances might a plant be more successful using wind pollination. Under what circumstances would animal pollination be better?

10. Which types of plants would you expect to grow and be successful in a hot, arid desert? In a tropical rainforest? What adaptations would you expect in each climate?

Chapter 23
Protostome Animals: Most Animals Form Mouth First

KEY CONCEPTS

1. Animals are multicellular, heterotrophic, eukaryotic organisms that develop from embryos.
2. Animals are divided into two subkingdoms—Parazoa (sponges) and Eumetazoa (all others).
3. Eumetazoa are grouped according to:
   - whether they have bilateral or radial symmetry;
   - presence or absence of a coelom or pseudocoel; and
   - whether the blastopore develops into a mouth (protostome) or an anus (deuterostome).
4. Parazoa are mostly sponges with neither symmetry nor organs. Cells of parazoans resemble protists but, unlike protists, parazoans develop from embryos and produce sperm and eggs.
5. Cnidarians and ctenophores are radially symmetrical and lack a true body cavity (coelom). Cnidarians are radially symmetrical acoelomates with true tissues but no organs. Ctenophores (comb jellies) are not radially symmetrical—they possess paired tentacles. Ctenophores also move through water by means of rows of cilia.
6. Bilaterally symmetrical acoelomates are all solid worms—some free-living, some parasites.
7. Platyhelminthes (flatworms, flukes, and tapeworms) are the simplest animals with heads. They have three body layers, true organs and tissues, a brain, and an organ for regulating water balance. They take in food and excrete wastes through the same structures.
8. Nemertea (ribbon worms) have most characteristics of flatworms and also a two-ended digestive system, a circulatory system, and a proboscis.
9. Pseudocoelomates have an internal body cavity between the endoderm and mesoderm.
10. Nematodes (roundworms) are tiny, cylindrical worms with a fluid-filled pseudocoel that acts as a skeleton, a circulatory system, sense organs, and a primitive brain.
11. Coelomates have many common properties but two distinct patterns of development.
12. Mollusks are soft-bodied coelomates with sophisticated organs for respiration, circulation, digestion, and excretion. All mollusks share a similar body plan, including a foot, a mantle cavity, and a visceral mass. Most have shells and a rasping tongue called a radula. They produce sexually, and their embryos develop into a characteristic larval stage.
13. Annelids are segmented worms consisting of a series of identical or near-identical sections.
14. Arthropods include the greatest number of species of all animal phyla. External adaptations include an exoskeleton and jointed limbs. Internal adaptations include circulatory, excretory, respiratory, reproductive, and nervous systems.
15. Arthropods are classified into three subphyla according to specialization of the appendages.

EXTENDED CHAPTER OUTLINE

PROGRESS AND THE BURGESS SHALE

1. About 530 million years ago, a mud slide the size of a city block buried tens of thousands of small marine animals. Geologic processes eventually lifted the rock—today it sits high in the Canadian Rockies and is called the Burgess Shale.
2. Charles Doolittle Walcott and his family first discovered the Burgess animals, and Walcott mistakenly assumed that the animals were all members of already well known phyla.
3. In the 1970s, three British paleontologists—Harry Whittington, Derek Briggs, and Simon Conway Morris—argued that the Burgess animals represented 25 basic body plans, of which only four survive in modern organisms.
4. In 1989, the paleontologist and popular writer Stephen Jay Gould wrote a book, called *Wonderful Life*, describing the animals of the Burgess Shale. Gould argued that each body plan in the Burgess Shale could have been classified as a phylum. Gould further held that the diversity of the Burgess Shale provided evidence that evolution is not a progression from simple to complex, and that life is less diverse today than during the Cambrian explosion.
5. Gould's arguments were not wholeheartedly accepted by other paleontologists and evolutionary biologists—no whole phylum of animals is known to have ever gone extinct.
6. The history of life suggests that diversity increases over time.
7. All living organisms, in all their current diversity, have descended, one generation at a time, from the same few ancestors. Increasing complexity may reasonably be described as progress, even though no goal is involved.

WHAT IS AN ANIMAL?

What features characterize the animals?

1. An animal is a multicellular, heterotrophic organism that develops from an embryo. All animals are eukaryotes and most reproduce sexually. Animals never show alternation of generations—with rare exceptions, the only haploid cells are the gametes.
2. Nearly all animals ingest their food.
3. Almost all animals contain many different kinds of specialized cells. Groups of similar cells, along with the matrix, form tissues—units of structure and function. Two or more kinds of tissue in turn can form an organ—a structural unit with a distinctive function.
4. Epithelium is a tissue that lines a surface; connective tissue is a network of loosely connected cells that help support other body parts.
5. Almost all animals coordinate movement in patterns known as behavior. All but the sponges coordinate such behavior through the functioning of a network of nerve cells (neurons).

How do zoologists classify the animals?

1. Zoologists are biologists who study animals.
2. Zoologists divide living animals into about 35 phyla that belong to two subkingdoms:
   - Eumetazoa include most of the 4 million named species of living animals.
   - Parazoa include 5000 species of sponges (the simplest animals) and one other species.

How are animals classified by the way they develop?

1. The Eumetazoa are divided into phyla according to how their embryos develop.
2. After fertilization, a zygote divides (mitosis) to produce a hollow ball of cells—a blastula—which folds in on itself to form a gastrula—three cell layers around a cavity (archenteron).

3. The innermost layer of the gastrula—endoderm—gives rise to the intestines and other digestive organs. The outermost layer—ectoderm—gives rise to skin, sense organs, and nervous system. The middle layer—mesoderm—gives rise to muscle, skeleton, and connective tissue.
4. The gastrula reveals the start of the adult body plan—a tube within a tube:
   - the inner tube is the digestive tract, with a mouth at one end and an anus at the other;
   - the outer tube is the body wall; and
   - the fluid-filled space between the two tubes, within which the internal organs hang, is the coelom in the most complex animals.
5. Coelomates are animals that have a coelom. The simplest animals are acoelomates—they have no cavity. Pseudocoelomates have an organ-containing cavity, but without the mesodermal lining of a true coelom. Almost all pseudocoelomates are nematodes.
6. The coelomates are further divided into two major groups, by what their blastopore becomes:
   - in protostomes, the mouth develops from the blastopore (most coelomates are protostomes);
   - in deuterostomes, the anus develops from the blastopore.

How do biologists use symmetry to classify animals?

1. Parazoa (sponges) lack symmetry, but Eumetazoa have one of two kinds of symmetry:
   - radial symmetry means the animal can be rotated around its central axis and looks the same;
   - bilateral symmetry means the right and left halves are almost mirror images of each other.
2. Most animals have bilateral symmetry, which gives a more complex organization.

THE PARAZOA HAVE NO ORGANS

1. The Parazoa contain two phyla—the Placozoa and the Porifera (sponges).
2. The Parazoa have no symmetry, and minimal organization and cell specialization. They have only simple connective tissues and no organs.
3. The Placozoa consists of a single species, *Trichoplax adhaerens*.
4. Most sponges are sessile—permanently anchored to rocks, logs, or coral.
5. Water, loaded with nutrients and organic matter, enters through pores in their outer layer, flows into the cavity, and exits through openings. It is pumped by the action of flagella on cells lining the central cavity.
6. Between the outer epidermal layer and the flagellated cells lining the cavity is a layer of mesenchyme, within which amebalike cells capture food and shuttle it to the outer cells.
7. Most sponges are extremely primitive. Each cell functions as an independent unit.
8. Although their cells are like those of protists, parazoans develop from embryos and produce sperm and eggs.

RADIALLY SYMMETRICAL ACOELOMATES

1. All radially symmetrical acoelomates are in two phyla—Cnidaria and Ctenophorae.
2. The Cnidaria are among the oldest animals in the fossil record.

Cnidarians

1. The four classes of cnidarians are Hydrozoa (hydroids), Scyphozoa (jellyfish), Anthozoa (corals and anemones), and Cubozoa.
2. Anthozoans have a flowerlike appearance—a cylindrical body with a crown of tentacles. Anthozoans have the most complex behaviors of the cnidarians.
3. All cnidarians contain two layers of cells that function as true tissues, but no organs.

4. Cnidarians exist as polyps, which resemble cylinders and are partly sessile, and medusae, which resemble bells and float about. Both forms have a mouthlike opening into a central cavity, use tentacles to catch food, and have an internal digestive cavity surrounded by an epidermis and an inner gastrodermis.
5. Most cnidarians' digestion is extracellular.
6. Cnidarians are named for the specialized stinging cells (cnidocytes) on their tentacles. Each cnidocyte can fire a tiny barbed spear called a nematocyst, which can discharge a poison.
7. After the nematocyst attaches to the prey, tentacles draw the prey mouthward, which requires coordination of the tentacles, controlled by simple muscle and nerve cells.

Ctenophores (comb jellies)

1. The ctenophores, or comb jellies, have rows of comblike cilia along their short bodies. By coordinated action of their cilia, ctenophores move forward, mouth first, capturing prey with their sticky tentacles, while slowly rotating.
2. Ctenophora are hermaphrodites.
3. Unlike the cnidarians, ctenophores are not radially symmetrical—they have paired tentacles.

BILATERALLY SYMMETRICAL ACOELOMATES

1. Cnidaria and Ctenophora are radially symmetric, but acoelomates may also have bilateral symmetry.

Platyhelminthes (flatworms)

2. Platyhelminthes (flatworms) are divided into three classes: Turbelleria (free-living flatworms), Trematoda (flukes), and Cestoda (tapeworms).
3. Flatworms have three body layers—an endoderm lining the digestive cavity, an outer ectoderm, and a mesoderm in between.
4. They have feeding, digestive, and reproductive organs, a brain, and a simple system for regulating water balance.
5. Flatworms lack a circulatory system, surviving by keeping diffusion distances small.
6. The single opening to the flatworm's gut cavity serves as both mouth and anus.
7. Turbellaria, the free-living flatworms, reproduce asexually or sexually.
8. A free-living flatworm's simple nervous system allows it to move efficiently toward food and away from light.
9. The classes of flatworms differ in how they eat and reproduce.
10. Flukes and tapeworms are parasites. Flukes take in food via a specialized mouth, and tapeworms lack a digestive system—they simply absorb digested food from the host.
11. Flukes and tapeworms illustrate many common adaptations of parasites:
    - rapid reproduction;
    - distinct stages that allow passage and dispersal through more than one host;
    - organs for attachment to their hosts;
    - specialized digestion (in the case of tapeworms, direct absorption); and
    - reduced sense organs.
12. Flukes attach to the host by a hook or sucker, and have a single-opening digestive system.
13. Many flukes are hermaphrodites.
14. Adult tapeworms have no digestive system.
15. A tapeworm maintains its easy life with a specialized attachment organ (a scolex) at its anterior end. Behind the scolex lies a set of repeated segments (proglottids). Each proglottid is a complete reproductive unit, with both male and female organs. Proglottids toward the rear of the tapeworm contain embryos, each within a separate shell, that pass, with the feces, to the outside world.
16. Platyhelminthes are the simplest animals with heads.

Nemertea (ribbon worms)

1. Ribbon worms are free-living, most residing in the sea, but a few are in fresh water or on land.
2. A ribbon worm has a characteristic snout, or proboscis—a long, sensitive, retractable, and sometimes venomous tube, which it uses for exploration, defense, and capturing prey.
3. Two important advances distinguish ribbon worms from flatworms:
   - a digestive tract with both a mouth and an anus, and
   - a circulatory system.
4. The one-way digestive tract allows continuous eating and more efficient nutrient extraction.
5. The ribbon worm's circulatory system is relatively simple—there is no pumping heart. The vessels and the body wall contract irregularly, sloshing blood through the system.

ASCHELMINTHS

1. The Aschelminths are united more by what they do not have than by what they do—they are a group of unrelated animals that fit nowhere else. They probably represent eight or nine phyla.
2. The Aschelminths lack a coelom, but some have a cavity called a pseudocoel because it lacks the mesodermal lining of a true coelom.
3. The pseudocoel:
   - is a distribution route for nutrients, gases, and wastes; and
   - serves as a hydrostatic skeleton.
4. Aschelminths are all tiny, have sticky glands for attachment, and sexes are usually separate.
5. Aschelminths have both a mouth and an anus.

Nematodes (roundworms)

1. Nematodes (roundworms) are slender, cylindrical, and unsegmented, and they are among the most numerous animals.
2. Many nematodes are parasites, and they infect virtually all plants and animals.
3. A nematode has a large pseudocoel between the body wall muscles and central tubular gut.
4. Nematodes lack a circulatory system; fluids move within the pseudocoel.
5. A hydrostatic skeleton allows the body to regain its original shape after muscle contraction.
6. A free-living nematode has sense organs that detect movement and chemicals, nerve cords, and a primitive brain—a ring of nerves just behind its mouth.
7. Nematodes reproduce sexually. In most species, fertilization is internal, requiring copulation.
8. The other seven phyla of Aschelminths include Rotifera, Nematomorphora, and Gastrotricha.

PROTOSTOME COELOMATES

1. An additional internal space (coelom or pseudocoel) brings several advantages:
   - it prevents muscle actions from disrupting the digestive and circulatory systems;
   - it can provide a hydrostatic skeleton against which muscles can work; and
   - it provides a protected space for the production of sperm and ova.
2. A pseudocoel arises from the space between endoderm and mesoderm, so that the gut is lined with a single layer of tissue—endoderm. A true coelom arises within the mesoderm, so that the gut is lined with endoderm surrounded by mesoderm.

How do protostomes differ from deuterostomes?

1. Nearly all coelomates have both a mouth and an anus. Those in which the mouth develops first are called protostomes; those in which the anus develops first are deuterostomes.
2. In protostomes, cell division after fertilization is usually spiral; in deuterostomes, it undergoes other patterns.

Mollusks share a common body plan

1. In diversity, mollusks are second only to arthropods.
2. Most mollusks live in protective shells.
3. All mollusks share a common body plan:
   - the intestinal tract, and excretory and reproductive organs, lie within the main body, called the visceral mass.
   - the foot, a muscular extension of the body that is used for sensing, grabbing, moving, digging, and holding on, is attached to the visceral mass;
   - the dorsal surface (top) of the visceral mass forms a mantle that secretes a shell;
   - folds of the mantle enclose a space—the mantle cavity—that contains breathing organs.
4. Shelled mollusks can withdraw into their mantle cavities, in some cases pulling behind them a protective flap (operculum) that resembles a trap door.
5. In aquatic mollusks, the mantle cavity contains gills—thin organs that transfer dissolved oxygen to the circulatory system. Air-breathing snails and slugs use this cavity as a lung.
6. Most mollusks have a radula, a rasping tongue covered with teeth made from chitin.
7. Mollusks have a specialized digestive tract, starting with a mouth and ending with an anus.
8. Mollusks have kidneylike nephridia—tubular excretory organs that remove nitrogenous wastes from the coelomic fluid and regulate water and salt concentration.
9. All mollusks have a circulatory system with a heart that receives oxygen-carrying blood from the gills and pumps it to other body tissues. Most mollusks have open circulatory systems, but cephalopods have a closed, well-developed circulatory system.
10. Mollusks reproduce sexually. Male female are often separate individuals, but some hermaphrodites exist, and in some species individuals change sex often. Cross-fertilization is the rule, but self-fertilization occurs.

Bivalves

1. Bivalves have two shells hinged by a ligament, with the foot protruding from between them.
2. Most bivalves are sedentary and attached to rocks.

Gastropods

1. The gastropods are the most diverse class of mollusks, and they differ from all others in having a mysterious 180° counterclockwise twist of the body—both the anus and the mantle cavity end up at the front, not far from the mouth.

Cephalopods

1. Cephalopods are the most active and intelligent mollusks. Of these, only the nautilus has an external shell. Surrounding the mouth of each species are a set of tentacles, evolved from the foot which the all-carnivorous cephalopods use to catch prey, pull along the sea bottom, and steer in open water.
2. An octopus' interesting behavior is controlled by the most sophisticated invertebrate brain. Information about the outside world comes from touch receptors on the tentacles, as well as from large and complex eyes.
3. Cephalopods respond to their environments with a complex set of color changes. Most cephalopods have specialized skin cells, usually with three different pigments.
4. Cephalopods are predatory, intelligent, and active.

What do all annelids have in common?

1. Earthworms and other annelids have long, segmented bodies consisting of identical (or nearly identical) sections called metameres.
2. Annelids have specialized excretory organs (nephridia), a closed circulatory system (with blood contained entirely within vessels—some larger vessels act as hearts to pump the blood), and no gills or lungs—gases move directly through the skin to the blood.
3. The three classes of annelids are Polychaeta, Oligochaeta, and Hirudinea.

Polychaete worms

1. Polychaetes are primitive marine annelids.
2. Each polychaete segment contains a pair of leglike paddles, called parapodia, which the worm uses to swim, crawl, or burrow, as well as to respire. Each parapodium is a bundle of bristles called setae.
3. A polychaete's well-developed head contains various sense organs, including two to four pairs of eyes.
4. Polychaetes have separate male and female sexes.

Oligochaetes

1. Oligochaetes include the common earthworms, and each segment lacks parapodia and contains just four pairs of bundled setae.
2. Oligochaetes are mostly hermaphrodites and cross-fertilize.

Hirudinea

1. Because most Hirudinea, or leeches, are parasites, they are more specialized than other annelids—their bodies are flattened rather than cylindrical, and they lack segments and setae. Many species have a sucker at each end of their bodies.
2. Leeches are hermaphrodites that cross-fertilize.
3. A leech's impressive capacity to consume blood rapidly depends on powerful muscles of the pharynx, sharp teeth, and a chemical that prevents blood clotting (an anticoagulant).

Why are arthropods so successful?

1. Arthropods are the most successful phylum of animals on the planet.
2. All arthropods have an exoskeleton secreted by the cells of the epidermis, and jointed limbs.
3. Jointed limbs, which give the arthropods their name, allow them to move with agility, despite the rigidity of their exoskeletons.
4. Arthropods have segmented bodies.
5. Over millions of years of evolution, legs have changed into mouthparts, antennae, gills, pincers, claws, and egg depositors. These specializations often develop from fused segments called tagmata.
6. Anterior segments of all arthropods are fused into a highly organized head. Other tagmata include the middle portion, called the thorax, and the rear portion, called the abdomen.
7. The exoskeleton, a hard external supporting framework, consists of layers of chitin and protein. It protects the animals from predators and parasites and gives mechanical support.
8. The hard exoskeleton does not grow, so many arthropods (spiders, for example) periodically must shed their exoskeletons and secrete new, larger ones in a process called molting. Other arthropods do all their growing as a soft larva that lacks an exoskeleton, then develop into the adult form in a process called metamorphosis.
9. Metamorphosis and molting divide the life cycle of an insect into many distinct stages, each of which may have different food and lifestyle.
10. Both arthropods and annelids have a tubular gut that stretches from mouth to anus; the arthropod circulatory system is open, and the respiratory systems vary.

11. Spiders and some other air-breathing arthropods use stacks of modified gills, collectively called a book lung, to pull oxygen directly from the air.
12. Most terrestrial arthropods take in air through regulated openings in the body wall called spiracles. Air passes from the spiracles to special ducts called tracheae. Each tracheae branches into tiny tracheoles, which deliver oxygen to individual cells throughout the body.
13. Reproduction in arthropods is almost always sexual. Male and female are usually separate.
14. Excretion in most terrestrial arthropods depends on unique organs called Malpighian tubules, which absorb fluid from the blood and convert the nitrogen-containing wastes into insoluble crystals of uric acid or guanine.
15. The nervous systems of arthropods are often complex, consisting of a double chain of ganglia running along the lower surface of the body. At the head, the chain curls upwards to form a brain—three pairs of fused ganglia.
16. While some arthropods (such as the spiders) have simple eyes, most arthropod species have compound eyes, made up of numerous simple light-detecting units, called ommatidia.

How do biologists classify arthropods?

1. Arthropods are classed into three subphyla based on the specialization of appendages (especially the first few segments).
2. In one subphylum—Chelicerata (chelicerates)—the first pair of appendages are mouthparts called chelicerae. The chelicerae serve as pincers or fangs, often associated with poison glands.
3. In the other two subphyla, Uniramia and Crustacea, the first pair (or the first few pairs) of appendages are antennae, and the next pair are jaws, or mandibles.
4. The Crustacea are almost entirely aquatic, while the Uniramia (including insects, centipedes, and millipedes) are mostly terrestrial. The crustaceans all have branched (or biramous) appendages, while the Uniramia all have unbranched (or uniramous) appendages.

Chelicerates

1. The chelicerates (including spiders and horseshoe crabs) are the only arthropods without antennae or jaws. The most anterior pair of appendages are the pincerlike chelicerae, and the second are the pedipalps, which serve different functions in different classes.
2. Spiders and other arachnids have no jaws and cannot chew. They digest only liquefied food. A spider injects a paralyzing poison through its chelicerae, tears and grinds a bit of its prey, then regurgitates digestive enzymes into the prey's body, which liquefies the prey.
3. Scorpions differ from most chelicerates in having a clearly segmented abdomen, which allows them to curl the abdomen, holding the stinger aloft.

Crustaceans

1. Crustaceans usually have three pairs of chewing appendages (including the mandibles), many pairs of legs, and, unlike other arthropods, two pairs of antennae.
2. Copepods are the most numerous crustaceans and, next to nematodes, are the second most abundant animals on Earth. They are a major component of plankton.

Unimaria

1. The Uniramia are the most diverse subphylum of arthropods. Almost all Uniramia take in air through tracheae, use Malpighian tubules to rid themselves of nitrogenous waste, and have unbranched (uniramous) appendages.
2. Insect bodies consist of three tagmata (fused segments): the head, thorax, and abdomen. The thorax consists of three segments, each of which carries a pair of legs, and in many insects, one or two pairs of wings.
3. Uniramia have one pair of antennae and one pair of mandibles.

VOCABULARY BUILDING

In your own words, first write a brief definition, then a full explanation for each of the following terms. Include examples where appropriate. Complete this section from your memory—you will not learn the terms by simply copying definitions from the textbook. Once you have finished, check your responses against the information in the chapter and make any necessary corrections.

matrix —

tissues —

organ —

epithelium —

connective tissue —

Eumetazoa —

Parazoa —

zygote —

blastula —

gastrula —

archenteron —

endoderm —

ectoderm —

mesoderm —

coelom —

coelomates —

acoelomates —

pseudocoelomates —

blastopore —

protostomes —

deuterostomes —

radial symmetry —

bilateral symmetry —

sessile —

mesenchyme —

cnidarians —

polyps —

medusae —

mesoglea —

cnidocytes —

nematocyst —

ctenophores —

Platyhelminthes (flatworms) —

Turbellaria —

Trematoda (flukes) —

Cestoda (tapeworms) —

scolex —

proglottids —

proboscis —

Nemertea (ribbon worms) —

Aschelminths —

pseudocoel —

nematodes (roundworms) —

mollusks —

visceral mass —

foot —

mantle —

mantle cavity —

operculum —

radula —

nephridia —

bivalves —

gastropods —

cephalopods —

metameres —

parapodia —

setae —

tagmata —

thorax —

abdomen —

exoskeleton —

molting —

metamorphosis —

book lung —

spiracles —

tracheae —

Malpighian tubules —

compound eyes —

ommatidia —

Chelicerates —

Uniramia —

Crustacea —

CHAPTER TEST

The following test has four parts. Complete as much of the exam as you can from memory. If you cannot answer a question, skip it. Once you complete all that you can, try to answer any questions you skipped. If you still cannot answer them, consult your textbook for the answers. Once you have completed all sections of the test, check your answers for Parts 1 - 3 against those in the back of this book. Highlight any incorrect answers then review that material in your textbook. Correct your answers for future reference.

Part 1: Multiple Choice

For each of the following, select all correct responses—more than one may be correct.

1. Which of the following is a characteristic of animals?
   a) multicellular
   b) develops from an embryo
   c) heterotrophic
   d) all of these

2. After fertilization, a zygote divides by mitosis to form a hollow ball called a(n):
   a) archenteron
   b) blastula
   c) gastrula
   d) embryo

3. Which of the following gives rise to the muscles and bones?
   a) ectoderm
   b) endoderm
   c) mesoderm
   d) all of these

4. Which of the following is not true of the Parazoa?
   a) They lack symmetry.
   b) They have no organs.
   c) They have no connective tissues.
   d) None—these are all true of the Parazoa.

5. Sponges are classified as:
   a) Cnidaria
   b) Porifera
   c) Eumetazoa
   d) Parazoa

6. Flatworms are classified as:
   a) bilaterally symmetrical acoelomates
   b) radially symmetrical acoelomates
   c) pseudocoelomates
   d) protostome coelomates

7. What distinguishes a pseudocoel from a true coelom?
   a) it lacks an open space
   b) it lies between the ectoderm and the endoderm
   c) it lacks a mesodermal lining
   d) all of these

8. Animals in which the mouth develops first are called:
   a) protostomes
   b) acoelomates
   c) Parazoa
   d) deuterostomes
9. A visceral mass, mantle, mantle cavity, shell, and foot are all characteristics of:
   a) ctenophores
   b) flatworms
   c) mollusks
   d) nematodes
10. Nephridia perform the same general function as what human structure?
    a) heart
    b) kidney
    c) lung
    d) stomach
11. Gastropods include which of the following?
    a) snails and slugs
    b) flatworms
    c) squids and octopods
    d) earthworms
12. Which of the following has an exoskeleton?
    a) cephalopods
    b) arthropods
    c) annelids
    d) mollusks
13. Which of the following is not an arthropod?
    a) Chelicerata
    b) Crustacea
    c) Pogonophora
    d) Uniramia
14. Arthropods are categorized according to:
    a) how they catch food
    b) specialization of their appendages
    c) the number of segments
    d) none of these
15. In flatworms, the proglottids contain:
    a) hearts
    b) lungs
    c) male and female reproductive structures
    d) None of these—proglottids are tiny mouths.

Part 2: Matching

For each of the following, match the correct term with its definition or example. More than one answer may be appropriate.

endoderm **ectoderm** **mesoderm**

1. ________________ This lines a true coelom.
2. ________________ This forms digestive organs.
3. ________________ This is the innermost layer.
4. ________________ This forms the skeleton.
5. ________________ This forms the skin and nervous system.

Cnidaria **Cestoda** **Chelicerata**

6. ________________ Reproduces by shedding proglottids into the host.
7. ________________ These are radially symmetrical acoelomates.
8. ________________ Some of these liquefy their prey with injected enzymes.
9. ________________ These are the only arthropods with no jaw or antennae.
10. ________________ These include jellyfish and sea anemones.

Part 3: Short Answer

Write your answers in the space provided or on a separate piece of paper.

1. List the two subkingdoms into which all animals are classified.

2. Differentiate between a tissue and an organ.

3. Differentiate between protostomes and deuterostomes.

4. How do Parazoa differ from protists?

5. For what do the ctenophores use their cilia?

6. How do flatworms survive without a circulatory system?

7. List five common adaptations of parasites that are found in tapeworms and flukes.

8. List two features that distinguish ribbon worms from flatworms.

9. What are two roles of a pseudocoel?

10. List three advantages of having a coelom.

11. What is the function of the mantle in most mollusks?

12. For what do cephalopods use their tentacles?

13. Explain the general body plan of annelids.

14. Which of the annelids are parasites?

15. What two major adaptations are shared by all arthropods?

Part 4: **Critical Thinking: Using Your Knowledge**

Answer each of these in essay form, using complete sentences and paragraphs. Provide as much information as you can. (For extra essay practice, write out answers to the Review and Thought Questions in your textbook.)

1. Having read about the Burgess Shale controversy, and with your understanding of the process of evolution, do you think evolution is or is not progressive? Define your use of the term "progressive," and back up your arguments.

2. Construct a table in which you list the different phyla discussed in Chapter 23 and distinguishing characteristics for each.

3. Discuss the needs of animals that likely led to the development of a true coelom. What led to the need for a coelom?

4. Explain the benefits and disadvantages of having an exoskeleton. Explain molting and metamorphosis.

5. Explain reasons that led to the success of arthropods. Why do they continue to be so successful today?

6. Explain what is meant by saying that the origin of all major animal phyla depended on the evolution of feces. What animals do you think would not exist today if the one-way digestive system never developed?

7. Some of the animals discussed in this chapter are hermaphroditic. Please explain this, and explain what advantages this provides them. Under what conditions would hermaphrodism be beneficial?

8. What are the advantages and disadvantages of being multicellular?

9. Overall, through evolution animals have become increasingly complex. Yet evolution is not a deliberate process. What do you think accounts for the increased complexity?

10. In animals, can complexity continue to increase indefinitely, or are there limits on how complex an animal can be? If so, what are some limits?

Chapter 24
Deuterostome Animals: Echinoderms and Chordates

KEY CONCEPTS

1. The deuterostomes consist of both invertebrates (echinoderms, arrow worms, and acorn worms) and vertebrates (fish, amphibians, reptiles, birds, and mammals).
2. Echinoderms are ancient, spiny-skinned animals that are radially symmetrical as adults but bilaterally symmetrical as larvae.
3. Arrow worms and acorn worms may resemble the ancestors of the vertebrates because they possess a postanal tail. In addition, the acorn worms have gill slits and a dorsal nerve cord like that of the chordates.
4. All chordates have a notochord, a dorsal hollow nerve cord, pharyngeal slits, and a postanal tail at some time in their lives.
5. As larvae, tunicates have a notochord, a dorsal hollow nerve cord, pharyngeal slits, and a postanal tail—all characteristics of chordates.
6. Cephalochordates have notochords both as larvae and as adults.
7. Vertebrates have vertebral columns, skulls, and a well-developed brain. Most vertebrate skeletons are made of bone.
8. Lampreys and hagfish (members of the Agnatha) are the only vertebrates that lack jaws.
9. Placoderms were the first vertebrates with jaws.
10. The Chondrichthyes have cartilaginous skeletons and no swim bladder.
11. Osteichthyes have bony skeletons, swim bladders, and paired fins.
12. Amphibians live a dual life—the larvae are usually aquatic and the adults are usually terrestrial.
13. The evolution of the amniotic egg enabled reptiles to dominate the Earth for millions of years. Reptiles' descendants, the birds and mammals, also have amniotic eggs.
14. The body temperatures of ectotherms fluctuate with air temperature; they regulate temperature somewhat by basking in the sun or moving to shade.
15. Endotherms maintain their body temperature within a few degrees of a constant temperature most of the time; they do this via metabolic processes.
16. Feathers, flight, and endothermy distinguish birds from reptiles.
17. Three characteristics that set mammals apart from other vertebrates are mammary glands, hair, and two sets of teeth.

EXTENDED CHAPTER OUTLINE

ARE WE UPSIDE DOWN?

1. In 1830, a public debate pitted French anatomist and paleontologist Baron Georges Cuvier against Étienne Geoffroy Saint-Hilaire.
2. Cuvier argued that all animals should be divided into four anatomical types; Saint-Hilaire maintained that all animals had the same body plan as vertebrates, but that we vertebrates are turned upside down.
3. Saint-Hilaire's view implied that vertebrates evolved from invertebrates.
4. Molecular biologists have identified genes that regulate the dorsal and ventral development of body plan, and one can override another, affecting the orientation of development.
5. If organisms as different as lobsters and humans share genes that help determine the development of the basic body plan, these organisms certainly share a common ancestor.

DEUTEROSTOME COELOMATES

1. Early development of deuterostome coelomates differs from that of protostomes in several ways, but in particular, in all deuterostomes, the blastopore develops into an anus instead of a mouth (which develops later).
2. The four phyla of deuterostomes are Echinodermata, Chaetognatha, Hemichordata, and Chordata.
3. The Chordata include the vertebrates.

Echinoderms

1. The 6000 species of echinoderms are divided into 6 classes of marine animals, and includes sea stars and sea urchins.
2. Echinoderms are an ancient lineage—fossils suggest they diverged from protostomes around 530 million years ago.
3. Most echinoderms have calcium-rich spines projecting from their skin.
4. As adults, all echinoderms are radially symmetrical, often with five nearly identical parts arranged around a central axis.
5. The larvae have bilateral symmetry.
6. The sexes of echinoderms are separate; gametes are released into open water so reproductive success relies on the sheer numbers of gametes.

Sea stars (starfish)

1. The sea star (starfish) is a typical echinoderm—a central region from which arms radiate outward. They have mouths under the center opening into a digestive tract.
2. Sea stars eat by extending their gut—turning the stomach inside-out
3. Sea stars' coeloms contain a unique set of canals—the water vascular system—that helps them control movement of their arms.
4. Starfish and some other echinoderms can regenerate arms and reproduce asexually.

Arrow worms and acorn worms

1. The 50 to 70 species of arrow worms are common marine organisms, and fossils date them back to late Cambrian times.
2. The body of an adult arrow worm extends beyond the anus, forming a tail.
3. The tail and other similarities between arrow worms and vertebrates leads to the belief that vertebrates are descended from an ancestor of arrow worms.

4. Most of the 65 species of acorn worms live in U-shaped burrows on the sea bottom.
5. Acorn worms are usually classed as hemichordates because they resemble chordates by:
   - having gill slits—holes that directly connect the throat to the outside, and
   - having the beginnings of a dorsal nerve cord (like our spinal cord), in addition to a ventral nerve cord.

What traits characterize the chordates?

1. The chordates include three subphyla—the tunicates and cephalochordates, with no backbones, and the vertebrates, which do have backbones.
2. Four hallmark characteristics are shared by all chordates at some point in their lives:
   - a notochord—a flexible rod running along the dorsal surface;
   - a dorsal, hollow nerve cord running between the notochord and the surface of the back;
   - pharyngeal slits (also called gill slits) that connect the inside of the intestinal tract to the outside; and
   - a segmented body and a postanal tail.
3. In many vertebrates the notochord disappears, replaced by a bony or cartilaginous backbone.
4. The dorsal nerve cord and its adult counterparts (brains and spinal cord) are partly hollow and contain fluid in their inner spaces.
5. Pharyngeal slits are present in all chordate embryos, but not in air-breathing adults. Aquatic chordates use them mostly for respiration.

What have we in common with tunicates?

1. Adult tunicates are sessile marine animals that do not look at all like other chordates. They have a tough outer coating (tunic) made of cellulose, which is rare in animals.
2. A tunicate larva has a notochord, dorsal hollow nerve cord, pharyngeal slits, and postanal tail.

What have we in common with cephalochordates?

1. Cephalochordates, or lancelets, include about 45 species of small, segmented, fishlike animals.
2. In cephalochordates, the notochord, present in both adult and larval forms, extends all the way through the head.

What distinguishes the vertebrates from other chordates?

1. Two important features distinguish vertebrates from other chordates:
   - a segmented vertebral column made of units called vertebrae (the vertebrae surround and protect the dorsal nerve cord, which becomes the spinal cord), and
   - a distinct head, with a cranium (skull) and a brain.
2. Vertebrates also have closed circulatory systems and characteristic internal organs.
3. The heart pumps blood through a system of fine capillaries to transport gases, nutrients and metabolic wastes.
4. The liver, kidneys, and other organs closely regulate and maintain a nearly constant internal environment for all cells.
5. Most modern vertebrates have bony skeletons, and the bone can grow.
6. The embryos of all vertebrates have skeletons made of cartilage, which is softer and more elastic than bone. Some cartilage remains in the adult skeleton, but in most adult vertebrates, most of the cartilage is replaced by bone (sharks and other cartilaginous fish are exceptions).
7. Vertebrates include four classes of fish (Agnatha, Placodermi, Chondricthyes, and Osteichthyes) and four classes of tetrapods (Amphibia, Reptilia, Aves, and Mammalia).

Agnatha have no jaws

1. Most jawless fish are extinct; they comprise some 60 families and are called ostracoderms.
2. Ostracoderm fossils indicate that vertebrates have been around as long as invertebrates.
3. The living jawless fish include only about 60 species of cyclostomes—soft-bodied fish with round, jawless mouths, which include lampreys and hagfish.
4. Almost all lampreys are parasitic; hagfish eat polychaete worms or scavenge dead fish.

Placoderms, now extinct, were armored fish with jaws

1. Freshwater fish developed the first vertebrate jaws at the beginning of the Devonian Period, about 405 million years ago.
2. Paleontologists sometimes call the Devonian the "Age of the Fishes."
3. Placoderms were the first vertebrates with jaws, and had less armor than ostracoderms.
4. Today, all vertebrates except the Agnatha have hinged jaws.

Chondrichthyes are fish with cartilaginous skeletons

1. In sharks, skates, and rays, bone does not replace cartilage during development. Instead, adult Chondrichthyes have lightweight cartilaginous skeletons.
2. Chondrichthyes also lack a swim bladder—a balloonlike organ that helps the bony fish float.
3. Although sharks and rays lack the bony scales of placoderms and ostracoderms, their skin is covered with small, toothlike scales.
4. Sharks have particularly well-developed senses that help them find their prey. These include:
    - keen vision and smell;
    - the lateral line system—a row of tiny sense organs along each side of the body that detect changes in water pressure; and
    - receptors that detect tiny electrical currents.

Oestichthyes are fish with bony skeletons

1. The bony fish, the Oestichthyes, are the most diverse of all the vertebrate classes.
2. Like the cartilaginous fish, bony fish first appeared in the Devonian Period, but in fresh water, rather than in oceans.
3. Bony fish have bony skeletons and thin, bony scales.
4. Most bony fish also have a swim bladder—an air-filled sac that helps them control their buoyancy.
5. Nearly all modern fish are ray-finned—they have paired fins supported by thin, bony rays derived from bony scales.
6. In contrast, the fins of lobe-finned fish are unusual—lungfish can use their sturdy fins to walk on land.
7. Biologists believe that the first amphibians must have resembled lobe-finned fish, whose fins gradually evolved into limbs.

WHAT ADAPTATIONS HAVE VERTEBRATES EVOLVED FOR LIFE ON LAND?

1. To colonize the land, vertebrates had to solve several problems:
    - They needed to obtain and conserve water in a comparatively dry environment.
    - They needed to extract oxygen from air rather than from water.
    - They needed strong skeletons to support their bodies.
    - They needed to control fluctuations in body temperature.
    - They still needed a watery environment for fertilization and early development.
2. Terrestrial vertebrates—amphibians, reptiles, birds, and mammals—evolved adaptations to solve the problems of land life.

Amphibians are terrestrial animals that begin their lives in water

1. The most important terrestrial adaptations—lungs and limbs—date to the Devonian Period.
2. Nearly all freshwater fish that survived the Devonian Period had a kind of lung—a simple sac enhanced with a rich blood supply.
3. The Devonian fish developed double circulation in which blood circulates between the heart and lungs and between the heart and the rest of the body. All terrestrial vertebrates have double circulation.
4. The Devonian freshwater vertebrates also developed legs, using their fins to walk.
5. Early amphibians were the first vertebrates to live most of their lives on land.
6. Today, three orders of Amphibia exist:
   - Urodela (salamanders),
   - Anura (frogs and toads), and
   - Gymnophiona (caecilians).
7. The Anurans are the most diverse amphibian order.
8. Anurans lack tails, but urodeles have them. Gymnophiona lack feet and resemble snakes or earthworms. They are tropical and live underground, rather like moles.
9. Amphibians can gulp air into their lungs, but they also absorb oxygen through their skins.
10. An amphibian's life cycle usually consists of two stages: the first stage is aquatic and the second terrestrial.
11. Amphibians are only partly adapted to life on land—the need for water dominates their lives.
12. Amphibians do not drink water; instead, they must absorb it through their thin, damp skins.
13. Most amphibians start their lives in water. The fertilized egg develops into an aquatic larva, or tadpole, which obtains oxygen through gills and uses its long tail to swim.
14. The tadpole becomes a terrestrial animal by losing its gills and developing lungs and limbs in a process called metamorphosis.
15. A few species have evolved direct development—the zygote develops directly into the terrestrial form, bypassing the larval form.
16. A few salamanders do not complete metamorphosis—they retain their gills and continue to live exclusively in water even after they are capable of reproducing.

Reptiles can live their entire lives on land

1. Reptiles, unlike amphibians, are entirely adapted to life on land.
2. Most of the 6000 species of living reptiles fall into three orders:
   - Crocodilia (crocodiles and alligators),
   - Chelonia (turtles and tortoises), and
   - Squamata (lizards), which are the most successful.
3. A fourth order of living reptiles has only one species—the lizardlike tuatara.
4. Only a fraction of the animals living in the Mesozoic Era—the Age of Reptiles—were actually dinosaurs.
5. The history of the reptiles began 300 million years ago in the late Carboniferous Period.
6. Reptiles were the first vertebrates to free themselves from the water—their dry, watertight skin helps conserve water.
7. Because they cannot breathe through their tough skin, they depend entirely on their lungs for oxygen.
8. Reptiles also have efficient kidneys to conserve water, and legs well-suited for life on land.
9. Reptiles also evolved the amniotic egg with the following components:
   - the amnion— a membrane that encloses the developing embryo in its own little pond;
   - a porous shell that allows gas exchange with the outside;
   - another membrane—the chorion;
   - the yolk—a rich food supply surrounded by another membrane, the yolk sac; and
   - the allantois—a fourth membrane that functions in respiration and excretion.

10. All reptiles and their descendants, including birds and mammals, have amniotic eggs.
11. In reptiles and birds, a shell forms around the egg before it emerges from the female, so fertilization must be internal.
12. All male reptiles except the tuatara have a penislike copulatory organ; snakes and lizards have a pair of hemipenes.
13. All reptiles of both sexes have a cloaca—a common opening shared by the digestive, urinary, and reproductive systems.

How do terrestrial vertebrates regulate body temperature?

1. Reptiles are ectotherms—they depend on external sources of heat to regulate their body temperature. Reptiles typically move into and out of the sun to adjust their temperature.
2. Mammals and birds are endotherms—they maintain a relatively constant body temperature by capturing the heat released from metabolism and releasing heat from the body as needed.

What distinguishes the birds from the reptiles?

1. Birds are wonderfully diverse, with some 8700 species.
2. Three features distinguish birds from reptiles:
   - feathers,
   - flight, and
   - endothermy.
3. Keeping a high body temperature and flying both require enormous amounts of energy.
4. Flight provides terrific protection from most predators.
5. In birds, fertilization is always internal and in most cases involves bringing the male and female cloacas together.
6. Feathers are made of keratin. Many biologists believe that feathers were originally an adaptation for warmth; flight came later.
7. Other adaptations lightened the bird body to improve flight, including:
   - loss of teeth,
   - hollowing of bones, and
   - reshaping of the breastbone.

What distinguishes mammals?

1. Mammals take their name from having mammary glands—milk-producing organs in the female that characterize the mammals. No other vertebrates have mammary glands.
2. Only mammals have hair and two sets of teeth (baby teeth and adult teeth).
3. Mammalogists divide the living mammals into three subclasses—prototherians, metatherians, and eutherians. These subclasses diverged during the Mesozoic Era.
4. Most mammals are eutherians, or placental mammals. In these, embryonic development takes place within the female's uterus. A placenta joins the lining of the uterus with the membranes of the fetus.
5. Metatherians, or marsupials, produce newborns that are undeveloped fetuses. A fetus pulls itself along the mother's fur until it finds a nipple, typically inside a furry pouch where the fetus completes its development.
6. The prototherians, or monotremes, are the strangest mammals—they lay eggs that are similar to those of reptiles and birds. Their digestive and reproductive systems open into a cloaca. The only living monotremes are the platypus and the spiny anteater.

VOCABULARY BUILDING

In your own words, first write a brief definition, then a full explanation for each of the following terms. Include examples where appropriate. Complete this section from your memory—you will not learn the terms by simply copying definitions from the textbook. Once you have finished, check your responses against the information in the chapter and make any necessary corrections.

dorsal —

ventral —

blastopore —

Echinodermata —

Chaetognatha —

Hemichordata —

Chordata —

gill slits (pharyngeal slits) —

notochord —

dorsal, hollow nerve cord —

bone —

cartilage —

tetrapods —

ostracoderms —

cyclostomes —

placoderms —

swim bladder —

double circulation —

metamorphosis —

amniotic egg —

amnion —

chorion —

yolk —

yolk sac —

allantois —

cloaca —

ectotherm —

endotherm —

mammary gland —

eutherians (placental mammals) —

placenta —

metatherians (marsupials) —

prototherians (monotremes) —

CHAPTER TEST

The following test has four parts. Complete as much of the exam as you can from memory. If you cannot answer a question, skip it. Once you complete all that you can, try to answer any questions you skipped. If you still cannot answer them, consult your textbook for the answers. Once you have completed all sections of the test, check your answers for Parts 1 - 3 against those in the back of this book. Highlight any incorrect answers then review that material in your textbook. Correct your answers for future reference.

Part 1: Multiple Choice

For each of the following, select all correct responses—more than one may be correct.

1. Which of the following is not an invertebrate?
   a) echinoderm
   b) arrow worm
   c) fish
   d) acorn worm
2. Which of the following characteristics distinguishes vertebrates from the other chordates?
   a) presence of gill slits at some developmental stage
   b) a segmented vertebral column
   c) a distinct head with a cranium
   d) a notochord in early development
3. Which of the following are hemichordates?
   a) echinoderms
   b) acorn worms
   c) reptiles
   d) arrow worms

4. Chordates are characterized by which of the following?
   a) notochord
   b) pharyngeal slits
   c) postanal tail
   d) all of these

5. Which of the following is found in bone?
   a) collagen
   b) calcium phosphate
   c) blood vessels
   d) all of these

6. Which of the following are jawless vertebrates?
   a) placoderms
   b) Agnatha
   c) Chondrichthyes
   d) None—all of these have jaws.

7. Which of the following is false about the Chondrichthyes?
   a) They have cartilaginous skeletons.
   b) They include the sharks and skates.
   c) They have a swim bladder.
   d) None—all of these are true about the Chondrichthyes.

8. In double circulation, blood circulates between:
   a) the heart and digestive system
   b) the heart and urinary system
   c) the heart and respiratory system
   d) the heart and nervous system

9. Which of the following is not an order of Amphibia?
   a) Chelonia
   b) Anura
   c) Squamata
   d) Gymnophiona

10. Frogs are:
    a) amphibians
    b) vertebrates
    c) chordates
    d) all of these

11. The first vertebrates to free themselves from an aquatic existence were the:
    a) reptiles
    b) amphibians
    c) birds
    d) mammals

12. In reptiles, the cloaca is a common passageway from which organ systems?
    a) reproductive
    b) digestive
    c) urinary
    d) all of these

13. Which of the following are endotherms?
    a) birds
    b) fish
    c) reptiles
    d) mammals
14. Which of the following provides nutrients to an embryo developing in an amniotic egg?
    a) allantois
    b) yolk
    c) amnion
    d) none of these
15. In which of the following subclasses are newborns really undeveloped fetuses?
    a) echinoderms
    b) placental mammals
    c) marsupials
    d) monotremes

Part 2: Matching

For each of the following, match the correct term with its definition or example. More than one answer may be appropriate.

amphibians **reptiles** **mammals**

1. ______________ This includes the monotremes.
2. ______________ This includes frogs, toads, and salamanders.
3. ______________ This includes turtles, crocodiles, and lizards.
4. ______________ These undergo metamorphosis.
5. ______________ These must breathe air at all developmental stages.
6. ______________ These have hair and two sets of teeth.
7. ______________ These have or had amniotic eggs.

ectotherms **endotherms**

8. ______________ These depend on external sources of heat.
9. ______________ These often bask in sun or move to shade to regulate temperature.
10. ______________ Mammals and birds are of this type.
11. ______________ These maintain a relatively constant internal body temperature.
12. ______________ These generate heat internally via metabolism.

Part 3: **Short Answer**

Write your answers in the space provided or on a separate piece of paper.

1. List the four phyla of deuterostome coelomates.

2. List the four hallmark characteristics of the chordates.

3. What are the main two characteristics that distinguish vertebrates from the other chordates?

4. The skeleton of all vertebrate embryos is composed of what?

5. What are the only jawless fish living today?

6. What is a swim bladder?

7. Explain metamorphosis.

8. What are the four orders of reptiles?

9. What era is considered the Age of the Reptiles?

10. What three features distinguish birds from the reptiles?

11. What are some adaptations that improved flight in birds?

12. List three traits that distinguish mammals from other vertebrates.

13. List the three subclasses of mammals, and provide examples of each.

14. Characterize the placoderms and list the ones that are alive today.

15. Who are the living prototherians (monotremes)?

Part 4: **Critical Thinking: Using Your Knowledge**

Answer each of these in essay form, using complete sentences and paragraphs. Provide as much information as you can. (For extra essay practice, write out answers to the Review and Thought Questions in your textbook.)

1. List the special aspects of terrestrial life that set it apart from aquatic life. What adaptations were necessary to prepare vertebrates to make the move from water to land? Who made this move?

2. Describe the amniotic egg—who has it, what are its advantages, why did it likely evolve?

3. Differentiate between ectotherms and endotherms. Who are they? Which type are we humans? What, if any, external heat sources affect our temperature? How do we maintain our temperatures in cold weather or in hot weather?

4. Differentiate between the three subclasses of mammals. How are they categorized? How does each group reproduce?

5. For each of the four phyla of deuterostomes, list the major classes, major adaptations, and what conditions might have led to the evolution of these adaptations. Also note any characteristics that link the various groups together, indicating common ancestry.

6. Why is egg-laying a better adaptation for birds than live births would be?

7. How do ectotherms, such as reptiles, survive in areas that have cold winters?

8. Marsupials are quite common in Australia and New Zealand, but the only native marsupial in North America is the opossum. How do you explain this?

9. If you discovered the fossil remains of a mammal, how might you determine if it was a herbivore, a carnivore, or an omnivore?

10. The embryos of all modern vertebrates have cartilage skeletons, but most are replaced by bone. What advantages are there to starting life with a cartilage skeleton before converting it to bone? What advantages are there of later having a bony skeleton instead of one made of cartilage?

Chapter 25
Ecosystems

KEY CONCEPTS

1. Ecology is the study of interactions of organisms with each other and with their physical environments.
2. A population is a group of individuals of the same species that inhabit a common area.
3. A community is an interacting group of species that inhabits a common area.
4. An ecosystem is a functional unit consisting of both biotic parts and abiotic parts.
5. All ecosystems together comprise the biosphere.
6. Producers harvest energy directly from sunlight or other inorganic molecules.
7. All producers are autotrophs—organisms that produce their own food. The majority are green plants that accomplish this task through photosynthesis.
8. Consumers obtain energy from eating producers or other consumers. Most are animals.
9. Decomposers, such as bacteria and fungi, obtain energy from eating dead organisms.
10. Consumers and decomposers are heterotrophs—they obtain food from other organisms.
11. The hierarchy of who eats whom is a food chain, and each level within it is a trophic level.
12. A pyramid of energy shows the amount of energy used at each trophic level, and it is always upright—producers always use far more energy than consumers, and each level of consumers use less energy than the level below.
13. The ten percent law states that each level of the energy pyramid provides the next level with only about ten percent of the energy it has assimilated.
14. Almost all energy for ecosystems comes from the sun.
15. Biogeochemical cycles are cycles by which materials in an ecosystem are recycled.

EXTENDED CHAPTER OUTLINE

ORGANIC ENTITY?

1. In the early 1960s, James Lovelock conceived an idea for an ingenious device that could detect tiny amounts of chemicals.
2. Entropy is a measure of the disorder in a system, which is an assemblage of interacting parts or objects.
3. If any system is left alone, its entropy, or disorder, tends to increase, according to the Second Law of Thermodynamics.
4. Organisms decrease local entropy by creating order, but in the universe as a whole, entropy still increases.
5. A reduction in local entropy, Lovelock argued, is a sure sign of life.
6. Lovelock envisioned the biosphere and atmosphere as a single immense entity, which he named Gaia, that acts to preserve itself through homeostasis (maintenance of a constant state).
7. In 1974, Lynn Margulis began enthusiastically promoting her own version of Gaia, which she viewed not as a superorganism, but rather as a giant ecosystem, and the highest level of organization of life.

WHAT IS ECOLOGY?

1. Ecology is the study of interactions of organisms with each other and with their physical environment.
2. A population is an interbreeding group of individuals.
3. A community is an interacting group of many species that inhabit a common area.
4. An ecosystem is a community of organisms together with the nonliving parts of the community's environment.

What is an ecosystem?

1. An ecosystem consists of both biotic (living) and abiotic (nonliving) parts.
2. An ecosystem has many of the properties of living organisms, but it also differs from organisms in important ways, such as:
   - ecosystems do not reproduce, and
   - ecosystems recycle.
3. Just as physical environments affect organisms, organisms affect their physical environments.
4. Each ecosystem, community, or population is part of a larger one.
5. All ecosystems together make up a much larger ecosystem called the biosphere (Earth for us), which differs from other ecosystems by having boundaries.

HOW DOES ENERGY FLOW THROUGH ECOSYSTEMS?

1. An ecosystem contains two groups of organisms—autotrophs and heterotrophs.
2. Autotrophs, or producers, make their own food. Most do this by harvesting energy directly from sunlight through photosynthesis, and are either green plants or photosynthesizing bacteria.
3. Heterotrophs consume other organisms, and most are animals, fungi, or protists.
4. Heterotrophs include:
   - consumers, usually animals, which obtain energy by eating other organisms (plant eaters are herbivores, meat eaters are carnivores, and omnivores eat both), or
   - decomposers, such as bacteria and fungi, which live on energy in the molecules of dead organisms.

5. Scavengers are consumers that eat dead organisms. What scavengers leave behind is decomposed by saprophytes, which include bacteria, fungi, and some plants.
6. A linear sequence revealing who eats whom is called a food chain, and usually has no more than four or five links.
7. A food web is a collection of all the interconnected food chains in an ecosystem.
8. A trophic level is an organism's position in a food chain. Herbivores are primary consumers, carnivores who eat herbivores are secondary consumers, and they are, in turn, eaten by tertiary consumers.
9. Omnivores are animals that consistently eat plants and animals.

How much energy flows through an ecosystem?

1. Ecologists can never completely describe a natural ecosystem because each is unique.
2. Ecologists may estimate the biomass of an ecosystem, which is the dry weight of all the organisms.
3. A graphic presentation of the total mass of organisms at each trophic level is called a pyramid of biomass.
4. A graph of the total numbers of organisms at each trophic level is called a pyramid of numbers.

Energy is continually dissipated in an ecosystem

1. Pyramids of biomass and numbers do not always have pyramidal shapes.
2. Productivity is a measure of the energy captured in chemical bonds of new molecules on a square meter of land per unit of time, usually a year.
3. The energy that drives the Earth's biosphere comes from the sun, and the amount of sunlight striking the Earth varies, except at the equator.
4. Primary productivity is the productivity of the first trophic level—the producers—and it varies tremendously from one ecosystem to another.
5. Marshes are the most productive ecosystems.
6. Biomass is the dry weight of all organisms living in the ecosystem.
7. Plant biomass is usually greater than herbivore biomass, which is usually greater than carnivore biomass.
8. A pyramid of biomass is a graph of the total biomass of organisms at each trophic level.
9. The biomass of consumers need not depend directly on the biomass of the producers.
10. Producers always process more energy than the consumers.
11. Energy becomes less and less available at successively higher trophic levels.
12. Producers convert only part of the light energy into chemical energy; the remainder is lost.
13. Primary consumers, in turn, use the chemical energy of the producers to make their own food, but, again, most of the energy is lost as heat.
14. A pyramid of energy is a graphic representation of the amount of energy used at each trophic level, and it is always upright—producers always process more energy than consumers.
15. The amount of sunlight reaching each square meter of land is about 1 to 2 million kcal per year.
16. Only 1% of the sunlight shining on Earth is available to producers.
17. According to the ten percent law, each trophic level passes on only about 10% of the energy that was in the previous trophic level.
18. Because the energy decreases with each trophic level, a food chain cannot have many trophic levels—usually not more than three or four.
19. Humans use about 40% of terrestrial primary productivity
20. Top carnivores consume a lot, so they typically must find food over wide areas and cannot specialize very much.

HOW DO ECOSYSTEMS RECYCLE MATERIALS?

1. Although organisms constantly dissipate energy, they recycle material. Materials such as carbon, oxygen, nitrogen, and phosphorus constantly change form but are rarely lost.
2. The cycle by which a specific material is recycled is called a biogeochemical cycle, and the material passes through numerous forms, both biotic and abiotic.
3. Oxygen makes up about 20% of the atmosphere and constantly cycles.
4. Animals and consumers take in oxygen from the atmosphere to burn carbon compounds, then release oxygen in carbon dioxide. Plants take up carbon dioxide and release oxygen back to the atmosphere.
5. Some biogeochemical cycles involve only local ecosystems, but others involve the whole biosphere.

What drives the water cycle?

1. All life is intimately connected to water.
2. Plants pull water from the soil and release it as vapor through tiny holes (stomata) in their leaves in a process called transpiration.
3. At any time, 97% of Earth's water is in the oceans, covering 70% of the Earth's surface.
4. Sunlight evaporates large quantities of water from the Earth's bodies of water, and plants transpire vast amounts—transpiration may account for up to 90% of all water that evaporates from the continents.
5. Water vapor easily condenses into rain, which falls back to the Earth's surface.
6. Of the water that evaporates into the atmosphere, just 17% comes from the land, but 25% of all rain falls on land.
7. More rain falls on the land than comes from the land; the excess water returns to the oceans as runoff in streams and rivers.

Where does carbon go?

1. Much of the carbon in the biosphere is in the oceans as bicarbonate ion, and in the atmosphere as carbon dioxide.
2. Bicarbonate ions tend to precipitate to the bottoms of oceans and lakes, where they form thick sediments that later turn into sedimentary rocks.
3. Organisms contain about 10% of the amount of carbon that is dissolved in the oceans.
4. Each year, terrestrial producers convert some 12% of atmospheric carbon dioxide into organic molecules, and this carbon then flows through the worldwide food web.
5. Organisms eventually return the carbon to the oceans and to the atmosphere—as carbon dioxide through respiration. (Even producers respire carbon dioxide.)
6. The largest reserves of carbon are in the fossil fuels—coal, oil, and methane—which result from incomplete decomposition of ancient organisms.

How have humans altered the carbon cycle?

1. During some periods, especially the Carboniferous Period, carbon accumulated because decomposers did not convert all the organic molecules back into carbon dioxide.
2. In the last two centuries, humans have begun returning these vast stores of fossil carbon to the carbon cycle by burning coal, oil, and gas.
3. Total carbon dioxide in our atmosphere has increased by 30% since 1750.
4. Higher levels of carbon dioxide, along with other gases, tend to trap heat, leading to the greenhouse effect.
5. Carbon dioxide buildup in the atmosphere is causing global warming. Of the ten hottest years on record, nine were between 1987 and 1997.

6. Global warming leads to increased melting of polar ice caps and flooding in some areas, but drought in others. Organisms are forced out of their natural habitats and into new ones.
7. In December 1997, representatives from 174 nations convened in Kyoto, Japan, and agreed to reduce carbon dioxide output. The Kyoto Protocol requires industrialized nations to decrease greenhouse gas emissions by 5% by the year 2010. Of all the industrialized nations, only Canada, Australia, and the United States (the largest emitter of these gases) have refused.

How can we get rid of carbon dioxide?

1. The obvious solution is to burn less fossil fuel.
2. Planting more trees will help remove carbon dioxide from the atmosphere (and increase our oxygen supply). Tropical forests are beginning to gain protection.
3. Carbon dioxide is now being buried under seas and oceans.
4. Another approach is to fertilize the oceans with iron, stimulating growth of photosynthetic plankton which will either become part of the food chain or die and sink to the bottom of the ocean, carrying the carbon dioxide with them.

Why is nitrogen both common and in short supply?

1. Molecular nitrogen (N_2) is the most abundant element in the atmosphere—78% of the air we breathe—yet most of this is not usable for organisms because N_2 is extremely stable.
2. Nitrogen is essential for synthesis of proteins and nucleic acids.
3. Most plants can take in nitrogen only as ammonia or nitrate; such nitrogen is called fixed nitrogen.
4. Consumers and decomposers obtain most of their nitrogen from proteins in plants and animals.
5. Lightning fixes atmospheric nitrogen during storms, then rain washes this to the Earth.
6. Nitrifying bacteria are the most important source for fixed nitrogen. These bacteria convert atmospheric nitrogen to ammonia or nitrate.
7. Some nitrifying bacteria are cyanobacteria; others form mutualistic relationships with plants, often living in root nodules.
8. The bacteria fix nitrogen for the plants' use, and the plants provide the bacteria with sugars that the bacteria need for energy.
9. Many ecosystems lose fixed nitrogen. For example, some bacteria obtain energy by converting ammonia to nitrate or nitrite, through a process called nitrification. The nitrate is highly soluble and rains wash it out of soils.
10. Denitrifying bacteria convert nitrate to atmospheric nitrogen, also causing loss of fixed nitrogen.
11. Animals and other organisms break down proteins and convert the nitrogen to ammonia; which aquatic organisms dilute with water and flush from their bodies. Terrestrial organisms convert the ammonia into less harmful urea or uric acid.
12. Decomposers obtain energy from urea and uric acid by converting them back to ammonia, which can then be used again by plants. This process is ammonification.
13. In most ecosystems, most fixed nitrogen moves locally from producers to consumers and decomposers, and back to producers, with only small amounts added through nitrification or lost by denitrification.

How have humans altered the nitrogen cycle?

1. In the last few decades, human activity has doubled the amount of fixed nitrogen entering the biosphere and accelerated the passage of fixed nitrogen through ecosystems.
2. The major source is fertilizers.
3. Even the crops—which have displaced native flora on one-third of the Earth's surface—harbor nitrifying bacteria that fix atmospheric nitrogen.

4. Burning fossil fuels, and trees and grasses (when clearing land), releases nitrogen as nitrous oxide and other gases.
5. All forms of fixed nitrogen are also greenhouse gases, contributing to global warming.
6. Nitric oxide dissolves in water and forms acid rain, which kills ponds, forests, and other habitats.
7. Extra nitrogen in soil leaches nutrients, depleting soil fertility and causing leaching of aluminum which, in turn, poisons plants, streams, and fish.
8. Excess nitrogen decreases species diversity in an ecosystem as some species grow unchecked, crowding out others.
9. Worldwide, 20% of all human waste ends up in rivers, and the nitrogen there flows into estuaries, creating dead zones as algal blooms lead to bacterial blooms and deoxygenation.
10. In the summer of 1999, researchers mapped a dead zone at the mouth of the Mississippi River that was the size of the area of Massachusetts.

VOCABULARY BUILDING

In your own words, first write a brief definition, then a full explanation for each of the following terms. Include examples where appropriate. Complete this section from your memory--you will not learn the terms by simply copying definitions from the textbook. Once you have finished, check your responses against the information in the chapter and make any necessary corrections.

system —

biosphere —

homeostasis —

superorganism —

symbiosis —

ecology —

population —

community —

ecosystem —

biotic —

abiotic —

autotrophs —

producers —

heterotrophs —

consumers —

herbivores —

carnivores —

scavengers —

saprophytes —

decomposers —

food chain —

food web —

trophic level —

omnivore —

productivity —

primary productivity —

biomass —

pyramid of biomass —

pyramid of energy —

ten percent law —

biogeochemical cycle —

transpiration —

fixed nitrogen —

ammonium ions —

nitrification —

denitrification —

urea —

uric acid —

ammonification —

CHAPTER TEST

The following test has four parts. Complete as much of the exam as you can from memory. If you cannot answer a question, skip it. Once you complete all that you can, try to answer any questions you skipped. If you still cannot answer them, consult your textbook for the answers. Once you have completed all sections of the test, check your answers for Parts 1 - 3 against those in the back of this book. Highlight any incorrect answers then review that material in your textbook. Correct your answers for future reference.

Part 1: Multiple Choice

For each of the following, select all correct responses. Some questions may have more than one correct answer.

1. A breeding group of individuals of the same species sharing the same location is a(n):
   a) community
   b) ecosystem
   c) population
   d) biosphere

2. All of the organisms living in one area, along with all the nonliving material in that area, form a(n):
   a) community
   b) ecosystem
   c) population
   d) biosphere

3. An interacting group of more than one species is a(n):
   a) community
   b) ecosystem
   c) population
   d) biosphere

4. Which of the following is a characteristic of an ecosystem that is not shared by an organism?
   a) An ecosystem uses energy to maintain a stable state.
   b) An ecosystem is not subject to natural selection.
   c) An ecosystem transforms chemicals.
   d) An ecosystem recycles its own materials.

5. All of Earth's ecosystems together form the:
   a) biosphere
   b) superorganism
   c) community
   d) all of these

6. Which of the following is a producer?
   a) green plant
   b) human
   c) cow
   d) scavenger

7. Most decomposers are:
   a) plants
   b) animals
   c) bacteria and fungi
   d) detritus

8. Organisms that produce their own food are called:
   a) heterotrophs
   b) producers
   c) autotrophs
   d) omnivores

9. Cows and hogs, which live on grain products, are examples of:
   a) heterotrophs
   b) herbivores
   c) carnivores
   d) omnivores

10. Organisms that eat both plants and animals are:
   a) herbivores
   b) carnivores
   c) omnivores
   d) consumers

11. Which of the following may be decomposers?
   a) bacteria
   b) saprophytes
   c) scavengers
   d) all of these

12. Which of the following is true of pyramids of energy?
   a) It perfectly corresponds to the pyramid of biomass.
   b) Producers always process far more energy than primary consumers.
   c) The shape is often inverted.
   d) All of these are true.

13. The productivity of a trophic level is:
   a) the dry weight of all organisms at that level
   b) the number of organisms at that trophic level
   c) the amount of energy captured in chemical bonds of new molecules at that trophic level
   d) the amount of energy produced by the organisms at that trophic level

14. In an ecosystem, who is usually the most mobile?
   a) producers
   b) primary consumers
   c) herbivores
   d) top carnivores

15. Biogeochemical cycles occur within which of the following?
   a) individual organism
   b) whole biosphere
   c) local ecosystems
   d) none of these

Part 2: Matching

For each of the following, match the correct term with its definition or example. More than one answer may be appropriate.

consumers **producers** **decomposers**

1. ________________ Most get their energy from sunlight.
2. ________________ Usually are bacteria and fungi.
3. ________________ Obtain energy by eating other organisms.
4. ________________ These are autotrophs.
5. ________________ Most of these are animals.

water **carbon** **nitrogen**

6. ________________ Respiration is the major mechanism for returning this to the atmosphere.
7. ________________ Plant transpiration is the major mechanism for returning this to the atmosphere.
8. ________________ Accumulation is associated with the greenhouse effect.
9. ________________ This is the most abundant element in the atmosphere.
10. ________________ Bacteria living in root nodules of plants fix this element, making it available to the plants.
11. ________________ Fossil fuels are a stockpile of this element.
12. ________________ This is a waste product from the breakdown of protein, and animals excrete it in the form of ammonia, urea, or uric acid.

Part 3: Short Answer

Write your answers in the space provided.

1. What is ecology?
2. How much of the sunlight shining on the Earth actually is available for primary consumers to use?
3. What is the most abundant element in the Earth's atmosphere?
4. Explain the roles of producers, consumers, and decomposers.
5. How are food chains and food webs related?

6. Who processes the most energy in a food chain?

7. Protists are usually where in a food chain?

8. From what trophic levels do omnivores eat?

9. Explain the ten percent law.

10. How much of the atmosphere is oxygen?

11. What is the relationship between animals and plants in the oxygen cycle?

12. What role do plants play in the water cycle?

13. Fossils are associated primarily with which biogeochemical cycle?

14. Nitrogen is an important component of what organic compounds?

15. What is the major source of increasing nitrogen levels?

Part 4: Critical Thinking: Using Your Knowledge

Answer each of these in essay form, using complete sentences and paragraphs. Provide as much information as you can. (For extra essay practice, write out answers to the Review and Thought Questions in your textbook.)

1. Consider the "ecosystem" in your home. List the biotic parts and provide examples of some of the abiotic elements in your ecosystem. How do you interact with other organisms in your ecosystem? How do you interact with your physical environment (abiotic parts)? How do the other organisms affect you? How does your physical environment affect you?

2. Trace the path of a molecule of each of the following as it moves through its biogeochemical cycle: oxygen, water, carbon, and nitrogen.

3. Consumers get more energy from less food if they "Eat low on the food chain." Explain why, and explain health benefits associated with this type of eating.

4. Explain ways in which ecosystems are both like and yet unique from organisms.

5. In what ways has human activity made it more difficult for top carnivores, such as tigers, to continue to thrive?

6. How can it be that there is too much nitrogen in our biosphere and atmosphere, yet we need to add it to our crops to help them grow?

7. Explain how clear-cutting the tropical rainforests impacts each of the biogeochemical cycles discussed in this chapter.

8. During the Carboniferous Period, a very wet time, as organisms died they often dropped into water, causing excessive amounts of carbon to be taken out of the atmosphere and trapped in the muck under water. It is, in fact, those very deposits that we now burn as fossil fuels, and which are contributing to the current buildup that us causing global warming. Although few disagree that global warming is occurring, some scientists say it is merely a natural reversal of a massive depletion that happened years ago. If this is a natural correction, should we be trying to stop it? How should this decision be made, and by whom?

9. What is your opinion of Martin's and Markel's idea that "fertilizing" the ocean can end the problem of global warming? Do you think such an endeavor should be done? What are the potential risks?

10. All ecosystems in the biosphere are intertwined. Food webs interconnect. Knowing this, explain how activities on a farm in the midwestern United States can alter the ecosystem found in a large city, such as Chicago. How can activities in Chicago affect the ecosystem "down on the farm?" Then, explain how human activity in the United States can affect ecosystems in Europe, or in the southern hemisphere. Finally, although you are merely one organism, what are some specific steps you not only can but WILL take to lessen your personal impact on the biosphere?

Chapter 26
Biomes and Aquatic Communities

KEY CONCEPTS

1. Climate determines the characteristics of an ecosystem.
2. Areas with heavier rainfall support forests; those with less rainfall support grasslands.
3. Latitude and the Earth's rotation influence the intensity and duration of solar radiation, wind, rain, and ocean currents. These factors, as well as terrain, help create distinct climates.
4. Desert plants and animals are all adapted to conserve water; the animals also have adaptations for staying cool when temperatures soar.
5. Tundra has a cold, dry climate and perpetually wet soil, and has a short growing season.
6. Temperate grasslands have hot summers, cold winters, and less rain than temperate forests.
7. Chaparral communities have hot, dry summers and mild, wet winters.
8. Savannas have abundant rain and hot climates, with regular dry seasons.
9. Taiga has cold winters and short, mild summers.
10. Temperate deciduous forests typically have warm, rainy summers and cold winters.
11. Tropical rain forests are hot and wet, with little variation from season to season. They also have the greatest species diversity.
12. Ocean ecosystems are more homogeneous than land communities.
13. Estuaries are productive and species rich, whereas ocean communities are, by comparison, unproductive and species poor.
14. Freshwater ecosystems are unusually susceptible to pollution.

EXTENDED CHAPTER OUTLINE

HOW MANY BIOMES CAN YOU SEE IN A DAY?

1. Each summer, in a 26-hour race from Death Valley to Mt. Whitney, runners pass through one of the most abruptly changing series of biological communities in the world.

What is a biome?

1. In each region, ecological succession leads to a characteristic climax community, or "biome."
2. A biome is a geographical region characterized by a distinctive landscape, climate, and community of plants and animals.
3. Biomes differ in the particular species that they contain.
4. Differences in the species in a biome result from the areas' unique evolutionary histories and geographical differences.

How do sunshine and the Earth's movements determine climate?

1. The most important determinants of a biome are temperature and moisture, which depend on sunlight intensity and patterns of wind, rain, and ocean currents; all of these depend on the Earth's movements and the distribution of mountains and valleys.
2. The most critical factor is sunlight.
3. Different latitudes have different climates for two reasons:
   - latitudes closest to the poles receive less solar energy than those near the equator, because the sun's rays arrive at a more oblique angle; and
   - latitudes closest to the poles experience more extreme seasons than those near the equator because the Earth's axis is tilted.
4. Differences in air temperature cause air to move in currents that transfer heat from the equator and toward the poles.
5. Air currents follow three rules:
   - hot air rises and cold air falls;
   - hot air holds more moisture than cold air; and
   - Earth's rotation twists the moving air.
6. Not much wind blows at the equator.
7. The movement of air from the tropics toward the equator creates the tradewinds—winds that blow from about 30° latitude steadily toward the equator.
8. Most surplus moisture falls as rain in the tropics—regions between the equator and 30° north and south latitudes—so the climate of the tropics is warm and wet.
9. Most of the moisture is gone by the time the equatorial air reaches 30° latitude, which has most of the world's great deserts.
10. Earth's rotation twists these air currents so that the resulting winds—westerlies—come from the west. Westerlies are characteristic of latitudes between 30° and 60° in both hemispheres.
11. As the air moves over the Earth's surface, it warms and picks up moisture again.
12. At about 60° latitude, the air again rises, cools, and rains or snows. Coniferous forests are found at this cool, wet latitude.
13. Near the poles, the air—again dry and cold—descends again and polar easterlies begin.
14. The effect of rotation on movement is the Coriolis effect, which accounts for the clockwise twist of the Northern Hemisphere wind current and the counterclockwise twist to Southern Hemisphere wind currents.
15. The Coriolis effect also shapes the ocean currents, so water also moves in great circles.
16. Ocean currents also depend on the placement of the continents.

How does topography influence climate?

1. Mountain ranges also influence the climate of the continents.
2. A rain shadow is the area adjacent to a mountain range, away from the prevailing winds, where little rain falls.
3. As we climb a mountain, the climate changes as it does when we travel from low latitudes to higher ones.
4. The tree line is the upper limit where subalpine trees grow.

HOW DO BIOMES DIFFER?

Desert ecosystems are surprisingly delicate

1. Deserts are relatively barren regions with lower productivity than the other biomes.
2. Deserts may be hot or cold, but all are dry, receiving less than 25 cm of rain per year.
3. The dominant plants of a desert depend heavily on its temperature.
4. Many desert plants have special adaptations to conserve water. For example, the small leaves of sagebrush minimize evaporation, while succulents (including the cacti) store water in fleshy tissue for future use.
5. Desert animals must tolerate dryness and hot summers—many are active only during the cool hours of dawn and dusk.
6. Many rodents estivate—retiring to underground burrows and entering a sleeplike state called a torpor—during the hottest and driest months of the summer.
7. A significant threat to American deserts is the off-road recreational vehicle, whose wheels destroy fragile desert plants and soils, as well as lizards, snakes, and tortoises.
8. Perhaps the biggest threat to desert-adapted organisms is conversion of desert into suburban developments and farms.
9. Unlike rainwater, irrigation water is loaded with dissolved minerals and salts (it comes from rivers and wells), and as it evaporates from hot soil, the minerals and salts are left behind as a deposit, poisoning the soil.

How can a dry biome be so wet?

1. The tundra—a vast, open land dominated by grasses and low shrubs—covers more than one-fifth of the Earth's land surface.
2. The climate is cold and dry, but the cool summers usually provide a growing season of one or two months.
3. Permafrost—permanently frozen ground less than a meter from the surface—underlies the soil during even the warmest summers.
4. Precipitation is low—less than 25 cm per year. Despite the low rainfall, the soil is wet, both because the permafrost prevents drainage and because the cold air curbs evaporation.
5. Bogs, ponds, and lakes punctuate the landscape.
6. Mammals in the tundra include caribou, polar bears, weasels, foxes, and lemmings.
7. In the summer, large numbers of birds arrive from southern latitudes to nest and raise young, and swarms of flies and mosquitoes abound.
8. Humans have had only limited impact on the tundra, but in the past 30 years that is changing with oil development, pipelines, and roads moving in.

Why are there no trees in grasslands?

1. Like temperate forests, temperate grasslands have well-defined seasons, with hot summers and cold winters.
2. The low annual rainfall, usually 25 to 75 cm per year, keeps grasslands from turning to deserts but cannot sustain trees.
3. Grasslands usually have a dry period, during which fires are common.
4. Grasslands extend over much of the Earth's surface, mostly in the interiors of the continents.

5. The types of grasses vary with the amount of rain and with the season.
6. Grasslands can support large numbers of mammals.
7. In North America today, humans have virtually eliminated the temperate grassland biome, replacing the native grasses with corn and wheat fields, and the native mammals with pigs and cows. Humans are now the top carnivores.

Chaparral always burns eventually

1. Chaparral predominates in five temperate regions: California, central Chile, the shores of the Mediterranean Sea, southwestern Africa, and southwestern Australia.
2. Chaparral is characterized by dense shrubs and scattered, broadleafed trees.
3. Each of the five chaparral regions has its own species, but all chaparral species share common adaptations to summer drought.
4. Wildfires are essential to maintenance of a chaparral—they kill trees that would shade out the chaparral and enable some seeds to sprout.
5. One of the greatest dangers to chaparral communities is overzealous fire prevention by humans. Inevitably a fire so big and hot that firefighters cannot put it out sweeps thousands of acres, incinerating the roots of the chaparral vegetation and destroying hundreds of suburban houses.

Savannas host some of the most spectacular grazing animals

1. Savannas are grasslands punctuated by solitary, small clumps of trees.
2. Like temperate grasslands, savannas have a regular dry season and periodic wildfires—conditions that limit the number of trees.
3. Savannas generally have more rain (usually 90 to 150 cm per year) and hotter climates than temperate grasslands.
4. Savanna grasses are tall and the trees tend to be short.
5. Animal life in the savannas varies greatly from continent to continent.
6. Hunters have driven many savanna species to the brink of extinction.
7. The major threat to the African savanna is the continuing conversion of savanna for agriculture.
8. The end of the savanna biome seems inevitable.

Taiga is the spruce-moose biome

1. Taiga refers to the broad band of sparse coniferous forest than extends across Canada, Alaska, Scandinavia, and Siberia.
2. In the taiga, winters are bitterly cold and summers are short.
3. The trees grow only 5 to 10 meters high, and the forests have only one or two species (often spruce or pine) dominating. Shrubs are relatively rare.
4. Moose, mice, squirrels, porcupines, and snowshoe hares are the most common herbivores in the taiga. Predators include wolves, grizzly bears, lynxes, and wolverines. Reptiles cannot survive the cold, birds migrate here in the summer only, and insects abound.
5. Because the taiga is so inhospitable, human populations are rare and have had relatively little impact, but logging has eliminated many of the original forests. This has also led to replacement of moose by deer, and a parasite carried by the deer is killing off the moose.
6. One of the greatest threats to the taiga comes from acid rain. Industry electrical power plants and automobiles produce oxides of sulfur and nitrogen that rise and travel hundreds or thousands of miles downwind. Eventually these oxides dissolve in rainwater or snow, making the precipitation acidic.

Why do the temperate deciduous forests lack top predators?

1. Temperate deciduous forests receive 80 to 140 cm of precipitation each year.
2. The particular type of forest vegetation depends on both temperature and rainfall.
3. Deciduous trees, which shed their leaves each autumn, characterize the temperate forest.
4. The plants of a temperate forest occupy four vertical layers:
   - trees,
   - shrubs,
   - herbs (nonwoody plants), and
   - ground cover.
5. Animal life is abundant in all the plant layers and in the soil beneath the litter.
6. Humans have cut down much of the forest for lumber and firewood, and to make way for farms, towns, and cities. Most of the new forests are small, isolated woodlots, and are relatively young and homogeneous.
7. This breakup of the forests eliminated the enormous ranges needed to support carnivores at the top of the food chain. As a result, destruction of the temperate deciduous forests, along with intensive hunting, has eliminated nearly all top carnivores in eastern American forests.

Tropical rain forests are the most productive and diverse biomes in the world

1. Most tropical rain forests are within 10° of the equator.
2. The weather in the tropical rain forests is more or less constant throughout the year, with average temperatures of about 27° C (80° F) all year.
3. Rainfall is heavy, ranging from 2 to 4.5 meters per year.
4. The limiting factor is usually light.
5. The forest floor is usually dim, with relatively little vegetation.
6. Soils of rain forests are surprisingly poor in nutrients.
7. The tropical rain forests are the second most productive communities in the world, second only to estuaries and marshes.
8. Tropical rain forests are first in biodiversity—at least half, and perhaps as many as 90% of all the Earth's species live in this biome.
9. More than two-thirds of the plants in a tropical forest are trees, most of which are flowering, broadleaved evergreens. Other tropical forest plants include:
   - lianas—vines that root in soil but climb into the canopy, and
   - epiphytes—plants, such as orchids, that grow entirely on other plants.
10. The vegetation forms discrete layers:
    - Most trees form a continuous layer—the canopy—30 to 40 meters above the ground.
    - Above the canopy are occasional tall, isolated emergents—trees with umbrella-shaped crowns growing to a height of 50 meters or more.
    - A third layer of shorter trees catch what light they can from gaps in the canopy.
    - Finally, a short shrub layer consists of dwarf palms and giant herbs with especially large leaves adapted to capture faint light that filters through upper stories.

How are tropical rainforests being destroyed?

1. Each year, estimates indicate that humans are logging or burning an area of tropical rain forest equal to the size of the state of Washington, but researchers believe this to be an underestimate. Some 1999 field surveys show the actual rate of destruction may be twice the estimates.
2. Rain forest destruction is an accelerating process.
3. Tropical nations export timber and plywood from rain forest trees, metals mined from rain forest lands, and beef cattle raised on clear-cut rain forest.

4. Once cleared, former rain forest land does not serve farmers or ranchers well—its soil, never rich to begin with, deteriorates rapidly. This leads to "slash and burn" farming. (The burning also contributes to carbon dioxide accumulation and global warming.)
5. The average cattle ranch in Central America lasts only six to ten years, and each hamburger made from these cattle costs five square meters of tropical rain forest, which includes about a half ton of diverse plants and animals.
6. Human activity threatens to eliminate the rain forest biome by the year 2030. This will include extinction of millions of plants and animals found nowhere else—one-fourth of the world's 250,000 plant species and one-fifth of the world's 20,000 named butterfly species. Most of the species that will be lost have not even been named.

WHAT ARE AQUATIC ECOSYSTEMS LIKE?

1. Most of Earth's surface is covered with water—about 70% by saltwater seas and oceans, and about 2% by freshwater lakes, ponds, rivers, and streams.
2. Aquatic communities in different geographical regions have plants and animals that resemble one another yet fall into just a few distinct types.
3. Ocean communities differ according to temperature, depth, and distance from shore.

How do oceanographers divide the ocean?

1. The oceans cover over two-thirds of the Earth's surface but contain only about 10% of the planet's species.
2. All the oceans are continuous with one another; consequently, the oceans provide only limited opportunity for adaptive radiation.
3. Oceanography is the study of seas and their ecosystems.
4. Biologists distinguish ocean communities according to ocean depth and distance from shore. The benthic division consists of all organisms that live on the ocean bottom, including those in the intertidal zone (between high tide and low tide) and those at the bottoms of deep canyons in the abyssal zone.
5. The pelagic division consists of organisms that live in open water, above the bottom. Microorganisms called plankton drift in the water, moving only where currents take them.
6. The pelagic division is further divided into:
   - the neritic zone—shallow coastal waters farther from shore, over the continental shelf, where algae are the primary producers, and
   - the oceanic zone—most of the world's oceans, located out beyond the continental shelf and over the deepest water, where photosynthetic cyanobacteria, diatoms, and dinoflagellates are the primary producers.
7. Off the coast of Peru, the Humboldt Current draws nutrients up from the bottom in a process called upwelling.
8. Upwelling creates a rich soup that supports a rate of photosynthesis six times that found in other parts of the open ocean.
9. A marine food chain typically contains four or five trophic levels. (In contrast, terrestrial food chains seldom have more than three trophic levels.)
10. In winter, the Humboldt Current reverses itself, preventing the nutrient rich cold waters from rising to the surface. Usually these changes, called "El Niño," are benign, but about every 25 years the current reversal lasts long enough to produce catastrophic effects.

Doesn't El Niño do more than kill fish and birds?

1. In the winter of 1997-1998 an area in the Pacific Ocean the size of the continental United States warmed by 3.5° C. This led to tremendous rainfall along the Pacific Coast from Canada south to Ecuador and Peru, while eastward, a severe drought occurred in Indonesia and Malaysia.
2. La Niña is a weather pattern in which areas of the Pacific Ocean turn cold, causing warm thunderstorms to return to Indonesia and bringing warmer winter weather to the southeastern United States and wetter, colder weather to the Pacific Northwest.

Why are estuaries and marshes valuable?

1. Estuaries are partly enclosed waters where freshwater streams or rivers meet the ocean.
2. Estuary water is less salty than the oceans, and its salinity is constantly changing.
3. Estuaries are among the most productive of all natural communities.
4. Estuaries provide breeding grounds for fish and shellfish, so destruction of estuaries inevitable leads to destruction of fisheries.
5. Primary producers in estuaries include plankton, attached algae, and large plants.
6. Because of the high productivity, estuaries are breeding and nursery grounds for many species of fish and shellfish.
7. Estuaries are especially vulnerable to human destructiveness—they are the first to receive pollution from streams and rivers, and they are prime targets for oceanfront development.
8. In the United States, state and federal governments are now struggling to clean up and restore estuaries.

Why are lakes, ponds, rivers, and streams susceptible to pollution?

1. Limnology is the study of freshwater lakes, ponds, rivers, and streams.
2. Many rivers and streams flow into lakes, bringing water and dissolved nutrients, so the character of lakes and ponds varies from region to region.
3. Primary production in a lake depends on the penetration of light. Both microorganisms and water plants use photosynthesis to capture light energy.
4. Most organic matter falls to the bottom as detritus, which is consumed by bacteria and invertebrates.
5. In the summer in temperate regions, warm air heats a layer of warm water as much as 20 meters thick. As winter approaches, the temperature of this layer falls until it is the same as the deeper layer. At that point, the waters mix—an event called the fall turnover.
6. A similar turnover occurs in the spring as the winter ice melts.
7. The two turnovers bring dissolved nutrients from the bottom sediments into the upper water, where photosynthesis occurs.
8. Fall and spring turnovers move oxygen from the surface down to bottom-dwelling heterotrophs.
9. A eutrophic lake is one with abundant minerals and organic matter. Organisms die and drop to the bottom, stimulating the lake-bottom decomposers to multiply. During the summer, decomposers deplete the deeper water's oxygen so the lake bottom can no longer support fish and other animals. As a result, a eutrophic lake's diversity is low.
10. In contrast, oligotrophic lakes have a limited supply of nutrients. These clear lakes have no permanent algal blooms so they contain more oxygen and support a more diverse community of organisms, including fish.
11. Eutrophication happens naturally over thousands of years; humans often accelerate the process by dumping nutrient-rich sewage and other wastes.
12. As nutrients and minerals accumulate, a lake becomes eutrophic. Productivity increases at first but diversity declines, and eventually the lake dies.

VOCABULARY BUILDING

In your own words, first write a brief definition, then a full explanation for each of the following terms. Include examples where appropriate. Complete this section from your memory—you will not learn the terms by simply copying definitions from the textbook. Once you have finished, check your responses against the information in the chapter and make any necessary corrections.

biome —

tradewinds —

westerlies —

polar fronts —

Coriolis effect —

rain shadow —

desert —

estivate —

temperate grassland —

chaparral —

savanna —

taiga —

coniferous forest —

temperate deciduous forest —

tundra —

permafrost —

tropical rain forests —

lianas —

epiphytes —

canopy —

emergents —

oceanography —

ocean ecosystems —

benthic division —

intertidal zone —

abyssal zone —

pelagic division —

plankton —

neritic zone —

oceanic zone —

upwelling —

estuaries —

limnology —

fall turnover —

spring turnover —

eutrophic —

oligotrophic —

CHAPTER TEST

The following test has four parts. Complete as much of the exam as you can from memory. If you cannot answer a question, skip it. Once you complete all that you can, try to answer any questions you skipped. If you still cannot answer them, consult your textbook for the answers. Once you have completed all sections of the test, check your answers for Parts 1 - 3 against those in the back of this book. Highlight any incorrect answers then review that material in your textbook. Correct your answers for future reference.

Part 1: Multiple Choice

For each of the following, select all correct responses—more than one may be correct.

1. The same biome in different areas (such as North America vs. Australia) may contain quite different species because of:
   a) differences in geography
   b) unique evolutionary histories
   c) drastically different climates
   d) none of these

2. Which of the following is not true about air currents?
   a) Hot air rises.
   b) Cold air is drier than hot air.
   c) Earth's rotation twists the moving air.
   d) All of these are true about air currents.

3. The Coriolis effect is responsible for which of the following?
   a) clockwise twist of Northern Hemisphere wind currents
   b) clockwise twist of Southern Hemisphere wind currents
   c) dryness of deserts
   d) coldness of polar fronts

4. Which of the following is not a plant layer in the temperate deciduous forest biome?
   a) trees
   b) ground cover
   c) herbs
   d) all of these are part of this biome

5. Which of the following prevents trees from becoming prominent in temperate grasslands?
   a) wildfires
   b) grazing
   c) lack of rain
   d) all of these

6. Which biome is always found on a western coast and has a "Mediterranean" climate?
   a) tundra
   b) taiga
   c) chaparral
   d) savanna

7. In tropical rain forests, the limiting abiotic factor is usually:
   a) rain
   b) temperature
   c) light
   d) carbon dioxide

8. The biome with the greatest species diversity is the:
   a) temperate deciduous forest
   b) savanna
   c) tropical rain forest
   d) chaparral

9. In the tropical rain forest, the tallest plants are:
   a) emergents
   b) those in the canopy
   c) shrubs
   d) lianas

10. Permafrost is associated with which biome?
   a) taiga
   b) tundra
   c) temperate grassland
   d) none of these

11. About how much of the Earth's surface is covered with saltwater?
    a) 2%
    b) 25%
    c) 55%
    d) 70%

12. Ocean ecosystems are distinguished by which of the following?
    a) climate
    b) depth
    c) distance from shore
    d) dominant plant life

13. All organisms that live on the ocean bottom are part of the:
    a) benthic division
    b) pelagic division
    c) neritic zone
    d) oceanic zone

14. In temperate lakes, what events bring dissolved nutrients from the bottom sediments into the upper waters?
    a) estivation
    b) turnovers
    c) El Niño
    d) eutrophication

15. Which of the following is true about a eutrophic lake?
    a) Nutrients are limited.
    b) They have little diversity.
    c) Algae are common on the surface.
    d) All of these are true.

Part 2: Matching

For each of the following, match the correct term with its definition or example. More than one answer may be appropriate.

temperate deciduous forest **tropical rain forest** **tundra** **savanna**

1. ____________________ This biome has the greatest species diversity.
2. ____________________ This biome has permafrost.
3. ____________________ Loss of this has eliminated most top carnivores.
4. ____________________ This is a tropical or subtropical grassland.
5. ____________________ These are the most productive communities.
6. ____________________ The climate here is cold and dry.

eutrophic lake	oligotrophic lake
7. ____________________	This lake has an abundance of organic matter.
8. ____________________	Diversity here is low.
9. ____________________	Surfaces of these tend to be clogged with algae.
10. ____________________	These lakes tend to be clear.
11. ____________________	These lakes contain more oxygen.
12. ____________________	These lakes have high productivity.

Part 3: Short Answer

Write your answers in the space provided or on a separate piece of paper.

1. Biomes are determined primarily by what?

2. List three rules that govern air movement.

3. Most of the great deserts are located at what latitude?

4. How do elevation and latitude relate to each other?

5. How do periodic fires benefit the chaparral? Why are big fires detrimental?

6. List two examples of adaptations that allow desert plants to conserve water.

7. The African plains are an example of what biome?

8. Why are the soils of tropical rain forests nutrient-poor?

9. Describe two unusual plant types unique to the tropical rain forest.

10. What is the dominant plant life found in the taiga?

11. List two reasons for the wet soil that is characteristic of the tundra.

12. What are two reasons that oceans resist temperature changes?

13. Distinguish between the two main divisions of ocean communities.

14. Why is the primary production of the oceanic zone low?

15. What accounts for the estuaries' high productivity?

***Part 4:* Critical Thinking: Using Your Knowledge**

Answer each of these in essay form, using complete sentences and paragraphs. Provide as much information as you can. (For extra essay practice, write out answers to the Review and Thought Questions in your textbook.)

1. Make a table that includes the following information for each of the biomes discussed in your textbook:
   - the features of the climate (temperature, rainfall, etc.);
   - characteristic plant life and any common adaptations;
   - characteristic animal life and any common adaptations;
   - any special features of the biome; and
   - specific examples of that biome.

2. For each of the biomes and aquatic ecosystems discussed in your textbook, discuss ways in which humans have had a negative impact. For each, do you think the situation will get worse or improve?

3. In the United States, there is a constant tug-of-war between those interested in preserving the environment and the natural biomes and aquatic ecosystems, and those in business and industry seeking new areas for development. If you had to decide the balance between these two sides, what issues would you need to consider? Select any one biome or aquatic ecosystem that you discussed in question #2, and present both sides of the argument of whether to conserve or develop an area. How would you resolve the conflict?

4. In several of the biomes discussed, top carnivores are disappearing. What are some of the specific reasons for this loss? What changes do you predict in that biome as a result of the loss of the top carnivores?

5. In North America, what was once temperate grassland is now most often wheat field. Wheat is a grassy plant, so why does mass planting of wheat not reproduce most of the features of the original grassland ecosystem?

6. The tundra has not been encroached upon very much until recent years, as oil drilling and pipelines have moved in. The oil companies maintain that they can do their business in these areas with no harm done to the ecosystem. What features of the tundra make it unlikely that this is possible?

7. A lot of tourists are interested in seeking out nature in its purest form, giving rise to a new industry called *ecotourism*, in which people are taken into uninhabited wilderness areas to admire nature in its purest state. The goal of many organizations providing these trips is to educate the public about the importance of conserving our natural habitats. In what ways might these trips, carefully regulated to minimize their impact on the environment, jeopardize the ecosystems which are being visited?

8. Each year we hear about brush fires in California raging out of control. Clearly, the homeowners would prefer not to have the fires, but in what ways are the fires essential for the survival of the ecosystem? How has human activity altered the beneficial nature of the fires? What steps can be taken to reestablish the natural balance?

9. As an increasing amount of our biomes are converted for human use, wildlife sanctuaries are being established to try to preserve native species. Although this approach may help preserve specimens, what are the limitations and why do these areas rarely exhibit the diversity and stability of the original ecosystems?

10. The tropical rain forests are disappearing at an alarming rate. Why are so many biologists disturbed by the destruction of this particular biome? Explain the reasons for this loss. The native people in the region have to earn a living, yet the way they are doing it is destroying their world. What remedies for this situation would you suggest?

Chapter 27
Communities: How Do Species Interact?

KEY CONCEPTS

1. Ecologists view the groups of species that compose communities as being both integrated and individualistic.
2. Consumers' adaptations for finding and eating food are very similar; the consumed have a varied repertoire of adaptational weapons to use against being consumed.
3. Species in a community interact as competitors, as predators (or parasites) and prey, and as symbionts.
4. Interactions between two species influence the evolution of both.
5. No two species can indefinitely occupy the exact same niche in the same habitat.
6. Organisms in natural environments compete for limited resources, and competition, through time, causes character displacement through natural selection.
7. Species diversity, a characteristic of biological communities, is influenced by many abiotic and biotic factors.
8. Over time, communities undergo predictable changes in a process called succession.
9. Succession in plant and animal communities usually implies a progression through a series of transitional stages to a climax community.
10. Succession occurs because pioneer and successional species improve environmental conditions for subsequent species.
11. The character of a climax community depends on climate, terrain, and history.
12. Factors that contribute to an ecosystem's resiliency include patchiness of the physical environment, primary productivity, and resistance to erosion.

EXTENDED CHAPTER OUTLINE

VOLCANO: AN ECOSYSTEM IS DESTROYED

1. On May 18, 1980, a midsized earthquake shook Mount St. Helens in Washington. Explosions from the volcanic eruption tore through 2.5 cubic kilometers of sliding debris, spewing rocks, ash, gas, and steam across adjacent valleys at nearly the speed of sound.
2. Although it was difficult to imagine that anything could survive, a vast amount of life did.
3. The Mount St. Helens ecosystem—almost completely destroyed—will rebuild itself.

ARE COMMUNITIES INTEGRATED OR INDIVIDUALISTIC?

1. Succession is the predictable process of change in the numbers and kinds of organisms in a community over time.
2. Frederic Clements argued that in every geographic area, defined by its climate and soil, succession tends toward a single, predictable community of plants—the climax community.
3. Succession is a developmental process.
4. Clements defined an integrated community as one that consists of characteristic assemblages of species that interact with each other in predictable ways.
5. Henry Gleason argued that communities are individualistic—composed of discrete populations that merely happen to occupy the same habitat.
6. Gleason and others argued that every species has an independent distribution and that every community is unique.
7. With the individualistic view, communities are not necessarily discrete entities with clear boundaries. Rather, species occupying the same community merely happen to share the same abiotic requirements.
8. The integrated and individualistic views of communities are not mutually exclusive ideas, but the extreme ends of a continuum.
9. Contemporary ecologists validate both ideas: a community is both an integrated set of interacting species and a group of organisms together because their ranges happen to overlap.

HOW DO SPECIES IN A COMMUNITY INTERACT?

1. The general attributes of a biome depend on abiotic factors such as temperature, moisture, sunlight, and nutrients.
2. Abiotic factors alone do not define a community—within each community, organisms interact with one another in many ways.
3. Species interactions include interactions between:
   - consumers and the consumed (predator-prey, herbivore-plant, paraite-host);
   - competitors; and
   - species that associate with each other.

Who eats whom?

1. Consumers come in many forms; the main ones (besides decomposers and scavengers) are:
   - herbivores, who consume plants;
   - parasites, who consume only part of the blood or tissues of the host and do not necessarily kill it; and
   - predators, who usually, but not always, kill their prey and eat most of it.
2. Ecologists rarely stress the parasite-predator distinction in herbivores, but it does apply.

How do plants and animals defend themselves?

1. Consumers typically have very similar adaptations for finding and eating their prey:
   - predators—swiftness, intelligence, acute senses, and sharp teeth; and
   - herbivores—patience and a good digestive system.

How do plants and animals defend themselves?

1. The consumed have an arsenal of different weapons for defending themselves against consumption. These include:
   - Camouflage—a species blends into the background, invisible to the predator.
   - Chemical defense—some plant species contain secondary plant chemicals—chemicals that are not essential to the plant's metabolism, but serve a defensive purpose; chemical defenses are found in both plants and animals.
   - Warning coloration—bright colors help the predator remember which species to avoid; these make chemical deterrents more effective (for example, the bright color of a bee reminds us of a past painful sting).
   - Batesian mimicry—harmless organisms often mimic the warning coloration of harmful species then, through deception, they avoid being consumed, too (for example, viceroy butterflies colored like bad-tasting monarch butterflies).
   - Müllerian mimicry—two or more equally dangerous species evolve similar colors or forms to represent similar dangers to their common predators; this helps predators learn the lesson fast—a single lesson protects it from more than one species.

Symbiosis: How do organisms live together?

1. In symbiosis, a species lives in an intimate association with another species.
2. Symbiosis may be cooperative or antagonistic.
3. In parasitism, one species benefits at another's expense.
4. Mutualism is a symbiotic relationship between two species that mutually benefits each.
5. Commensalism is an intimate relationship between two species that benefits one species (the commensals) but neither helps nor harms the second species.

Do organisms coevolve?

1. The mutualistic relationship demands special adaptations in each species.
2. The interdependent evolution of two or more species whose adaptations appear to be selected by mutual ecological interactions is called coevolution.
3. Demonstrating that coevolution has occurred in individual cases is difficult.
4. All species that have ecological interactions may exert some selective pressure on one another, and virtually every species in a community interacts (however indirectly) with every other species.
5. Coevolved relationships can be very specific or very diffuse. Diffuse coevolution is further support for the idea that all species in a community affect one another.

How does competition affect organisms?

1. Some species are competitors—they use a resource in a way that limits the availability of that resource to others.
2. The competitive exclusion principle says that when two species compete directly for exactly the same limiting resources (those in short supply), the more efficient species will eliminate the other species.
3. Although one species will win, the same species may not win under all conditions.
4. The place in which an organism lives, along with the environmental conditions that characterize that place, is the organism's habitat.

5. The way an organism uses its environment is called its niche.
6. No two species can simultaneously exploit the same resources in the same place.
7. No two species can occupy the exact same niche in the same habitat indefinitely.

How do competing species coexist?
1. In resource partitioning, species divide the niche.
2. A species' fundamental niche is its potential ability to utilize resources; the resources that the species actually uses in a particular community are its realized niche is.
3. Competition does occur in natural populations and can even help determine species distributions.

Character displacement: Does competition influence evolution?
1. Any change in morphology, life history, or behavior that results from competition is called character displacement.
2. Character displacement can lead to speciation, as in the Galápagos finches.

WHAT DETERMINES BIODIVERSITY?
1. Both resource partitioning and character displacement increase the number of species within an ecosystem.
2. The number of species in an ecosystem is called species richness.
3. Relative abundance is a measure of diversity that takes into account how common or rare each species is.
4. Biotic interactions, such as character displacement, may also increase the number of species in a community.
5. Predators can maintain diversity by preventing one species from successfully competing with another.
6. Predators that promote diversity are called keystone species.
7. Species diversity depends heavily on abiotic components:
   - Diversity generally increases from the poles to the equator, and from higher to lower elevations.
   - Physically varied areas allow more species to have unique niches than flat, relatively featureless landscapes.
   - The larger the size of an ecosystem, the more species it is likely to contain.
8. According to the MacArthur-Wilson theory of island biogeography, a new island, devoid of already established species, should have a high rate of immigration and a low rate of extinction, since there are fewer species to disappear. As more species arrive, competition increases and the rate of extinction also increases as fewer species can become established.
9. The theory of island biogeography also predicts that islands closer to the mainland will have more immigration and, therefore, more species than more distant lands.
10. Finally, this theory formalized the earlier belief that the number of species that can inhabit a habitat is proportional to the area of the habitat.
11. Habitat islands are small habitats surrounded, and thus isolated, by different habitats. Many habitat islands are too small to support sufficient populations for breeding or predation.
12. A metapopulation is a group of local populations that alternately go extinct and recolonize vacant habitats.

COMMUNITIES CHANGE OVER TIME
1. The characteristics of a community depend on the geography and form of its habitat as well as on the history of the community.

Why do communities change?

1. Succession is the change in the numbers and kinds of organisms in a community over time.
2. Primary succession is the invasion of a completely new environment; secondary succession is the sequence of stages in a community that has suffered serious damage.
3. A pioneer community is the first in a succession; a climax community is long-lived and occurs at the end of a succession. Intermediate stages are called successional communities.
4. The changes that occur during succession are cumulative and directional.
5. Although climax communities change over time, they do so at an extremely slow rate.
6. Pioneer communities consist of species that:
   - flourish in disturbed areas,
   - breed rapidly, and
   - are short-lived.
7. Climax communities are characterized by species that:
   - breed slowly,
   - gradually take over in undisturbed areas, and
   - are long-lived.

Why does succession occur?

1. Species present at early successional stages may facilitate succession, tolerate succession, or inhibit succession.
2. Pioneer organisms inevitably facilitate succession because no other organism can colonize a pile of sand or a piece of bare rock until the path has been established.
3. Pioneer or successional species change their environments in ways that allow colonization by succeeding species.

What determines which community is the climax community?

1. The climax community is the one that persists the longest without changing.
2. The character of a climax community depends on the climate, terrain, and history.

Why do communities stay the same for long periods?

1. The properties of an ecosystem are more stable than the individual organisms that compose the ecosystem.
2. Stability can be defined as simply lack of change, in which case the climax community would be considered most stable because it changes the least over time.
3. Alternatively, stability can be defined as the speed with which an ecosystem returns to a particular form following a major disturbance—this kind of stability is called resiliency.
4. With this latter definition of stability, the climax community would be most fragile and least stable because they can take hundreds of years to reestablish.
5. In general, diversity, by itself, does not ensure stability.

Can ecosystems bounce back?

1. Ecologists are especially interested to know what factors contribute to the resilience of communities because climax communities all over the world are being severely damaged or destroyed by human activities.
2. An environment that varies from place to place supports more kinds of organisms than one that is more uniform, because a local population that goes extinct is quickly replaced by immigrants from an adjacent community.
3. Factors that contribute to an ecosystem's resilience include resistance to erosion and increased primary productivity.
4. Partly damaged ecosystems often support greater plant growth than mature ones because younger trees grow more vigorously than trees in a mature climax forest.

VOCABULARY BUILDING

In your own words, first write a brief definition, then a full explanation for each of the following terms. Include examples where appropriate. Complete this section from your memory—you will not learn the terms by simply copying definitions from the textbook. Once you have finished, check your responses against the information in the chapter and make any necessary corrections.

succession —

climax community —

integrated —

individualistic —

parasite —

predator —

camouflage —

secondary plant compounds —

warning coloration —

mimicry —

Batesian mimicry —

Müllerian mimicry —

symbiosis —

parasitism —

mutualism —

commensalism —

commensals —

coevolution —

competition —

competitive exclusion principle —

limiting resource —

habitat —

niche —

resource partitioning —

fundamental niche —

realized niche —

character displacement —

species richness —

relative abundance —

keystone species —

theory of island biogeography —

habitat islands —

metapopulation —

succession —

primary succession —

secondary succession —

pioneer community —

climax community —

successional communities —

stable —

resilience —

CHAPTER TEST

The following test has four parts. Complete as much of the exam as you can from memory. If you cannot answer a question, skip it. Once you complete all that you can, try to answer any questions you skipped. If you still cannot answer them, consult your textbook for the answers. Once you have completed all sections of the test, check your answers for Parts 1 - 3 against those in the back of this book. Highlight any incorrect answers then review that material in your textbook. Correct your answers for future reference.

Part 1: Multiple Choice

For each of the following, select all correct responses—more than one may be correct.

1. By definition, a parasite:
   a) kills its host
   b) consumes its host's whole carcass
   c) consumes only part of its host
   d) is larger than a predator
2. Which of the following consumers is often smaller than the organism it consumes?
   a) predator
   b) herbivore
   c) parasite
   d) none of these
3. Which of the following is not likely an adaptation found in predators?
   a) warning coloration
   b) swiftness
   c) intelligence
   d) acute senses
4. Which of the following works as a deterrent primarily through deception?
   a) warning coloration
   b) secondary plant chemicals
   c) Batesian mimicry
   d) Müllerian mimicry
5. Which of the following employs Müllerian mimicry?
   a) a walking stick, an insect that looks like a twig
   b) the brightly colored and poisonous coral snake
   c) the palatable viceroy butterfly that resembles the toxic monarch butterfly
   d) wasps and bees
6. In which type of symbiosis do two organisms associate in a way that one benefits and the other neither benefits nor is harmed?
   a) mutualism
   b) parasitism
   c) commensalism
   d) none of these
7. Bees collect nectar from flowers for food and, as they fly to other flowers, they pollinate the plants. This is an example of:
   a) mutualism
   b) parasitism
   c) commensalism
   d) none of these

8. The competitive exclusion principle states that:
   a) Two species cannot utilize the same resources.
   b) Two species with different abiotic needs cannot occupy the same location.
   c) When two species compete directly for the same resource, the more efficient species eliminates the other.
   d) When two species share the same location, they start using the same resources, and differences in their abiotic needs disappear.

9. When species occupy the same habitat and compete for the same resource, the habitat is often split into smaller sections in which individual species stay, reducing direct competition. This is called:
   a) mutualism
   b) symbiosis
   c) habitat division
   d) resource partitioning

10. The resources that a species actually uses in a particular community are referred to as its:
    a) habitat
    b) fundamental niche
    c) realized niche
    d) home

11. Character displacement, which led to speciation in the Galápagos finches, refers to:
    a) changes in structure, behavior, or life history due to competition
    b) individuals from one species moving to different locations to establish identical communities
    c) individuals from one species moving to different locations to undergo speciation
    d) forcing individuals into small areas within their realized niche due to competition

12. Which of the following can increase the number of species within an ecosystem?
    a) character displacement
    b) resource partitioning
    c) increased direct competition
    d) all of these

13. Which of the following areas would likely demonstrate the greatest species diversity?
    a) small ecosystems
    b) flat landscapes
    c) areas near the equator
    d) low elevations

14. Which of the following would undergo primary succession?
    a) a forest after a fire
    b) a deserted cornfield
    c) rock exposed when a glacier recedes
    d) all of these

15. Which of the following would be typical of pioneer species?
    a) short life
    b) slow breeders
    c) seeds do not disperse easily
    d) small size

Part 2: Matching

For each of the following, match the correct term with its definition or example. More than one answer may be appropriate.

parasitism **mutualism** **commensalism**

1. ________________ Both species benefit.
2. ________________ One species benefits; one is harmed.
3. ________________ One species benefits; the other is not affected.
4. ________________ A type of symbiosis.
5. ________________ Requires special adaptation in both species.

pioneer community **climax community**

6. ________________ Short-lived stage.
7. ________________ Consists mostly of annuals.
8. ________________ The final stage in succession.
9. ________________ Seeds disperse easily, which is how they enter the area.
10. ________________ This is the most persistent stage.

Part 3: Short Answer

Write your answers in the space provided or on a separate piece of paper.

1. In what way are a community's species both integrated and individual?
2. List three types of interspecific interactions.
3. List five types of organisms that are consumers.
4. Differentiate between Batesian and Müllerian mimicry.
5. Are any of the symbiotic relationships harmful to both species?
6. What is coevolution?
7. Explain the competitive exclusion principle.
8. Differentiate between a habitat and a niche.

9. How does resource partitioning increase the number of species in an area?

10. Differentiate between species richness and relative abundance.

11. How can predators maintain diversity?

12. List examples (not from the book) of parasitism, mutualism, and commensalism.

13. Pioneer and successional communities change over time. Do climax communities?

14. What is resilience?

15. Why would an organism occupy a realized niche instead of a fundamental niche?

Part 4: Critical Thinking: Using Your Knowledge

Answer each of these in essay form, using complete sentences and paragraphs. Provide as much information as you can. (For extra essay practice, write out answers to the Review and Thought Questions in your textbook.)

1. Explain the process of succession. Trace the development of two ecosystems, one arising after an earthquake lifts previously unexposed rock, and the other arising from the ash after a volcanic eruption.

2. In what ways are farming and succession incompatible? How do farmers deal with this?

3. Humans control their environment more than most other species. Give examples of limiting resources that are used by humans. Are we in danger of depleting any of our resources? Are humans taking any actions to avoid doing so? Are you personally?

4. Explain the views of communities as being either integrated or individualistic, and why these two extreme positions are not considered to be mutually exclusive. Give some specific examples to illustrate each position.

5. Increased development of native habitats is leading to an increase in habitat islands. What are the potential consequences of this fragmentation of habitats?

6. Differentiate between a parasite and a predator. Both consume parts of other organisms, but in different ways. How does their approach to consumption relate to the adaptations these types of organisms have?

7. Describe and provide examples of various weapons used by prey species (the consumed) to avoid being consumed. For at least three pairs, examine the relationship between predator and prey adaptations and, using specific examples, explain how coevolution is at work.

8. Years ago, a misguided program led to the extermination of wolves and mountain lions living near the Grand Canyon. As a result, the deer in the area began to starve. Explain this link.

9. Describe the outcomes predicted by the competitive exclusion principle when two or more species compete for the same resource.

10. Many people are concerned about human impact on the ecology of the world. Yet it can be argued that human encroachment into various ecosystems is merely a sign that we are the best competitors and that we are seeing natural selection in effect—other species disappear through competitive exclusion. Is human "dominance" on the planet just a natural process? If we continue losing other species by competitive exclusion, what will happen to us, and why?

Chapter 28
Populations: Extinctions and Explosions

KEY CONCEPTS

1. Humans drive other species to extinction by destroying their habitats. One way this happens is by fragmenting the habitat into small areas with a lot of edge and little core.
2. According to the extinction vortex hypothesis, loss and degradation of habitat reduces population size and changes the population's structure. This decrease in population size leads to decreased genetic diversity, increased inbreeding, and decreased reproduction, all of which lead ultimately to extinction.
3. Small populations are more susceptible to extinction due to chance events, such as natural disasters or harsh weather.
4. We can preserve many more species if we set aside larger preserves and connect them with protected corridors.
5. All populations can grow exponentially. The larger a population, the faster it can grow.
6. The carrying capacity of the environment limits the size of a population.
7. Both density-dependent and density-independent factors limit population growth.
8. Population cycles probably depend on specific interactions between species.
9. Differences in survivorship curves lead to differences in age distributions.
10. Even if human couples began to reproduce only at the replacement level, the human population would continue to grow for another 50 years.
11. Cities depend on energy subsidies and agriculture, which also depends on energy subsidies. Countries with the highest standards of living make the greatest demands on the world's resources.
12. The green revolution increased the world food supply and, thus, also the planet's carrying capacity for humans. But today, population growth is exceeding the food supply.
13. Water and soil pollution come in the form of thermal pollution, biodegradable pollutants, nondegradable pollutants, and poisons.
14. Air pollution includes visible particles of soot and other carcinogens, invisible products of combustion and agriculture (which will raise the temperature of the Earth), and chlorofluorocarbons (which are damaging the ozone layer).

EXTENDED CHAPTER OUTLINE

FIGHTING FOR LIFE

1. In general, symmetry in organisms implies good health, and asymmetry is often an indicator of environmental stress, such as pollution.
2. Michael Soulé coined the term, and produced a model for, the extinction vortex. By this model, when a population shrinks due to hunting or habitat reduction, the number of reproducing individuals also diminishes, which leads to inbreeding and reduced reproduction, which in turn eventually lead to extinction.
3. Soulé founded the field of conservation biology.

HOW DO POPULATIONS GO EXTINCT?

1. Factors that lead to extinction can be categorized as:
   - deterministic events, such as habitat destruction or excessive hunting or trapping, or
   - stochastic events, which occur by chance.

How do deterministic factors drive species to extinction?

1. Population extinction occurs at the local level, and can occur by competition or habitat destruction.
2. Humans drive other species to extinction primarily by habitat destruction.

How does fragmentation contribute to habitat destruction?

1. Fragmentation occurs when habitats are divided into smaller areas with a lot of edge space and little central core.
2. If a population from one habitat disappears, the habitat is quickly recolonized by species from adjacent habitats.
3. Fragmentation reduces the total area of a habitat, but the number of species in a habitat is dependent on the area of the habitat.
4. Due to the edge effect, at the edges of a habitat:
   - different plants grow because of different lighting;
   - safe nesting areas may be absent; and
   - predators and other species from adjacent habitats may invade and kill residents directly or by competition.
5. Due to the edge effect, some habitat cores are too small to sustain the native species.

Draining the gene pool

1. According to the extinction vortex hypothesis, loss and degradation of habitat reduces the size of a population, which leads to inbreeding.
2. Inbreeding leads to a loss of genetic diversity and decreased birth rate (due to genetic defects).
3. This, in turn, further reduces population size.

How do stochastic factors drive species to extinction?

1. Chance (stochastic) events are far more likely to lead to extinction in populations that are already small and stressed.

How can we preserve ecosystems and the species that live in them?

1. Unless something drastic is done soon, approximately half of all the world's species will go extinct within the next decade or two.
2. Humans have converted fully one third of the Earth's surface to agriculture, and covered more with cities, suburbs, roads, and development.
3. Most of the remaining land is among the planet's least productive.

4. As of this writing, 29 of 63 tropical nations have already cut or burned nearly 80% of their forests.
5. As of 1998, 20 tropical nations had agreed to try to preserve 10% of their countries' land in parks and preserves, but many conservation biologists feel certain that, due to fragmentation and the edge effect, this is too little too late and we will still likely lose half of all species.
6. Conservation biologists also recommend corridors which are strips of protected habitat that connect the various parks and preserves and allow safe movement of species between them. This would benefit small, slow-moving organisms that have to migrate, for example, to breeding grounds, and also large predators who require large areas for hunting.
7. Michael Soulé put the solution very clearly: "Our generation is the key. Whatever we save in the next few decades is all future humanity will have to use and to appreciate. If we don't do it, it will be too late."

HOW DO POPULATIONS GROW?

1. The sizes of populations within a community fluctuate—changes in abiotic conditions or in biotic interactions may dramatically influence them.
2. Natural populations usually remain rather stable over the long term.

What is exponential growth?

1. Exponential growth is a growth pattern in which a population doubles in some constant period of time (called the doubling time).
2. Exponential growth means that the larger the population, the faster the population increases.
3. A graph of the increasing number looks like the letter J and is therefore called a J-shaped curve. At the bottom of the J-shaped curve, the population size (N) increases rather slowly, then it accelerates. When there are no limits to population increase, N increases faster.
4. The growth rate (r) is the difference between the average birth rate and the average death rate.
5. When a population's birth rate is at its maximum and the death rate is at its minimum, r is as large as it can be. This is called r_{max}, or the intrinsic rate of natural increase.
6. The change on the total population size per unit of time is G, and is determined by this equation: $G = rN$
7. The larger a population, the faster it grows.

What limits the size of a population?

1. Natural populations never achieve their intrinsic rate of natural increase.
2. The whole growth curve consists of three parts—an initial "J" phase in which the growth rate is accelerating; then a deceleration phase; and finally a gradual leveling off. The whole curve now looks more like a tilted and stretched "S".
3. The larger the population size (N), the faster growth slows down.
4. The carrying capacity (K) is also used to refer to the actual number of organisms (population density) that a habitat can support indefinitely.
5. The logistic growth equation is $G = r N (K - N)/K$ and it merely describes the restricted growth of many natural populations.
6. When population size reaches the carrying capacity, the environment cannot support any more growth so the population stops growing and reaches zero population growth.

What determines whether a population grows?

1. The environment's carrying capacity for a particular species depends on the species' needs.
2. Change in a population size results from the combination of four processes: birth, death, emigration, and immigration.

3. The net change in population size is the sum of births and immigration, minus deaths and emigration.
4. As a population's density approaches the carrying capacity, its members must increasingly compete for limited reasons.
5. Density-dependent factors, such as predation, parasitism, disease, competition, and emigration, limit growth in proportion to population density—the denser the population, the more slowly it grows.
6. Some populations are limited by density-independent factors, such as natural disasters, and harsh weather.

Do natural populations actually grow exponentially?

1. Population density may temporarily overshoot carrying capacity, then drop far below the carrying capacity, and then cycle back up to *K* again.
2. When a population size exceeds carrying capacity, the population must decrease.
3. Many populations fluctuate for a while, then gradually level off at the carrying capacity.
4. Species that can grow rapidly tend to overshoot carrying capacity more than slow-growing species.

Why do some populations cycle up and down?

1. Predators help regulate herbivore populations, but other factors (the supply of plant food, defensive behavior by plants, and self-regulation) often play even more important roles.
2. Just as prey do not necessarily regulate the numbers of predators, plant growth does not necessarily regulate the number of herbivores.
3. Density-independent factors such as weather also cause irregular population fluctuations.
4. Many populations show extremely regular fluctuations in their densities.
5. Cycling in natural populations remains hard to explain.

WHAT FACTORS INFLUENCE THE VALUES OF *r* AND *K*?

1. Species differ greatly in their intrinsic rates of increase, r_{max}.
2. A species' life history traits include average number of offspring produced, age of first reproduction, and life span.
3. Some species are especially well-adapted to have high values of r_{max}. These species have evolved to reproduce as quickly and frequently as possible. Such species are referred to as *r*-selected species, and they tend to have short life spans, early first reproductions, and many offspring.
4. Unlike *r*-selected species, *K*-selected species have adaptations that increase their ability to maintain populations as close to the carrying capacity (*K*) as possible.
5. In general, *K*-selected species live long lives, reproduce late, and produce few offspring.
6. Most species fall somewhere between the two extremes. About half of all species show no tendency in either direction.
7. Both *r*-selected and *K*-selected species live in about every habitat.

How do life strategies affect survivorship and age distributions?

1. Some species invest energy in the production and care of a few large offspring; in others, most of the offspring die young.
2. A survivorship curve shows the fraction of a population that is alive at successive ages.
3. The opposite of survivorship is mortality—the fraction that dies at a given age.
4. With a convex (type I) survivorship curve, survivorship starts out high, decreases slowly with age to a certain point, then decreases more rapidly.
5. Organisms with this type of survivorship curve tend to survive long enough to reproduce and care for their young. Humans in developed countries show this type of curve.

6. Some organisms experience a steady rate of death that is not affected by age. These species show a diagonal (type II) survivorship curve—a straight, declining line.
7. Many species have the greatest chances of dying early. Their survivorship curves are called concave (or type III).
8. The most common natural pattern is increased mortality of both the young and old, with decreased mortality at midlife.
9. The age structure of a population is the fraction of individuals of various ages.
10. A cohort is a set of individuals that enter the population or are born at the same time.
11. The age distribution in a population also depends on the birth rate. Populations that are reproducing quickly (humans in India) have a greater fraction of young individuals than in a population of slower growth (humans in the United States).
12. The age structure of the human population as a whole varies from country to country.
13. The future age structure of a given population can be predicted by knowing:
    - the present age structure;
    - the mortality of each cohort;
    - the age structure of immigrants and emigrants; and
    - the fertility of each cohort.

Demography: How do scientists measure populations?

1. Demography is the statistical study of populations.
2. Mortality is expressed in terms of death rate—the number of individuals per 1000 who die annually; fertility is expressed in terms of birth rate—the annual number of births per 1000.
3. The annual rate of increase is the actual percentage by which a population increases each year. The annual rate of increase for the entire world in 1989 was about 1.66%.
4. Completed family size is the average number of children that reach reproductive age born to each family.
5. Replacement reproduction is the family size at which each couple is replaced by just two descendants. At the replacement level, a population neither grows or shrinks.

Why is population momentum important?

1. The population of the United States continues to grow even though completed family size is below replacement level. This paradox is called population momentum.
2. In a rapidly growing population, a large percentage of individuals will be young—about 20% of the world's population today is under age 15. Even if people in this group only replace themselves, they will increase the world population by 20%.

WHAT IS THE EARTH'S CARRYING CAPACITY FOR HUMANS?

1. Whether we have yet exceeded the Earth's carrying capacity for humans is a matter of debate.

How well is our energy supply keeping pace with human consumption?

1. Most humans live in cities and suburbs and consume tremendous amounts of energy each year, most derived from agriculture. Thus agriculture provides energy to, or subsidizes, cities and suburbs.
2. Agriculture requires tremendous energy to grow and harvest food, thus agriculture is subsidized by fossil fuels.
3. The overall energy flow through an urban area may be 1000 times more than that of a pond or meadow.
4. Humans exceed limits on local carrying capacity because we can tap into resources worldwide.

5. Ecological footprint refers to the area of land needed to provide all the essential resources to sustain a single person.
6. The average human in the United States uses more than 10 times the energy used by an average human in the world as a whole. To fulfill these needs, each person in the United States requires the use of 5.1 hectares.
7. Population size must not be overlooked in considering impact. Saudi Arabia has a population growth rate 5.5 times that of the U.S., but their small population size means that they will add 100,000 people per year, compared to the U.S. adding 400,000 per year with its larger population.

How well has agriculture met the human demand for food?

1. In the 1960s, in recognition of the seriousness of world hunger, agricultural researchers bred rice, wheat, and other crops that could produce up to 10 times more food as older varieties. The project is referred to as the Green Revolution
2. Crop increases temporarily outstripped human population growth.
3. These "miracle" crops require a very high energy and resource investment—primarily from fossil fuels.
4. Old crop varieties have lower yields and are disappearing.
5. Currently, enough food is produced to feed the world, but getting it delivered remains a challenge.
6. Each day, 150,000 people die. Of that group, 70,000 people starve to death. Each year, 10 million children and 10 to 15 million adults starve to death.
7. Agriculture is suffering due to increased unpredictability of crop yields.
8. The Green Revolution increased the world's food supply, and thus Earth's carrying capacity for humans. Today, though, population growth exceeds the increased food supply.

HOW IS HUMAN OVERPOPULATION POLLUTING THE BIOSPHERE?

1. Humans are not only using up limited resources—we are also ruining renewable resources by dumping industrial waste, garbage, sewage, and other waste products.

Pollution has degraded the water

1. Pollution is most apparent in freshwater communities, but marine ecosystems also suffer from pollution.
2. Historically, most people have disposed of sewage and other waste by dumping it into the nearest body of water.
3. Thermal pollution occurs when factories dump hot or warm water into rivers or lakes, such as when a stream is used to cool a nuclear power plant. Thermal pollution destroys existing aquatic communities.
4. Sediments from erosion accumulate in water and damage aquatic ecosystems by blocking light, burying organisms that live on the bottom, and filling in bodies of water.
5. Human activities enormously accelerate erosion.
6. Biodegradable pollutants include carbon dioxide, sewage, fertilizers, and other wastes that organisms can consume. Nitrogen and phosphorus in fertilizers used in agriculture and home gardening wash into rivers and lakes and promote eutrophication. Raw sewage can have the same effect.
7. Nonbiodegradable pollutants accumulate as solid wastes that are usually mixed with biodegradable solid wastes in landfills. Even many biodegradable wastes do not break down in landfills.
8. Most wastes dumped at sea are merely dumped overboard.

9. The hardest pollutants to deal with are poisons. These include heavy metals such as mercury and lead, reactive gases from smog, radioactive substances, pesticides, and an unknown but growing number of toxic chemicals used in industry and agriculture.
10. A huge number of compounds are estrogenics—compounds that mimic the effects of estrogen. These interfere with development and reproduction and cause death for some organisms.
11. The effects of many toxic chemicals are amplified at higher levels in the food chain—a phenomenon called biomagnification. Organic compounds such as DDT accumulate in tissues, especially in the fats of animals. The highest concentrations are in top carnivores.
12. Pesticides change whole communities, with major effects on the food chain's higher levels.
13. Many of these substances increase people's susceptibility to cancer and other diseases and lead to long-term changes in brain function.

Pollution has also degraded the air

1. Industrialization has often meant a darkening of the air and surrounding landscape, as particles from incompletely burned coal and oil spewed from factories and power plants.
2. Smoke and soot are merely the most visible products of burning of fossil fuels, and the most easily removed.
3. Fossil fuels are mostly hydrocarbons. The major products of combustion are carbon dioxide and water, which increases the amount of carbon dioxide in the atmosphere.
4. Carbon dioxide causes the atmosphere to warm—the greenhouse effect.
5. Nitrogen and sulfur oxides may travel thousands of miles with air currents before dissolving in falling rain. The resulting so-called acid rain may devastate ecosystems a continent away from the source of the pollution.
6. The energy of sunlight can convert nitrogen oxides in the atmosphere to the still more reactive compounds of photochemical smog.
7. Smog strains the lungs and heart, and may cause lung disease, heart disease, and cancer.
8. The ozone layer absorbs most of the ultraviolet light from the sun, shielding living things from potentially dangerous amounts of ultraviolet light.
9. Chlorofluorocarbons are believed to have created the growing "ozone hole" over Antarctica.
10. Where the ozone has thinned, ultraviolet light reaching the Earth's surface increases mutations in DNA, destroys protein molecules, and increases chemically reactive molecules in the atmosphere. For humans, the most obvious effect of increased ultraviolet light is higher rates of skin cancer.
11. Human population growth has crowded out whole ecosystems, fragmented others, driven thousands of species to extinction, polluted air and water, and altered the global climate.
12. Many researchers optimistically believe we can clean up the mess, but it will take tremendous time, effort, money, and support.
13. We cannot bring back the extinct species. Biologists estimate that the evolution of replacement species will take 5 to 10 million years.

VOCABULARY BUILDING

In your own words, first write a brief definition, then a full explanation for each of the following terms. Include examples where appropriate. Complete this section from your memory—you will not learn the terms by simply copying definitions from the textbook. Once you have finished, check your responses against the information in the chapter and make any necessary corrections.

extinction vortex —

conservation biology —

deterministic —

stochastic —

habitat destruction —

fragmentation —

edge effects —

corridors —

doubling time —

exponential growth —

J-shaped curve —

growth rate (*r*)

birth rate —

death rate —

r_{max} (intrinsic rate of natural increase) —

carrying capacity (*K*) —

logistic growth equation —

S-shaped curve —

zero population growth —

emigration —

immigration —

density-dependent factors —

density-independent factors —

overshoot —

life history traits —

r-selected species —

K-selected species —

survivorship curve —

mortality —

convex (type I) survivorship curve—

diagonal (type II) survivorship curve —

concave (type III) survivorship curve—

age structure —

cohort —

fertility —

demography —

annual rate of increase —

completed family size —

replacement reproduction —

population momentum —

subsidizes —

energy subsidy —

ecological footprint —

biomagnification —

chlorofluorocarbons —

ozone layer —

CHAPTER TEST

The following test has four parts. Complete as much of the exam as you can from memory. If you cannot answer a question, skip it. Once you complete all that you can, try to answer any questions you skipped. If you still cannot answer them, consult your textbook for the answers. Once you have completed all sections of the test, check your answers for Parts 1 - 3 against those in the back of this book. Highlight any incorrect answers then review that material in your textbook. Correct your answers for future reference.

Part 1: Multiple Choice

For each of the following, select all correct responses—more than one may be correct.

1. Which of the following is typical of unchecked exponential population growth?
   a) slow growth at first, then a growth spurt, followed by slow growth again
   b) slow growth at first, followed by increasingly rapid growth with time
   c) rapid growth at first which slows with time
   d) rapid growth followed by a crash in population size

2. Carrying capacity is affected by:
   a) biotic factors only
   b) abiotic factors only
   c) both biotic and abiotic factors
   d) genetic reproductive capacity of the species only

3. Which of the following will increase population size?
   a) increased birth rate
   b) decreased death rate
   c) emigration
   d) all of these

4. Which of the following is a density-dependent factor that can affect population size?
   a) predation
   b) drought
   c) disease
   d) fire

5. Which of the following is not typical of *r*-selected species?
   a) numerous and small offspring
   b) relatively short-lived species
   c) large body size
   d) little energy invested in rearing young

6. Which of the following is not typical of *K*-selected species?
   a) reproduce late in life
   b) larger body size
   c) provide extended parental care
   d) all of these are characteristic of *K*-selected species

7. Which type of survivorship curve is typical for humans in developed countries?
   a) concave survivorship curve
   b) convex survivorship curve
   c) diagonal survivorship curve
   d) none of these

8. Most real survivorship curves are:
   a) concave
   b) convex
   c) diagonal
   d) a combination of these three

9. Completed family size refers to:
   a) the total number of children a set of parents has
   b) the average number of children that reach reproductive age born to each family
   c) two—the number needed to replace the parents
   d) none of these

10. The current growth spurt in human population size began with the invention of:
    a) agriculture
    b) antibiotics
    c) air travel
    d) closed sewers and refrigeration

11. What is the greatest risk of our reliance on modern crops?
    a) We cannot produce enough food for everyone.
    b) Today's crops are less nutritious.
    c) Genetically identical crops are quite susceptible to devastation by pests.
    d) None of these—modern crops are stronger than older varieties.

12. Which of the following were disadvantages of the programs of the Green Revolution?
    a) The crops relied heavily on fossil-fuel fertilizers and irrigation.
    b) It decreased the genetic diversity in agricultural crops.
    c) It had no effect on the world's food supply.
    d) It was energetically expensive.

13. When it comes to feeding all the people of the world, what is currently the biggest problem?
    a) inability to deliver the food to all who need it
    b) poor nutrient value of food supply
    c) not enough food to feed all the hungry people
    d) all of these

14. Due to biomagnification, the greatest amount of toxic chemicals are found in:
    a) producers, such as plants, that take them in from the soil and water
    b) herbivores, who consume the plants
    c) top carnivores, who are at the top of the food chain
    d) All of these contain equal amounts of toxins.

15. Which of the following results from the burning of fossil fuels?
    a) acid rain
    b) greenhouse effect
    c) smog
    d) all of these

Part 2: Matching

For each of the following, match the correct term with its definition or example. More than one answer may be appropriate.

***r*-selected species**	***K*-selected species**
1. ______________	These reproduce frequently and prolifically.
2. ______________	These tend to be present in all ecosystems.
3. ______________	These are small and short-lived.
4. ______________	These tend to reproduce late in life.
5. ______________	Populations are maintained near carrying capacity.

convex survivorship curve	**diagonal curve** **concave curve**
6. ______________	The chance of dying does not change with age.
7. ______________	Typical of humans in developed countries.
8. ______________	Typical of *r*-selected species.
9. ______________	Chance of dying increases after reproductive years.
10. ______________	Chance of dying is greatest for the youngest.

Part 3: Short Answer

Write your answers in the space provided or on a separate piece of paper.

1. Explain exponential population growth.

2. List four processes that determine a population's size.

3. List three examples of density-dependent factors that affect population size.

4. List three examples of density-independent factors that affect population size.

5. The population size of *r*-selected species is likely controlled by what factors?

6. List three examples of *r*-selected species and three examples of *K*-selected species.

7. What four bits of information allow us to predict a population's future age structure?

8. What is demography?

9. Differentiate between completed family size and replacement reproduction.

10. Roughly what is today's world population of humans?

11. Explain population momentum.

12. What is meant by saying that agriculture subsidizes cities and suburbs?

13. List three ways in which city dwellers live subsidized existences.

14. Explain biomagnification.

15. Explain how heavy automobile usage in some countries can have a negative impact on the entire planet.

Part 4: Critical Thinking: Using Your Knowledge

Answer each of these in essay form, using complete sentences and paragraphs. Provide as much information as you can. (For extra essay practice, write out answers to the Review and Thought Questions in your textbook.)

1. Using specific examples, explain how humans have avoided the normal restraints on population growth. Provide specific examples of density-dependent factors that have become more important as the population continues to soar. What do you predict will ultimately happen to human population growth?

2. Discuss the Green Revolution. What was the motivation behind it? What were the outcomes? In your opinion, was the plan a success or failure (support your position).

3. When the Green Revolution was conceived, it seemed like a good idea. What issues did the Green Revolution fail to consider that led to our current state?

4. Where do you think human society would be today—three decades later—had the Green Revolution not occurred, if we were, instead, still relying on older crop varieties with relatively low yields? How do you think our population would have been affected? How about the environment? What global problems might have been worse than they are now, and which might have been better?

5. Knowing now that the Green Revolution was a bit shortsighted, we are now faced with a new agricultural revolution—genetically engineered crops that have higher yields, disease and herbicide resistance, etc. Many people are concerned about possible environmental consequences. Review some of the arguments both pro and con. What issues need to be considered as we advance in this area? What impact do you predict this will have on the human population and Earth's carrying capacity? Are we moving in the right direction?

6. With what you have learned about factors that influence population size and growth, the nature of the human population, and the impact of humans on the biosphere, write an essay about what you imagine life will be like on planet Earth in the year 3000. Consider such aspects as population size, density-dependent factors, biomagnification, and pollution. Will we humans change our behaviors or continue on the same path we currently follow?

7. Many states have learned that deer hunting is essential for maintaining a healthy and balanced ecosystem. What factors have led to the need to kill part of the deer population in many areas, in order to keep it healthy. What role are the hunters taking on?

8. Consider the area in which you live. What are some examples of habitat fragmentation, even on a very small scale, that are evident there? What organisms are directly affected, and how? Who is indirectly affected, and how?

9. In nature, we often see that a population's behavior is different when the population is small than it is when the population is large. Large populations experience more stress. What signs indicate that the human population is experiencing a lot of stress? In what ways might the large size of the human population be altering human behavior?

10. Evaluate your actions on this single day in your life. Which of your actions will have a harmful effect on the environment? Were you aware of the impact at the time of your action? What have you done today to minimize your impact on the world? How can you change your behavior to be more "environmentally friendly"? What changes in your day-to-day behavior are you willing to make, even if they inconvenience you, knowing that your children, your grandchildren, and all future generations of your family will be living in the world you leave behind?

Chapter 29
The Ecology of Animal Behavior

KEY CONCEPTS

1. Any behavior is likely to include both innate and learned aspects.
2. All behavior develops under the influence of both genetic inheritance and environmental experience.
3. Individual behavior patterns affect an individual's ability to survive in a specific environment—thus the genetic component of any behavior pattern is subject to natural selection.
4. Ecology—an individual's relationship to its environment—determines the adaptive value of behavior.
5. A modal action pattern is a stereotypical behavior triggered by a sign stimulus, which can be seen, heard, felt, or smelled.
6. Behaviors that seem to be controlled by genes alone are rare. Even behavioral traits that are strongly influenced by genes are, like most other traits, probably polygenic.
7. Many behaviors may be strongly influenced by genes but modifiable through learning.
8. Many complex behaviors can develop normally only during certain sensitive periods in development. Even complex social behavior can have a sensitive period.
9. Through optimal foraging, animals may efficiently maximize the amount of food they get for the energy expended.
10. Animals compete with others of their own species for mates, resources, and rank through agonistic encounters.
11. Competition for mates over evolutionary time results in sexual selection—the differential ability of individuals with different genotypes to acquire mates.
12. Animals defend territories for raising young, mating, or foraging—as long as defending the territory will increase their reproductive potential.
13. Sociality allows animals to spot both prey and predators more easily, but groups of animals are more visible to predators, and individuals in groups must constantly compete for food, mates, and breeding areas.
14. Kin selection seems to explain many cases of altruism, but not all, and the phenomenon of altruism continues to puzzle and fascinate biologists. One explanation for the adaptive value of altruism is inclusive fitness, which includes the concept of kin selection.
15. Haplodiploidy in some social insects probably facilitates kin selection.
16. Kin selection may explain altruism in other animals besides the social insects. But other explanations, such as reciprocal altruism, are possible as well.

EXTENDED CHAPTER OUTLINE

Why is sociobiology controversial?

1. E.O. Wilson, an eminent Harvard professor, wrote an ambitious textbook on animal behavior called *Sociobiology: The New Synthesis*.
2. The firestorm of criticism ignited by Wilson's book left the lay public with the impression that the field of sociobiology provided support for racism, sexism, classism, and Nazism.
3. Sociobiology, the study of behavior from an evolutionary perspective, got such a bad name that biologists renamed it behavioral ecology.
4. Until about 1972, two separate disciplines for the study of animal behavior existed:
   - the European science of ethology (the systematic study of animal behavior from a biological point of view), and
   - the American science of experimental psychology.
5. Experimental psychologists focused on learned behavior and were primarily concerned with immediate mechanisms—what physiological processes controlled behavior in animals, and what animals were used as substitutes for humans.
6. Experimental psychologists tended to think of behavior as being mostly learned.
7. Ethologists focused on how behavior helped animals survive.
8. Konrad Lorenz and Niko Tinbergen assumed that most animal behavior was adaptive and emphasized innate behavior—behavior that is not learned.
9. Psychologists, most interested in human behavior, were suspicious of ethology because:
   - they believed human behavior was so malleable that genetics and evolution were not relevant, and
   - ethology brought images of past attempts to prove one race or group superior to others.

HOW IS BEHAVIOR ACQUIRED?

1. Behavior, like other characteristics, develops from the interplay of genes and environment, so natural selection can act, through the genes, on behavior.
2. An individual's ecological environment determines the adaptiveness of a behavioral trait.

What is innate behavior?

1. Innate behavior is behavior that an animal engages in regardless of previous experience.
2. Learning is a change in an animal's behavior in response to a specific previous experience.
3. All animals can learn, and every animal has innate tendencies to learn different things, so learned and innate behavior are not completely separable.

What stimulates innate behavior?

1. A highly stereotyped innate behavior is called a modal action pattern. The pattern can be partly flexible and partly fixed.
2. Modal action patterns serve a purpose only when performed at the proper time and place.
3. Some signal in the environment stimulates each modal action pattern. The key aspect of an object that triggers a modal action pattern is called a sign stimulus.
4. Pheromones—chemicals made by one organism that influence the behavior of another organism of the same species—can also activate modal action patterns.
5. An animal reacts to only a few aspects of its environment.
6. A supernormal stimulus is a stimulus even more stimulating than anything normally encountered in nature.

How much can animals learn?

1. Animals can learn a lot from their environment and from each other.
2. Even behavior strongly influenced by genes is still open to modification through learning.

Can old dogs learn new tricks?

1. Timing is critical for many kinds of learning.
2. The period of time during which an animal can learn a particular behavior pattern is called the sensitive period, similar to critical periods in morphological development.
3. Sensitive periods exist for a variety of different kinds of behavior.
4. Imprinting is the process by which an animal learns such behavior during a sensitive period.
5. Just as development of normal vision depends on visual experience during a critical period, development of normal social behavior depends on experience at sensitive periods.
6. Appropriate later experience can in some cases overcome deficits in early experience.
7. When two species differ in their behavioral tendencies, these differences may result from the genetic differences between the species.

How do genes influence behavior?

1. Behavior is the result of both genetic influences and experience.
2. The influence of genes on behavior is hard to determine because behavior, like most traits, is polygenic.

HOW DOES ECOLOGY INFLUENCE BEHAVIOR?

How do animals make the best choices?

1. When animals forage, they may efficiently maximize the amount of food they get for the energy expended, also weighing into the equation other costs, such as the risk of being eaten. Ecologists called this process optimal foraging.

How do competition and predation influence foraging behavior?

1. In the presence of predators, animals must alter behavior to increase their chances of survival.
2. The presence of competing species can also change feeding behavior.

How do animals express aggressive feelings?

1. Reproduction and survival necessitate both competition and cooperation.
2. Contests are examples of agonistic behavior, which includes all aspects of competitive behavior in a species—aggression, aggressive displays, appeasement, and retreat.
3. Some animals use agonistic interactions to establish a dominance hierarchy—a ranking of individuals that fixes which animals have first access to resources or mates.
4. Dominance hierarchies establish rules that allow animals to live in the same territory with limited resources.

Territoriality

1. In many species, animals engage in competitive behavior to establish and defend a territory —an area occupied by an individual or group from which others of that species are excluded.
2. Some species use territories exclusively for mating; more commonly, species also use territories for foraging and raising young.
3. Establishing and defending a territory necessarily involves agonistic behavior.
4. Animals use a wide variety of signals to mark their territories.
5. Defending a territory requires energy expenditure.

How does female choice influence the evolution of males?

1. The long-term result of competition for mates is sexual selection—the differential ability of individuals with different genotypes to acquire mates.
2. Some biologists distinguish sexual selection from natural selection, saying natural selection favors adaptive traits, but sexual selection often leads to "maladaptive" traits.
3. Natural selection can act both on inherited variation in structures that are important in mating and on inherited mating behaviors.

How do animals compete for resources or mates?

1. Animals of the same species may cooperate or compete with one another, according to the demands of their current environment and the constraints of their evolutionary histories.
2. Mating rituals allow females to assess the health and vigor of potential mates.

SOME ANIMALS LIVE IN SOCIAL GROUPS

How is social behavior advantageous?

1. Living in groups affords distinct advantages—large groups of animals more quickly spot predators and other dangers than small groups of animals or solitary ones.
2. Sometimes groups of animals can cooperatively defend themselves.
3. Group living can be an advantage in finding food as well—predators that hunt cooperatively, such as lions and wolves, can kill prey far larger than themselves.
4. Social living also has disadvantages—animals living in groups must compete with each other for food, mates, and breeding areas.
5. Whether animals live in groups depends on the ecological costs and benefits.

Honeybees are an extraordinary example of social behavior

1. An insect colony can be viewed as a superorganism, in which thousands or even millions of individuals function as a single unit.
2. Social insects form enormous colonies.
3. Social organization allows social species to take advantage of environmental resources that are unavailable to more solitary animals.
4. All of the social insects have three traits in common:
   - they cooperate in raising their young;
   - few individuals reproduce, as sterile workers assist and defend the fertile queen; and
   - generations overlap, so that offspring help their mothers.

Is altruism adaptive?

1. Altruism is behavior that benefits others at a cost to the animal that performs the behavior.
2. Social insects demonstrate the most extreme example of altruism.
3. We would expect natural selection to favor sacrifice by parents to help their offspring, but to act against the sacrifice of offspring to help their parents.

Does altruism arise by natural selection?

1. Darwin argued that a family (the queen and her offspring, for example) is the unit of selection, not the individual. By self-sacrifice, workers perpetuate genes they share with their sisters, one of whom will become a queen.
2. W.D. Hamilton suggested the theory of kin selection as a more general answer to the question of altruism in the social insects. By this theory, an individual increases its reproductive output by helping relatives (who share its genes) to reproduce. Siblings are just as related to one another as they are to their parents—they share one-half of their genes.
3. Fitness is a measure of selective advantage—one genotype's contribution to the next generation, relative to the contribution of other genotypes.
4. Inclusive fitness is the sum of an individual's genetic fitness (which includes that of its own direct descendants) plus all of its influence on the fitness of its other relatives. It is the total success in passing alleles to the next generation, either by the efforts of a parent or by the altruistic acts of a relative.

Why might intricate societies have evolved among certain insects?

1. Hymenopterous social insects have an unusual kind of sex determination called haplodiploidy—males develop from unfertilized eggs and are therefore haploid, while females are diploid. Other than providing sperm in a single mating event, males contribute nothing to the economy of the hive.
2. Half your alleles come from your mother, half from your father. Each of your brothers and sisters also share one-half of their alleles with each of your parents.
3. The chance that you and a sister each inherited a given allele from your mother is 0.5 x 0.5 = 0.25. Similarly, the chance that you and a sister each inherited the same allele from your father is 0.25. So the chance that you and your sister share any allele is 0.25 + 0.25 = 0.5, exactly the same as the chance that you share an allele with your mother or father.
4. J.B.S. Haldane summarized these relationships by saying, "I would gladly lay down my life for two of my brothers or eight of my first cousins."
5. The queen passes half her alleles to her female offspring. However, since males are haploid, a male passes on all his alleles—his daughters inherit all of his genes. The (female) workers share 75% (rather than 50%) of their alleles with one another.
6. Because the queen mates with many different males, not all of her offspring share the same father. However, the average relatedness will be greater than 50%.
7. If the workers were to have their own offspring, they would pass on 50% of their alleles. They can therefore pass on more of their genes by putting their energies into making sisters than they can by reproducing and making daughters.

Does inclusive fitness always explain altruistic behavior?

1. Grazing animals, birds, and other animals that live in groups often help protect one another.
2. A few studies suggest kin selection as an explanation for altruism.
3. In the last 20 years, many biologists, anthropologists, and other enthusiasts have tried to extend the concept of kin selection to virtually all altruistic behavior, but not all examples of altruism conform to the kin selection model.
4. Robert Trivers argued that animals capable of recognizing one another might engage in altruism on the understanding that the recipient would return the favor some time in the future. Trivers called this idea reciprocal altruism, since the recipient reciprocates.

VOCABULARY BUILDING

In your own words, first write a brief definition, then a full explanation for each of the following terms. Include examples where appropriate. Complete this section from your memory—you will not learn the terms by simply copying definitions from the textbook. Once you have finished, check your responses against the information in the chapter and make any necessary corrections.

sociobiology

behavioral ecology —

ethology —

experimental psychology —

conditional reflex —

innate —

learning —

modal action pattern —

sign stimulus —

pheromones —

supernormal stimulus —

sensitive period —

imprinting —

optimal foraging —

agnostic behavior —

dominance hierarchy —

sexual selection —

territory —

altruism —

kin selection —

inclusive fitness —

haplodiploidy —

reciprocal altruism —

CHAPTER TEST

The following test has four parts. Complete as much of the exam as you can from memory. If you cannot answer a question, skip it. Once you complete all that you can, try to answer any questions you skipped. If you still cannot answer them, consult your textbook for the answers. Once you have completed all sections of the test, check your answers for Parts 1 - 3 against those in the back of this book. Highlight any incorrect answers then review that material in your textbook. Correct your answers for future reference.

Part 1: Multiple Choice

For each of the following, select all correct responses—more than one may be correct.

1. Innate behavior is governed by which of the following?
   a) genes
   b) learning
   c) previous experience
   d) all of these

2. A modal action pattern is an example of:
   a) conditioned reflex
   b) innate behavior
   c) learned behavior
   d) supernormal stimulus

3. Which of the following is not true of a modal action pattern?
   a) It is triggered by a sign stimulus.
   b) It is a highly stereotyped behavior.
   c) It is less flexible than a reflex.
   d) It requires coordination of muscles and nerves.

4. A supernormal stimulus is one that:
   a) occurs with the highest frequency in nature
   b) is more stimulating than any natural stimuli
   c) eventually is learned and elicits no response
   d) only works in laboratory-raised animals not exposed to natural stimuli

5. Most behavior traits in humans are:
   a) polygenic
   b) polymorphic
   c) controlled by single genes
   d) not influenced by genes

6. Which of the following determines the adaptive value of a behavior?
   a) animals' interactions with each other
   b) animal's interactions with their environment
   c) both a and b
   d) neither a nor b

7. A sensitive period can best be described as the period during which:
   a) an innate behavior is learned
   b) an innate behavior is first expressed
   c) a behavior pattern can be learned
   d) a behavior pattern cannot be modified

8. Which of the following is primarily associated with a sensitive period?
   a) innate behavior
   b) modal action pattern
   c) perception of sign stimuli
   d) imprinting

9. Optimal foraging is characterized by:
   a) maximizing both food intake and energy expenditure
   b) maximizing food intake while minimizing energy expenditure
   c) minimizing food intake while maximizing energy expenditure
   d) minimizing both food intake and energy expenditure

10. Which of the following can have an impact on foraging behavior?
   a) predation
   b) competition
   c) mutualism
   d) all of these

11. Which of the following is not an example of agonistic behavior?
    a) clashing of antlers by two deer
    b) mating rituals
    c) a mongoose attacking a cobra
    d) These are all examples of agonistic behavior.
12. Animals defend territories for which of the following reasons?
    a) mating
    b) foraging
    c) raising their young
    d) all of these
13. For honeybees, all the workers in the hive are:
    a) sisters
    b) males
    c) mothers
    d) reproducers
14. Which of the following characterizes altruism?
    a) Both individuals benefit directly from the behavior of one individual.
    b) The behavior benefits one at the expense of the individual who performs it.
    c) The behavior benefits the performer at the expense of another individual.
    d) Two individuals perform a behavior together that benefits them both.
15. Which of the following is true of haplodiploidy?
    a) males develop from unfertilized eggs
    b) males and females are both haploid
    c) females are diploid
    d) males are haploid

Part 2: Matching

For each of the following, match the correct term with its definition or example. More than one answer may be appropriate.

innate behavior | **learned behavior**

1. ____________________ This is based in genetics.
2. ____________________ All animals are capable of this.
3. ____________________ Involves modal action patterns.
4. ____________________ May involve sign stimuli.
5. ____________________ Experience shapes this type of behavior.

sexual selection | **kin selection**

6. ____________________ This is an explanation of altruism.
7. ____________________ This is the differential ability to acquire mates.
8. ____________________ This can lead to maladaptations.
9. ____________________ Relatives are assisted to pass on shared genes.

***Part 3:* Short Answer**

Write your answers in the space provided or on a separate piece of paper.

1. What was the central disagreement between experimental psychologists and ethologists?

2. What determines the adaptive value of behavior?

3. Explain modal action patterns.

4. Why is it difficult to pinpoint the gene responsible for a particular behavior?

5. What is the significance of sensitive periods?

6. Differentiate between mating rituals and sexual selection.

7. List four general categories of agonistic behavior.

8. Explain a dominance hierarchy. Can you think of any examples of this in humans?

9. On what aspects of behavior can natural selection work?

10. Under what circumstances are animals most likely to expend energy to defend territories?

11. List some advantages and disadvantages of social behavior.

12. List three traits that are common to all social insects.

13. Provide two examples of altruistic behavior and explain why you think they occur.

14. What is meant by inclusive fitness?

15. What is meant by reciprocal altruism?

Part 4: Critical Thinking: Using Your Knowledge

Answer each of these in essay form, using complete sentences and paragraphs. Provide as much information as you can. (For extra essay practice, write out answers to the Review and Thought Questions in your textbook.)

1. Discuss some behavior pattern that you display on a regular basis. What aspects of that behavior are likely determined genetically? What aspects more likely result from learning?

2. Describe various types of interspecific and intraspecific interactions that might affect behavior. Provide specific examples.

3. Explain the social structure within a honeybee hive, as well as the types of behavior seen within each group. Discuss haplodiploidy and the concept of kin selection.

4. Does kinship selection occur in humans? Use examples to explain your answer.

5. News media often report on "heroes" who risk their lives to save someone else. Are these examples of altruistic behavior? If so, does the kin selection concept account for such deeds? Why do humans engage in energy-costing activities if their own lives are not in danger?

6. Do humans exhibit optimal foraging behavior? Defend your answer.

7. In what ways, if any, do humans utilize dominance hierarchies? How are they established? What types of agonistic behaviors are involved?

8. In humans, children are abused and die from abuse more often at the hand of a stepfather than at the hand of any other relative. What might explain this?

9. Many species exhibit very elaborate courtship behavior. Why are these behaviors, which often make the suitor more obvious to predators, so elaborate?

10. Some species engage in social hunting part of the time but hunt solo at other times. What circumstances might lead to this switching?

Chapter 30
Structural and Chemical Adaptations of Plants

KEY CONCEPTS

1. Flowering plants are largely terrestrial organisms that rely on below-ground roots and above-ground shoots to obtain water, minerals, carbon dioxide, and sunlight.
2. Terrestrial plants have evolved structures and mechanisms to:
   - perform photosynthesis,
   - obtain water and carbon dioxide,
   - support themselves mechanically, and
   - transport minerals and nutrients throughout the plant.
3. Terrestrial plants have:
   - roots for absorbing water and nutrients from soil,
   - broad, thin leaves for collecting solar energy, and
   - stomata for collecting carbon dioxide and releasing water and oxygen.
4. Most of the parts of a plant—shoots (with stems and leaves) and roots—are composed of four tissue types.
5. The vascular system consists of xylem, which carries water and minerals from the roots, and phloem, which carries sugar and other nutrients from the leaves.
6. Xylem consists of two kinds of nonliving cells: narrow tracheids and wider vessel elements.
7. Phloem is made of living cells called sieve tube members, which accommodate flow of materials through the phloem.
8. A plant is supported against the pull of gravity by turgor pressure within its cells and by structural materials, such as cellulose and lignin, within the cell walls of support cells.
9. Flowering plants fall into two classes that have characteristic differences in their embryos, flowers, leaves, roots, and vascular systems. These two classes are monocots and dicots.
10. Water from the soil enters the root via the root hairs, travels through the loose parenchymal cells of the cortex, then moves through the endodermis into the vascular tissue.
11. Leaves, the major sites of photosynthesis, consist of an outer layer of epidermis and a central region of mesophyll through which run the vascular bundles (veins). The epidermis is dotted with stomata.
12. Plants produce secondary plant compounds, which are chemicals not needed for the major metabolic pathways of a plant. Many secondary plant compounds repel or poison insects and other herbivores; some other compounds provide structural support while others attract pollinators and other animals.

EXTENDED CHAPTER OUTLINE

SHOULD WE EAT OUR VEGETABLES?

1. In 1983, Bruce Adams, a biochemist at the University of California, Berkeley, asserted that enormous numbers of ordinary vegetables are loaded with mutagens—chemicals that cause DNA to mutate.
2. Ames listed scores of toxic chemicals found naturally in ordinary foods.
3. Many plants respond to any damage by secreting high levels of toxins, many of which are mutagenic, teratogenic, or carcinogenic.
4. A large part of Ames' 1983 paper was devoted to natural plant compounds, such as vitamin E, vitamin C, and β-carotene, that seem to prevent cancer. These compounds (antioxidants) prevent formation of free radicals—compounds that damage DNA—which are formed during normal metabolic processes.
5. Later, Ames increasingly focused on free radical production during normal metabolism as the ultimate cause for all degenerative diseases associated with aging.

WHAT ARE THE PHYSICAL CONSTRAINTS ON PLANTS?

1. During photosynthesis, plants combine the hydrogen from water molecules (H_2O) with carbon dioxide (CO_2) to build the simple sugar glucose.
2. Sunlight provides the energy needed for this transformation.
3. Early plants may have descended from green algae, or chlorophytes, which can live in fresh water and even on land.
4. Recognizable plants evolved in the Silurian Period and diversified throughout the 100 million years of the Devonian and Carboniferous Periods.
5. Plants have become the dominant feature of land life.

How did terrestrial plants evolve?

1. For the aquatic ancestors of terrestrial plants the land offered distinct advantages:
   - plenty of light and nutrients, and
   - far more carbon dioxide than was available at sea.
2. Photosynthetic organisms needed new mechanisms to obtain water and carbon dioxide and to support their bodies against gravity.
3. Terrestrial plants evolved roots that pull water from moist soil and tiny holes in their leaves (stomata) that open and close like tiny mouths to collect carbon dioxide from the air.
4. Leaves are solar collectors loaded with chlorophyll—the green pigment used to capture the energy of sunlight.
5. Stems support leaves, flowers, and fruits and provide an avenue for transporting materials.
6. Most terrestrial plants are vascular plants with roots, stems, and leaves.

What are the roles of shoots and the roots?

1. Most plants consist of:
   - an above-ground shoot system—stems, photosynthetic leaves, and flowers and other reproductive organs; and
   - an underground root system which anchors the plant and absorbs water and minerals from the soil.
2. Materials travel between shoot and root in the plant's vascular system—the network of vessels in which fluids move through the plant.
3. The vascular system is composed of two tissues: xylem and phloem.
4. Xylem transports water and minerals; phloem transports sugars produced by photosynthesis.
5. Xylem and phloem associate together in vascular bundles.

6. The cell wall is the most important plant structure for supporting the plant, and it consists of cellulose, other carbohydrates, and specialized proteins.
7. All plants have a primary cell wall outside the plasma membrane. Some cells have a thicker secondary cell wall between the cell membrane and the primary cell wall.
8. A polymer called lignin adds strength to the secondary wall, and in some cases to the primary wall.
9. Most of the internal space of plant cells consists of a large central vacuole filled with a watery fluid (cell sap). The pressure of this fluid against the cell wall (turgor pressure) helps support plants against the pull of gravity.
10. Some plants (herbs) are entirely soft and green and depend mainly on turgor pressure for support. Herbs are highly susceptible to wilting, and include all plants with no above-ground woody parts.
11. Woody plants, such as trees, shrubs, and other perennial plants, have woody parts that provide support independent of turgor pressure.

HOW ARE FLOWERING PLANTS STRUCTURED?

1. Four tissue systems occur in all organs of a plant:
   - the dermal tissue makes up the outer, protective covering of the plant (the equivalent of our skin);
   - the vascular tissue system which conducts materials around the plant;
   - ground tissue, in which the vascular tissue is embedded; and
   - undifferentiated meristem tissue.
2. The dermal tissue system is the protective covering of the plant. It consists of the epidermis (primary tissue) and the periderm (secondary tissue).
3. The epidermis of the shoot secretes a cuticle that prevents water loss. The epidermis is a complex tissue, consisting of multiple cell types, each with a distinct function.
4. The ground tissue system consists of three main types of tissue:
   - parenchyma,
   - collenchyma, and
   - sclerenchyma.
5. Cells of parenchyma are living cells that have only a thin primary cell wall. Parenchymal cells make up most of the plant's soft tissues and are frequently full of starch.
6. Collenchymal cells have a thick primary cell wall and often provide mechanical support.
7. Sclerenchyma consists of one of two cell types—fibers or sclereids. Fibers and sclereids have primary and secondary cell walls and provide mechanical support.
   The vascular tissue consists of xylem and phloem, and other associated tissues. The conductive functions of xylem and phloem were discussed above, and these tissues also provide mechanical support.
8. Dermal, ground, and vascular tissues all derive from meristems—areas of rapidly dividing cells.
9. Meristems at the tips of roots and shoots are called apical meristems, and growth from apical meristem tissue is termed primary growth.
10. Many plants also possess lateral meristems, which are cylinders of actively dividing cells within the roots and stems. Growth at lateral meristems, called secondary growth, increases the thickness of a shoot or a root.

How is the vascular system organized?

1. Xylem caries water and dissolved substances in two kinds of cells—tracheids and vessel elements.

Xylem carries fluids in two kinds of nonliving cells

1. Tracheids are long, thin, spindle-shaped cells. Water and dissolved substances move from cell to cell through pits—regions that lack secondary cell walls and water-tight lignin. Pits form pairs between adjacent cells; water flows through the pits form one tracheid to the next.
2. In a simple pit pair, a region of secondary wall is absent; in the bordered pit pair, an adjacent secondary wall overarches the pit membrane and reinforces the wall of the tracheid.
3. In conifers, bordered pits have an additional feature that allows each pit to function as a valve—a thickened region of the cell wall called the torus.
4. A larger, more efficient kind of conduction cell, called a vessel element, evolved in flowering plants. Vessel elements are shorter, broader, and less tapered than tracheids and connect end to end to form long open channels.
5. The end of each vessel element, called a perforation plate, contains one or more holes through which water easily flows. Vessel perforations lack primary and secondary walls.

The conducting elements of phloem are living cells

1. In the phloem, the sieve tube members, which form the sieve tube, are the cells that actually conduct fluid.
2. The wall between adjoining sieve tube members is called a sieve plate and it has more and bigger pores than the side of the sieve tube, allowing more fluid to flow through a tube rather than in and out of it.
3. Although sieve tube members are living cells, they are highly specialized and lack nuclei, ribosomes, and vacuoles.
4. Next to each sieve tube member is a companion cell—a long, nucleated, fully functional parenchymal cell that supplies the sieve tube member with proteins and energy-rich molecules.

How do monocots and dicots differ?

1. When a seed first sprouts, it has a root and a shoot. The very first leaves, quite simple in form, are called cotyledons, or seed leaves.
2. Monocotyledons (monocots) have one cotyledon, while dicotyledons (dicots) have two. Most monocots are herbaceous and include all grasses; dicots include most trees and shrubs, as well as many herbs.
3. The flower parts of a dicot usually come in multiples of four or five, while those of a monocot come in multiples of three.
4. In monocots, the leaf veins generally run parallel to one another, while those of a dicot usually form a complex netlike pattern.
5. Inside a dicot's stem, the vascular bundles lie in the form of a ring, while those of a monocot are scattered throughout the stem.
6. Inside the vascular cylinder of a dicot's root, xylem and phloem often form a star shape that runs the length of the root. In many monocot roots, this area, instead, has a nonconducting core—the pith—with bundles of xylem and phloem lying in distinct bundles around it.
7. Dicot roots have lateral meristems and secondary growth, while monocots have no lateral meristems and thus are incapable of secondary growth.
8. In many dicots, the entire root system connects to a single vertical root—the taproot.

HOW DO TERRESTRIAL PLANTS OBTAIN WATER AND MINERALS?

1. Water-dissolved nutrients flow from the soil into the vascular system of the roots, then move up through the stem to the plant's leaves and other organs.
2. More than 90% of the water that enters a plant departs through the stomata in the leaves as water vapor.
3. Plants need large amounts of nitrogen, sulfur, phosphorus, magnesium, calcium, and potassium; they also need small amounts of micronutrients such as molybdenum, copper, zinc, manganese, boron, iron, and chlorine. Plants obtain nutrients almost entirely as ions dissolved in the water.

How do roots carry water?

1. Most water enters a root through root hairs, which are extensions of single epidermal cells, and which greatly increase the root's surface area.
2. The epidermis and the root hairs secrete a slimy substance called mucigel, which lubricates the root tips to allow them to move between soil particles.
3. Most of a root's cortex consists of parenchymal cells—large cells with large vacuoles which, in many plants, contain stored starch.
4. The innermost layer of the cortex—the endodermis—differs from the rest of the cortex.
5. Water flows from the soil to the epidermis by two paths, and then from the epidermis through the cortex to the endodermis.
6. The first path is the symplast, which runs through the cytoplasm of the parenchymal cells of the cortex. Water moves from cell to cell via plasmodesmata.
7. The second path involves the apoblast, which lies in the material of the cell walls of the cortex. Water and dissolved nutrients soak into the cellulose fibers there.
8. Each cell of the endodermis secretes a waxy, water-impermeable substance called suberin into its cell walls, on four of the six sides. Presence of this waxy wall, called the Casparian strip, prevents water from going between the cells.
9. The endodermis is a selective barrier to the entrance of water and dissolved substances from the soil into the xylem.
10. Within the cylinder formed by the endodermis is the stele, which contains the root's vascular system. At the center of the stele are the xylem and phloem; surrounding them is a sheath of parenchymal cells called the pericycle.

Stems carry water and minerals between roots and leaves

1. A node is the region of a stem to which a leaf attaches, and the regions between nodes are called internodes.
2. Each leaf-node-internode segment represents a module of the plant that can be repeated over and over.
3. Like primary roots, primary stems consist of dermal, ground, and vascular tissues.
4. The stem's epidermis, unlike that of the roots:
   - lacks hairs,
   - has stomata, and
   - has a thick cuticle—a waterproof layer of polyester and wax that encases the stem and prevents water loss.
5. Inside the stem's epidermis is a ring of cortex surrounding a core of vascular tissue and pith.
6. The stems of most plants do not have an endodermis.
7. In a cross section, the vascular bundles of a monocot stem appear to be scattered. As it ascends the stem, though, a given bundle gradually moves toward the center, periodically branching to connect to other bundles or to leaves.

What do leaves do?

1. Most leaves have a thin flat part (the blade), and a stalk (the petiole) which connects the leaf to the stem.
2. The arrangement of leaves on a stem is characteristic of each species, with three principal patterns:
   - In the alternate or spiral—the most common arrangement—the bases of the petioles spiral up the stem.
   - In opposite arrangements, there are two leaves arising from a single node.
   - In the whorled arrangement, three or more leaves arise at each node.
3. Like stems and roots, leaves consist of dermal, ground, and vascular tissue systems.
4. Both the upper and lower surfaces of a leaf contain specialized pores called stomata, through which carbon dioxide, oxygen, and water vapor enter and exit the leaf.
5. Surrounding each stoma are two guard cells—specialized epidermal cells that serve as a valve. Stomata close when the plant cannot afford to lose water and at night when it is not consuming carbon dioxide. They can also regulate the plant's temperature.
6. The ground tissue of the leaf consists of mesophyll—green parenchymal cells that are responsible for most of a plant's photosynthesis.
7. Most leaves have two kinds of mesophyll—palisade parenchyma and spongy parenchyma.
8. Cells of palisade parenchyma are elongated and packed with chloroplasts; cells of spongy parenchyma are irregular but rounded, and separated by numerous air spaces, which allow oxygen and carbon dioxide to diffuse in and out, and water vapor to leave.
9. In general, the tops of leaves are more specialized for light absorption and the bottoms for exchange of gases and water.
10. The leaf's veins are its larger vascular bundles. Vascular bundles of a vein are continuous with those of the petiole and with the leaf traces of the stem. Inside the leaf, the veins branch into finer and finer vascular bundles.
11. In monocots, leaf veins usually run parallel to one another. In dicots, the leaf usually has only one or a few large veins which branch into fine networks.
12. Aside from photosynthesis, other leaf functions include
    - protecting buds and flowers, and
    - storing food for the embryo and young plant.

HOW DO PLANTS DEFEND THEMSELVES AGAINST HERBIVORES?

1. Plants are good to eat and they can't get away.
2. Even other plants parasitize plants.

Mechanical defenses

1. The plant's waxy cuticle excludes both bacteria and fungi fairly well.
2. Hairy leaves or very sticky leaves discourage caterpillars and other tiny herbivores.
3. Spines, thorns, and prickles fend off larger mammalian herbivores.

Chemical defenses

1. Plants' most effective defenses are probably chemical. Most of these are secondary plant compounds. Primary compounds are essential for plants' day-to-day functioning, but secondary plant compounds are not.
2. Secondary plant compounds are diverse chemicals that function in very specific ways.
3. The coevolution of plants and their attackers has led to a balanced ecosystem in which plants have so far managed to stay ahead.

4. Most secondary plant compounds fall into three groups:
   - terpenes,
   - phenolic compounds, or
   - alkaloids.
5. Terpenes are lipids and the largest class of secondary compounds. Vast numbers of terpenes seem to exist solely to drive away or poison herbivores.
6. Mixtures of terpenes are called essential oils, which repel insects.
7. The terpene gossypol also works as a contraceptive in human males.
8. Phenolic compounds are aromatic substances with a variety of roles in plants. Some repel herbivores and pathogens, others attract pollinators or fruit dispersers. Lignin is a phenolic compound that also contributes to mechanical support. Some phenolic compounds poison almost all nearby plants. Some regulate gene expression in nitrogen-fixing bacteria. Others rapidly accumulate in the plant in response to bacterial or fungal infections.
9. Tannins, commonly found in woody plants, are phenolic polymers that reduce growth and survivorship in many herbivores.
10. The pigments (anthocyanins) responsible for most of the red, pink, purple, and blue colors of flowers, fruits, leaves and other plant parts are also phenolic compounds.
11. Alkaloids are nitrogen-containing secondary plant compounds found in 20% to 30% of all vascular plants. They include many of the most powerful drugs known to humans, including nicotine, atropine, cocaine, morphine, the hallucinogen psilocybin, and the toxin strychnine.
12. Many alkaloids are highly toxic and these compounds may also be highly teratogenic.
13. Other secondary plant compounds include the mustard oil glycosides.
14. Herbivores that specialize on noxious plants often avoid competition with other herbivores.
15. Many compounds that originally evolved as repellents to herbivores now act as attractants for certain groups of animals.

VOCABULARY BUILDING

In your own words, first write a brief definition, then a full explanation for each of the following terms. Include examples where appropriate. Complete this section from your memory—you will not learn the terms by simply copying definitions from the textbook. Once you have finished, check your responses against the information in the chapter and make any necessary corrections.

antioxidant —

free radicals —

stomata (stoma) —

shoot system —

root system —

vascular system —

xylem —

phloem —

vascular bundles —

cellulose —

primary cell wall —

secondary cell wall —

lignin —

central vacuole —

cell sap —

turgor pressure —

herbaceous plants —

woody plants —

dermal tissue system —

epidermis —

periderm —

cuticle —

dermal tissue system —

epidermis —

periderm —

cuticle —

ground tissue system —

parenchyma —

collenchyma —

sclerenchyma —

vascular tissue system —

meristems —

apical meristems —

primary growth —

lateral meristems —

secondary growth —

tracheids —

pits —

simple pit pair —

bordered pit pair —

torus —

vessel elements —

perforation plate —

sieve tube members —

sieve tubes —

sieve plate —

companion cell —

cotyledons —

monocotyledons (monocots) —

dicotyledons (dicots) —

pith —

taproot —

cortex —

root hairs —

mucigel —

endodermis —

symplast —

apoblast —

suberin —

Casparian strip —

stele —

pericycle —

node —

internode —

blade —

petiole —

alternate —

opposite —

whorled —

mesophyll —

palisade parenchyma —

spongy parenchyma —

veins —

secondary plant compounds —

primary compounds —

terpenes —

essential oils —

gossypol —

phenolic compound —

tannins —

anthocyanins —

alkaloids —

mustard oil glycosides —

CHAPTER TEST

The following test has four parts. Complete as much of the exam as you can from memory. If you cannot answer a question, skip it. Once you complete all that you can, try to answer any questions you skipped. If you still cannot answer them, consult your textbook for the answers. Once you have completed all sections of the test, check your answers for Parts 1 - 3 against those in the back of this book. Highlight any incorrect answers then review that material in your textbook. Correct your answers for future reference.

Part 1: Multiple Choice

For each of the following, select all correct responses—more than one may be correct.

1. A compound that causes birth defects is called a:
   a) mutagen
   b) teratogen
   c) carcinogen
   d) none of these

2. Which of the following is the most important reason for having root hairs?
   a) better anchoring of the plant in the soil
   b) more carbon dioxide can be released
   c) more surface area for uptake
   d) more surface area for increased photosynthesis

3. Stomata have what function?
   a) They collect oxygen.
   b) They collect water.
   c) They collect carbon dioxide.
   d) They absorb sunlight.

4. Which of the following transports water and minerals from the roots to the shoots?
   a) xylem
   b) central vacuole
   c) phloem
   d) all of these

5. Which of the following helps physically support a plant?
   a) turgor pressure
   b) lignin
   c) cell wall
   d) all of these

6. Which of the following makes up most of the soft tissues of plants?
   a) cuticle
   b) parenchymal cells
   c) epidermis
   d) phloem

7. Which of the following is not associated with xylem?
   a) tracheid
   b) vessel element
   c) pits
   d) sieve tubes

8. Monocots and dicots differ in which of the following?
   a) number of flower parts
   b) leaf vein pattern
   c) vascular system arrangement
   d) all of these

9. What is the function of the slimy mucigel secreted by the epidermis and root hairs?
   a) lubrication so tips can move through the soil
   b) water-proofing
   c) protection
   d) defense against herbivores

10. What function do guard cells serve?
   a) They act as valves controlling the stomata.
   b) They secrete secondary plant compounds.
   c) They regulate entrance of water into the roots.
   d) They store water for dry periods.

11. Which is the largest group of secondary plant compounds?
   a) phenolic compounds
   b) alkaloids
   c) tannins
   d) terpenes

12. Drugs such as nicotine and morphine are from what group of secondary plant compounds?
   a) terpenes
   b) alkaloids
   c) phenolic compounds
   d) These chemicals are not secondary plant compounds.

13. What is the function of phloem?
   a) It transports water from the roots to the shoots.
   b) It absorbs energy from sunlight.
   c) It transports the products of photosynthesis throughout the plant.
   d) It brings carbon dioxide into the plant at the stomata.

14. Central vacuoles in plant cells contain cell sap, which:
   a) stores the plant's starch
   b) stores carbon dioxide
   c) provides turgor pressure to support the plant
   d) conducts photosynthesis

15. Which of the following is not associated with monocots?
   a) trees
   b) flower parts in multiples of three
   c) leaf veins usually parallel
   d) lateral meristems

Part 2: Matching

For each of the following, match the correct term with its definition or example. More than one answer may be appropriate.

roots **stems** **leaves**

1. ________________ Contain stomata for taking in carbon dioxide.
2. ________________ Have a thick cuticle to prevent water loss.
3. ________________ Casparian strips are found here.
4. ________________ Contain xylem and phloem.
5. ________________ Most photosynthesis occurs here.

monocot **dicot**

6. ________________ A flowering plant is one of these.
7. ________________ A tree or shrub is one of these.
8. ________________ Flower parts usually come in multiples of three.
9. ________________ Vascular bundles are scattered throughout the stem.
10. ________________ Capable of secondary growth from lateral meristems.

Part 3: Short Answer

Write your answers in the space provided.

1. What does Bruce Ames believe causes all the degenerative diseases associated with aging?

2. From what ancestors do biologists believe early terrestrial plants may have evolved?

3. What plant adaptations did moving onto land require?

4. How do the functions of xylem and phloem differ?

5. Compare the positions and structures of the primary and secondary cell walls.

6. What effect does secondary growth have on a plant?

7. What are the cells that conduct fluid in phloem called?

8. List three differences between monocots and dicots.

9. In what type of plant would you be most likely to find a taproot?

10. How do plants obtain their nutrients?

11. What is the internal organization of a typical root?

12. Most of a root's cortex is composed of what type of cells?

13. What are the three main patterns of leaf arrangement?

14. List three mechanical defenses employed by plants.

15. What are the two roles played by lignin?

Part 4: Critical Thinking: Using Your Knowledge

Answer each of these in essay form, using complete sentences and paragraphs. Provide as much information as you can. (For extra essay practice, write out answers to the Review and Thought Questions in your textbook.)

1. Discuss the three groups of secondary plant compounds and provide an example from each group. What are some of the functions of these compounds?

2. Trace the path of water as it moves into, through, and out of a plant.

3. Explain adaptations that allow a plant to stand upright against the force of gravity. How might these adaptations have occurred?

4. Compare and contrast monocots and dicots.

5. Explain the general organization of a flowering plant.

6. After thoroughly reviewing the vascular system of terrestrial plants, what advantages, if any, are there to having two types of conductive tissue—xylem and phloem, instead of one type?

7. Explain the functioning of the stomata. What would happen to a plant if its stomata remained open all the time? What if they got "stuck" shut?

8. What happens to old, limp carrot sticks when they are soaked in water, and why?

9. If you pinch off the tip of a straight stem, why do many plants grow branches there?

10. After transplanting a plant, why should you remove most, but not all, of the leaves?

Chapter 31
What Drives Water Up and Sugars Down?

KEY CONCEPTS

1. Water moves from the soil, through the plant, then into the air via the plant's vascular system.
2. Roots use ionic pumps to draw water from the soil.
3. Water and minerals from the soil travel up through the plant in the xylem.
4. The driving force for the ascent of water is transpiration—passage of water from the leaves to the air. Transpiration pulls water up the stem in continuous columns.
5. Upward transport depends on water's adhesive and cohesive properties. The tensile strength of water is sufficient to keep it from breaking under the tension generated by transpiration.
6. Stomata regulate the rate of transpiration by opening and closing. Guard cells open in response to light and low carbon dioxide by pumping potassium ions from adjacent cells. This draws water into the guard cells and opens the stoma. In the presence of abscisic acid, guard cells lose potassium ions, water leaves the cells and they relax, closing the stoma.
7. The products of photosynthesis—sugars—move from sources (usually leaves or storage areas) to sinks (usually roots or storage areas).
8. Sugars move by way of the phloem—a cylinder of living cells just under the outer bark, or cork, of trees and shrubs.
9. Mesophyll cells pump sugars by active transport from cell to cell, then into the phloem.
10. The osmotic flow of water drives the transport of photosynthetic products in the phloem.
11. Crassulacean acid metabolism (CAM) gives some plants another way of saving water.

EXTENDED CHAPTER OUTLINE

THE DEVASTATING DRY LEAF CREATURE

1. In the 1970s, the California wine industry bloomed. Everyone was planting vineyards, mostly with vines on AXR1 rootstock, making AXR1 a vast monoculture.
2. If *Phylloxera*—a sucking insect that feeds on roots and leaves—could adapt to that one rootstock, the insects would eventually feast on hundreds of thousands of acres of grapes.
3. *Phylloxera* do not kill the plants themselves, rather they cluster in a gall on the roots and their feeding leaves the roots susceptible to fungal and bacterial infections which can kill the plants.
4. By 1985, researchers at the University of California Department of Viticulture knew that *Phylloxera* could kill vines growing on AXR1.
5. Not until 1989, however, was a press release issued warning growers to stop using AXR1.
6. Many factors contributed to the disaster, and blame can be found on all parts.

WHAT DRIVES WATER UP?

What is the route of water from soil to air?

1. Water enters the roots via the root hairs, moves through the cortex, crosses the endodermis, and enters the xylem.
2. Once in the xylem, water ascends from the roots to the leaves through parallel and usually interconnected xylem vessels in the stem.
3. These vessels branch into leaf traces, and water flows through petioles into the leaves.
4. The water then enters cells of the mesophyll and evaporates at the cell surface. The water vapor diffuses through the air spaces of the mesophyll and exits the leaf through the stomata. Transpiration is the passing of water vapor from leaf to air. The rate of transpiration is extraordinary—a leaf can transpire its own weight in water in less than an hour.
5. Plant scientists early in this century proposed three possible mechanisms for water transport:
   - water is pushed up from the roots;
   - capillary action in the xylem pulls water in; or
   - evaporation at the leaves creates a suction that pulls water up the xylem.

Root pressure: Can the roots push water to the tops of tall trees?

1. Water tends to move from areas of purer water to areas of high concentrations of solutes (dissolved materials in the water).
2. Osmosis is the tendency for water to move across a membrane in response to a difference in concentration of solutes.
3. The pressure exerted against the cell wall by osmosis is called turgor pressure. At the same time, the rigid cell wall pushes back with wall pressure.
4. Root xylem usually has a higher concentrations of solutes than the soil does; water from the soil therefore flows into the roots.
5. The pressure that the water inflow creates in the roots is called root pressure.
6. In small plants, root pressure can push water all the way up the stem and out of tiny holes at the margins of the leaves, in a process called guttation. Guttation occurs on cool, humid mornings, when the air is too saturated for water to evaporate from the leaves.
7. Many plants do not develop any root pressure at all.
8. The water in the xylem is sucked upward by a vacuum.
9. The rate of water flow due to root pressure is too slow to account for the more rapid transport of materials from roots to leaves. The driving force of water transport must therefore come from some source other than root pressure.

Can capillary action raise water to the tops of tall trees?

1. Water is cohesive—it sticks to itself and columns of water do not break up as they move through narrow tubes (capillaries).
2. Ordinary capillary action is not enough to drive water up very high.
3. Water is highly adhesive—it tends to cling (via hydrogen bonds) to the surfaces of carbohydrates and other polar substances.
4. Although root pressure and capillary action are not enough to drive water to the tops of tall trees, both effects are important to water movement in plants.

The driving force for the ascent of water is transpiration

1. The now accepted explanation for the ascent of water is called the cohesion–tension theory, first proposed in 1914 by Irish botanist Henry Dixon. The theory states that transpiration in the leaves pulls water up the stem in continuous columns.

How can transpiration draw water to the tops of tall trees?

1. Water molecules move into dry air as effectively as they move into cells with high concentrations of solutes.
2. The tendency for water molecules to move along gradients of concentration or pressure is called water potential, which is always relative to something else. The greater the gradient, the greater the water potential.
3. The osmotic pressure that results from the difference in water concentration between humid air and the liquid water inside a plant is enough to support a 300-meter column of water—nearly three times the height of the tallest trees.
4. Transpiration of water from the air spaces of the leaves to the atmosphere could provide the driving force for the ascent of water.
5. The driving force for all this movement is the tendency for liquid water to evaporate.

Is the flow of water through the stem coupled to transpiration?

1. Fluid moves more quickly as transpiration increases during the day.

Is water cohesive enough to sustain the transpiration pull in a tall plant?

1. The cohesion–tension theory depends on water columns in the xylem staying together.
2. Opposing forces pull the column—transpiration pulls upward as gravity pulls downward.
3. The great cohesiveness of water allows columns to hold together under tremendous pull.

Is water in the xylem really under tension?

1. The greater the rate of transpiration, the greater the tension in the xylem.

STOMATA REGULATE TRANSPIRATION BY OPENING AND CLOSING

1. Stomata are the gates through which carbon dioxide enters and water (and oxygen) leave plants. Far more water escapes than carbon dioxide enters, however.
2. All plants closely regulate water loss by opening and closing their stomata in response to environmental changes.
3. Closing the stomata reduces water loss, but it also prevents the plant from collecting carbon dioxide, therefore limiting photosynthesis.
4. The trade off between saving water and maintaining photosynthetic productivity is called the transpiration-photosynthesis compromise.
5. Stomata usually open in response to light and to low levels of carbon dioxide.
6. Stomata close when the air is hot and dry, and excessive transpiration threatens to wilt the plant.

How do stomata open and close?

1. Two guard cells, which can change shape, surround each stoma.
2. As guard cells gain water, they become turgid and bulge outward, opening the stoma.
3. When water exits the guard cells, they become flaccid and close the gap between them.
4. Guard cells regulate the flow of water by actively pumping potassium ions (K^+) from adjacent cells of the epidermis.
5. By osmosis, water rapidly follows these potassium ions, the cells become turgid, and the stoma opens.
6. Stomata close in response to water stress as a result of the action of the plant hormone abscisic acid. As water becomes less available, the concentration of abscisic acid increases.
7. Abscisic acid inhibits the flow of potassium ions into the guard cells, closing stomata.

How can plants save water without limiting photosynthesis?

1. The transpiration-photosynthesis compromise precludes most plants living in dry deserts.
2. Crassulacean acid metabolism (CAM) plants can collect all the carbon dioxide they need by opening their stomata only at night, when temperatures are cool and transpiration is low.
3. CAM plants have evolved a biochemical trick for storing carbon dioxide collected at night for use in photosynthesis during the day.
4. In CAM plants, carbon dioxide enters the stomata at night and forms an organic acid (crassulacean acid).
5. During the day, this acid, which is stored in the vacuole, gradually diffuses back into the cytoplasm, where it releases carbon dioxide for photosynthesis.

WATER IN THE XYLEM CARRIES MINERALS FROM THE SOIL

Most plants obtain both water and essential minerals from the soil

1. Researchers have determined which nutrients are needed by plants through hydroponic culturing (growing plants in water instead of soil) and varying the amount of each nutrient to observe the effects.

Most plants obtain both water and essential minerals from the soil

1. Both water and minerals usually come from the soil surrounding a plant's roots.
2. Soils consist of particles of weathered rocks and partially decayed organic matter.
3. Soils are classified according to the size of their particles:
   - clays consist of only small particles, less than 2 μm;
   - silts consist of medium-sized particles, from 2 μm to 20 μm;
   - sands consist of large particles, from 20 μm to 200 μm; and
   - loams consist of mixtures of all three particle sizes.
4. The ideal loam is about 20% clay, 40% silt, and 40% sand.
5. In most soils, bacteria and fungi quickly convert dead organisms into small molecules. The residue of such decay is a black or brown material—humus—that decays more slowly.

Why do roots need oxygen from the soil?

1. Plants actively transport ions into their root cells and concentrate them there, especially in the xylem.
2. Such active transport requires energy—root cells use large amounts of ATP.
3. Due to the high ATP demand, roots depend on oxygen in the soil (for oxidation).

Why do some plants obtain minerals from animals?

1. Many plants live in areas where minerals, especially nitrogen, are in short supply.
2. Plants such as the Venus flytrap solve this problem by trapping and digesting insects for their nitrogen compounds, phosphates, and other ions.

HOW DO PLANTS TRANSPORT THE PRODUCTS OF PHOTOSYNTHESIS?

1. Roots depend on the products of photosynthesis, but cannot themselves photosynthesize.

Which way does sugar move?

1. Products of photosynthesis move entirely within the sieve tubes of the phloem, moving from the leaves to the roots or to other areas where sugar is needed or stored.
2. In woody stems, functional sieve tubes are confined to the inner bark.
3. Removing a strip of bark around a tree trunk, a technique called girdling, interrupts the phloem and blocks the transport of photosynthetic products—girdling always kills the tree.
4. Sugars are said to move from source (the site of production) to sink (the site of storage or consumption).
5. A source may be either a site of photosynthesis, such as the leaves, or a storage site that releases sugars by breaking down starch.
6. Sinks include growing tips, young leaves, fruits, and storage organs.

What is inside a sieve tube?

1. Phloem contains a concentrated solution of sugars.
2. Sucrose (table sugar) may make up as much as 30% of the phloem contents.
3. Phloem sap also contains a few minerals and plant hormones.

The osmotic flow of water drives translocation in the phloem

1. Translocation occurs much faster than can be explained by diffusion alone.
2. The pressure flow hypothesis was first advanced in 1926 by Ernst Munch, a German plant physiologist and artist.
3. This hypothesis states that the concentration of sugar is higher in the phloem near a source than near a sink. More water therefore flows into the sieve tubes near a source and pushes the contents toward a sink.

What draws water and sugar into the phloem?

1. Cells of the mesophyll are connected to one another by plasmodesmata, which allow sugars to flow freely from cell to cell.
2. There are no plasmodesmata between the mesophyll cells and the cells of the phloem, so sugar must therefore always pass through two cell membranes—that of a mesophyll cell and that of a phloem cell. Both passages require energy and are selective.
3. Sugar flows into companion cells—specialized cells adjacent to the sieve elements.
4. The active pumping of sucrose is coupled to a pH gradient across the cell membrane, created by using ATP to pump out hydrogen ions.
5. Hydrogen ions flow back into the companion cell via a transport protein that also carries sucrose.
6. Once the sieve tube contains a high concentration of sucrose, osmotic pressure forces water into the phloem from the surrounding cells, driving the sugar away.
7. Although phloem transport can occur in either direction, most researchers now think that the flow within a single sieve tube goes in a single direction at a given time.

VOCABULARY BUILDING

In your own words, first write a brief definition, then a full explanation for each of the following terms. Include examples where appropriate. Complete this section from your memory—you will not learn the terms by simply copying definitions from the textbook. Once you have finished, check your responses against the information in the chapter and make any necessary corrections.

transpiration —

osmosis —

turgor pressure —

root pressure —

guttation —

cohesive —

capillary —

adhesive —

cohesion–tension theory —

water potential —

transpiration-photosynthesis compromise —

abscisic acid —

crassulacean acid metabolism (CAM) —

clays —

silts —

sands —

loams —

humus —

girdling —

source —

sink —

pressure flow hypothesis —

companion cells —

CHAPTER TEST

The following test has four parts. Complete as much of the exam as you can from memory. If you cannot answer a question, skip it. Once you complete all that you can, try to answer any questions you skipped. If you still cannot answer them, consult your textbook for the answers. Once you have completed all sections of the test, check your answers for Parts 1 - 3 against those in the back of this book. Highlight any incorrect answers then review that material in your textbook. Correct your answers for future reference.

Part 1: Multiple Choice

For each of the following, select all correct responses—more than one may be correct.

1. How does *Phylloxera* kill grape vines?
   a) The insects secrete a toxin that kills the plants.
   b) The insects secrete plant hormones that destroy the roots.
   c) The insects increase the plants' susceptibility to infection.
   d) The insects eat all the grapes.

2. Water travels upward through a plant within the:
   a) phloem
   b) cortex
   c) guard cells
   d) xylem

3. The tendency for water to move across a membrane in response to a difference in the solute concentration is called:
   a) osmosis
   b) guttation
   c) transpiration
   d) adhesiveness

4. The main force driving the ascent of water through a plant is called:
   a) osmosis
   b) guttation
   c) transpiration
   d) adhesiveness

5. Which of the following happens at the stomata?
   a) Carbon dioxide enters.
   b) Oxygen enters.
   c) Water leaves.
   d) All of these happen.

6. Which of the following will likely cause stomata to open?
   a) increased light
   b) dry air
   c) high temperatures
   d) low levels of carbon dioxide

7. Which of the following is a consequence of closing the stomata?
   a) increases the plants' carbon dioxide levels
   b) reduces water loss
   c) impairs photosynthesis
   d) all of these

8. Which of the following plays a role in closing the stomata?
   a) abscisic acid
   b) potassium ions
   c) guard cells
   d) all of these

9. Which of the following is true of CAM plants?
   a) They do not undergo photosynthesis.
   b) They do not transpire.
   c) They only open their stomata at night.
   d) They are well adapted for desert conditions.

10. Soils that consist of mid-sized particles are called:
    a) clays
    b) loams
    c) sands
    d) silts

11. An ideal loam consists of:
    a) 33% clay, 33% silt, and 33% sand
    b) 20% clay, 60% silt, and 20% sand
    c) 20% clay, 40% silt, and 40% sand
    d) 40% clay, 20% silt, and 40% sand

12. The products of photosynthesis move through the plant within the:
    a) phloem
    b) cortex
    c) guard cells
    d) xylem

13. Girdling kills a tree because it:
    a) restricts growth
    b) interrupts the downward flow of photosynthetic products
    c) releases toxic chemicals
    d) blocks transpiration

14. Which of the following are sources?
    a) leaves
    b) roots
    c) fruits
    d) storage organs that are not releasing energy-rich molecules

15. Which of the following is involved in moving products of photosynthesis through a plant?
    a) a gradient in sucrose concentration
    b) osmosis
    c) movement of water
    d) all of these

Part 2: Matching

For each of the following, match the correct term with its definition or example. More than one answer may be appropriate.

cohesion-tension theory **transpiration-photosynthesis compromise**

pressure flow hypothesis

1. ____________ Explains movement of photosynthetic products.
2. ____________ Transpiration in the leaves pulls water up the stem.
3. ____________ Explains movement of water from root to leaves.
4. ____________ Refers to the balance between water conservation and photosynthetic productivity.
5. ____________ Explains low crop yields during droughts.
6. ____________ Water moves as a continuous column.

stomata open **stomata close**

7. ____________ The plant has a low level of carbon dioxide.
8. ____________ The air is hot and dry.
9. ____________ Carbon dioxide enters.
10. ____________ Water is conserved.

Part 3: Short Answer

Write your answers in the space provided or on a separate piece of paper.

1. What economic interests might have played a role in the University of California's three-year delay before issuing a warning for growers to stop using AXR1 rootstock?

2. What substances move through the xylem?

3. What causes guttation?

4. Differentiate between water's adhesiveness and its cohesiveness, and explain how these properties aid movement of water up through the plant.

5. Explain the cohesion-tension theory.

6. Warm air in spring causes bubbles to form from the gases in the xylem, which breaks the water column. How do woody plants avoid this problem?

7. How does tension in the xylem relate to the diameter of a tree trunk?

8. Explain the transpiration-photosynthesis compromise.

9. Briefly explain how CAM plants have adapted to hot, dry environments.

10. What is humus?

11. List the three types of soils in order of decreasing particle size.

12. Why do roots require oxygen?

13. Why do carnivorous plants eat animals?

14. What visible sign indicates that girdling interferes with the downward flow of nutrients?

15. List some of the contents in plant sap.

Part 4: Critical Thinking: Using Your Knowledge

Answer each of these in essay form, using complete sentences and paragraphs. Provide as much information as you can. (For extra essay practice, write out answers to the Review and Thought Questions in your textbook.)

1. In your opinion, did the University of California act properly in dealing with the Phylloxera crisis? Should the press release warning have been issued earlier? What were the economic concerns, and do you believe that the delayed response was prudent? Would you have handled this situation differently, and, if so, how?

2. The situation with the AXR1 rootstock arose because vineyards relied almost exclusively on this one crop. In what other areas of agriculture do you see the potential for similar disasters?

3. Explain the various properties of water and the forces that are involved in moving water from the soil into and through the plant, ultimately releasing it to the air by transpiration.

4. Thoroughly explain the pressure flow hypothesis and all forces at work in moving nutrients from the sites of photosynthesis to other parts of the plant.

5. What is the nature of the soil at your home? Consider each of the three soil types. Why is none of them alone ideal for plant growth? How does particle size affect a plant's uptake of water, minerals, and oxygen?

6. What role can humus play in improving poor soil?

7. Why is it best to recut stems of cut flowers an inch from the end, and under water?

8. In how many ways, and why, can drought harm farm crops?

9. Nitrogen is required for healthy plants. What essential organic molecules in the plant depend on an adequate supply of this nutrient?

10. Epiphytes are "air" plants—they seem to require only air to grow, often perched high up on trees in rain forests and having no roots in the soil. How do these plants receive all the materials they require to survive?

Chapter 32
Growth and Development of Flowering Plants

KEY CONCEPTS

1. Stamens produce haploid microspores which develop into male gametophytes, known as pollen grains.
2. Carpels produce haploid megaspores which develop into female gametophytes, called embryo sacs. Each embryo sac has one egg.
3. Flowering plants undergo double fertilization—the fusion of two sperm nuclei with two cells of the embryo sac to form:
   - a diploid zygote, and
   - a triploid endosperm.
4. A seed contains a dormant embryo with one or two cotyledons surrounded by a seed coat.
5. At germination, the seed swells with water and the radicle breaks out of the seed coat to grow down into the soil. Development resumes with a burst of metabolic activity.
6. After germination in cereals, the aleurone layer secretes enzymes (such as α-amylase) that release nutrients to the growing embryo.
7. Light, water, temperature changes, stomach acid, and other environmental cues can all trigger germination.
8. Plants repeatedly generate modular structures from apical meristems at shoot and root ends.
9. The development of form in plants depends on oriented cell division and cell expansion.
10. Growing root tips include three zones:
    - the slimy root cap,
    - the apical meristem, and
    - the elongation zone.
11. Each leaf starts as a tiny extension of the apical meristem—through apical dominance, terminal buds suppress the growth of axillary buds.
12. Although many flowers are missing one or more parts, flowers generally develop from four whorls of primordia which form (from the outside in):
    - the sepals,
    - the petals,
    - the stamens, and
    - the carpels.
13. Homeotic selector genes help regulate development of flower parts.
14. Secondary growth, which depends on vascular cambium, changes with the seasons, resulting in characteristic growth rings in trees and other woody perennials.
15. Meristems allow plants to reproduce asexually through vegetative reproduction that does not involve fertilization or seed production. Cutting, grafting, and tissue culture allow artificial propagation of plants.

EXTENDED CHAPTER OUTLINE

THE ORIGIN OF CORN

1. Columbus left Spain in 1492, searching for a shortcut to Asia, but when he reached the New World, one of his most valuable discoveries was corn.
2. Within 50 years of Columbus' arrival in the West Indies, corn grew around the world.
3. What we know as corn is the product of intense artificial selection—modern corn is almost unable to reproduce by itself and is entirely domesticated.
4. Most people now agree that corn is about 10,000 years old and descended from some form of teosinte (grass) from central Mexico.
5. Corn's artificially induced anatomy has made it entirely dependent on humans for reproduction, while nearly every other species of plant on Earth can reproduce and grow with no help at all.

HOW DO PLANT EMBRYOS ESTABLISH A BODY PLAN?

1. A flowering plant's life cycle consists of two phases:
   - the diploid sporophyte, which produces haploid spores by meiosis, and
   - the haploid gametophyte, which produces haploid gametes by mitosis.
2. Cycling between these two phases is called alternation of generation.

How do plants make sperm?

1. Within the anther (the pollen-producing part of the stamen) are four pollen sacs.
2. Each pollen sac contains diploid cells that undergo meiosis to form haploid microspores, each of which develops into a male gametophyte.
3. In most plants, each microspore undergoes mitosis to produce two haploid cells—a generative cell and a vegetative cell (which completely surrounds the generative cell).
4. The generative and vegetative cells are enclosed together within a common wall to form a pollen grain.
5. Anthers release pollen which is then carried by the wind, insects, or other agents to the stigma of a flower.

How do plants make egg cells?

1. The pistil is the female reproductive part, and it contains one or more carpels. The ovary produces the female gametophytes and is located in the swollen base of the pistil.
2. The stigma receives the pollen, and the stigma is attached to the ovary via the style.
3. The walls of the ovary produce ovules which, after fertilization, form seeds.
4. One diploid cell in each ovule undergoes meiosis to form four haploid cells. One of these—the megaspore—develops into the embryo sac, which is the mature female gametophyte.
5. The diploid tissue surrounding the megaspore becomes the nucellus. The diploid tissue of the ovule forms the integuments and ultimately becomes the seed coat.
6. The megaspore undergoes mitosis to form eight haploid nuclei which form seven cells—the central cell contains two nuclei, and one of the remaining cells is the egg.

How does fertilization occur in plants?

1. Once pollen lands on the stigma, the vegetative cell forms a pollen tube through which the sperm passes.
2. The generative cell divides to form two sperm cells, and these both fuse with the egg cell.
3. When the pollen tube arrives at the embryo sac, the sperm are released.

4. Flowering plants differ from all other organisms by undergoing double fertilization—the simultaneous fusion of two sperm cells with two cells from the embryo sac. One sperm fuses with the egg, forming a diploid zygote; the other sperm fuses with the central cell (with its two nuclei) forming a triploid nucleus that gives rise to the endosperm.
5. The triploid endosperm provides nourishment and hormones for the growing embryo.
6. A developing seed consists of the embryo, endosperm, nucellus, and integument. In most species, the endosperm is used up during development so that the mature seed has only the other three components.

How does a plant form a seed?

1. The first cell division in a plant embryo divides the zygote into two unequal-sized cells:
   - the basal cell divides to form a column of cells called a suspensor, which attaches the embryo to surrounding tissue and nourishes it; and
   - the smaller cell forms the embryo, one end forming the root apical meristem while the other forms the shoot apical meristem.
2. The fundamental shoot-root polarity of the embryo is established early in development.
3. The embryo continues to divide, forming three regions:
   - tissues that will form the epidermis,
   - the ground tissue system, and
   - the vascular system.
4. Continued division of the shoot-end cells produces the cotyledons. Dicots have two cotyledons; monocots have one.
5. Cotyledons absorb nutrients from the endosperm for later distribution to the developing plant.
6. Above the attachment of the cotyledons the axis is called the epicotyl; below the attachment it is called the hypocotyl.
7. The first bud of the embryo—the plumule—includes the epicotyl and first true leaves.
8. The radicle, at the hypocotyl's end, is a primordial root containing the root's apical meristem.
9. The seed dries, the seed coat hardens, and the embryo enters a dormant stage in which growth and development are suspended.

How does the embryo resume development?

1. Resumption of growth after the dormant stage is called germination.
2. The first stage of germination is imbibition—the swelling of the seed with water.
3. When the embryo is hydrated, it awakens and undergoes a burst of metabolic activity.
4. The radicle and hypocotyl are the first parts of the embryo to break out of the seed, and the radicle immediately turns downward into the soil, forming a functioning root system.
5. In many dicots, the top of the hypocotyl forms a hook that pushes up through the soil and pulls the cotyledons and epicotyl behind it. In these plants, most of the stem derives from the epicotyl.
6. In other plants, the hypocotyl doesn't grow, the cotyledons stay below ground, and the epicotyl grows, pushing the plumule and apical meristem through the soil.

How does the embryo obtain nourishment?

1. Germination triggers a huge increase in biochemical activity—the energy and building blocks are derived from starches, oils, and proteins stored in the seed.
2. Breakdown of these stored materials requires enzymes that are not present (or are insufficient) in the ungerminated seed. In cereals, these enzymes are secreted by the aleurone layer—an outer layer of cells surrounding the endosperm.

3. One enzyme—α-amylase—helps break down starches stored in the endosperm.
4. The embryo itself stimulates the aleurone layer to make digestive enzymes—the chemical signal that triggers this production is the plant hormone gibberellin.

What environmental cues trigger germination?

1. Light is important for normal development of a plant. Germinating seedlings with insufficient light suffer etiolation—they are tall and spindly, have poor leaf development, and little or no chlorophyll production.
2. Researchers have succeeded in growing plants from lupine seeds found in 10,000-year-old frozen sediments.
3. Seeds do not have a built-in clock that determines the time of germination; they depend on environmental cues.
4. Plants avoid premature germination through various mechanisms, including:
   - hard seed coats that prevent water and oxygen from entering (some of these seeds require scarification to break the seed coat), or
   - chemical inhibitors, which are washed away by rainfall.
5. The most widespread of the chemical inhibitors is abscisic acid, which inhibits translation of messenger RNA in seeds, blocking synthesis of enzymes needed for germination.

HOW DO PLANTS GENERATE A SHOOT AND A ROOT?

1. Apical meristems are located just behind the tip of both the shoot and the root. They contain groups of undifferentiated cells that generate the specialized cells of the shoot and root.
2. An apical meristem can repeatedly generate the same set of plant parts, which serves as a construction module that may be repeated any number of times.
3. A shoot apical meristem can generate a stem and leaf; the area where the leaf attaches is the node, and the area between nodes is the internode.
4. Between nodes are repeating modules, each consisting of a stem segment, leaf, node, and bud.
5. A bud contains another shoot tip, with its own apical meristem, and a few tiny leaves below the meristem.
6. Action of the apical meristems is responsible for the plant's primary growth in both the shoot and the root.
7. Each plant species has a characteristic pattern of modular repetition.

How do meristem cells elongate and differentiate?

1. Undifferentiated meristem cells can stop dividing and start to differentiate into various cell types. They grow by taking water into a central vacuole.
2. These elongated, differentiated cells are arranged in columns.
3. Plant cells cannot move to form new associations during development—each cell has its place.
4. Cell expansion depends mainly on the osmotic influx of water into the central vacuole; due to the orientation of cellulose fibers, the cell wall tends to be weakest in one direction, so expansion occurs there, causing elongation rather than general enlargement.
5. Plant hormones dramatically influence growth of an entire plant. For example:
   - ethylene causes longitudinal orientation of cellulose fibers, leading to seedlings that are short and fat, and
   - gibberellin stimulates transverse orientation of cellulose fibers, leading to seedlings that are tall and thin.

Root tips grow and mature in three zones

1. Primary root growth occurs almost exclusively near their tips.
2. In general, the growth zone has three overlapping regions:
   - the root cap,
   - the meristem, and
   - the elongation zone.
3. The root cap, at the tip of the root, houses many Golgi complexes and vesicles, which contribute to making mucigel—a polysaccharide slime that lubricates the path of the growing root. The root cap also protects the apical meristem.
4. The apical meristem contains most of the dividing cells in the root.
5. The pattern of cell division in the meristem produces a multitude of cells already organized to resemble mature roots.
6. Cells in the elongation zone are increasingly large; those farthest away from the meristem are the most mature and differentiated.
7. Appearance of abundant root hairs marks the beginning of the differentiation zone—root cells with root hairs are fully mature.

Primary growth of shoots is more complicated than that of roots

1. The apical meristem of the shoot must produce stem, leaves, branches, and flowers, so primary growth here is more complex than in the roots.
2. The stem apical meristem also contains dividing cells arranged to elongate and push the meristem upward. Again, the cells farthest away from the meristem are the most mature and differentiated.
3. Each leaf originates as a leaf primordium—a tiny extension of the apical meristem.
4. A bud consists of a shoot tip and several young shoot modules, each containing a node, a short internode, and one or more leaf primordia.
5. Terminal buds lie at the tips of shoots; axillary buds are just above nodes and can form branches.
6. Many species show apical dominance, in which the presence of terminal buds suppresses development of axillary buds (via a plant hormone secreted from the terminal buds). Such plants tend to grow taller rather than wider.
7. Removing the terminal bud allows axillary buds to develop.

How does a flower derive from a specialized bud?

1. Some buds are specialized to form flowers or flowering shoots.
2. In most species, flowers contain both carpels (female reproductive structures) and stamens (male reproductive structures).
3. A flower consists of up to four sets of parts arranged in whorls; the two inner whorls are fertile—they produce the gametes. The whorls are arranged as follows:
   - the outermost whorls are the sepals—green parts at the base of the flower;
   - the next whorl forms the petals;
   - the third whorl inward has the male stamens; and
   - the innermost whorl has the female pistils, each containing one or more carpels.
4. Not all flowers have all four whorls:
   - Those that have all four whorls are called complete flowers.
   - Those lacking any of the whorls are incomplete flowers.
   - Those that contain both male and female structures (stamens and carpels) are perfect flowers.
   - Those that lack either stamens or carpels are imperfect flowers.

5. Corn flowers are incomplete and imperfect. Tassel-like flowers lack carpels, axillary flowers on the ears lack stamens.
6. Corn is an example of a monoecious plant—both male and female reproductive structures are carried on the same plant. In dioecious species, the male and female parts are on separate plants.

How do genes regulate the development of flowers?

1. Homeotic mutations are mutations that influence the location of organs.
2. Homeotic selector genes carry sequences of DNA that seem to regulate development of a great diversity of organisms.
3. Despite great differences between plants and animals, both share common strategies and even common molecular designs for regulating the development of form.
4. Plant homeotic selector genes probably serve the same roles in all or almost all plants.
5. Evidence indicates that meristems develop into leaves unless regulatory genes instruct them to form something else.

Secondary growth depends on two kinds of lateral meristems

1. Secondary growth refers to thickening of plant parts and occurs in two kinds of lateral meristems (cambia):
   - vascular cambium produces secondary xylem and phloem; and
   - cork cambium produces cork—the waterproof, insectproof covering of woody stems and roots.
2. Only the tips of woody stems and branches lack cambia.
3. Bark includes everything between the vascular cambium and the stem surface.
4. The vascular cambium lies between the xylem and phloem and adds secondary xylem and phloem to the primary vascular tissue.
5. In woody plants, this process continues each year, and the xylem accumulates as wood.
6. As a tree trunk expands, the mature epidermis is torn and destroyed, but just inside it is the cork cambium, which produces rows of protective cork cells.
7. Cork cells secrete suberin and other protective secondary compounds in their cell walls, then die, forming protective bark.
8. Secondary growth changes with the seasons in such a way that summer wood is much denser than spring wood, producing visible layers known as growth rings.
9. Each growth ring represents one year of growth, thus revealing the age of the tree.

MERISTEMS MAY PRODUCE WHOLE PLANTS

1. In vegetative reproduction, plants reproduce asexually, such as when strawberry plants reproduce by runners.
2. In some species, a stem cutting taken from the main plant and placed in water generates new roots and, when planted, grows into a clone—a mature plant that is genetically identical to the parent plant.
3. The plant hormone auxin promotes root development in plant cuttings.
4. In grafting, a cutting (the scion) from one woody plant is grafted onto the root (rootstock) of another plant. For example, in winemaking, French grape varieties are universally grafted onto American rootstock.
5. The trick to grafting is binding the scion to the rootstock so that their cambia come together. If this is accomplished, secondary growth establishes continuity between the vascular parts.

6. In tissue culture, a callus (undifferentiated tissue) is derived from the cut surface of a plant or from a single cell.
7. In the proper chemical environment, the callus becomes a plantlet—a tiny plant with both shoot and root. The chemical environment must include an energy supply (usually sucrose), nutrients, and suitable plant hormones, because the callus lacks these and does not undergo photosynthesis.

VOCABULARY BUILDING

In your own words, first write a brief definition, then a full explanation for each of the following terms. Include examples where appropriate. Complete this section from your memory—you will not learn the terms by simply copying definitions from the textbook. Once you have finished, check your responses against the information in the chapter and make any necessary corrections.

sporophyte —

gametophyte —

alternation of generations —

pollen grains —

stamens —

microspores —

pistil —

carpels —

ovary —

style —

stigma —

ovules —

nucellus —

integument —

megaspore —

embryo sac —

pollen tube —

double fertilization —

endosperm —

suspensor —

cotyledons —

dicots —

monocots —

epicotyl —

hypocotyl —

plumule —

radicle —

dormant stage —

germination —

imbibition —

aleurone layer —

α-amylase —

etiolation —

scarification —

apical meristems —

node —

internode —

bud —

root cap —

mucigel —

elongation zone —

leaf primordium —

terminal buds —

axillary buds —

apical dominance —

whorls —

sepals —

petals —

complete flower —

incomplete flower —

perfect flower —

imperfect flower —

monoecious —

dioecious —

homeotic selector genes —

secondary growth —

cambia —

vascular cambium —

cork cambium —

cork —

bark —

growth ring —

vegetative reproduction —

grafting —

CHAPTER TEST

The following test has four parts. Complete as much of the exam as you can from memory. If you cannot answer a question, skip it. Once you complete all that you can, try to answer any questions you skipped. If you still cannot answer them, consult your textbook for the answers. Once you have completed all sections of the test, check your answers for Parts 1 - 3 against those in the back of this book. Highlight any incorrect answers then review that material in your textbook. Correct your answers for future reference.

Part 1: Multiple Choice

For each of the following, select all correct responses—more than one may be correct.

1. Pollen sacs are located within the:
   a) carpel
   b) anther
   c) sepal
   d) stigma

2. Pollen grains are received by the:
   a) carpel
   b) anther
   c) sepal
   d) stigma

3. Which of the following develops into the embryo sac?
   a) microspore
   b) megaspore
   c) nucellus
   d) pollen grain

4. The vegetative cell of the pollen grain forms which of the following?
   a) pollen tube
   b) embryo
   c) endosperm
   d) egg

5. The endosperm is:
   a) haploid
   b) diploid
   c) triploid
   d) none of these

6. The primordial root that contains the root's apical meristem is called the:
   a) cotyledon
   b) hypocotyl
   c) radicle
   d) node

7. The first stage of germination is:
   a) dormant stage
   b) fertilization
   c) pollination
   d) imbibition

8. What does the enzyme α-amylase do?
   a) helps break down stored starches in the endosperm
   b) helps break dormancy at germination
   c) assists fertilization
   d) forms pollen grains

9. Undifferentiated cells that give rise to the specialized cells of the shoot and root are located in the:
   a) nodes
   b) apical meristems
   c) axillary buds
   d) all of these

10. Leaves attach to the stem at the:
    a) nodes
    b) apical meristems
    c) axillary buds
    d) internodes

11. What is the function of mucigel?
    a) It nourishes the embryo.
    b) It lubricates the path of growing roots.
    c) It triggers germination.
    d) It inhibits germination.

12. Apical dominance can be overcome by:
    a) removing all axillary buds
    b) removing the apical meristem
    c) removing the apical buds
    d) removing the terminal buds

13. The innermost whorl of complete flowers contains:
    a) sepals
    b) petals
    c) stamens
    d) carpels

14. An imperfect flower is lacking:
    a) pollen
    b) stamens
    c) carpels
    d) petals

15. The wood of woody plants is formed by accumulation of:
    a) cork cambium
    b) xylem
    c) starch
    d) phloem

Part 2: Matching

For each of the following, match the correct term with its definition or example. More than one answer may be appropriate.

microspore **megaspore**

1. ____________ This develops into the embryo sac.
2. ____________ This gives rise to the male gametophyte.
3. ____________ This produces two haploid cells.
4. ____________ This undergoes three rounds of mitosis, producing eight haploid nuclei but only seven cells.
5. ____________ This produces the generative cell.

tissue culture **grafting** **stem cutting**

6. ____________ This is a type of vegetative reproduction.
7. ____________ Roots grow when the sample is placed in water.
8. ____________ A scion is joined to a rootstock.
9. ____________ A callus is provided a proper chemical environment.
10. ____________ Secondary growth connects the xylem and phloem.

Part 3: Short Answer

Write your answers in the space provided or on a separate piece of paper.

1. List the components of a pollen grain.
2. Explain double fertilization in flowering plants.
3. What structure attaches the embryo to the surrounding tissues, like an umbilical cord?
4. The attachment point of the cotyledons divides the axis into what two regions?
5. What is imbibition?
6. What is the first part of the embryo to emerge from the seed coat?
7. What are two common mechanisms used to prevent premature germination?

8. Explain etiolation.

9. How does the apical meristem construct the plant parts?

10. How does cell expansion cause elongation?

11. Explain how ethylene can lead to short, fat seedlings.

12. List the three regions of the growth zone in the root.

13. Explain apical dominance.

14. List in order, from outside to inside, what the four whorls consist of.

15. What is meant by corn being monoecious and having incomplete and imperfect flowers?

Part 4: Critical Thinking: Using Your Knowledge

Answer each of these in essay form, using complete sentences and paragraphs. Provide as much information as you can. (For extra essay practice, write out answers to the Review and Thought Questions in your textbook.)

1. Modern corn is an unusual species—a product of years of artificial selection which is entirely dependent on humans for its reproduction. Can you think of any potential risks from having the world so reliant on this major food source?

2. Thoroughly describe the processes by which plants form egg and sperm cells.

3. Explain the relationship between the environmental conditions and germination.

4. Thoroughly describe how primary growth occurs in the shoot and the root. How does secondary growth occur?

5. Why are fires essential to maintaining the health of forest and grassland biomes?

6. Thoroughly discuss the process of fertilization in plants.

7. What are some ways in which seed scarification occurs in nature?

8. Discuss the formation of cork, bark, and growth rings.

9. You started some tomato plants from seeds in your house to extend your growing season, but the young plants are growing tall and spindly, and look very pale. Some are even dying. What, if anything, can you do to save your plants?

10. I plant lima beans each year. As soon as most of the plants emerge, split seed pod still attached, squirrels and rabbits eat the top. A few make it past this stage and two leaflike structures open, a beautiful healthy green, but that part gets eaten. All of these seedlings have good root growth, but the plants still die. Why are they dying, and what could I do to help them survive?

Chapter 33
How Do Plant Hormones Regulate Growth and Development?

KEY CONCEPTS

1. Hormones direct development and coordinate cellular and biochemical activities in many cells throughout the plant.
2. Auxin has many effects, including:
   - coordination of phototropism;
   - stimulation of cell elongation;
   - formation of roots;
   - development of fruit; and
   - inhibition of growth and development of axillary buds.
3. Auxin, with gibberellin, helps determine whether cells differentiate into xylem or phloem. In large concentrations, auxin and its synthetic analogs are potent herbicides.
4. Gibberellins promote:
   - germination;
   - breakdown of starches and sugars;
   - cell division;
   - cell elongation, especially in stems; and
   - gene expression.
5. Cytokinins, in conjunction with auxin:
   - stimulate cell division and growth;
   - delay senescence;
   - stimulate growth in axillary buds; and
   - determine how cells differentiate.
6. Ethylene is a gas that promotes:
   - abscission of leaves and fruit;
   - fruit ripening;
   - the triple response in seedling; and
   - release of more ethylene.
7. Abscisic acid promotes dormancy in buds and seeds, and helps stimulate abscission, antagonizing the effects of gibberellins and auxin.
8. Studies suggest that other plant hormones, such as systemin, salicylic acid, and jasmonic acid, differ significantly from animal hormones.
9. Like other organisms, plants are highly responsive to changes in their environment—they may alter their shape, structure, and the exact positions of the leaves and flowers to increase their ability to survive.
10. In gravitropism, cells in the roots somehow detect the direction of gravity. If the root is horizontal, calcium ions and auxin seem to facilitate the turning of the root downward.
11. Phototropism and solar tracking depend on a pigment other than chlorophyll.
12. Plants detect passing of seasons by measuring the length of the night by means of two pigments, called phytochromes.

EXTENDED CHAPTER OUTLINE

A LIFE LIVED FULL

1. In the 1880s, Charles Darwin and his son performed experiments to explain phototropism—the tendency of plants to bend toward light. They showed that something in the tip of oat seedlings causes this response.
2. Over four decades later, Frits Went and his father demonstrated that the "something" is a chemical substance in the tip of the plant. The substance was named auxin.
3. Kenneth Thimann took a job as an instructor at California Institute of Technology and was recruited to help identify the chemical nature of auxin, which he did.
4. Later, at Harvard, Thimann noticed large amounts of auxin are fatal to plants, and synthetic auxins are especially lethal because the plants lack enzymes needed to break them down.
5. During World War II, Thimann's synthetic auxins were developed for use as herbicides, but they were never used against the Japanese—the atomic bomb made them unnecessary.
6. The herbicides were heavily used after the war—road crews used them along highways, and corn and wheat farmers increased their yields by 30%.
7. The herbicides were revived during the Vietnam War—forests were sprayed heavily and the trees dropped their leaves to reveal targets for American and South Vietnamese aircraft.
8. The most commonly used herbicide was "Agent Orange," which was produced in bulk quantities that were heavily contaminated with the poison dioxin. Later, American service personnel who were exposed to the dioxin attributed a variety of illnesses to Agent Orange. The chemical companies settled out of court for $180 million.

HOW DO PLANTS DETECT ENVIRONMENTAL CHANGES AND HOW DO THEY COORDINATE THEIR RESPONSES TO THOSE CHANGES?

1. Plants change shape throughout their lives.
2. Plants are active and responsive to their environments:
   - A plant shaded by other plants grows toward any available light.
   - The roots of a plant growing in dry soil grow downward until they reach water.
   - Plants adjust the position of their leaves or flowers to track the sun from hour to hour.
   - Flowers open and close according to time of day and other variables.
   - Some plants forcefully eject seeds from their fruits.
   - Depending on environmental cues, the same cells that give rise to leaves and stems may also give rise to flowers or roots.
3. A plant's ability to respond to its environment is especially remarkable because the cell walls make its cells far more rigid than those of animals.

How do plants respond to the pull of gravity?

1. Turning toward or away from any stimulus is called a tropism.
2. A plant's response to gravity is called gravitropism:
   - roots, especially primary roots, grow down;
   - shoots, especially primary shoots, grow up; and
   - branches are usually horizontal.
3. Large cells in the root cap contain organelles called amyloplasts, which contain several starch grains. These organelles are denser than others and settle to the bottom of the cells.
4. Many plant physiologists suspected that amyloplasts were the root's statoliths (gravity detectors), but more recent research has not supported this hypothesis.

5. Auxin plays a role in gravitropism, and other suspected signals are ethylene and abscisic acid; but the initial trigger seems to be calcium ions.
6. When a root is horizontal, calcium ions accumulate along the lower surface, triggering a similar accumulation of auxin in these areas. Then auxin inhibits cell elongation at the lower edge, but cells continue to elongate along the top. The continued growth on the top, in the absence of this expansion on the bottom, causes the root tip to turn downward.

What pigment facilitates phototropism?

1. Plants respond to light, or the lack of it, in many ways, including phototropism, solar tracking (the turning of leaves to follow the sun) and nighttime sleep movements.
2. Two classes of proteins—cryptochromes and phototropin—somehow respond to blue light by stimulating transport of auxin from the illuminated to the shaded side of growing shoots.

How do plants detect changes in season or daylight?

1. In some species, the time of flowering depends on maturity or size; in other species, flowering always occurs around a certain date.
2. For any species, the flowering date varies with latitude and altitude.
3. Flowering response is an example of photoperiodism—the response to the relative lengths of day and night throughout the year.
4. Short-day plants flower when days are short and nights are long; long-day plants require long days and short nights; day-neutral plants do not depend on day length for flowering.
5. Researchers now know that plants actually measure the length of the night, not the day.
6. The pigment involved in measuring night length is phytochrome.
7. Phytochrome regulates many processes that depend on the timing of dark and light, such as flowering, germination, and leaf formation.
8. Phytochrome exists in two interconvertible forms: the red-absorbing form, P_r, and the far-red-absorbing form, P_{fr}.
9. Hendricks and his colleagues explained the effect of light and dark on flowering as follows:
   - sunlight contains more energy in the red than in the far-red part of the spectrum;
   - during daylight, the red component of sunlight converts P_r to P_{fr}; and
   - at night, P_{fr} converts back to P_r.
10. P_{fr} promotes flowering in long-day plants and inhibits flowering in short-day plants.
11. The action spectrum for other light-regulated changes suggests that phytochrome is important in numerous developmental processes which vary between species, but include:
    - seed germination,
    - elongation of new seedlings,
    - beginning of chlorophyll synthesis, and
    - production of enzymes for photosynthesis.
12. In almost every case, the developmental change controlled by phytochrome depends on the accumulation of P_{fr}. Red light (or daylight) promotes these processes by converting P_r to P_{fr}; far-red light inhibits them by allowing P_{fr} to be converted to P_r. The longer the night, the lower the levels of P_{fr} become.

HORMONES COORDINATE ACTIVITIES IN CELLS THROUGHOUT THE PLANT BOTH DURING DEVELOPMENT AND IN RESPONSE TO ENVIRONMENTAL CUES

1. A hormone is an organic compound produced in a tissue or organ and transported to another tissue or organ, called the target, where it produces one or more special effects.
2. Plant hormones move by diffusion and by transport in vascular tissues.
3. Plant hormones exert their effects at very low concentrations.

4. Plant physiologists have identified five families of plant hormones: auxins, gibberellins, cytokinins, ethylene, and abscisic acid.
5. The same hormones affect different tissues differently.
6. Hormones have different effects at different concentrations.
7. Hormones interact with each other in complex ways.

Auxin coordinates many aspects of plant growth and development

1. Briggs concluded that phototropism depends on auxin distribution within the tip.
2. Auxin acts at the level of the individual cells, stimulating cell elongation in shoots and roots.
3. Auxin also affects gene expression in target cells.
4. The main event in cell elongation is breaking the bonds in the cell's rigid wall, allowing the cell contents to expand. This is done by expansin, secreted by the cells in response to auxin.
5. Auxin's concentration varies from place to place within a plant, creating an auxin gradient from the top to the bottom of a shoot, and a gradient is also formed in the roots. The gradient is essential for gravitropism.
6. The relative concentrations of auxin and gibberellin in vascular tissue help establish the developmental fate of individual cells—which become xylem, and which become phloem.
7. Auxin inhibits abscission of leaves and fruit.
8. Auxin inhibits development of the axillary buds, and is responsible for apical dominance.
9. High concentrations of auxin are very effective as herbicides, and synthetic forms are especially effective because the plants cannot clear them away.

How do gibberellins act on target cells?

1. Plants have more kinds of gibberellins than of any other hormone, and most plant species have several different kinds of gibberellins.
2. Gibberellins generally affect overall growth of intact plants more than other plant hormones.
3. Like the auxins, gibberellins seem to promote cell elongation, especially in the stems, by loosening cell walls, but with a more dramatic effect.
4. Gibberellins stimulate the breakdown of starches and sucrose to monosaccharides (simple sugars), thus increasing the availability of energy and also the solute concentration inside the cell. This latter effect causes more water to flow into the cells, increasing their elongation.
5. In some cases, gibberellins can stimulate cell division.
6. Gibberellins can change the pattern of expression of specific genes.
7. Gibberellins appear to be important in the normal growth of shoots, the germination of seed, and the mobilization of food reserves from the endosperm or cotyledon.

Cytokinins stimulate growth and differentiation

1. In addition to stimulating cell division, cytokinin and auxin together help determine the developmental fate of plant tissue—for example, whether a callus culture continues to divide without differentiation or will differentiate into shoots or roots.
2. Cytokinin and auxin also work in opposition to one another:
   - Auxin promotes root initiation but cytokinin opposes it.
   - Auxin promotes bud dormancy and inhibits dormancy bud initiation, but cytokinin relieves dormancy and promotes new buds.
   - Auxin originates in the shoot tips and moves downward, but cytokinin originates in the roots and moves up through the xylem.
3. Application of cytokinin to a cut leaf delays senescence—the breakdown of cellular components leading to cell death.
4. During the normal life cycle of a plant, cytokinin transported to leaves by the xylem appears to prevent senescence from occurring prematurely.
5. Naturally occurring cytokinins are present and work at exceedingly low levels.
6. Some plant scientists think that it is incorrect to call cytokinin a hormone.

How did researchers identify ethylene as a plant hormone?

1. Ethylene is a simple molecule, with just two carbon atoms and four hydrogen atoms.
2. Unlike other hormones, ethylene is a gas and is not transported in the vascular system.
3. Among the chemical changes that ethylene starts or speeds up during fruit ripening are:
   - breakdown of chlorophyll, changing the fruit's color;
   - breakdown of starches, making the fruit sweeter; and
   - breakdown of cellulose, making the fruit softer.
4. Ethylene also triggers increased production of more ethylene, so ripening often spreads rapidly within a single fruit and between fruits.
5. Ethylene is present almost everywhere in a plant.
6. Under the right conditions, ethylene causes abscission—dropping of leaves and fruit.
7. In many plants, ethylene causes a triple response—stems thicken, stop elongation, and begin to grow horizontally.

Abscisic acid promotes dormancy in buds and seeds

1. Abscisic acid takes its name from its accumulation during abscission of leaves and fruit, although it actually plays only a minor role in this process.
2. Abscisic acid plays a major role in suspending the development of buds and seeds.
3. Abscisic acid is primarily an inhibitor—it opposes the effects of auxins and gibberellins.
4. Abscisic acid inhibits transcription of some genes and stimulates transcription of others.
5. Abscisic acid also coordinates the responses to a variety of environmental stresses.

Are there other plant hormones?

1. Wounding a plant leaf can evoke both local and long-distance effects, suggesting the involvement of one or more plant hormones.
2. In response to leaves being nibbled, plants may secrete systemin, which stimulates synthesis of jasmonic acid, which, in turn, stimulates transcription of "systemic wound response genes) that promote various defensive responses.
3. Plants can develop resistance to viruses, bacteria, or fungi at the wound site or farther away.
4. Signaling molecules may mediate both local and long-range effects.
5. Salicylic acid, which we know as aspirin, also:
   - helps prevent plant infections,
   - promotes flowering, and
   - promotes defensive responses.
6. In heat-generating (thermogenic) flowers, salicylic acid stimulates a mitochondrial "short circuit" that generates heat, increasing the flower's temperature by 14°C (25°F). The heat vaporizes foul-smelling chemicals that attract pollinators.

VOCABULARY BUILDING

In your own words, first write a brief definition, then a full explanation for each of the following terms. Include examples where appropriate. Complete this section from your memory—you will not learn the terms by simply copying definitions from the textbook. Once you have finished, check your responses against the information in the chapter and make any necessary corrections.

phototropism —

auxin —

tropism —

gravitropism —

amyloplasts —

statoliths —

photoperiodism —

phytochrome —

hormone —

target —

gibberellins —

cytokinins —

senescence —

ethylene —

abscission —

abscisic acid —

CHAPTER TEST

The following test has four parts. Complete as much of the exam as you can from memory. If you cannot answer a question, skip it. Once you complete all that you can, try to answer any questions you skipped. If you still cannot answer them, consult your textbook for the answers. Once you have completed all sections of the test, check your answers for Parts 1 - 3 against those in the back of this book. Highlight any incorrect answers then review that material in your textbook. Correct your answers for future reference.

Part 1: Multiple Choice

For each of the following, select all correct responses—more than one may be correct.

1. The growth of a plant toward light is called:
   a) gravitropism
   b) photoperiodism
   c) phototropism
   d) none of these

2. Which of the following is responsible for apical dominance?
   a) auxins
   b) gibberellins
   c) ethylene
   d) abscisic acid

3. The relative concentrations of which of the following determine whether cells become xylem or phloem?
   a) auxins
   b) gibberellins
   c) ethylene
   d) abscisic acid

4. Gibberellins cause cell elongation primarily in the:
   a) roots
   b) leaves
   c) shoots
   d) flowers

5. Senescence refers to:
   a) dropping of leaves
   b) dropping of flowers
   c) dropping of fruit
   d) breakdown of cell components leading to death

6. Which of the following promotes cell division and growth in axillary buds?
   a) auxins
   b) gibberellins
   c) ethylene
   d) cytokinin

7. During ripening, fruit changes color as a result of breakdown of:
   a) cellulose
   b) starches
   c) chlorophyll
   d) ethylene

8. Which of the following is primarily responsible for the ripening of fruit?
   a) auxins
   b) gibberellins
   c) ethylene
   d) abscisic acid

9. Which of the following is not a part of the triple response caused by ethylene?
   a) abscission occurs
   b) stems thicken
   c) elongation stops
   d) stems grow horizontally

10. Which of the following plays a major role in causing plants to drop their leaves and fruit?
   a) auxins
   b) gibberellins
   c) ethylene
   d) abscisic acid

11. Which of the following promotes flowering and initiates defensive responses?
   a) salicylic acid
   b) gibberellins
   c) auxin
   d) abscisic acid

12. Which of the following triggers the gravitropism response?
    a) auxins
    b) calcium ions
    c) salicylic acid
    d) ethylene

13. The response to the relative lengths of day and night is called:
    a) phototropism
    b) gravitropism
    c) photoperiodism
    d) none of these

14. Short-day plants will most likely flower during which of the following times?
    a) spring
    b) early summer
    c) late summer
    d) fall

15. Which of the following is the light detector involved in plants' sensitivity to the seasons?
    a) auxins
    b) salicylic acid
    c) ethylene
    d) phytochrome

Part 2: Matching

For each of the following, match the correct term with its definition or example. More than one answer may be appropriate.

auxin **cytokinin** **ethylene**

1. ____________________ Stimulates cell elongation and growth.
2. ____________________ Opposes some effects of auxin.
3. ____________________ Inhibits development of axillary buds.
4. ____________________ Helps determine the developmental fate of tissue.
5. ____________________ Promotes fruit ripening.
6. ____________________ Prevents premature senescence.

gravitropism **phototropism** **photoperiodism**

7. ____________________ Involves movement of auxin.
8. ____________________ Triggered by calcium ions.
9. ____________________ Regulated by phytochromes.
10. ____________________ Plants respond to amount of light (or dark).
11. ____________________ Causes roots to grow downward.
12. ____________________ Causes plants to grow toward a light source.

Part 3: Short Answer

Write your answers in the space provided or on a separate piece of paper.

1. Chemically, Agent Orange is related to what plant hormone, and why was it used in the Vietnam War?

2. By what mechanisms do plant hormones move?

3. Explain how auxin regulates phototropism.

4. Which of the plant hormones discussed has the greatest effect on the overall growth of intact plants?

5. How does gibberellins' breakdown of starches and sucrose enhance cell elongation?

6. How do auxin and cytokinin work in opposition to each other at axillary buds?

7. Differentiate between senescence and abscission.

8. Which of the plant hormones discussed has the simplest composition, and what is it?

9. How can burning incense or building bonfires hasten fruit ripening?

10. Explain the validity of the saying that "One bad apple spoils the whole barrel."

11. Abscisic acid opposes the actions of what other hormones?

12. List two effects of salicylic acid in plants.

13. List three examples of how plants respond to their environment.

14. Explain long-day and short-day plants.

15. Explain the relationship between sunlight, P_r, and P_{fr}.

Part 4: **Critical Thinking: Using Your Knowledge**

Answer each of these in essay form, using complete sentences and paragraphs. Provide as much information as you can. (For extra essay practice, write out answers to the Review and Thought Questions in your textbook.)

1. Provide a list of the effects and interactions for all plant hormones discussed in this chapter.

2. Explain how P_r and P_{fr} regulate photoperiodism in plants. Include a discussion of how this relates to short-day, long-day, and day-neutral plants.

3. Trace the history of the usage of synthetic auxins. Have all of these uses been beneficial? Are there any health risks to using large amounts of these herbicides?

4. In what ways are plant and animal hormones similar, and in what ways do they differ?

5. What change in the growth pattern would you expect if plants were grown during a space mission?

6. Explain the mechanisms by which auxins can, on one hand, benefit plants, yet can also be used as herbicides.

7. Your flowering houseplants have not bloomed for some time, so you decide to help the process. You install a grow light so the plants get natural daylight during the day, and artificial light most of the night. Your plants grow, but still do not bloom. What is wrong?

8. What do you predict would happen to the growth and development of a plant if you placed a seed in a clear container that was loosely filled with moist soil, then flipped the container upside-down every day so it spent every other day in the opposite vertical position? Why?

9. Why are some herbicides fine for use on soybean crops but not usable for corn crops?

10. You are having friends over in two days for your famous spaghetti with tomato sauce. The tomatoes you had on hand were old and getting mushy, so you bought new ones, but these are not ripe enough to add the robust tomato flavor your sauce requires. How can you remedy this situation in time to save dinner?

Chapter 34
Form and Function in Animals

KEY CONCEPTS

1. Because of gravity, large animals must have proportionately stronger and thicker bones than small animals. Similarly, fast runners need stronger bones than their sedentary counterparts.
2. Metabolic rate is partly a function of size.
3. Many organs of the body rely on enormous surface areas to function most efficiently. These include the lungs, intestines, circulatory system, and even webbed feet.
4. Countercurrent systems enable animals to establish steep and permanent gradients of temperature or concentration.
5. The style of locomotion partly determines the overall shape of an animal.
6. Most swimmers undulate through the water.
7. Snakes generally undulate like fish, but many also move by rectilinear movement and sidewinding.
8. Flying animals have many specializations, including powerful wings; a light, rigid skeleton; hollow bones; reduced jaws and legs; and reproductive structures that shrink when not needed.
9. Runners increase overall speed by increasing stride length and stride rate. Long legs, specialized hips and shoulders, and a flexible spine all increase stride length. Light legs and muscle attachments close to the joint increase stride rate.
10. Animals move by contracting pairs of antagonistic muscles which pull against some type of skeleton.
11. Tough collagen fibers arranged in parallel or in layers give bone and connective tissues much of their strength.
12. Osteoclasts and osteoblasts continually shape and reshape bones as an organism grows, heals, and changes habits.
13. In the human body, some 600 skeletal muscles, most of them arranged in antagonistic pairs, work our bodies by pulling against bone and against each other.
14. Smooth muscle and cardiac muscle are not attached to bones. They are also not under conscious control—rather, they are under control of the autonomic nervous system.
15. Skeletal muscles contain two types of fibers:
    - glycolytic (fast) fibers deliver great power quickly, but have poor endurance; and
    - oxidative (slow) fibers generate less power, but have greater endurance.
16. The striped appearance of striated muscle comes from the ordered arrangement of striped sarcomeres; the striping in sarcomeres comes from the arrangement of actin and myosin filaments.
17. Muscle contracts when actin and myosin filaments slide past one another—a movement powered by ATP.
18. All organisms resist changes through homeostasis and negative feedback mechanisms.
19. Many animals conform to their environment—a phenomenon called tolerance.

EXTENDED CHAPTER OUTLINE

TYRANNOSAURUS WRECKS: *T. REX* TREKS, BUT SLOWLY

1. Because dinosaur skeletons share many features with modern reptiles, early paleontologists assumed that dinosaurs resembled modern reptiles in a lot of ways. In the last 20 years, all these assumptions have been called into question.
2. Zoologist R. McNeill Alexander concluded that although many dinosaurs had long legs that enabled them to run fast, their legs were not strong enough to carry them at high speeds.
3. At best, Alexander estimated a *T. rex* could run no more than about 18 to 22 mph.
4. Farlow and Robinson confirmed Alexander's calculations.

WHY ARE ANIMALS SHAPED THE WAY THEY ARE?

1. Form follows function—form evolves in response to the demands placed on a body part.
2. Form can change over time.
3. Most form takes shape during development from the embryo to the adult, resulting from interactions between genes and the environment.

How are the parts laid out?

1. All vertebrates and their embryos share a common body plan—a tube within a tube. The inner tube is the intestinal tract, and the intestines and associated organs lie within a cavity—the coelom.
2. In vertebrates, the coelom has two distinct parts:
   - the thoracic cavity, enclosing the heart and lungs; and
   - the abdominal cavity, housing the intestinal tract, the liver, pancreas, and kidneys.
3. The backbone (spinal column) runs down the back, with the skull and tail attached at the ends. Just behind the skull, the pectoral girdle supports the forelimbs; the ribs are behind that, and, finally, the pelvic girdle supports the pectoral fins or hind limbs.

What are animal bodies made of?

1. Groups of similar cells form tissues, which include cells and matrix—materials secreted by the cells.
2. Four classes of tissue exist:
   - Epithelial tissues tend to be flat sheets of tightly interconnected cells. These tissues cover and line all our body parts. Endothelial tissue, which lines the heart and blood vessels, is similar but has a different origin.
   - Connective tissues hold parts together, provide mechanical support, and include cartilage and bone. These tissues often have individual cells dispersed in a flexible matrix.
   - Muscle tissue comes in three types—skeletal, cardiac, and smooth. All muscle cells contract.
   - Nervous tissue contains neurons, which conduct electrochemical signals, and glial cells, which support the neurons.
3. Organs are two or more tissues that make distinct structures with specific functions.
4. Organs are grouped into organ systems.

How are size, surface area, and shape related?

1. The most obvious differences between animals are differences of size and shape.

How does size affect shape?

1. For each animal, there is an appropriate range of sizes, and evolution of a different size requires a change in form.
2. As the linear dimensions of an animal double, its area increases four-fold. In contrast, the volume and weight increase eight times. The result is that the bones and muscles bear a proportionately heavier load.

3. In general, the heavier the animal, the thicker its bones must be to support its weight.
4. Ants and other small animals are saved during falls by their relatively large surface area—wind resistance, which is proportional to surface area, slows any falling object until the object reaches a terminal velocity.
5. Small animals are more suited to life in trees and other high places. While most primates live in trees, large apes, such as humans, gorillas, baboons, and chimpanzees, are more likely to live safely on the ground.
6. Large herbivores that eat leaves from trees do not climb the trees—they have evolved long necks or trunks for reaching up into the trees.

Why does metabolic rate depend on size?

1. Size influences metabolic rate—small, warm-blooded animals have much greater metabolic rates than larger animals. This means they also need more calories and oxygen.
2. Why larger animals have slower metabolisms is not entirely understood, but larger animals have proportionately smaller surface areas—they lose less heat than do small animals.

How do animals use expanded surface areas?

1. Animals of the same size can vary enormously in shape.
2. One change that occurs in the evolution of an organ is a change in surface area.
3. Webbed feet increase surface area without increasing overall size.
4. The wings of bats and birds are limbs whose surface area has increased enormously, whether by means of extended bones, long feathers, or wide membranes.
5. Another use for increased surface area is absorption of nutrients.
6. The lungs, digestive tract, and circulatory system all have enormous surface areas for moving nutrients and wastes to and from the cells of the body.
7. The rate at which oxygen and nutrients diffuses across a membrane depends on:
   - how easily the material diffuses across that membrane, and
   - the difference in the amounts of the material on each side of the membrane (gradient).

Bulk flow

1. Diffusion works only over small distances.
2. Larger animals use bulk flow, where a force pushes a substance from one area to another.

How do animals use countercurrent systems?

1. Bulk flow and diffusion can work together in arrangements called countercurrent systems, in which fluids or gases run past each other and exchange heat or materials.
2. Countercurrent mechanisms let animals have steep temperature or concentration gradients.
3. Since the mucous membranes in the nose are moist, evaporation from this large surface area cools the inside of the nose. Dogs depend on cool noses to cool themselves off on hot days.
4. To keep the whole body cool would require more water than is available to most desert animals. The solution is to allow the body to heat up while keeping the head cool.
5. To do this, gazelle and antelope run warm blood from the heart into a network of small arteries that run through a large sinus filled with cool, venous blood from the nose.
6. Running the warm, arterial blood past the cold, venous blood is an example of a countercurrent system—a system in which fluids or gases are run past each other so that they can exchange something.
7. Dolphins and whales use countercurrent systems to prevent the cold blood from their flippers from chilling the rest of the body. The veins that carry the cold blood run adjacent to and surround the arteries that carry warm blood from the heart to the flippers. The cold blood is thus warmed, and the warm blood cooled.

8. This type of countercurrent heat exchanger is also found in the legs of wading birds and the testes of mammals.
9. Fish use a countercurrent system to extract oxygen from water passing through their gills.

HOW DOES STYLE OF LOCOMOTION INFLUENCE SHAPE?

1. To propel themselves forward, animals must push against whatever is available.
2. Each style of movement uses the limbs in different ways.

What kinds of adaptations do swimmers have?

1. Animals exclusively adapted to swim rapidly in water all have the same oblong shape; a swimmer's tail is broad and flat, which forces the water against the tail; and fast vertebrate swimmers rely on webbed flippers and feet.

How do snakes crawl?

1. Snakes usually undulate across the ground in the same way that eels move through water.
2. Snakes on very smooth surfaces cannot undulate, so they resort to rectilinear movement, in which they progressively bunch up and relax their ventral scales.
3. Most snakes are capable of rectilinear movement, but desert species are especially good at sidewinding, which keeps much of the snake's body off the hot ground.

What kinds of adaptations do flyers and gliders have?

1. True flying demands extreme specializations that can limit ground activities.
2. All vertebrate flyers have adapted forelimbs that are used as wings. Wings are virtually useless for anything but flying, so most vertebrate flyers walk on two legs instead of four and use their hindlimbs the way others use their forelimbs.
3. Other adaptations found in flyers include:
    - extremely light bodies;
    - hollow bones and light skeletons;
    - lack of teeth;
    - reduced legs;
    - their reproductive organs may shrink when not breeding;
    - a diet that avoids leaves and grass (which allows a smaller digestive tract);
    - a compact and stiff body; and
    - a rigid skeleton.

How do animals walk and run?

1. Most terrestrial vertebrates walk as if their legs are upside-down pendulums.
2. When running, the legs act more like pogo sticks.

What kinds of adaptations do runners have?

1. For terrestrial vertebrates, speed is determined by two factors—stride length and stride rate.
2. Stride length increases over the course of evolution:
    - when the leg length increases in proportion to other parts of the body, or
    - if the shoulder is made highly flexible.
3. Large animals contract their muscles more slowly than smaller ones.
4. Some animals also extend stride length by alternately flexing and extending their spines.
5. Muscles that attach far from a joint have more leverage and thus can produce more power. Conversely, those that attach close to the joint move much faster but are less powerful.
6. One important way to increase stride rate is to keep the legs light.

HOW DO ANIMALS USE SKELETON AND MUSCLES TO MOVE?

1. Muscles and bones work together to move the body.

How do muscles and skeletons work together?

1. All animals move by contracting muscles against a rigid framework—the skeleton—which can be made of bone or other materials.
2. Earthworms have a hydroskeleton—a rigid, fluid-filled internal space called a coelom.
3. Arthrodpods—the most numerous animals—have exoskeletons. These are external skeletons that resemble a hollow cylindrical tube.
4. Gram for gram, exoskeletons can support more body weight than vertebrate endoskeletons.
5. The same mechanical principles apply to invertebrates and vertebrates, and in both cases muscles can pull but not push.
6. Vertebrates have endoskeletons.

Naming the parts of the vertebrate endoskeleton

1. Vetebrates are defined by their backbone—a column of hollow bone segments called vertebrae. The skull—the bony case that encloses and protects the brain—lies at the forward end of the skeleton.
2. Two bones are attached at a joint.
3. Some joints are immovable, such as those between the skull bones, which are called sutures.
4. Three kinds of connective tissue connect the bones and muscles of a joint:
   - cartilage cushions the joint (skeletons of sharks and rays are entirely made of cartilage, as are fetal mammals);
   - ligaments attach bone to bone and surround the joint;
   - tendons attach muscles to bones.
5. The structure of a joint determines how the adjacent bones can move.
6. Bone consists of three major tissue types:
   - The outer layer is compact bone tissue.
   - The centers and ends of most mammalian bones are spongy bone tissue.
   - Inside the spongy bone is red bone marrow, which contains an aggregation of undifferentiated stem cells from which blood cells develop.

What forces act on bones and connective tissue?

1. Ligaments and tendons must resist tension—the pulling action of two opposing forces.
2. Bones must resist:
   - tension;
   - compression—the pushing action of two opposing forces; and
   - shear—twisting action created by forces that are not opposite one another.
3. Connective tissue consists largely of flat, irregularly shaped cells called fibroblasts embedded in an extracellular matrix.
4. Fibroblasts secrete a fibrous protein called collagen, and other substances.
5. Collagen, which strengthens cartilage, ligaments, tendons, and bones, is the most abundant protein in the extracellular matrix.
6. Collagen molecules form triple-stranded helices which assemble into cross-linked fibrils which, in turn, form large, interconnected fibers.
7. Fibrils resist tension by assembling in parallel along lines of stress; they resist shear by assembling in perpendicular layers.
8. Bone derives additional strength from crystals of calcium phosphate embedded in the extracellular matrix. These crystals allow the bone to resist compression and shear.

How does bone respond to use?

1. Osteoblasts are specialized cells that make bone matrix.
2. Osteoclasts are similar cells that digest collagen and bone matrix.
3. In living bone, osteoblasts and osteoclasts are constantly at work.
4. Normally, the rate of bone formation equals the rate of bone absorption, but bones of those who exercise become thicker and stronger, while bones of those who are inactive become thinner, weaker, and decalcified.
5. Osteoporosis is a condition in which the bones become extremely thin and weakened and can easily break, due to the loss of bone mineral deposits that accompanies age and menopause.

How do skeletal muscles move bones?

1. The human body contains more than 600 skeletal muscles—muscles attached to bone.
2. Together, the skeletal muscles are the largest tissue in the vertebrate body.
3. Muscles can only exert force in one direction—they can pull, but not push.
4. Most body movement thus depends on the working of antagonistic pairs of muscles which pull in opposite directions.

How do smooth and cardiac muscles differ from skeletal muscle?

1. Cardiac muscle pumps blood through the heart; smooth muscle lines the walls of hollow organs and blood vessels.
2. Smooth muscle lacks the striations seen in skeletal and cardiac muscle.
3. Striations reflect the regular arrangement of protein filaments responsible for muscle contraction.
4. Humans and other animals have voluntary control of their skeletal muscles—these muscles are also called voluntary muscles. Cardiac and smooth muscle are involuntary muscles.
5. Skeletal muscles are generally under the control of a system of nerves called the motor axis, but cardiac and smooth muscle are instead under control of the autonomic nervous system.
6. The autonomic nervous system has two distinct sets of nerves:
   - the sympathetic nervous system brings about the "fight or flight" response, slowing digestive activities and speeding up the heart; and
   - the parasympathetic nervous system saves resources by slowing the heart and increasing digestive activity.

How do the two kinds of muscle fibers work?

1. Every skeletal muscle consists of many parallel muscle fibers, which are giant cells.
2. Thick fibers, called glycolytic fibers, derive most of their energy from glycolysis and have few mitochondria. Thinner fibers, called oxidative fibers, derive most of their energy from respiration and thus have many mitochondria.
3. Oxidative fibers depend on a continuing supply of oxygen, so they also contain high levels of myoglobin—an iron-containing muscle protein that pulls oxygen from the blood. The myoglobin also gives these fibers a reddish color (in contrast to the pale or whitish color of glycolytic fibers).
4. Glycolytic fibers are also called fast fibers because they can deliver a lot of power very quickly. However, they can only be used for short bursts of activity.
5. Oxidative fibers power muscles over longer periods—they have greater endurance.
6. Muscles grow or atrophy according to how much they are used.

How are contractile proteins arranged?

1. A muscle fiber from a skeletal muscle consists of many myofibrils, which are thin threads that run the full length of the muscle fiber. Each myofibril consists of a string of smaller subunits called sarcomeres.
2. Each sarcomere is marked by light and dark stripes of varying widths. The bands of each sarcomere are perfectly aligned with those of all other sarcomeres within the myofibril, and this precise arrangement produces the characteristic striations.

How do striated muscles contract?

1. When a muscle fiber contracts, the sarcomeres contract.
2. Each sarcomere contains hundreds of tiny, parallel filaments which come in two sizes:
   - actin filaments are thin, and
   - myosin filaments are thick.
3. Actin filaments are paired and attach at one end to a Z-band (at the end of the sarcomere). Myosin filaments run parallel to, and alternate with, the actin filaments.
4. During contraction, each pair of actin filaments slides past the myosin filaments.
5. This model of contraction is called the sliding filament model.
6. During contraction, thick and thin filaments are linked to each other by tiny cross-bridges that are extensions of the thick myosin filaments.

INTRODUCTION TO PHYSIOLOGY: HOMEOSTASIS AND TOLERANCE

1. Physiology refers to function—how the body's parts work individually and together.
2. Homeostasis is the tendency of a living organism to maintain a relatively constant internal environment.
3. Tolerance refers to an organism's ability to change with its changing environment.

Homeostasis: How do organisms self-regulate?

1. To maintain a relatively constant condition, organisms need to separate their inside world from the outside world. Your body's barrier is your skin.
2. The cell membrane forms a division between the intracellular environment and the extracellular environment.
3. Technically, the inside of your digestive tract is also the outside of the body—it directly communicates with the outside world.
4. Homeostasis always involves two processes:
   - detecting change, and
   - counteracting the change (through negative feedback).
5. A set point is your body's normal value for some condition, or the desired state.
6. In vertebrates, the hypothalamus detects temperature changes.
   - If the body is too cold, the hypothalamus stimulates heat-seeking behavior, shivering, increased heat production by the cells, and raising of the hair or feathers.
   - If the body is too hot, the hypothalamus causes fur or feathers to lay down, sweating, and decreased activity to generate less heat.
7. Many aquatic animals use countercurrent heat exchangers to keep parts warm in cool water.

How much change can animals tolerate?

1. Animals often give in to the environmental change, rather than trying to resist it.
2. Excess heat kills by disrupting the three-dimensional shape of proteins, including enzymes, rendering them nonfunctional, and by disrupting cell membranes.
3. Some animals have adaptations that allow them to tolerate extreme heat, while others are better able to tolerate extreme cold.
4. Many animals conform to their environment.

VOCABULARY BUILDING

In your own words, first write a brief definition, then a full explanation for each of the following terms. Include examples where appropriate. Complete this section from your memory—you will not learn the terms by simply copying definitions from the textbook. Once you have finished, check your responses against the information in the chapter and make any necessary corrections.

coelom —

thoracic cavity —

abdominal cavity —

tissues —

matrix —

epithelial tissues —

connective tissues —

muscle tissue —

skeletal muscle tissue —

smooth muscle tissue —

cardiac muscle tissue —

nervous tissue —

neurons —

glial cells —

organs —

organ systems —

bulk flow —

countercurrent systems —

turbinate bones —

countercurrent heat exchanger —

rectilinear movement —

sidewinding —

hydrostatic skeleton —

exoskeleton —

endoskeleton —

antagonistic muscles —

backbone —

vertebrae —

skull —

joint —

sutures —

cartilage —

ligament —

tendon —

compact bone tissue —

spongy bone tissue —

stem cells —

red bone marrow —

tension —

compression —

shear —

fibroblasts —

extracellular matrix —

collagen —

fibrils —

fibers —

osteoblasts —

osteoclasts —

osteoporosis —

striations —

voluntary muscle —

involuntary muscle —

motor axis —

autonomic nervous system —

sympathetic nervous system —

parasympathetic nervous system —

muscle fibers —

glycolytic fibers —

oxidative fibers —

myoglobin —

myofibrils —

sarcomeres —

thin filaments (actin) —

thick filaments (myosin) —

sliding filament model —

cross-bridges —

physiology —

homeostasis —

tolerance —

negative feedback —

set point —

CHAPTER TEST

The following test has four parts. Complete as much of the exam as you can from memory. If you cannot answer a question, skip it. Once you complete all that you can, try to answer any questions you skipped. If you still cannot answer them, consult your textbook for the answers. Once you have completed all sections of the test, check your answers for Parts 1 - 3 against those in the back of this book. Highlight any incorrect answers then review that material in your textbook. Correct your answers for future reference.

Part 1: Multiple Choice

For each of the following, select all correct responses—more than one may be correct.

1. Homeostasis refers to:
   a) the function of a structure
   b) the structure of an organ
   c) maintenance of a constant internal environment
   d) none of the above

2. Which of the following is part of the vertebrate coelom?
   a) thoracic cavity
   b) spinal column
   c) abdominal cavity
   d) tail

3. Which of the following utilizes an increased surface area?
   a) wings
   b) webbed feet
   c) lungs
   d) all of these

4. A countercurrent system can be used to establish which of the following gradients?
   a) temperature
   b) concentration
   c) pressure
   d) all of these

5. To minimize contact with the hot ground, desert snakes are especially good at which type of locomotion?
   a) sidewinding
   b) rectilinear movements
   c) undulation
   d) none of these

6. Which of the following is not an adaptation you would expect in flyers?
   a) light for their size
   b) highly flexible skeleton
   c) compact body
   d) lack of teeth

7. For terrestrial vertebrates, speed is determined by:
   a) stride length
   b) stride rate
   c) metabolic rate
   d) age

8. The pushing action of two opposite forces is called:
   a) shear
   b) tension
   c) compression
   d) all of these

9. Connective tissue gets its strength primarily from:
   a) actin
   b) calcium phosphate crystals
   c) collagen
   d) myosin

10. Muscles are usually attached to bones by:
    a) ligaments
    b) tendons
    c) cartilage
    d) sutures

11. Which of the following is voluntary muscle?
    a) smooth muscle
    b) cardiac muscle
    c) skeletal muscle
    d) all of these

12. Which of the following is an example of smooth muscle?
    a) wall of the intestine
    b) wall of the blood vessels
    c) wall of the heart
    d) all of these

13. Which of the following slows the heart?
    a) sympathetic nervous system
    b) parasympathetic nervous system
    c) both of these
    d) neither of these

14. Which type of muscle is striated?
    a) smooth
    b) skeletal
    c) cardiac
    d) all of these

15. Which of the following is a characteristic of glycolytic fibers?
    a) few mitochondria
    b) deliver a lot of power in a short time
    c) used for short bursts of activity
    d) all of these

Part 2: Matching

For each of the following, match the correct term with its definition or example. More than one answer may be appropriate.

small animals **large animals**

1. ____________________ Have larger relative surface area.
2. ____________________ Have higher metabolic rates.
3. ____________________ Require more calories and oxygen.
4. ____________________ These fall harder.
5. ____________________ Lose a smaller proportion of their metabolic heat.

smooth muscle **skeletal muscle** **cardiac muscle**

6. ____________________ This is involuntary.
7. ____________________ This is striated.
8. ____________________ This is under conscious control.
9. ____________________ This is controlled by the autonomic nervous system.
10. ____________________ This can remain contracted indefinitely.

Part 3: Short Answer

Write your answers in the space provided or on a separate piece of paper.

1. On what does the form of an organ depend?

2. What shape are animals that are exclusively adapted to swimming rapidly?

3. List three types of movements used by snakes.

4. Why do most vertebrate flyers walk on two legs?

5. List three ways in which stride length can be increased.

6. Differentiate between tendons and ligaments.

7. List an advantage and a disadvantage of having an exoskeleton.

8. Differentiate between tension and shear.

9. In addition to collagen, what provides bone's strength?

10. How does exercise affect bones?

11. Why do skeletal muscles tend to work in antagonistic pairs?

12. Which muscles are involuntary, and why?

13. Differentiate between the action of glycolytic and oxidative fibers.

14. What is myoglobin?

15. Cross-bridges are part of what?

Part 4: Critical Thinking: Using Your Knowledge

Answer each of these in essay form, using complete sentences and paragraphs. Provide as much information as you can. (For extra essay practice, write out answers to the Review and Thought Questions in your textbook.)

1. Explain the relationship between size and surface area.

2. Explain, and give examples of, a countercurrent system.

3. Discuss the sliding filament model of muscle contraction.

4. Heat stroke, a fatal condition if it is not corrected, results from the failure of normal heat loss mechanisms to work sufficiently. Heat is a catalyst—it increases the rate of metabolic reactions. Knowing this, describe what happens during heat stroke and why it is fatal.

5. Discuss as many ways as you can in which the skin is essential for homeostasis.

6. Explain how homeostasis is involved with hunger, thirst, blood pressure. What signals trigger a response, and what responses correct a change away from the set point for each?

7. Blood is also categorized as a connective tissue. In what ways is blood a typical connective tissue, and in what ways is it unique from the other connective tissues?

8. If you place all your food supplies and cooking utensils up high, and stretch to reach them each day, will your arms get longer over time? Explain your answer.

9. When you buy ground meat at the grocery store, the outer meat is bright red, but the inner meat is darker. What accounts for this color difference?

10. Explain ways in which tolerance is evident in your life.

Chapter 35
How Do Animals Obtain Nourishment from Food?

KEY CONCEPTS

1. Animals depend on food for energy and raw materials.
2. During digestion, an animal hydrolyzes macromolecules and lipids into their component building blocks and absorbs the resulting small molecules.
3. In vertebrates, digestion and absorption occur in a digestive tract that starts at the mouth and ends at the anus. A two-ended digestive tract allows the specialization of different regions for specific processes that take place in a fixed order.
4. Food must provide enough energy to supply an animal with its basic metabolic needs and to support its various activities.
5. Animals usually regulate their food intake to match their activities, but this regulation can go awry either because of an inadequate food supply or a failure in normal regulation.
6. Malnourishment in humans usually results from an imbalance of amino acids in dietary protein, but may also result from deficiencies in minerals and vitamins.
7. Digestion begins in the mouth with division of the food, start of enzyme hydrolysis, formation of a bolus, and movement of the bolus into the gut.
8. The stomach mechanically disrupts food while hydrochloric acid and pepsin begin to hydrolyze proteins. Food leaves the stomach as semiliquid chyme.
9. The length of the small intestine allows sufficient time for digestion and absorption. Specializations that increase the surface area increase the efficiency of absorption.
10. Digestion depends on enzymes produced by the pancreas and bile produced by the liver.
11. The large intestine removes ions and water.
12. Bacteria grow in both the colon and the cecum; in some species, bacteria in the cecum digest cellulose.
13. Nerve cells and hormones coordinate the actions of the parts of the digestive system.
14. The cells of the gut wear out and must be replaced. Radiation or cancer-causing mutations can interfere with normal cellular repair and replacement.

EXTENDED CHAPTER OUTLINE

Another American shot heard around the world

1. In 1822, a shotgun accidentally discharged into the belly of 19-year-old Alexis St. Martin, immeasurably adding to the scientific understanding of digestion.
2. Beaumont found a rib and part of a lung protruding from the wound, as well as a portion of St. Martin's stomach, which was punctured.
3. St. Martin recovered amazingly, but lived the rest of his life with a hole that led from the outside directly into his stomach.
4. Beaumont realized that St. Martin's stomach provided "an excellent opportunity for experimenting upon the gastric fluids and the process of digestion.
5. Beaumont listed 51 conclusions derived from his experiments and observations.

WHY MUST ANIMALS EAT?

1. The digestive system derives both energy and raw materials from food.
2. Digestive systems perform the same basic task—to break down the complex molecules of food into simpler molecules that can be absorbed and incorporated.
3. Energy comes from complex molecules that must go through hydrolysis before they can enter the energy pathways.
4. Each molecule undergoes a series of enzyme-catalyzed conversions to simple molecules.
5. Although an animal can make many building blocks needed to produce its own large molecules, food provides a ready supply, including some that the animal cannot make itself.
6. Herbivores eat only plants; carnivores eat only animals, with some species specializing on particular parts, like the blood or the skin. Other animals only consume detritus—remains of plants and animals broken down to smaller fragments by the action of microorganisms.
7. Humans are omnivores—our diets range widely.

How much must an animal eat?

1. Food must provide energy for all of an animal's activities. The amount required depends on the animal's energy requirements and differs with size, species, and activities.
2. Energy consumption is roughly proportional to weight. For an average college student, daily energy expenditure, in kilocalories, is roughly 15 to 30 times the body weight in pounds.
3. Every animal requires a certain amount of energy just to stay alive and awake. This minimal energy requirement is called the basal metabolic rate.
4. For humans, the basal metabolic rate is about 1400 kcal per day for a 120-pound woman and about 1700 kcal per day for a 160-pound man.
5. The basal metabolic rate is usually slightly more than half the energy used in ordinary activities.
6. Individual animal species have widely varying basal metabolic rates.

How does an animal know how much food to eat?

1. Most animals adjust food intake according to their activity levels.
2. The hypothalamus controls appetite.
3. A hormone called orexin triggers feeding behavior.
4. Fat cells in the body of a well-fed animal secrete a hormone called leptin which signals the hypothalamus to suppress feeding behavior.

What are the consequences of too much or too little food?

1. If an animal takes in more food energy than it uses, the excess is stored as glycogen or as fat. This situation—called overnourishment—is common in human industrialized societies.
2. If an animal takes in less food energy than it needs, it must derive the extra energy from its own body—animals begin to break down their muscles and other tissues.

3. Undernourishment—taking in too few calories—occurs in modern human populations mostly in times of drought, war, or other catastrophes.
4. Malnourishment—deficiency in one or more essential nutrients—is quite common: it affects at least 10% of the world's population, and is an increasing problem.
5. Animals can usually adjust their food intake to match their activities.

What are the causes of malnourishment?

1. Malnourishment may result from a deficiency of protein, minerals, or vitamins. Each type of deficiency leads to a recognizable set of symptoms.
2. Protein deficiency is the principal cause of human malnourishment.
3. Protein deficiency disease—kwashiorkor—is characterized by lethargy, severe anemia, change in hair color, inflammation of the skin, and a pot belly, and is especially prevalent in children just after weaning, when their diet switches from milk to a single kind of starch.
4. All animal cells depend on dietary protein to provide the essential amino acids—those which an animal cannot produce itself.
5. Because animals do not store amino acids between feedings, each meal must provide the proper proportions of essential amino acids needed to make proteins in that animal.
6. Proper balance requires mixing protein sources. An appropriate mixture—one providing enough of each essential amino acid—is said to contain complementary proteins.
7. Animals require inorganic ions for a wide variety of structures and processes. Some (calcium, phosphorus, potassium, sulfur, sodium, chlorine, and magnesium) are required in relatively large amounts. Others are needed only in trace amounts.
8. Animals appear to have some mechanisms for ensuring that they take in the proper minerals.
9. Vitamin deficiency can cause disease. Thiamine deficiency can cause beriberi.
10. Ascorbic acid (vitamin C) is the specific substance needed to cure scurvy.
11. Chemically, the 13 vitamins fall into two broad groups: water-soluble vitamins and fat-soluble vitamins. Each water-soluble vitamin is a precursor of a particular coenzyme—a small organic molecule that binds to an enzyme and plays a role in catalysis. Fat-soluble vitamins, including vitamins A, D, E, and K, play diverse roles.
12. Substances that are essential (needed but not made in the body) for humans are not necessarily essential for all animals.

HOW DO SINGLE CELLS AND INVERTEBRATES DIGEST?

1. The most nourishing parts of food consist of carbohydrates, protein, and lipids.
2. To obtain this nourishment, an animal must:
   - digest the food—convert it into smaller molecules, such as glucose, amino acids, fatty acids, and glycerol; and
   - absorb the smaller molecules—take them up into cells and into the circulation.
3. Digestion is a chemical process that involves splitting bonds between building blocks of macromolecules and lipids.
4. Each splitting reaction is a hydrolysis—it is accompanied by addition of a water molecule.
5. None of the hydrolysis reactions requires energy—all are exergonic and spontaneous.
6. Specific enzymes participate in each reaction. Proteases work on proteins, glycosidases work on polysaccharides, lipases work on lipids, and nucleases work on nucleic acids.

Harsh conditions help disrupt tissue interactions

1. Freeing food molecules to interact with enzymes requires harsh conditions.
2. Humans use such mechanisms as cooking, marinating in an acid solution, and chewing.
3. Harsh conditions prepare the consumable tissue, but may destroy tissues of the consumer.
4. To avoid such "self" digestion, animals perform digestion extracellularly—a membrane separates the digestive enzymes and the digesting food from most of the cellular proteins.

Digestion by individual cells

1. Pinocytosis is a form of endocytosis—the process by which a cell traps extracellular materials into a vesicle (endosome) by infolding of the plasma membrane.
2. The endosome then fuses with a lysosome, full of hydrolytic enzymes, and the resulting small molecules move into the cytosol.
3. Phagocytosis involves larger particles than pinocytosis, and the plasma membrane expands to surround the material.
4. Phagocytosis is usually used to remove cellular debris, rather than for feeding.

Digestion in cavities with one opening

1. The simplest digestion chamber has a single opening which serves as both mouth and anus.
2. Food enters the chamber, secreted enzymes digest it, nutrients enter the surrounding cells by endocytosis, and undigested materials are expelled from the chamber.

What are the advantages of a two-ended digestive tract?

1. In more complex animals with two-ended digestive systems, food enters the mouth and passes through an extended tube (the digestive tract or gut), then undigested and unabsorbed material leaves through the anus.
2. The remnants of the food, with bacteria that inhabit the tract, form feces.
3. Animals can process food more efficiently with a two-ended digestive system.
4. Some invertebrate carnivores perform extracellular digestion by using their victims' own bodies as a preliminary digestion vessel, secreting enzymes into its body to liquefy the meal.

HOW DOES THE VERTEBRATE DIGESTIVE TRACT FUNCTION AS A "DISASSEMBLY LINE?"

1. The digestive system performs four functions:
   - movement—agitating food in specialized regions and pushing it through the system;
   - secretion—production of lubricants, enzymes, and detergents;
   - digestion—breakdown of large molecules into their component building blocks; and
   - absorption—taking up small molecules into cells or into the circulation.

How is the vertebrate digestive tract organized?

1. The general organization of the vertebrate digestive system has five specialized divisions:
   - mouth and throat (headgut);
   - esophagus and stomach (foregut);
   - small intestine (midgut);
   - pancreas and biliary system; and
   - large intestine (hindgut).
2. The lumen of the gut (the inner space) is continuous.
3. The lining of the gut lumen is an epithelium—a tightly connected sheet of cells.

The human digestive system and some vertebrate variations

What does the headgut (mouth and throat) contribute to digestion?

1. Food enters the digestive tract through the mouth and throat.
2. All vertebrates except lampreys and hagfish have jaws, and nearly all have teeth.
3. Mammals are the only animals that can actually chew their food, and their teeth are highly specialized. Carnivorous mammals have well-developed incisors and canine teeth; herbivorous mammals may lack canines and have incisors adapted for cutting vegetation.
4. Presence of food in the mouth stimulates the first secretions of the digestive tract.
5. Salivary glands secrete both enzymes and mucus—a viscous, slippery substance that coats food particles and lubricates their movement.

6. Grinding of the teeth exposes the food to salivary amylase, an enzyme that hydrolyzes polysaccharides into shorter fragments.
7. The tongue shapes the chewed food and mucus into a ball—the bolus.
8. The tongue and other muscles push the bolus back into the pharynx (throat), which is a common entryway for food and air.
9. The bolus then passes into the esophagus, then into the stomach.
10. If food is too large to enter the esophagus, it may block the glottis (airway) and the resultant choking can be fatal.
11. Animals can voluntarily initiate chewing and swallowing, but both are involuntary reflexes.
12. Swallowing involves the coordinated response of smooth and striated muscles in the pharynx, the sphincters, and esophagus, and inhibition of breathing muscles.

What does the foregut (esophagus and stomach) contribute to digestion?

1. Peristaltic waves—coordinated contraction of smooth muscles—carry the bolus down the esophagus to the stomach.
2. Some animals have a dilated crop at the end of the esophagus that can store food for later.
3. The stomach, the system's most dilated and muscular structure, accumulates food, disrupts it mechanically, and begins protein hydrolysis.
4. Sphincters at the start and end of the stomach prevent backflow into the esophagus and prevent digesting material from moving on too quickly.
5. Presence of food in the stomach stimulates secretion of hydrochloric acid and pepsinogen. The acid alters pepsinogen's structure so that it becomes an active enzyme called pepsin, which begins to hydrolyze proteins.
6. The mixture of the bolus with the gastric secretions results in a semi-liquid substance called chyme, which enters the small intestine when peristaltic waves open the sphincter.
7. In birds, food passes quickly from the esophagus through an unspecialized stomach into the gizzard, a region specialized for grinding food. The inner surface of the gizzard is coated by a horny, abrasive material and small stones.
8. The stomachs of cows and ungulates have special adaptations that enable them to extract energy efficiently from hard-to-digest grasses. These animals are ruminants—herbivores that can regurgitate partly digested food (cud) for further chewing.
9. A ruminant's stomach contains four distinctive chambers; bacteria and protists housed here produce enzymes that break down cellulose.

What does the midgut (small intestine) contribute to digestion?

1. Chyme next enters the small intestine, which is the longest division of the digestive tract and accomplishes two tasks:
   - completion of digestion of macromolecules; and
   - absorption of most useful products of digestion.
2. Complete digestion requires a relatively long time for enzyme action, and absorption requires a large surface area.
3. In herbivores and other ruminants, undigested cellulose interferes with the digestion and absorption, so their small intestines are proportionately longer than in omnivores.
4. The small intestine's structure reflects its specialization—a layer of connective tissue surrounds the epithelium and contains blood vessels, nerves, and lymphatics. This layer is surrounded by a thin layer of smooth muscle. Together, these structures form the mucosa, which is highly folded into fingerlike projections called villi to increase the surface area.
5. The mucosa is surrounded by another layer of connective tissue, then two layers (circular and longitudinal) of smooth muscle.
6. Finally, another layer of connective tissue (the serosa) surrounds the entire tract, and forms the mesenteries that anchor the tract to the abdominal wall.

7. Movements through the digestive tract are performed by the two layers of smooth muscle:
   - Segmentation churns the chyme more or less in the same place, mixing the chyme with secretions and enhancing absorption by forcing contact with the wall.
   - Peristaltic waves propel the material onward.
8. Intestinal epithelial cells have extensions called microvilli, which increase the surface area.
9. All molecules and ions move out of the lumen by passing through the epithelium.

What do the pancreas, liver, and gallbladder contribute to digestion?

1. Digestion within the small intestine depends on secretions by the pancreas and liver.
2. Most digestion occurs in the duodenum, which is the first part of the small intestine. The remainder of the small intestine is specialized for absorption.
3. A common duct carries materials from the pancreas, liver, and gallbladder to the duodenum.
4. Pancreatic enzymes do not work at acidic pH, so the pancreas also secretes bicarbonate ion to neutralize the acidic chyme as it enters the duodenum.
5. Proteases chop peptide fragments (produced by pepsin) into small lengths of amino acids, then peptidases complete the hydrolysis to free amino acids, which can be absorbed.
6. Pancreatic amylase finishes digestion of polysaccharides begun by salivary amylase.
7. Enzymes on the surface of the microvilli hydrolyze the disaccharides into simple sugars ready for immediate absorption. But cellulose remains intact in most animals.
8. Nucleases produced by the pancreas hydrolyze the DNA and RNA that are consumed.
9. Lipids are not water-soluble, so they usually aggregate into droplets whose oily interiors are not accessible to hydrolytic enzymes.
10. The liver secretes a detergent solution called bile, and it moves to the gallbladder for storage.
11. Bile breaks up fat droplets in the chyme, then lipases, secreted by the pancreas into the duodenum, hydrolyze the lipid molecules into fatty acids, glycerol, and other molecules.
12. Absorption of small molecules across the epithelium of the small intestine depends on both passive diffusion and specific protein molecules on the plasma membrane. Some of these finish the final steps of digestion; others are transport molecules.
13. The overall process of absorption is so efficient that normally, by the time the chyme appears in the large intestine, no nourishing molecules remain—only water, ions, cellulose, and unabsorbed molecules, along with mucus and digestive secretions.

What does the hindgut (large intestine) contribute to digestion?

1. Most of the length of the large intestine is specialized to recover water and dissolved ions.
2. The three major sections of the colon are the ascending colon, the transverse colon, and the descending colon, which ends in a straight portion called the rectum, which stores feces.
3. Water is removed from both small and large intestines by pumping ions across the epithelium.
4. If the colon is irritated, its contents may move too fast for efficient absorption, causing diarrhea; if the contents move too slowly, too much water is absorbed causing constipation.
5. Undigested cellulose provides roughage (fiber, bulk) that stimulates peristalsis. Insufficient roughage also leads to constipation.
6. The colon contains large amounts of bacteria that live on unabsorbed molecules and produce some vitamins.
7. The cecum, at the junction of the small and large intestine, diverts from the main tract.
8. The human cecum does not appear to serve any purpose; the appendix which projects off it is now believed to have some role in the immune system.
9. The cecum of other animals, especially herbivores, may be relatively large.

HOW DOES AN ANIMAL COORDINATE PROCESSES IN INDIVIDUAL PARTS OF THE DIGESTIVE SYSTEM?

1. The digestive system must coordinate processes performed by cells that are far apart.

Hormones and nerve cells coordinate the actions of the gut

1. In multicellular organisms, chemical signals are responsible for almost all intercellular communication and are especially important to coordinating the responses of different organs to changes in their environments.
2. A hormone is a substance that is made and released by cells in a well-defined organ or structure and moves throughout the organism, exerting specific effects on specific cells in other organs or structures, called target organs.
3. Both nerve cells and hormones use chemical signals to coordinate muscle activity, secretion, digestion, and absorption.
4. Part of the control of the digestive system depends on a complex network of nerve cells within the digestive tract itself that continuously receives information about the status of the tract from specialized receptors on the epithelial membranes.
5. The nerve network processes this information and sends appropriate commands to cells within the tract.
6. The intestinal nerve network also receives information from the brain.

How does the gastrointestinal tract replenish itself when cells die?

1. Movement of food particles and chyme subject the gastrointestinal tract to tremendous wear and tear.
2. As cells die, new ones replace them, but they must be the correct type in the correct place.
3. All cells of the gut epithelium derive from common stem cells.
4. Chemical signals called growth factors regulate the stem cells' division and differentiation.
5. Exposure to radiation can destroy stem cells' ability to divide, leading to gastrointestinal disorders.
6. Regulation of cell division can fail, and this failure is more common in the large intestine, causing small wartlike growths called polyps. This abnormal division can become excessive and lead to colon cancer, the second most common type of cancer in the United States.

VOCABULARY BUILDING

In your own words, first write a brief definition, then a full explanation for each of the following terms. Include examples where appropriate. Complete this section from your memory—you will not learn the terms by simply copying definitions from the textbook. Once you have finished, check your responses against the information in the chapter and make any necessary corrections.

herbivores —

carnivores —

omnivores —

basal metabolic rate —

malnourishment —

kwashiorkor —

essential amino acids —

digest —

absorb —

digestive tract —

gut —

anus —

feces —

movement —

secretion —

digestion —

absorption —

lumen —

epithelium —

salivary glands —

mucus —

amylase —

bolus —

pharynx —

sphincter —

esophagus —

peristaltic waves —

crop —

stomach —

chyme —

gizzard —

ruminants —

villus —

duodenum —

microvilli —

pancreas —

bile —

gallbladder —

large intestine (colon) —

rectum —

diarrhea —

constipation —

cecum —

homeostasis —

hormone —

target organs —

CHAPTER TEST

The following test has four parts. Complete as much of the exam as you can from memory. If you cannot answer a question, skip it. Once you complete all that you can, try to answer any questions you skipped. If you still cannot answer them, consult your textbook for the answers. Once you have completed all sections of the test, check your answers for Parts 1 - 3 against those in the back of this book. Highlight any incorrect answers then review that material in your textbook. Correct your answers for future reference.

Part 1: Multiple Choice

For each of the following, select all correct responses—more than one may be correct.

1. Herbivores eat which of the following?
   a) animals only
   b) plants only
   c) animals and plants
   d) everything

2. Energy consumption is roughly proportional to:
   a) height
   b) weight
   c) age
   d) height and age

3. Consumption of too few calories is called:
   a) malnourishment
   b) undernourishment
   c) overnourishment
   d) none of these

4. Kwashiorkor results from:
   a) protein deficiency
   b) vitamin C deficiency
   c) vitamin B_1 (thiamine) deficiency
   d) too much fat

5. Which of the following are water-soluble vitamins?
   a) A
   b) B complex
   c) C
   d) D

6. Hydrolysis involves which of the following?
   a) breaking bonds between building blocks of macromolecules
   b) addition of amino acids
   c) addition of water
   d) absorption of water

7. Compared to pinocytosis, the process of phagocytosis involves:
   a) smaller particles
   b) removal of cellular debris
   c) feeding
   d) larger particles

8. Feces contains significant amounts of which of the following?
   a) body wastes
   b) undigested material
   c) bacteria
   d) all of these

9. The lumen of the gut is lined with:
   a) smooth muscle
   b) epithelium
   c) connective tissue
   d) serosa

10. Salivary amylase begins the hydrolysis of:
    a) proteins
    b) lipids
    c) carbohydrates
    d) nucleic acids

11. The crop, found in some animals, is an extension of the:
    a) stomach
    b) small intestine
    c) esophagus
    d) pancreas

12. Which of the following is the correct order of structures through which food passes?
    a) mouth, pharynx, stomach, esophagus, small intestine, large intestine, anus
    b) mouth, esophagus, pharynx, small intestine, stomach, large intestine, anus
    c) mouth, esophagus, stomach, large intestine, small intestine, pharynx, anus
    d) mouth, pharynx, esophagus, stomach, small intestine, large intestine, anus

13. To be absorbed, all carbohydrates must be broken down to:
    a) amino acids
    b) nucleotides
    c) simple sugars
    d) fatty acids

14. Bile aids the digestion of:
    a) fats
    b) carbohydrates
    c) proteins
    d) nucleic acids

15. The large intestine is specialized for absorption of:
    a) fats
    b) water
    c) ions
    d) nutrients

Part 2: Matching

For each of the following, match the correct term with its definition or example. More than one answer may be appropriate.

amylase **pepsin** **lipase**

1. ________________ Secreted by salivary glands.
2. ________________ Secreted by the pancreas.
3. ________________ Activated in the stomach by hydrochloric acid.
4. ________________ Breaks down fats.
5. ________________ Breaks down proteins.
6. ________________ Breaks down carbohydrates.

stomach **small intestine** **large intestine**

7. ________________ Stores food after ingestion.
8. ________________ Completes digestion.
9. ________________ Absorbs mostly water.
10. ________________ Starts digestion of protein.
11. ________________ Receives secretions from the liver and pancreas.
12. ________________ Contains bacteria that produce some vitamins.

Part 3: Short Answer

Write your answers in the space provided or on a separate piece of paper.

1. Differentiate between undernourishment and malnourishment.

2. How do animals protect their own cells from digestive enzymes?

3. Explain the process of pinocytosis, and how it differs from phagocytosis.

4. What is the main advantage of a two-ended digestive tract?

5. What is unique about the digestive process in spiders?

6. What are the four main functions performed by the digestive system?

7. What are the specialized functions of the different types of human teeth?

8. What are the functions of the salivary gland secretions?

9. The pharynx is a shared pathway for what?

10. What is the role of hydrochloric acid in the stomach?

11. What is a gizzard?

12. What is the purpose of the villi in the small intestine?

13. Explain the cause of diarrhea and constipation.

14. List three enzymes found in the digestive system.

15. From where does the nerve network in the digestive tract receive information?

Part 4: Critical Thinking: Using Your Knowledge

Answer each of these in essay form, using complete sentences and paragraphs. Provide as much information as you can. (For extra essay practice, write out answers to the Review and Thought Questions in your textbook.)

1. Discuss factors that affect basal metabolic rate and how it relates to total energy consumption.

2. Differentiate between overnourishment, undernourishment, and malnourishment. What are some ways that social structure can affect who suffers from each of these conditions?

3. Which condition of the previous conditions is the most common in the United States, and why? Obesity is on the rise in the United States, and yet a lot of these same people are malnourished. How can that be possible?

4. Trace the path of the components of a brownie (eggs = protein, butter = lipid, and sugar = carbohydrate) through the human digestive system, from the mouth through the anus. Explain what happens to the food in each location, including specific chemicals involved in hydrolysis.

5. Discuss specializations found in herbivores, carnivores, and omnivores.

6. How can strict vegetarians get all the nutrients they need each day?

7. Vitamin C has been touted as a miracle cure alleged to cure colds and numerous other ailments. Due to this, a lot of people consume massive doses of this nutrient on a regular basis. Some critics of this approach suggest that these individuals might just as well flush the extra money they are spending down the toilet. Why?

8. Some of the newer weight-loss products purportedly allow you to lose weight by blocking up to one-third of the absorption of fat from your digestive tract. How will that lead to weight loss? What effect might this have on normal digestive functioning? Some critics are concerned that extended use of these products might lead to malnourishment. Why?

9. Why do we need to have two layers of smooth muscle in the digestive tract?

10. A person who has had his or her gallbladder removed is often advised to limit their fat intake. Why? What symptoms might they expect to suffer if they consume a very high fat meal?

Chapter 36
How Do Animals Coordinate Cells and Organs?

KEY CONCEPTS

1. Nerve cells and hormones both contribute to homeostasis—the maintenance of a constant internal environment—and in animals, this involves coordinating the activity of many cells and organs.
2. Hormones are chemical signals made in specific organs or cells that travel through the entire body and evoke characteristic responses in target cells and organs.
3. In both invertebrates and vertebrates, specific hormones regulate growth and development.
4. Insulin, made when blood glucose is high, stimulates glucose absorption from the blood. Glucagon, made when blood glucose is low, promotes production of more blood glucose from stored glycogen.
5. Hormones produced in the digestive system help coordinate absorption of food molecules within the digestive system and in the rest of the body.
6. The adrenal cortex produces two hormones in response to hormone signals from the brain. The adrenal medulla produces epinephrine, which is the major mediator of the "fight or flight" reaction.
7. The hypothalamus is the major mediator of information between the brain and the endocrine system. Hormones made by the hypothalamus regulate hormone production elsewhere, especially in the anterior pituitary, which also produces hormones that regulate hormone production elsewhere.
8. Hormones are chemically diverse, including steroids, amines, and peptides. The major distinction among hormones is between those that are lipid-soluble and those that are water-soluble.
9. Hormone action begins with binding to a receptor. Molecules that resemble a natural hormone can bind to the same receptor and initiate or prevent the same response.
10. Epinephrine binds to different receptor types in different tissues.
11. Hormones from the thyroid gland regulate metabolism, growth, and calcium levels in the blood. Parathyroid hormone also contributes to calcium regulation.
12. The gonads secrete sex hormones.
13. Prostaglandins are paracrine factors that exert their efforts through second messengers.

EXTENDED CHAPTER OUTLINE

Environmental effects on sexual development

1. Western men of the 1990s make half as many sperm as men did in the 1930s.
2. Some researchers blame this trend on a new class of environmental pollutants called environmental estrogens (estrogenics)—natural or synthetic substances that mimic the physiological effects of the hormone estrogen.
3. Other researchers have found other compounds that interfere with the actions of testosterone.
4. Estrogens are a group of steroid hormones found in vertebrates that are responsible for development of secondary sexual characteristics.
5. Androgens are another group of steroid hormones, responsible for initial development of male genitalia and secondary sexual characteristics.
6. Environmental estrogenics include hundreds of different compounds, such as the pesticides atrazine, chlordane, and DDT. Most are weak compared to natural estrogens.
7. Combinations of different compounds can have powerful effects—when two different PCBs are combined, their estrogenic effects turn male turtles (in eggs) into females.

HOW DO CHEMICAL SIGNALS COORDINATE RESPONSES IN MANY CELLS?

1. The function of each cell depends on the animal's ability to regulate its own internal environment, which is the external environment of the animal's cells.
2. Homeostasis is the maintenance of a constant internal environment.
3. In vertebrates, extracellular fluids are constantly renewed by the blood, which carries nutrients and oxygen to individual cells, waste products to the kidneys and lungs, and chemical signals that mediate responses to changes in the internal or external environment.
4. Animals also use the electrical activity of nerve cells to coordinate events in physically separated organs.

How do organisms send chemical signals?

1. Biologists distinguish three types of chemical signals:
   - pheromones—secreted by one organism and influencing the behavior or physiology of another organism of the same species;
   - hormones—made and released by a well-defined organ or structure, which travel throughout the circulation and exert specific effects on specific cells in other organs or structures; and
   - paracrine signals—acting only on cells in their immediate vicinity.
2. Paracrine signals include neurotransmitters, prostaglandins, and growth factors.
3. Chemical signals are used by both the nervous and the endocrine systems.
4. Hormones are produced by endocrine cells or organs, specialized for secreting hormones into the general circulation. In contrast, exocrine cells or organs secrete their products to be carried to their targets via ducts.
5. Hormones reach essentially all tissues of the body, but nerves carry signals only to particular targets.
6. A single hormone signal can elicit different responses in different targets.

HOW DO HORMONES COORDINATE PHYSICALLY DISTANT ORGANS?

1. Among the most dramatic examples of hormone action is the transformation of a wormlike caterpillar into a beautiful butterfly or moth.

How do hormones control insect metamorphosis?

1. In butterflies and moths, the fertilized egg first develops into a larva (caterpillar), specialized for eating. The larva molts several times as it grows.
2. The final larva undergoes metamorphosis into a pupa, a dormant stage enclosed in a silken cocoon.
3. When the pupa emerges, it is an adult and strikingly different from pupal or larval forms.
4. At least three hormones are required:
   - juvenile hormone causes molting and prevents metamorphosis,
   - ecdysone promotes metamorphosis, and
   - prothoracicotropin (PTTH) stimulates ecdysone production.

How do hormones help keep blood glucose at a nearly constant level?

1. Shortly after a meal, the body is in the absorptive state. Cells take up glucose, make glycogen, and increase fat and protein synthesis.
2. The absorptive state results in a drop in the blood concentration of glucose and other fuel molecules.
3. The body then enters the postabsorptive state—the liver breaks down glycogen to increase the blood glucose and other cells again derive energy from internal sources.
4. Insulin and glucagon are the two hormones that regulate the switching between the absorptive and postabsorptive state.
5. Insulin stimulates cells throughout the body to take up glucose, thus decreasing the blood glucose concentration. It also stimulates the liver, muscles, and fat to increase synthesis of fats, protein, and glycogen.
6. Glucagon, along with a low level of insulin, triggers the postabsorptive state—cells reduce their glucose uptake and begin to break down glycogen, fats, and proteins.
7. Clusters of endocrine cells in the pancreas, called pancreatic islets or islets of Langerhans, secrete hormones as follows:
   - α (alpha) cells produce glucagon, which promotes glycogen breakdown in muscles;
   - β (beta) cells produce insulin, which promotes cellular uptake of glucose from the blood; and
   - δ (delta) cells produce somatostatin, which decreases activity in the gut and inhibits insulin and glucagon synthesis.
8. Specialized cells of the stomach lining make gastrin, which stimulates hydrochloric acid production; specialized cells in the small intestine make secretin and cholecystokinin.
9. In diabetes mellitus, failure to produce or respond to insulin causes a permanent postabsorptive state, producing excess blood glucose, which leads to glucose in the urine.
10. People with insulin-dependent diabetes mellitus (IDDM, Type 1, or juvenile onset diabetes) fail to make any insulin through an autoimmune disorder in which their immune systems attacked and destroyed the insulin-producing cells of the pancreas.
11. Most people with diabetes have Type 2 or non-insulin dependent (NIDDM) diabetes, in which they make insulin but their target cells fail to respond to it.

How do hormones coordinate the response to stress?

1. Stress—the response to a noxious or potentially noxious stimulus—is part of life.
2. Two major hormones prepare the body to deal with stress:
   - cortisol helps the body cope with stress by increasing the amount of energy molecules and building blocks available for use.
   - epinephrine (adrenaline) increases the heart rate, increases production of glucose from glycogen, and dilates blood vessels.
3. Both epinephrine and cortisol are made in the adrenal gland.
4. The adrenal glands, next to the kidneys, have an outer cortex and an inner medulla.
5. The adrenal cortex makes two major hormones:
   - cortisol, which mediates an animal's response to stress; and
   - aldosterone, which helps regulate kidney function.
6. The hypothalamus controls the production of the hormones from the adrenal cortex by releasing corticotropin-releasing hormone (CRH), which in turn causes the anterior pituitary gland to secrete adrenocorticotropic hormone (ACTH), which then triggers cortisol release from the adrenal cortex.
7. The adrenal medulla operates independently of the brain.
8. Epinephrine is the main instigator of the body's "fight or flight" reaction, which prepares the animal for immediate action and energy expenditure when faced with stress or danger.
9. Transmitting a danger signal to the adrenal medulla involves the central nervous system, which interprets the danger, and the peripheral nervous system, which transmits information to the adrenal medulla.

How do the pituitary gland and hypothalamus regulate the production of hormones by other endocrine organs?

1. The pituitary gland is about the size of a pea and is located at the base of the brain.
2. The posterior pituitary is part of the brain itself, and consists of terminals of nerve cells that originate in the hypothalamus.
3. The posterior pituitary secretes two main hormones:
   - *vasopressin* (antidiuretic hormone, ADH) stimulates water reabsorption in the kidney;
   - *oxytocin* stimulates uterine contractions and milk release at childbirth.
4. The anterior pituitary is not part of the brain—it is a separate gland—and it produces the following peptide hormones:
   - *corticotropin* (adrenocorticotropic hormone, ACTH) stimulates production of steroids in the adrenal cortex;
   - *endorphin* is a natural pain suppressor;
   - *thyroid-stimulating hormone* (TSH) stimulates production of thyroxin in the thyroid;
   - *growth hormone* (GH) stimulates tissue and skeletal growth;
   - *prolactin* stimulates milk production;
   - *melanocyte-stimulating hormone* (MSH) stimulates pigment production;
   - *follicle-stimulating hormone* (FSH), in females, stimulates estrogen production and maturation of the follicle during the menstrual cycle; and
   - *luteinizing hormone* (LH which, in females, induces ovulation and stimulates estrogen production; in males, it increases testosterone production.
5. The hypothalamus produces at least seven hormones that control release of hormones by the anterior pituitary. Five of these are releasing hormones, each of which stimulates the release of a specific hormone; two are inhibiting hormones that inhibit hormone release.
6. Hypothalamic hormones reach the anterior pituitary through special blood capillaries.
7. Regulation of hormone release often involves negative feedback—the use of information about the output of a system to reduce further output.

8. The hypothalamus and anterior pituitary also participate in coordinating responses to neural information about the physical and emotional state of the body.
9. Stress increases production of corticotropin-releasing hormone (CRH) from the hypothalamus, which in turn leads to increases in ACTH (anterior pituitary) and cortisol (adrenal cortex).
10. High levels of cortisol heighten the body's reaction to stress, increasing heart output, lung ventilation, blood flow to the muscles, and glycogen production in the liver.
11. The hypothalamus integrates information from the endocrine system and the nervous system.
12. Information from the endocrine system also affects the functioning of the brain. Hormonal changes can lead to altered behavior.
13. Communication between the endocrine and nervous systems runs in both directions.

HOW DO HORMONES FIND THEIR TARGETS?

1. Hormones fall into two classes:
   - lipid-soluble signals (steroids), whose nonpolarity allows them to pass directly through the plasma membrane of the target cell; and
   - water-soluble signals (peptides), which cannot pass through the plasma membrane.
2. Action on a target cell begins when the signal molecule binds to a specific receptor protein.
3. Lipid-soluble signals bind to specific receptors inside their target cells; water-soluble signals bind to specific receptors on the surface of their target cells.
4. Lipid-soluble signaling molecules include the steroids, such as:
   - glucocorticoids (such as cortisol),
   - mineralocorticoids (such as aldosterone), and
   - sex steroids (estrogen, progesterone, and testosterone).
5. Water-soluble chemical signals include amines, the simplest chemicals that act as hormones.
6. Peptides, which are water-soluble, are the largest class of hormones.
7. The action of water-soluble signals depends entirely on their interactions with receptors on the surface of their target cells; the receptors, in turn, stimulate synthesis or release of intracellular "second messengers" that influence cellular activity.
8. For all kinds of chemical signals, signaling between cells involves six steps:
   - a cell must make a signaling molecule;
   - the signaling molecule must be released from the cell;
   - the molecule must travel to its target cells;
   - the target cells must detect the signal;
   - the target cells must respond to the signal, and
   - something must end the signaling process.

How do target cells detect chemical signals?

1. A signaling molecule forms noncovalent bonds at the binding site of a receptor.
2. Molecules that resemble a natural hormone can bind to the same receptor and initiate the same response.
3. Other molecules that are chemically similar may bind to the receptor and thereby prevent the binding of a natural signal.

Hormones may target more than one kind of receptor

1. For virtually every type of signaling molecule, researchers have found several types of receptors, which are deployed differently in different types of target cells.

HOW MANY OTHER ENDOCRINE ORGANS ARE THERE IN A MAMMAL?

Thyroid and parathyroid glands

1. The thyroid gland, in the neck, produces thyroid hormone which regulates growth and metabolism. Hyperthyroidism causes uncontrolled, rapid metabolism; hypothyroidism leads to lethargy, weight gain, and cold intolerance.
2. The thyroid also produces calcitonin, which stimulates removal of calcium from the blood.
3. The parathyroid gland produces parathyroid hormone, which has the opposite effect of Calcitonin; it increases the concentration of calcium ions in the blood.

Testes and ovaries

1. The gonads produce the sex steroid hormones:
   - testes produce testosterone; and
   - ovaries produce estrogen and progesterone.
2. Sex steroids are responsible for development and maintenance of secondary sex characteristics, and for development of the germ cells. Progesterone is essential for maintaining pregnancy.

PARACRINE SIGNALS ACT OVER SHORT DISTANCES

1. Paracrine signals affect only cells in the immediate vicinity of the signaling cell.
2. Paracrine signals include:
   - neurotransmitters—signals by which nerve cells communicate,
   - prostaglandins—derivatives of an unsaturated fatty acid, and
   - growth factors—signals which stimulate cell division.
3. A series of enzymes converts arachidonic acid, a minor component of membrane lipids, into prostaglandins.
4. Aspirin inhibits this pathway, so it acts as both an anti-inflammatory agent and an anticlotting agent. By this latter action, small amounts of aspirin significantly diminish the risk of a heart attack.

VOCABULARY BUILDING

In your own words, first write a brief definition, then a full explanation for each of the following terms. Include examples where appropriate. Complete this section from your memory—you will not learn the terms by simply copying definitions from the textbook. Once you have finished, check your responses against the information in the chapter and make any necessary corrections.

estrogenics —

hormone —

target cell/organ —

internal environment —

homeostasis —

pheromone —

paracrine signal —

neurotransmitter —

prostaglandin —

growth factor —

endocrine system —

endocrine cell/organ —

exocrine cell/organ —

absorptive state —

postabsorptive state —

insulin —

diabetes mellitus —

adrenal cortex —

adrenal medulla —

epinephrine —

thyroid gland —

parathyroid gland —

testes —

ovaries —

pituitary gland —

posterior pituitary —

hypothalamus —

anterior pituitary —

releasing hormone —

inhibiting hormone —

median eminence —

negative feedback —

lipid-soluble signal —

water-soluble signal —

receptor —

steroid —

amine —

peptide —

binding site —

second messenger —

CHAPTER TEST

The following test has four parts. Complete as much of the exam as you can from memory. If you cannot answer a question, skip it. Once you complete all that you can, try to answer any questions you skipped. If you still cannot answer them, consult your textbook for the answers. Once you have completed all sections of the test, check your answers for Parts 1 - 3 against those in the back of this book. Highlight any incorrect answers then review that material in your textbook. Correct your answers for future reference.

Part 1: Multiple Choice

For each of the following, select all correct responses—more than one may be correct.

1. Which of the following is true of phytoestrogens?
   a) They mimic estrogens.
   b) They are synthetic.
   c) They are made by plants.
   d) All of these.
2. Substances secreted by one organism that influence the behavior or physiology of another organism of the same species are called:
   a) hormones
   b) paracrine signals
   c) growth factors
   d) pheromones
3. Which of the following is not a type of paracrine signal?
   a) prostaglandins
   b) growth factors
   c) hormones
   d) neurotransmitters
4. Which of the following occurs during the absorptive state?
   a) glycogen synthesis
   b) synthesis of fats
   c) glucose uptake by the cells
   d) all of these

5. Which of the following is true of the postabsorptive state?
   a) Glycogen is broken down in the liver.
   b) Insulin controls this phase.
   c) Glucose is removed from the blood.
   d) All of these are true.
6. Which of the following is an effect of glucagon?
   a) It signals the absorptive state.
   b) It inhibits glycogen synthesis.
   c) It mimics the effects of insulin.
   d) All of these are effects of glucagon.
7. Which of the following is true of Type I diabetes mellitus?
   a) People with this fail to make insulin.
   b) The immune system of a person with this attacks and destroys the insulin-producing cells.
   c) This usually starts early in life.
   d) All of these are true.
8. Insulin is produced by which cells in the pancreas?
   a) alpha cells
   b) beta cells
   c) delta cells
   d) none of these
9. Epinephrine is produced by the:
   a) pancreas
   b) adrenal cortex
   c) adrenal medulla
   d) pituitary
10. Which of the following hormones is involved in the regulation of calcium?
    a) parathyroid hormone
    b) calcitonin
    c) thyroid hormone
    d) aldosterone
11. Which of the following hormones is especially important for maintaining pregnancy?
    a) estrogen
    b) testosterone
    c) progesterone
    d) oxytocin
12. Which of the following is not true of the posterior pituitary?
    a) It consists of nerve fibers that originate in the hypothalamus.
    b) It secretes vasopressin.
    c) It secretes oxytocin.
    d) It secretes releasing hormones.
13. Secretion of which of the following is expected to increase when experiencing stress?
    a) corticotropin-releasing hormone (CRH)
    b) adrenocorticotropic hormone (ACTH)
    c) cortisol
    d) all of these

14. Which of the following is true of water-soluble signal molecules?
    a) They cannot enter cells.
    b) They rely on the function of second messengers within the cells.
    c) They do not bind to receptors.
    d) All of these are true.

15. Which of the following is true of paracrine signals?
    a) cAMP is a paracrine signal.
    b) They act only on cells that are near to their site of release.
    c) They travel through the blood.
    d) All of these are true.

Part 2: Matching

For each of the following, match the correct term with its definition or example. More than one answer may be appropriate.

hypothalamus **anterior pituitary** **posterior pituitary**

1. ____________________ This is part of the brain.
2. ____________________ This is controlled by the hypothalamus.
3. ____________________ Secretes releasing and inhibiting hormones.
4. ____________________ Secretes growth hormone and luteinizing hormone.
5. ____________________ Secretes vasopressin and oxytocin.

water-soluble signals **lipid-soluble signals**

6. ____________________ Includes steroids.
7. ____________________ Includes peptides.
8. ____________________ These enter cells.
9. ____________________ These rely on the action of second messengers.
10. ____________________ These include estrogen and testosterone.

Part 3: Short Answer

Write your answers in the space provided or on a separate piece of paper.

1. What two functions are served by the sex hormones estrogen and testosterone?

2. List three synthetic estrogenics.

3. List three types of chemical signals and briefly explain each of them.

4. List major events during the postabsorptive state.

5. Differentiate between insulin-dependent and non-insulin-dependent diabetes mellitus.

6. List the cell types in the endocrine pancreas, and what each type secretes.

7. List one hormone from each part of the adrenal gland that is secreted in response to stress.

8. What is hyperthyroidism, and what are its effects?

9. Explain the relationship between the posterior pituitary and the hypothalamus.

10. Explain the roles of the gonadotropins secreted by the anterior pituitary.

11. Briefly explain negative feedback.

12. What structure integrates information from both the endocrine and the nervous systems?

13. List three categories of steroids.

14. List the six steps required for signaling between cells.

15. What is the most widely used second messenger?

Part 4: Critical Thinking: Using Your Knowledge

Answer each of these in essay form, using complete sentences and paragraphs. Provide as much information as you can. (For extra essay practice, write out answers to the Review and Thought Questions in your textbook.)

1. Explain the controversy about estrogenics. Give examples of ways in which you are exposed to these compounds. Do you think they are having ill effects on the world's health? What should we do about these compounds?

2. Explain the relationship between the hypothalamus and both parts of the pituitary gland. How are the parts of the pituitary controlled, and what do their secretions, in turn, control?

3. Differentiate between the functioning of lipid-soluble chemical signals (steroids) and water-soluble chemical signals. Explain the function of second messengers.

4. Make a chart. First list all the endocrine organs discussed, then list the hormones secreted by each of these organs and the hormones' major effects.

5. Assume you have just eaten a big meal before going to bed for the night. Discuss the relationship between insulin and glucagon in maintaining your blood glucose levels from when you stop eating until you arise and have breakfast the next morning.

6. DDT was shown to cause serious reproductive problems in eagles, almost to the point of causing extinction. This led to a ban on the use of DDT as a pesticide in the United States, but not in other countries. How effective is this ban? Does it protect people in other countries? Does it protect people in the United States?

7. In the United States, Japanese beetle traps are available which claim to contain a Japanese beetle pheromone attractant. The concept is that the trap will draw the beetles to it and then can be destroyed. Is this an effective means of keeping your property Japanese beetle-free?

8. If you were treating a patient with Type II diabetes mellitus, why would insulin injections likely not be the best treatment? How could diet be used to treat the disorder?

9. A patient with Type I diabetes took a larger does of insulin than prescribed and became very disoriented, weak, and confused. What is causing the symptoms, and how could you treat this?

10. We often refer to getting an "adrenaline rush" form a scary event. Describe what you feel like at that moment. What causes those feelings? How is this tied into neural and hormonal control of homeostasis?

Chapter 37
How Do Animals Move Blood Through Their Bodies?

KEY CONCEPTS

1. In mammals, the heart pumps blood through two different routes. One circuit passes through the lungs; the other passes through all the other tissues in the body.
2. The four-chambered hearts of birds and mammals keep deoxygenated blood from mixing with oxygenated blood, guaranteeing that the body tissues receive abundant oxygen.
3. Circulating blood maintains the extracellular fluid by removing wastes and delivering nutrients.
4. While invertebrate blood lacks cells, the blood plasma of vertebrates contains numerous erythrocytes, leukocytes, and platelets.
5. Arteries carry blood away from the heart and into smaller vessels called arterioles; arterioles branch into extensive capillary beds. Capillaries converge to form venules, which then converge to form veins that return the blood to the heart.
6. Debris, such as dead and foreign cells, in the extracellular fluids flows into the lymphatic system and through the lymph nodes. There the material is filtered from the lymph and ingested by phagocytic white blood cells (leukocytes).
7. The wall of a capillary is composed of only a single layer of endothelial cells, allowing easy movement of materials. Walls of larger vessels are thicker and the endothelium is surrounded by elastic fibers, collagen, and smooth muscle.
8. Oxygen in the lungs moves into the blood in the capillaries. The newly oxygenated blood in the capillaries enters the pulmonary veins, which return the blood to the left side of the heart.
9. Systole is the period during which the heart muscle is contracted; the period between contractions, when the heart is at rest, is called diastole.
10. One-way valves in the heart—atrioventricular and semilunar—ensure that the blood flows in only one direction, from atria to ventricles, then out the ventricles.
11. Cells in the atria contract almost simultaneously as a single unit, as do those of the ventricles.
12. Electrical signals for these contractions are conducted between the atria and ventricles by the atrioventricular (AV) node.
13. A pacemaker—the sinoatrial (SA) node—sets the rhythm of the heart's contractions.
14. The amount of blood pumped by the heart (cardiac output) is a function of how fast the heart beats (pulse) and how much blood is pumped out with each contraction (stroke volume). This amount varies depending on activity, mood, and other factors.
15. The neurotransmitters epinephrine and norepinephrine, which mediate the fight-or-flight response, increase the amount of blood pumped through the heart. Acetylcholine reduces blood flow through the heart.
16. The body regulates blood flow to different parts by means of:
    - local control over vasodilation and vasoconstriction, which are mediated by local changes in the concentration of oxygen, carbon dioxide, and other molecules; and
    - central control mediated by the nervous and endocrine systems.

EXTENDED CHAPTER OUTLINE

CHARLES DREW AND THE BATTLE OF BRITAIN

1. Dr. Charles Drew was a leading authority on blood storage and processing. In 1940, he led an effort to develop better ways to collect, store, and ship blood plasma to Britain for wounded soldiers.
2. Drew developed a foolproof system that prevented bacterial contamination, and a new invention to speed blood processing.
3. Drew was named Associate Director of a new blood bank for the U.S. Army and Navy, but resigned because the Red Cross, which ran the facility, refused to accept blood from African-Americans, including Drew.
4. Dr. Drew did tremendous work to advance the acceptance of African-American doctors both in communities and in professional societies.

WHY DO ANIMALS NEED BLOOD?

1. The "internal" environment of an animal is actually the "external" environment of its cells—the extracellular fluid.
2. Maintaining a constant internal environment (homeostasis), required for life, requires constant exchange of materials between the extracellular fluid and the animal's external environment.
3. Blood carries oxygen from the lungs and nutrients from the gut to the extracellular fluid, while also moving carbon dioxide to the lungs and nitrogenous wastes to the kidneys for disposal.
4. Blood's other functions include;
   - distribution of hormones,
   - transport of molecules and cells of the immune system, and
   - conduction of heat throughout the body.

What is the connection between blood and extracellular fluid?

1. The extracellular fluid contains many dissolved substances, some of which establish a stable environment.
2. All cells in the body are bathed in extracellular fluid.
3. Among many other substances, blood transports to and from the extracellular fluid:
   - nutrient molecules from the gut,
   - oxygen from the lungs, and
   - carbon dioxide from the cells.

What does blood contain?

1. The fluid part of the blood is called plasma.
2. Plasma is chemically quite similar to extracellular fluid, except plasma has more protein.
3. Invertebrate blood consists only of plasma containing a balance of nutrients, wastes, and stabilizing buffers.
4. Vertebrate blood contains plasma and vast numbers of blood cells of three types:
   - erythrocytes (red blood cells),
   - leukocytes (white blood cells), and
   - platelets.
5. Red blood cells are by far the most numerous in mammalian blood (about 5 billion cells per ml of blood).
6. Vertebrate blood's red color comes from hemoglobin, an oxygen-binding protein whose iron-containing heme groups are red when oxygen is attached.

7. Invertebrate blood has fewer cells and its oxygen-binding protein molecules are extracellular.
8. Leukocytes contribute to the body's defenses against infection and tumors.
9. Platelets are small cell fragments that help blood coagulate (clot).
10. In mammals, platelets and erythrocytes lose their nuclei as they develop, so only the leukocytes are true cells.

HOW DID SCIENTISTS DISCOVER THAT THE BLOOD CIRCULATES?

1. Galen (129–199 A.D.) was the greatest contributor to the understanding of circulation.
2. Galen demonstrated that arteries carry blood rather than air, and that nerves from the brain control body movements.
3. Galen's understanding was only partially accurate. He thought blood moved due to the pulsing of the arteries. He thought the heart acted as a sort of furnace that heated the blood.
4. In 1628, William Harvey published the first accurate description of how blood circulates.
5. His ideas were quite radical, so he worked alone until he was certain, then published his work.
6. He discovered the general path of blood and that the heart pushes, not pulls, the blood.
7. The mammalian heart consists of four chambers—two ventricles and two atria.
8. No connection was visible between the arteries and veins so nobody—including Harvey—guessed at the circular path of the blood.
9. Italian microscopist Marcello Malpighi discovered the capillaries that link arteries and veins.

Why do many animals need a blood pump?

1. Animals less than 1 mm in diameter get by without a circulatory system, because their needs are met by diffusion.
2. All larger animals require a circulatory system with a pump.
3. Hearts come in many forms and some animals pump blood by the squeezing of blood vessels during body movements.

How does the heart pump blood through the body?

1. Humans and other mammals have a double circulation:
    - pulmonary circulation, carrying deoxygenated blood from the heart to the lungs, and
    - systemic circulation, delivering oxygenated blood from the heart to the rest of the body.
2. The blood, heart, and blood vessels make up the cardiovascular system; the route of the blood is called circulation.
3. The heart pushes, or pumps, the blood through the circulatory system.
4. Birds and mammals have a single four-chambered heart, but other animals have hearts with two to five chambers, and some have more than one heart.
5. In humans, the fist-sized heart is in the chest, just behind the sternum.
6. A fibrous sac called the pericardium surrounds the heart and bathes its surface with a lubricating fluid.
7. The myocardium is the muscular wall of the heart, and it forms the four chambers.
8. The largest vessel is the aorta—the artery that carries the blood from the left ventricle to the rest of the body.
9. The mammalian heart consists of two separate halves. Each side has a relatively thin-walled receiving chamber, called the atrium, and a thicker-walled sending unit called the ventricle.
10. On the right side (pulmonary side), deoxygenated blood returns from the body via the venae cavae—the two largest veins. This blood enters the right atrium, moves into the right ventricle, then is ejected through the pulmonary artery, which carries the deoxygenated blood to the lungs.

11. On the left side, oxygenated blood returns from the lungs through the pulmonary veins, entering the left atrium. It passes into the left ventricle, then is ejected, via the aorta, to the rest of the body.

How do birds and mammals prevent oxygen-rich blood from mixing with oxygen-poor blood?

1. In many animals, freshly oxygenated blood mixes with deoxygenated blood from the tissues.
2. The four-chambered bird and mammalian heart prevents this mixing by segregating the deoxygenated and oxygenated blood.
3. Reptilian and amphibian hearts contain only three chambers—two atria and one ventricle—so mixing of the blood does occur, but the ventricle's structure minimizes this.
4. The hearts of bony fish have only two chambers—one atrium and one ventricle. The ventricle pumps blood to the gills to pick up oxygen, and from there the blood flows to the tissues.

HOW DO THE BLOOD AND THE EXTRACELLULAR FLUID EXCHANGE SMALL MOLECULES AND IONS?

1. In an open circulatory system, blood travels from the heart through an artery and into a large open space that may occupy as much as 40% of the body's volume, bathing the animals tissues.
2. In a closed circulatory system, the blood surges through a continuous circuit, enclosed within vessels. The system takes up only about 5 to 10% of the total body volume, and the tiny capillaries allow close contact of blood and almost all cells.
3. A closed circulatory system requires a heart that can generate more force than needed in an open system, but the system also provides greater homeostatic control.
4. The disadvantage to a closed system is that the blood and cells are not in direct contact.
5. Capillaries are loosely joined, with small gaps between them, so plasma can ooze out to form the extracellular fluid.
6. This fluid collects in the lymphatic system and is eventually returned to the blood, becoming plasma again.
7. Lymph contains less protein than plasma because the blood cells and large proteins cannot fit through the gaps in capillary walls.

Arteries, veins, and capillaries

1. In vertebrates, arteries carry blood away from the heart and veins carry blood back to it.
2. Arteries branch many times into smaller vessels called arterioles; from these the blood enters the very fine capillaries.
3. No cell in the body is more than three or four cells away from a capillary.
4. Capillaries are the immediate source of extracellular fluid. They leak constantly, seeping nutrients and oxygen into the surrounding space, and also absorb cell wastes.
5. The absorbed fluid, with the blood cells and dissolved substances, moves into the venules, which converge to form the larger veins for the return trip to the heart.

The lymphatic system

1. Some fluid reenters through the capillaries of the lymphatic system, which provides a secondary route for returning extracellular fluid to the blood.
2. The lymphatic system also carries waste proteins and particles too large to enter capillaries.
3. The lymphatic system also contains specialized leukocytes called lymphocytes, which are involved in the immune response.
4. Fluid, dead or foreign cells, and unneeded proteins form the lymph, which moves passively through the lymphatic system.

5. Lymph nodes contain:
   - filterlike tissue that separates cells and debris from the lymph,
   - phagocytic cells that engulf material, and
   - immune cells that monitor the lymph for signs of infection.

The structure of blood vessels

1. Blood vessels are hollow tubes.
2. Surrounding the space inside (the lumen) is a thin layer of cells called the endothelium.
3. The endothelium of capillaries is a single layer thick, which allows easy passage of substances through the wall.
4. The linings of the arteries and veins are thick and relatively impermeable, and the surrounding walls contain layers of elastic fibers, collagen, and smooth muscle.
5. Veins generally have thinner and less muscular walls than do arteries.
6. The elastic walls of arteries and large veins allow the vessels to narrow or expand in response to changes in blood pressure and to signals from the autonomic nervous system supplying the smooth muscle in the walls.
7. Collagen fibers stiffen the vessel walls, and if the collagen sheath fails, ballooning occurs, and the result is called an aneurysm.

How does the blood acquire oxygen?

1. A failure of the blood supply to the heart is called a heart attack; failure of the supply to the brain is a stroke.
2. In air-breathing vertebrates, oxygen enters the body and then the blood by way of the lungs.
3. In most tissues, oxygen moves out of the capillaries into the body's tissues; but in the lungs, oxygen moves into the capillaries and binds to the hemoglobin in the red blood cells.
4. The blood in the lungs also releases carbon dioxide.
5. The refreshed blood returns to the heart through the pulmonary veins—the only veins in the body that carry oxygenated blood.

WHAT MAKES THE HEART BEAT?

1. In a resting human adult, the heart beats about 70 times per minute.
2. Each beat consists of a cycle of contractions by the different chambers of the heart.

Systole and diastole

1. During half of the cycle, both the atria and the ventricles are relaxed—this is diastole.
2. The contraction period is systole.
3. Blood pressure is measured in mm Hg (millimeters of mercury), and is expressed by two numbers:
   - the first—systolic blood pressure—is the highest pressure and occurs when the ventricle contracts; and
   - the second—diastolic pressure—is the lowest, and occurs when the heart is relaxed.
4. Heart valves ensure one-way flow of blood through the heart.
5. Atrioventricular valves ensure that blood cannot move from the ventricle back into the atrium, and semilunar valves ensure that blood that has left the ventricles cannot return.
6. Smaller valves in the veins ensure that blood flows in a single direction through the circulatory system.
7. A heartbeat has two sounds:
   - the first ("lub") is that of the atrioventricular valves shutting, and
   - the second ("dup") is that of the semilunar valves shutting.
8. A defective valve can cause a hissing sound called a heart murmur.

What coordinates the beating of the heart muscle?

1. The muscle fibers of the ventricles must all work together to pump blood.
2. Fibrillation refers to continuous disorganized contractions. Ventricular fibrillation, also known as cardiac arrest, accounts for one-fourth of all deaths in the United States. It is fatal because the ventricles stop pumping blood, cutting off their own supply, and the muscle weakens and dies from lack of oxygen.
3. Although both are striated, heart muscle is organized differently than skeletal muscle.
4. Heart (cardiac) muscle fibers have many cells, each with a single nucleus.
5. Heart muscle cells are connected through special structures called intercalated discs, each of which contains:
   - desmosomes, which reinforce the muscle against mechanical stress, and
   - gap junctions, which allow ions and small molecules to pass freely from cell to cell.
6. When one muscle fiber in a chamber is stimulated to contract, all the other cells in that chamber are also stimulated.
7. Atrial and ventricular cells are connected by special tissue called the atrioventricular (AV) node, which conducts electrical signals from the atria to the ventricles.
8. About 1% of the myocardial cells are capable of rhythmic spontaneous contractions.
9. The pacemaker—a group of these cells—sets the pace of contraction for the rest of the heart. In mammals, the pacemaker is near the top of the right atrium and is called the sinoatrial (SA) node.
10. Electrical signals travel from the SA node to the AV node, which relays the signal down into the ventricles.

Blood pressure depends on the action of the heart and the properties of blood vessels

1. Each beat of the pulse corresponds to an expulsion of blood from the heart (systole).
2. The pressure within the vessels decreases as the arteries divide.
3. The pressure in the capillaries is rather constant and drives some of the fluid from the capillaries into the extracellular space.
4. Blood in veins has even less pressure.

NERVES, GLANDS, AND OTHER TISSUES REGULATE THE FLOW OF BLOOD

1. The cardiac output (amount of blood pumped from the heart each minute) depends on:
   - heart rate—the number of beats per minute, and
   - stroke volume—the amount of blood ejected by each contraction of the ventricles.
2. In the resting adult, cardiac output is about 5 liters per minute—about the total blood volume.
3. Both heart rate and stroke volume increase during exercise; other factors also affect them.

Regulation by nerves and hormones

1. Sympathetic nerves mobilize the body in times of stress, speeding up the heart and narrowing capillaries.
2. Sympathetic nerves act on the heart's pacemaker by secreting norepinephrine.
3. Parasympathetic nerves slow the heart by secreting acetylcholine.
4. Norepinephrine and acetylcholine are neurotransmitters—molecules that transmit signals from nerve cells.
5. Epinephrine (adrenaline) is closely related to norepinephrine and also speeds the heart. Both of these chemicals increase stroke volume by causing smooth muscle in the veins to contract, increasing blood pressure, which sends more blood into the heart (thus more is ejected).

How does the body deliver oxygen where it is needed?

1. Changes in the diameters of specific arterioles route blood to the specific organs or muscles that are most active at a given time.
2. One form of control on blood flow is highly localized and depends on the concentrations of oxygen, carbon dioxide, hydrogen ions, and other molecules that indicate metabolic activity.
3. Smooth muscles at the entrance of each capillary regulate the blood flow into that vessel.
4. Effects of the sympathetic nervous system vary from place to place in the body.
5. Decreased sympathetic nerve activity allows vasodilation—arterioles open so flow increases.
6. Endothelium-derived relaxation factor—nitric oxide—comes from the endothelium and causes vasodilation.
7. The sympathetic nervous system can rapidly override local control by releasing norepinephrine, which causes vasoconstriction—contraction of the smooth muscles surrounding the arterioles.
8. Many substances influence vasodilation and vasoconstriction. For example, histamine, released by damaged tissues, dilates capillaries and increases their leakiness.

VOCABULARY BUILDING

In your own words, first write a brief definition, then a full explanation for each of the following terms. Include examples where appropriate. Complete this section from your memory—you will not learn the terms by simply copying definitions from the textbook. Once you have finished, check your responses against the information in the chapter and make any necessary corrections.

extracellular fluid —

plasma —

erythrocytes —

leukocytes —

platelets —

hemoglobin —

cardiovascular system —

circulation —

pulmonary circulation —

systemic circulation —

heart —

sternum —

pericardium —

myocardium —

aorta —

venae cavae —

pulmonary artery —

pulmonary veins —

atrium —

ventricle —

sinuses —

open circulatory system —

closed circulatory system —

arteries —

veins —

arterioles —

capillaries —

venules

lymphatic system —

lymphocytes —

lymph —

lymph nodes —

lumen —

endothelium

aneurism —

heart attack —

stroke —

diastole —

systole —

atrioventricular valves —

semilunar valves —

fibrillation —

ventricular fibrillation —

atrioventricular (AV) node —

pacemaker —

sinoatrial (SA) node —

cardiac output —

heart rate —

stroke volume —

sympathetic nervous system —

parasympathetic nervous system —

norepinephrine —

acetylcholine —

neurotransmitters —

epinephrine —

vasoconstriction —

vasodilation —

histamine —

CHAPTER TEST

The following test has four parts. Complete as much of the exam as you can from memory. If you cannot answer a question, skip it. Once you complete all that you can, try to answer any questions you skipped. If you still cannot answer them, consult your textbook for the answers. Once you have completed all sections of the test, check your answers for Parts 1 - 3 against those in the back of this book. Highlight any incorrect answers then review that material in your textbook. Correct your answers for future reference.

Part 1: Multiple Choice

For each of the following, select all correct responses—more than one may be correct.

1. The fibrous sac that surrounds the heart is the:
   a) sternum
   b) myocardium
   c) pericardium
   d) endocardium

2. The largest vessel is the:
   a) aorta, which is an artery
   b) aorta, which is a vein
   c) pulmonary artery
   d) pulmonary vein

3. Deoxygenated blood that has been through your body returns to what heart chamber?
   a) right ventricle
   b) left ventricle
   c) right atrium
   d) left atrium

4. In which of the following does deoxygenated blood mix with oxygenated blood?
   a) birds
   b) amphibians
   c) reptiles
   d) bony fish

5. Which of the following is a function of blood?
   a) transports hormones
   b) transports nutrients
   c) transports wastes
   d) all of these

6. Which of the following carries most of the oxygen in the blood?
   a) plasma
   b) erythrocytes
   c) leukocytes
   d) platelets

7. In mammals, which of the following do not contain nuclei?
   a) erythrocytes
   b) leukocytes
   c) platelets
   d) These are all cells, so they all contain nuclei.

8. The smallest blood vessels are the:
   a) arterioles
   b) capillaries
   c) venules
   d) veins
9. Which of the following is not part of a blood vessel wall?
   a) endothelium
   b) collagen
   c) elastic fibers
   d) skeletal muscle
10. A failure of the blood supply to the brain is:
    a) an aneurysm
    b) a heart attack
    c) a stroke
    d) a fibrillation
11. Cardiac output depends on:
    a) heart rate
    b) stroke volume
    c) neither of these
    d) both of these
12. Which of the following increases the heart rate?
    a) norepinephrine
    b) epinephrine
    c) acetylcholine
    d) all of these
13. Which of the following helps to regulate the flow of blood to different parts of the body?
    a) local chemical concentrations
    b) the nervous system
    c) the endocrine system
    d) all of these
14. The rate at which the heart contracts is determined primarily by:
    a) the SA node
    b) the parasympathetic nervous system
    c) the AV node
    d) the atrioventricular valve
15. Oxygenated blood from the lungs returns to the heart through the:
    a) pulmonary artery
    b) vena cava
    c) pulmonary veins
    d) aorta

Part 2: Matching

For each of the following, match the correct term with its definition or example. More than one answer may be appropriate.

pulmonary circulation **systemic circulation**

1. ____________ This involves the right side of the mammalian heart.
2. ____________ The aorta is part of this.
3. ____________ This delivers oxygenated blood.
4. ____________ The ventricle in this route has a thicker muscle wall.
5. ____________ The artery in this route carries deoxygenated blood.

arteriole **capillary** **vein**

6. ____________ Returns blood to the heart.
7. ____________ Exchange of materials occurs here.
8. ____________ These dilate or constrict to adjust blood flow.
9. ____________ The walls of these are a single-cell layer thick.
10. ____________ These are the smallest vessels.

Part 3: Short Answer

Write your answers in the space provided or on a separate piece of paper.

1. Where does the pulmonary artery begin and end, and what type of blood does it carry?
2. What are the functions of the atria and the ventricles?
3. List, in order, the types of vessels blood passes through as it leaves and then returns to the heart.
4. What is the main difference between the blood of invertebrates and vertebrates?
5. List three types of blood cells and their main functions.
6. List two molecules that bind oxygen.
7. Explain the difference between an open and a closed circulatory system.

8. List two types of organisms that have open circulatory systems.

9. List two characteristics of capillaries that allow easy exchange across their walls.

10. List two functions of the lymphatic system.

11. Why is elastic necessary in the walls of the blood vessels?

12. What causes the two heart sounds?

13. What is ventricular fibrillation, and why is it fatal?

14. Define cardiac output and the two factors that determine it.

15. Histamine causes much grief among allergy sufferers. Does it provide any benefit?

Part 4: **Critical Thinking: Using Your Knowledge**

Answer each of these in essay form, using complete sentences and paragraphs. Provide as much information as you can. (For extra essay practice, write out answers to the Review and Thought Questions in your textbook.)

1. Starting at the vena cava, trace the path of blood as it moves through the pulmonary and systemic routes of circulation. Indicate when the blood is deoxygenated and oxygenated, and discuss what happens to the blood as it leaves each side of the heart.

2. Discuss how oxygen enters the blood, is transported by the blood, and is delivered to the tissues.

3. Explain how the heart regulates its own contraction, and other factors that influence it.

4. After women have had complete mastectomies (removal of the breast and some lymph nodes), they often have problems with swelling in their arms. Why?

5. Varicose veins, often visible in the back of the legs, occur when valves in the veins don't function properly. If these valves aren't working, what happens? Why are these veins so visible? What types of work might make you more susceptible to varicose veins, and how could you prevent them?

6. At birth, some humans are said to be "blue babies" because they have a bluish appearance. The defect is a hole in the wall between the two atria. What happens as a result of this "hole in the heart" to cause the blue color?

7. Why is atrial fibrillation usually not serious, but ventricular fibrillation is fatal?

8. Explain how high blood pressure can cause each of the following:
   - edema (swelling)
   - stroke
   - heart attack

9. Anemias cause fatigue because the blood is delivering less oxygen to the cells. How many specific disorders of the blood can you think of that would cause this decreased oxygen?

10. The atria contract as if they are a single chamber, and so do the ventricles. What would happen if all four chambers contracted at the same time, as a single unit?

Chapter 38
How Do Animals Obtain and Distribute Oxygen?

KEY CONCEPTS

1. To participate in biochemical reactions, oxygen must first dissolve in water.
2. Oxygen is far more plentiful in air than in water, but both land and aquatic organisms depend on oxygen that has dissolved in water.
3. Most gas molecules bounce off a solution's surface, but some are captured and dissolved.
4. The amount of dissolved gas in a solution depends on temperature and partial pressure.
5. The rate at which a gas moves into a cell or organism depends on the surface area through which the gas can move.
6. Evaginated breathing structures are called gills; invaginated breathing structures are lungs.
7. Dissolved gas can diffuse into an organism through gills.
8. Oxygen-depleted water must be replenished by movement. Fish move water over their gills by pumping it or by swimming.
9. Gills do not have enough mechanical support to remain functional in air.
10. Almost all air-breathing vertebrates depend on moist lungs to acquire oxygen.
11. More efficient oxygen extraction requires a large surface area and a rich blood supply.
12. The mammalian respiratory system consists of finely divided airways leading to hundreds of millions of blind sacs, called alveoli, where oxygen is absorbed.
13. Mammals pull air into their lungs by expanding the thoracic cavity; they push air out of the lungs by relaxing the muscles of the thoracic cavity.
14. The respiratory tract traps dust and bacteria in mucus, which cilia propel towards the mouth. Macrophages also engulf debris in the airways.
15. Alveoli's cellular structure promotes diffusion of oxygen and carbon dioxide.
16. A detergent (called surfactant) in the alveoli reduces the surface tension and allow the lungs to expand freely.
17. Air flows through a bird's lungs in a single direction for very efficient oxygen extraction.
18. In insects, oxygen diffuses to tissues through air tubes called tracheae.
19. Oxygen-binding proteins greatly increase the oxygen-carrying capacity of the blood.
20. Blood carries carbon dioxide, a waste product, from the tissues to the lungs.
21. In humans, low oxygen increases BPG, which increases the efficiency of oxygen delivery.
22. Hemoglobin's affinity for oxygen depends on the partial pressure of oxygen, pH, and the amount of carbon dioxide and BPG.

EXTENDED CHAPTER OUTLINE

Stanton Glantz and the tobacco industry

1. The National Cancer Institute awarded Stanton Glantz, professor of medicine at the University of California, a three-year grant to study the effects of public policy on tobacco use.
2. Glantz published a series of papers in JAMA describing how the tobacco industry had for 30 years carefully concealed its knowledge that tobacco products are both deadly and addictive.
3. Glantz received an unsolicited box of copies of internal memos and research reports from Brown and Williamson tobacco company, which showed that the company and their legal advisors understood the hazards of smoking since the 1960s.
4. The University of California posted the documents on the Internet and Glantz published scientific papers analyzing the documents in JAMA. Within weeks, the House Appropriations Committee canceled Glantz's grant.

Why was the tobacco industry fighting so hard?

1. Until the late 19th century, tobacco smoking was considered a dirty habit of only a few.
2. Addiction increased, especially during WWI when soldiers were given cigarettes to calm their nerves and pass time between battles.
3. In the 1880s, a physician had already published a report suggesting smoking caused cancer—nearly all patients with cancer of the mouth, throat, and lungs were smokers.
4. In the 1920s, Lucky Strike cigarettes were marketed directly to women as "healthy."
5. Major advertising campaigns were launched, touting Lucky Strikes as a "torch of liberty."
6. During WWII, smoking was again promoted to soldiers and Hollywood embraced it as glamorous, seductive, and romantic.
7. In 1964, the Surgeon General of the United States issued a report stating that smoking is hazardous to health, and sales leveled. Two years later, Congress required manufacturers to label cigarettes as hazardous, and sales dropped.
8. Between 1964 and 1979, the percentage of smokers dropped from 42% of all adults to 32%. Today's numbers are even lower.
9. In 1988, The Surgeon General issued a report stating that smoking is also highly addictive—nicotine causes dependency in the same way as heroin and cocaine.
10. Safe cigarettes never materialized—filters, low tar, and low nicotine forced smokers to inhale more deeply and to smoke more to get the same "dose" of nicotine.
11. All tobacco smoke interferes with the healthy functioning of the lungs and the cardiovascular system.
12. The United States military, whose officers once ordered soldiers to take a rest with the words, "Rest! Light 'em up!" has instituted a massive program to eliminate all tobacco use from the armed forces.

HOW DO ANIMALS EXCHANGE CARBON DIOXIDE AND OXYGEN?

1. Animals usually get most of their energy from cellular respiration, which requires oxygen, so animal life requires the means to acquire and distribute oxygen.
2. Birds, mammals, reptiles, and amphibians obtain oxygen in the lungs and distribute it with the blood; fish and other aquatic animals extract oxygen with gills; insects use networks of trachea to distribute oxygen directly to the tissues.

Can animals take oxygen directly from the air?

1. Oxygen is far more plentiful in air than in water.
2. Land animals must dissolve the oxygen in water before they can use it.

How do gases exert pressure?

1. When gas molecules hit the surface of a liquid, most bounce back into the space above, but some are "captured" (dissolved) by the liquid. Each molecule striking the surface pushes on the surface, or exerts a bit of pressure.
2. The total pressure depends on the number of molecules striking the surface and the speed with which each molecule strikes it (which varies with temperature).
3. Pressure is measured in torr or in mm Hg (millimeters of mercury).
4. Each type of molecule independently contributes to the total pressure.
5. In air, about 21% of the molecules are oxygen, 0.03% are carbon dioxide, and almost all the rest are nitrogen.
6. The concentration of a type of molecule in a gas is the partial pressure of that gas—the pressure exerted by that type of molecule.
7. The total pressure is the sum of all the partial pressures.

How much dissolved gas will a solution hold?

1. In time, gas molecules enter and leave the solution at the same rate and are in equilibrium.
2. The equilibrium concentration of a type of molecule in a solution depends directly on its
   - concentration in the gas phase, which is proportional to its partial pressure; and
   - temperature—gas molecules are less soluble at higher temperatures.

Why is surface area important to gas exchange?

1. The rate at which molecules of a gas can enter a solution depends on three factors:
   - temperature,
   - partial pressure, and
   - surface area.
2. Both the molecules of a gas and the molecules dissolved in a solution move by a process called diffusion—the spontaneous movement of a substance from a region of high concentration to a region of low concentration.
3. The total amount of oxygen that moves across a boundary in a given time is proportional to the difference in concentrations on the two sides of the boundary.
4. Large surface areas also speed the exchange of molecules.
5. Each cell must have a large enough surface area to permit enough oxygen to enter to support the needed level of respiration and ATP synthesis.
6. Animals employ two general strategies to increase total surface area for gas exchange:
   - evagination, forming gills, or
   - invagination, forming lungs.

HOW DO AQUATIC ORGANISMS OBTAIN DISSOLVED OXYGEN?

1. Aquatic animals larger than about 1 mm obtain dissolved oxygen through gills.
2. Oxygen diffuses from the surrounding water across the expanded surface area into the animal's internal environment.
3. To gather more oxygen, the animal must continually replace the oxygen-depleted water with new, oxygen-rich water. Some animals do this by moving the gills through the water, but most move the water over the gills.

How do the gills obtain enough oxygen to power a fast-swimming fish?

1. The oxygen requirements of a fast-swimming fish are greater than those of most invertebrates and other slower-moving fish, so they have enormous surface areas on their gills.
2. Fish adjust their behavior to maintain oxygen supplies.
3. The gills of fish are highly organized:
   - several gill arches lie on each side of the head, all covered by a bony flap (operculum);
   - each arch carries two rows of gill filaments; and
   - each filament carries many rows of parallel, platelike lamellae.
4. Water passes through the lamellae in a single direction, and a dense network of capillaries exposes the circulating blood to the oxygen in the water.

Why can't land animals breathe with gills?

1. Gills collapse when exposed to air because they lack mechanical rigidity.
2. In air, external gills quickly dry and lose the water into which oxygen must dissolve.
3. Spiders solve this problem with moist internal gill-like plates in a breathing apparatus called the book lung.

HOW DO AIR-BREATHING ORGANISMS OBTAIN OXYGEN?

1. Air-breathing organisms almost all (except spiders) obtain oxygen from invaginated surfaces that do not easily collapse.
2. Lungs are localized organs of gas exchange always associated with the circulatory system. Insects depend on tracheae—a complex set of tubes—which also arise by invagination and carry air through the animal's body.
3. Lungs and tracheae lie protected within the body.

What qualities enable the lungs to exchange gases?

1. The surface area of the lungs limits animals' oxygen-gathering ability.
2. Birds and mammals, endotherms that require the most energy and thus the most oxygen, have elaborately folded lungs.
3. The surface area of a human lung is about the size of a tennis court.
4. The ability of a lung to extract oxygen depends not only on the surface area, but also on the blood supply that carries oxygen away from the lungs.
5. The membrane between the blood and the gases must be extremely thin for exchange.

By what path does air enter the lungs?

1. Lungs are part of a respiratory system that consists of all structures responsible for exchange of gases between the blood and the external environment.
2. The vertebrate respiratory system consists of the lungs, airways, and muscles that move air.
3. Air enters the body through the nose and mouth, which lead to the pharynx then the glottis.
4. The epiglottis prevents food from entering the airways.
5. The glottis leads to the larynx, or voicebox, which contains the vocal cords, which produce sound by vibrating as air passes across them.
6. The larynx carries air to the trachea, or windpipe, which enters the chest and splits into the bronchi, which lead to the lungs.
7. Bronchi branch into smaller and smaller tubes, the smallest of which are the bronchioles.
8. Alveoli are tiny hollow sacs within the lungs, where gas exchange occurs.
9. The surface area of the alveoli in a healthy person is about 135 m^2, more than 80 times the outer surface area of the whole body!

How do we breathe?

1. Tidal volume is the amount of air moved into and out of the lungs with each normal breath. For a healthy adult human, tidal volume at rest is about half a liter (500 ml).
2. The dead space is the volume of air that does not come in contact with the alveoli, and thus cannot undergo gas exchange.
3. The dead space volume includes the air in the conducting tubes—trachea, larynx, bronchi, and bronchioles.
4. In a resting adult human, each breath brings into the lungs about 350 ml of fresh air (500 ml tidal volume – 150 ml dead space = 350 ml).
5. Mammals, birds, and most reptiles pull air into the lungs using muscles within the chest cavity. Amphibians and some reptiles push or gulp air into the lungs using the muscles and valves of their mouths.
6. Humans and other mammals expand the chest cavity to create a vacuum, reducing the pressure inside the lungs so that air is drawn inward.
7. Ventilation (the flow of air into and out of the lungs) depends on muscles of the chest cage, which surrounds the thoracic cavity.
8. The diaphragm, a sheet of muscle beneath the lungs, separates the thoracic cavity from the abdomen. Muscles and elastic connective tissue run between the ribs, forming the sides of the cavity.
9. During inspiration, the diaphragm contracts and moves downward, and the rib muscles contract, moving the ribs outward. Together, these actions expand the chest, decreasing the pressure within the alveoli compared to outside so that air is drawn inward.
10. During expiration, the diaphragm relaxes into a steep dome and the ribs resume their lowered position. Together, these passive actions decrease the size of the chest, increasing the pressure within the alveoli compared to outside, which forces air out of the lungs.

What keeps the airways moist and clean?

1. Most of the cells that line the trachea, bronchi, and bronchioles have fringes of constantly beating cilia, while other cells secrete mucus which keeps the surface of the airways wet.
2. Mucus also traps airborne particles, often carrying bacteria, which are then moved by the beating cilia into the throat, where they can be swallowed and digested.
3. The airways also contain macrophages, which are scavenger cells that engulf accumulated debris in the airways.

How does tobacco smoke damage the lungs?

1. The three most dangerous substances in tobacco smoke are nicotine, carbon monoxide, and tar.
2. Tar immediately interferes with lung function.
3. Lungs secrete mucus, which traps foreign particles; cilia move the mucus and particles upward to the throat to be swallowed; and macrophages (large white blood cells) remove bacteria, small particles, and viruses from the inner surfaces of the lungs.
4. Tobacco smoke paralyzes the macrophages and cilia so they cannot clear the lungs.
5. One cigarette paralyzes the cilia for an hour; further smoking kills them.
6. Smoke causes mucus to accumulate in the lungs.
7. The only way to clear the lungs is for the smoker to cough frequently ("smoker's cough").
8. Smokers have frequent respiratory infections, especially bronchitis and pneumonia, and these chronic infections, with continual coughing, further damage the lungs.
9. The alveoli are tiny air sacs in the lungs that provide a huge surface air for gas exchange.
10. Coughing from chronic bronchitis ruptures the alveolar walls.
11. Nitric acid and sulfuric acid in burning tobacco weaken the alveolar walls.

12. In emphysema, so many alveoli break down and become useless that the lungs can no longer absorb enough oxygen to support life. People with advanced emphysema rely on oxygen tanks, and ultimately even that is insufficient.
13. Tobacco smoke contains at least 50 known carcinogens—chemicals that cause cancer—and smokers' lungs cannot clean themselves, so smoking causes lung and other types of cancer.
14. In time, small tumors develop which eventually metastasize (spread) to other body areas.

How do alveoli overcome the surface tension that resists expansion?

1. A liquid's resistance to an increase in surface area is called surface tension.
2. To overcome this tension, alveoli produce surfactant—a detergent that reduces surface tension and allows expansion.
3. Infants born prematurely often lack surfactant—they breathe with great difficulty and are said to be suffering from hyaline membrane disease.

How do birds breathe so well?

1. Among all the vertebrates, birds are the most efficient at extracting oxygen:
    - As a bird takes a breath, air passes into the posterior air sac, which does not allow gas exchange.
    - When the bird exhales, the contents of the posterior sac move into the lungs.
    - On the next breath, air moves out of the lung into the anterior air sac.
    - On the second exhalation, the contents pass back out into the atmosphere.
2. Thus, in birds, air moves through the lungs in a single direction, while the blood in the lungs flows in the opposite direction. As the blood acquires more oxygen, it flows into areas of the lung that have still more.
3. Countercurrent exchange is diffusion between two flows that move in opposite directions.
4. This arrangement maintains the concentration difference even as materials move.

How do insects distribute oxygen?

1. In all vertebrates and most invertebrates, the circulatory system distributes dissolved oxygen:
    - in most invertebrates and fish, oxygen moves through the gills to the circulation; and
    - in most vertebrates, oxygen moves from the lungs to the circulation.
2. Insects do not distribute dissolved oxygen—they distribute air through a system of tiny ventilation pipes called tracheae.
3. Air enters the body through openings called spiracles, which lead directly to the tracheae.
4. With this separate system to distribute oxygen, insects are able to tolerate the sluggishness of their open circulatory systems.

How do the lungs foster gas exchange?

1. The primary function of the epithelial cells in the alveoli is to promote gas exchange.
2. Fine capillaries surround the alveoli, carrying blood that binds oxygen and discharges carbon dioxide.
3. As oxygen in the inhaled air enters the alveoli, it dissolves in the liquid that coats the epithelium and diffuses first across the epithelial layer, then across the endothelium (the epithelial cells that form the capillaries), and enters into the blood plasma (the liquid part of the blood).
4. Oxygen molecules then diffuse from the plasma into the red blood cells where they can bind to hemoglobin.
5. Carbon dioxide follows the reverse path, but is mostly carried in the plasma instead of in the red blood cells.

HOW DOES OXYGEN GET TO THE TISSUES?

1. In most animals, blood carries oxygen to tissues throughout the body.
2. Most animals have special oxygen-binding proteins, such as hemoglobin and hemocyanin in the blood, and myoglobin in the muscle.

How do red blood cells transport oxygen?

1. In invertebrates, oxygen carriers are dissolved directly in the blood; in vertebrates, hemoglobin molecules are packed into red blood cells.
2. Hemoglobin consists of four polypeptide chains, each of which contains a small organic molecule called heme. In the middle of each heme is an iron atom.
3. One oxygen molecule can bind to each heme, so one hemoglobin molecule can bond to four oxygen molecules.
4. Hemoglobin with oxygen bound to it is called oxyhemoglobin, and hemoglobin lacking oxygen is called deoxyhemoglobin.

How does hemoglobin load and unload oxygen?

1. The partial pressure of oxygen in the alveoli is about 100 mm Hg, while that of the deoxygenated blood entering the lungs is about 40 mm Hg.
2. With the huge surface area, oxygen quickly reaches equilibrium so the blood leaving the lungs has an oxygen partial pressure of about 100 mm Hg.
3. Tissue cells have a partial pressure of about 46 mm Hg, so oxygen easily and quickly diffuses out of the blood and into the cells.
4. Hemoglobin changes its affinity for oxygen—each time an oxygen binds to a hemoglobin, the hemoglobin molecule becomes more likely to bind more oxygen. The molecule actually changes its shape.
5. Likewise, when hemoglobin begins to unload its oxygen, it loses its affinity for oxygen and starts to lose it more easily.

What determines hemoglobin's affinity for oxygen?

1. Oxyhemoglobin is redder than deoxyhemoglobin.
2. In adult humans, the percent of oxyhemoglobin increases from a partial pressure of about 10 mm Hg to one of about 60 mm Hg. Above that level, the partial pressure increases little and the hemoglobin is said to be saturated with oxygen. As blood moves from lungs to heart to resting muscle it gives up about 25% of its oxygen.
3. Very active cells may exceed their normal oxygen supply, then turn to glycolysis for ATP production.
4. Glycolysis produces lactic acid, which lowers the pH. The lowered pH decreases hemoglobin's affinity for oxygen, so the oxygen is more easily released.
5. Very active cells also release large amounts of carbon dioxide, and when hemoglobin binds with carbon dioxide, its affinity for oxygen is lowered so areas generating a lot of carbon dioxide (through aerobic processes) receive more oxygen.
6. In many mammals, hemoglobin binds with BPG, which also decreases hemoglobin's affinity for oxygen, so more is released.
7. In summary, hemoglobin's affinity for oxygen changes according to:
   - partial pressure of oxygen,
   - pH,
   - amount of carbon dioxide, and
   - BPG.

How does blood carry carbon dioxide?

1. Lungs also dispose of carbon dioxide—as oxygen enters, CO_2 leaves.
2. As oxygen enters the blood in the lungs, carbon dioxide leaves. The difference in partial pressure for carbon dioxide is small—46 mm Hg in the blood and 40 mm Hg in the alveoli—but it readily diffuses to equilibrium.
3. Once CO_2 dissolves in water, it can chemically combine with water to form carbonic acid:

$$CO_2 + H_2O \rightarrow H_2CO_3$$

4. Carbonic acid then dissociates to give a hydrogen ion (H^+) and a bicarbonate ion (HCO_3^-):

$$H_2CO_3 \rightarrow H^+ + HCO_3^-$$

5. Most (about two-thirds) of the CO_2 carried by the blood is in the form of bicarbonate in red blood cells and plasma.
6. Hemoglobin can also bind and transport carbon dioxide.
7. A small amount—about 5%—dissolves directly in the blood.

WHAT FACTORS CONTRIBUTE TO OXYGEN HOMEOSTASIS?

How can blood deliver oxygen at high altitude?

1. At high altitudes, the difference in the oxygen concentrations in the lungs and muscles is not as great as at sea level.
2. When a person climbs a mountain, an increase occurs in the level of BPG, which binds to hemoglobin and increases its ability to deliver oxygen efficiently.
3. At high altitude, the number of red blood cells increases, thus increasing the total amount of hemoglobin and the oxygen-carrying capacity of the blood.
4. The stimulus for increased red blood cell production is a hormone—erythropoietin—which is made in the kidneys in response to low oxygen delivery.

How do animals regulate breathing in response to changes in oxygen and carbon dioxide delivery?

1. As tissues use oxygen for respiration, they also produce carbon dioxide as a waste product—increased respiration always leads to an increase in CO_2.
2. In mammals, it is an increased CO_2 concentration in the blood that signals the need for deeper or more rapid breathing.
3. Coordination of breathing depends on the breathing center—a nerve complex in part of the brain called the medulla—that regulates the rate of breathing.
4. Nerves in the breathing center trigger contraction of the diaphragm and inspiratory muscles.
5. As the lungs expand, stretch receptors report back to the breathing center and the contraction signal ends, allowing expiration to occur.
6. The medulla also controls the circulatory system, changing heart output according to needs.

VOCABULARY BUILDING

In your own words, first write a brief definition, then a full explanation for each of the following terms. Include examples where appropriate. Complete this section from your memory—you will not learn the terms by simply copying definitions from the textbook. Once you have finished, check your responses against the information in the chapter and make any necessary corrections.

partial pressure —

equilibrium —

surface-to-volume ratio —

gills —

lungs

operculum —

book lung —

respiratory system —

pharynx —

larynx —

vocal cords —

trachea —

bronchi —

bronchioles —

alveoli —

tidal volume —

dead space —

residual —

ventilation —

thoracic cavity —

diaphragm —

inspiration —

expiration —

cilia —

mucus —

emphysema —

surfactant —

spiracles —

heme —

oxyhemoglobin —

deoxyhemoglobin —

breathing center —

medulla —

CHAPTER TEST

The following test has four parts. Complete as much of the exam as you can from memory. If you cannot answer a question, skip it. Once you complete all that you can, try to answer any questions you skipped. If you still cannot answer them, consult your textbook for the answers. Once you have completed all sections of the test, check your answers for Parts 1 - 3 against those in the back of this book. Highlight any incorrect answers then review that material in your textbook. Correct your answers for future reference.

Part 1: Multiple Choice

For each of the following, select all correct responses—more than one may be correct.

1. According to its own internal memos, since when has the Brown and Williamson tobacco company known the addictive nature of smoking?
   a) since the 1880s
   b) since the 1960s
   c) since the 1980s
   d) The company still does not believe smoking is addictive.

2. Nicotine causes addiction in the same way as:
   a) cocaine
   b) heroin
   c) other less serious drugs
   d) Nicotine is not addictive.

3. What percentage of lung cancer deaths are caused by smoking?
   a) 10%
   b) 50%
   c) 77%
   d) 90%

4. What is the size limit for animals to receive adequate oxygen by diffusion alone?
   a) 1 cell
   b) No animals get enough from diffusion alone.
   c) 1 inch
   d) 1 millimeter

5. Which of the following paralyzes cilia and macrophages in the lungs?
   a) nicotine
   b) tar
   c) carbon dioxide
   d) carbon monoxide

6. Emphysema is characterized by which of the following?
   a) paralyzed cilia
   b) paralyzed macrophages
   c) accumulation of mucus
   d) destruction of alveoli

7. About what percentage of the air we breathe is oxygen?
   a) 5%
   b) 21%
   c) 78%
   d) 100%

8. The amount of dissolved gas is dependent on:
   a) temperature
   b) concentration in the gas phase
   c) partial pressure of the gas
   d) all of these

9. How do fish move water across their gills?
   a) by swallowing water
   b) by pumping water over their gills
   c) by swimming through the water
   d) all of these

10. In vertebrates, the common passage for both food and air is the:
   a) trachea
   b) pharynx
   c) esophagus
   d) larynx

11. Sound is produced by air moving across the vocal cords, which are located within the:
   a) trachea
   b) pharynx
   c) esophagus
   d) larynx

12. Which of the following muscles contract during normal expiration when at rest?
   a) diaphragm
   b) muscles between the ribs (intercostal muscles)
   c) all of these
   d) None of these; the muscles relax during normal expiration.

13. Insects bring air into their bodies through the:
    a) gills
    b) spiracles
    c) mouth
    d) nose

14. Which of the following is true about hemoglobin?
    a) It is packed into vertebrates' red blood cells.
    b) It binds and carries oxygen throughout the body.
    c) One hemoglobin molecule can bind four oxygen molecules.
    d) All of these are true.

15. The stimulus for increased red blood cell production, to increase the blood's oxygen-carrying capacity, is:
    a) erythropoietin
    b) hemoglobin
    c) hemocyanin
    d) BPG

Part 2: Matching

For each of the following, match the correct term with its definition or example. More than one answer may be appropriate.

lungs **gills**

1. ____________________ These are evaginated breathing surfaces.
2. ____________________ These are always internal.
3. ____________________ These contain alveoli.
4. ____________________ These are always associated with the circulation.
5. ____________________ Oxygen must first be dissolved before uptake occurs.

oxygen **carbon dioxide**

6. ____________________ Accounts for 21% of the air we breathe.
7. ____________________ Most is transported bound to hemoglobin.
8. ____________________ Has a small but sufficient concentration gradient between the blood and alveoli.
9. ____________________ Leaves the blood in areas with a low pH.
10. ____________________ Released by active cells.

***Part 3:* Short Answer**

Write your answers in the space provided or on a separate piece of paper.

1. In the history of smoking, what two events caused cigarette sales to level off and decline?

2. List two ways that exercise reduces hemoglobin's affinity for oxygen.

3. How does tar cause smoker's cough?

4. Briefly explain three mechanisms that normally help keep the lungs clean.

5. Explain how the lungs are damaged in emphysema.

6. What is the typical lifespan of someone with untreated lung cancer? What if the disease is aggressively treated?

7. Explain the relationship between the partial pressure and total pressure of gases in a mixture.

8. Explain three factors that determine the rate at which gas molecules enter a solution.

9. Explain the breathing apparatus used by spiders.

10. What is the function of the epiglottis?

11. Explain dead space.

12. Explain how air is brought into the lungs.

13. How do oxygen and carbon dioxide cross the walls of the alveoli and blood vessels?

14. Explain hyaline membrane disease.

15. What are two mechanisms by which blood can deliver adequate oxygen at high altitude?

***Part 4:* Critical Thinking: Using Your Knowledge**

Answer each of these in essay form, using complete sentences and paragraphs. Provide as much information as you can. (For extra essay practice, write out answers to the Review and Thought Questions in your textbook.)

1. In your opinion, should the government put restrictions on cigarette sales and smoking? If the decision was yours, what restrictions would you allow and which would you not allow?

2. How would you resolve the issue of health concerns from both first- and second-hand smoke vs. the economic concerns of many states that depend on the tobacco industry?

3. In the United States, some major lawsuits are being won against the tobacco industry on behalf of diseased and dying smokers. Keep in mind the following information:
   - Tobacco companies have known since the 1960s about the addictive potential of nicotine, and are accused of increasing the nicotine concentration in their products.
   - The government, media, and society actively promoted smoking earlier in our history.
   - The people suing are primarily lifetime smokers who started, and became addicted, decades ago when smoking was quite common and popular.
   - Warning labels have appeared on cigarette packages since the mid-1960s.

4. Should smokers be allowed to sue tobacco manufacturers?

5. Should states be able to sue tobacco industries to regain lost revenue from supporting ill smokers with tax dollars?

6. Describe and trace the path of oxygen through the respiratory systems of fish, insects, birds, and humans. Draw comparisons between each system, and explain the need for unique features in each.

7. In the United States, many smokers are irritated by bans on smoking in public places, such as restaurants, taverns, and offices, claiming that they are tax-paying American citizens and have the right to smoke if they want to. How would you address their concerns?

8. Describe the processes involved in both inspiration and expiration, and relate them to tidal volume and dead space.

9. What are the advantages to having most of the oxygen carried bound to hemoglobin, which can change its affinity for oxygen?

10. The disease cystic fibrosis causes excessive secretion of mucus. Explain why this disease can severely impair lung function.

Chapter 39
How Do Animals Manage Water, Salts, and Wastes?

KEY CONCEPTS

1. The extracellular space lies between the external environment and the inside of cells.
2. Organisms must maintain a balance of water and solutes in cells and in extracellular fluid.
3. Movement of water, ions, and molecules between the outside environment, the extracellular space, and the insides of the cells depends on the osmolarity of fluids in each space.
4. Freshwater animals rid themselves of excess water by excreting dilute urine; marine and terrestrial animals tend to excrete salts and other wastes in minimal amounts of water.
5. Land animals acquire water from water in beverages, water in food, and the metabolism of food. Animals lose water in urine, feces, and evaporation from moist surfaces.
6. Cells and organisms rid themselves of nitrogenous wastes by means of excretion—aquatic organisms excrete dilute ammonia; mammals excrete urea; and insects, snails, birds, and most reptiles excrete uric acid.
7. As blood enters the glomerulus, all of the constituents of the blood except the cells and large molecules filter into Bowman's capsule and the renal tubule.
8. Renal tubules return water and valuable ions and molecules from the filtrate to the blood.
9. The parts of the renal tubule vary in their structure and function:
   - the proximal tubule is lined with microvilli which pump ions across the epithelium;
   - the descending loop of Henle is smooth and does not actively pump ions; and
   - the ascending loop of Henle and the distal tubule are lined with microvilli.
10. Active pumping of ions by the cells in the wall of the renal tubules creates osmotic gradients that pull water from the filtrate.
11. Urea stays in the renal tubule until it reaches the far end of the collecting duct, where it accumulates in surrounding tissues and draws water from the remaining fluid in the tubule.
12. A nephron forms urine with high concentrations of urea, as well as various ions, ammonia, organic acids and bases, and foreign molecules. The nephron normally absorbs glucose.
13. An increase in the concentration of salt or protein wastes in the blood causes an increase in the blood flow in the renal arteries.
14. When salt concentrations in the blood increase, the brain's hypothalamus detects the increase and triggers the sensation of thirst.
15. A decrease in blood pressure or in sodium causes the kidneys to secrete renin, which forms angiotensin. Angiotensin constricts arterioles, increasing blood pressure and causing the kidneys (and therefore the body) to retain both salt and water.
16. Aldosterone prevents water loss and salt loss by stimulating sodium reabsorption in the kidneys, the sweat glands, and the intestines.
17. When the concentration of salt in the blood rises, the posterior pituitary secretes ADH, which increases water permeability in the kidney's collecting ducts. By causing the body to retain water, ADH can increase blood pressure by as much as 40 points.
18. The heart produces atrial natriuretic factor, which is a hormone that increases water loss and lowers blood pressure.

EXTENDED CHAPTER OUTLINE

WILL CORN CHIPS RAISE YOUR BLOOD PRESSURE?

1. In the spring of 1996, Julian Midgley, Andrew Matthew, and their colleagues wrote that for most people an extremely low-salt diet would not lower blood pressure and in some people could actually cause a heart attack.
2. Even with recent declines in salt consumption, the average American consumes twice the recommended daily intake of salt.
3. The minimum sodium requirement for normal day-to-day health is 115 mg, yet most of us take in nearly 4000 mg each day. Government nutritionists, in an attempt to recommend something reasonable, suggest no more than 2400 mg per day.
4. A study of 32 countries found a consistent relationship between sodium consumption and high blood pressure (hypertension).
5. Some researchers argue that a closer examination of people's behavior usually shows that those who consume a lot of sodium also drink more alcohol and tend to be overweight.
6. Researchers have long known that only a small percentage of people with high blood pressure benefit from a low-salt diet—in about half of all people with hypertension, which is 5 to 10% of all Americans, the kidneys cannot compensate for large amounts of sodium, leading to high blood pressure.
7. As we age, our sodium sensitivity increases and our kidneys lose some of the ability to regulate salt and fluid levels in the blood.
8. Midgley, Matthew, and their colleagues concluded that a low-sodium diet lowers blood pressure only in people over age 45 who have high blood pressure; and the drop is modest.
9. A super-low-sodium diet may be dangerous, and can increase the risk of heart attack by four times in men with high blood pressure
10. Too much salt burdens the kidneys and may possibly contribute to osteoporosis later in life.

WHY DO ANIMALS NEED TO REGULATE WATER AND SALT?

1. Water makes up more than half of the body's weight.
2. Most biochemical reactions occur in water and water participates in many reactions.
3. Cells and organisms strictly regulate the solution composition inside and outside the cells.
4. A cell needs not only the right amount of water, but also the right amounts of ions and other molecules dissolved in that water.
5. Organisms must be able to detect deviations from the norm and take corrective action.
6. This is the essence of homeostasis—the tendency to maintain a stable internal environment.
7. The kidneys:
   - regulate the volume and composition of the blood and extracellular fluid;
   - maintain constant salt concentrations in the body fluids; and
   - regulate the disposal of water, wastes, and toxins.

Which way does water flow?

1. In organisms with circulatory systems, the extracellular space is continuous with the liquid part of the blood.
2. The extracellular space contains a dilute solution of salts and organic molecules, whose concentrations differ from those in the cytoplasm, but the total concentration of solutes is the same inside and outside the cell.
3. In general, water can flow freely from between the extracellular space and the cytoplasm.
4. Osmolarity is roughly the sum of the concentrations of all ions and molecules in a solution.
5. When a pathway is open, water flows passively from a solution with lower osmolarity to a solution of higher osmolarity, but energy is used for active transport to establish the gradient.

Water and salt balance in aquatic animals

1. Two solutions that have the same osmolarity are said to be isotonic, and there will be no net flow of water between the two solutions.
2. A solution that has higher osmolarity than another one, such as a saltier solution, is hypertonic and will tend to take in water.
3. Freshwater animals are saltier than their environments, so their body fluids are hypertonic, and they must fight the inflow of water with water-resistant skins.
4. A solution that is more dilute than another is hypotonic and will tend to lose water.
5. Freshwater animals produce dilute urine by pumping ions out of it, then back into the body.
6. Excreting large amounts of urine results in loss of essential molecules and ions, so they must be able to reabsorb these.

How do saltwater fish hold on to water?

1. Saltwater fish must counteract the tendency of water to flow outward. They do this by pumping ions out of their body fluids, into their urine and back to the sea.
2. Saltwater fish must continually compensate for water loss. They do this by drinking lots of seawater, which also brings in more salt. Excess salt is removed, mostly through the gills.
3. Marine mammals don't usually drink salt water—they derive most of their water from foods they eat and as a byproduct of metabolism.

How do land animals balance water and salt?

1. The major problem of life on land is getting and keeping water.
2. Although some can get water through their skin or directly from the air (insects), most land animals derive water from three sources: drinking, eating, and metabolism.
3. The average adult human drinks about 1200 ml of fluids each day and obtains another
4. 1000 ml in food. Metabolism yields another 350 ml, for a total of about 2.5 L per day.
5. Normally, the body loses the same amount of water it takes in.
6. Water leaves land animals through urine (most in humans is lost this way), feces, and evaporation (from the outer body surface and respiratory organs).
7. Desert animals, with access to almost no water, conserve water meticulously.
8. In all animals, water intake and loss may vary greatly from day to day.
9. Animals must regulate their water flow, usually by adjusting drinking and urine production.
10. Under most circumstances, our water content varies by less than 1%.
11. Dehydration occurs when too much water flows out of cells; water intoxication occurs when water accumulates in the tissues.

HOW DO ANIMALS DISPOSE OF NITROGEN-CONTAINING WASTES?

1. In land animals, water and salt regulation is tightly connected to disposal of various wastes, especially nitrogenous wastes—products of the breakdown of proteins, nucleic acids, and other nitrogen-containing compounds.
2. Cells and organisms rid themselves of nitrogenous wastes by means of excretion—a disposal process by which cells pass materials across a cell membrane.
3. Elimination is the disposal of remains of digested food.
4. Fish and most aquatic animals accomplish excretion by converting excess nitrogen to ammonia, which is toxic to animals. Aquatic animals can dilute it; land animals cannot.
5. Land mammals convert ammonia to urea, a relatively nontoxic compound.
6. Insects, land snails, most reptiles, and birds require even less water than mammals, and they dispose of nitrogenous wastes by producing a nearly insoluble compound called uric acid.

HOW DO KIDNEYS WORK?

1. In vertebrates, the major organs of excretion are the kidneys.
2. Kidneys are paired structures lying below the stomach and liver.
3. Blood, carrying wastes from all over the body, arrives at each kidney by a renal artery and leaves through the renal vein.
4. The kidneys remove wastes and water from the blood and produce urine that flows down to the bladder through a pair of tubes called ureters.
5. The bladder drains to the outside of the body through the urethra.
6. In female mammals, the urethra empties directly to the outside, but in male mammals, the urethra joins the reproductive tract, serving as a common exit for both urine and sperm.
7. In most birds, reptiles, and amphibians, the ureter runs directly into a cloaca—a common exit for the digestive, excretory, and reproductive systems.

How does a kidney filter the blood?

1. A kidney has an outer region, called the cortex, and an inner region, called the medulla.
2. A single human kidney contains about a million glomeruli—tangled networks of capillaries that are derived from the renal arteries.
3. Surrounding each glomerulus is a bulb called Bowman's capsule, which leads to a long, narrow tube, called the real tubule.
4. The glomerulus, Bowman's capsule, and the tubule together form a nephron, which is the functional unit of a kidney.
5. Blood enters each glomerulus and filters through pores into Bowman's capsule.
6. Water, urea, and other small molecules from the blood pass freely into Bowman's capsule, forming the filtrate, which passes through the renal tubule.
7. The filtrate's composition is similar to that of the blood plasma, but with fewer proteins.

How do the tubules recycle essential molecules and ions and recapture water?

1. In humans, the total volume of filtrate is almost 200 L per day, but only about 1.5 L leaves the body as urine.
2. The filtrate flows down the length of the tubule and the cells of the tubule selectively remove salt, water, glucose, and other ions and molecules and return them to the blood. Water follows passively.
3. This process is called reabsorption.
4. In humans, the plasma is filtered about 60 times each day.

What does the structure of a tubule suggest about its workings?

1. The three main regions of a tubule are:
   - the proximal tubule, which lies just next to Bowman's capsule, in the cortex;
   - the loop of Henle, which descends into the medulla then ascends into the cortex; and
   - the distal tubule, which carries urine to a collecting duct, then to a ureter and the bladder.
2. The proximal tubule has brush borders—fingerlike microvilli—that help pump materials across the epithelium that lines the tubule.
3. Most of the sodium ions are removed from the lumen here. Chloride ions and water follow the sodium.
4. The descending loop of Henle and the deepest part of the ascending limb lack brush borders and have far fewer mitochondria because they do not actively pump materials.
5. The ascending limb of the loop of Henle and the distal tubule do have brush borders and lots of mitochondria, and is also involved in pumping sodium.

How do the tubules regulate the ion content of the lumen?

1. The proximal tubule actively pumps sodium ions out of the filtrate in the lumen; other ions follow. The ion movement causes water to follow out of the lumen and into the capillaries.
2. By the time the filtrate enters the loop of Henle, about 75 percent of the water has returned to the blood and extracellular fluid.
3. The deep part of the medulla, at the bottom of the loop of Henle, is very salty, so water in the lumen moves out through the wall of the tubule by osmosis, concentrating the filtrate as it moves to the bottom of the loop.
4. Active pumping of ions by the cells in the wall of the renal tubules creates osmotic gradients that pull water from the filtrate.

What happens to the urea?

1. Urea enters the tubule in the original filtrate, and cannot move out of the lumen of the tubule.
2. Urea diffuses out at the far end of the collecting duct, raising the ion concentration of the fluid surrounding the deep part of the loop of Henle.
3. The high urea content and saltiness of this fluid pull water and salt passively out of the lumen of the deepest part of the descending limb.

How does the kidney form concentrated urine?

1. The nephron also controls the concentrations of other substances in the blood and urine by moving substances in both directions across the tubular epithelium.
2. Reabsorption is transport of substances (such as salt, water, glucose, and other nutrients) from the filtrate back to the blood.
3. Secretion is transport of substances (such as potassium, hydrogen ions, ammonia, and organic acids and bases) into the filtrate, to be excreted in the urine.

How do the kidneys of nonmammalian vertebrates meet the challenges of different environments?

1. The structure and function of nephrons is similar in all vertebrates, but with notable exceptions.
2. Fish have special cells in their gills that can pump chloride and other ions. The direction of the pumping depends on their environment.
3. The kidneys of most birds resemble those of mammals, but marine birds, who drink salt water, have special salt glands above their eyes that excrete excess salt. Some reptiles also have salt glands.

HOW DO ANIMALS ADJUST WATER AND SALT IN RESPONSE TO CHANGING CONDITIONS?

1. Maintaining a constant level of water and salt concentration in the blood means that the kidneys must constantly adjust urine concentration and flow.

How does the brain control fluid intake?

1. Animals control how much water is in the body by controlling how much we drink.
2. In mammals, the thirst center is in the hypothalamus, which has osmoreceptors—cells that detect changes in the salt concentration. When the blood gets too "salty," these receptors stimulate the sensation of thirst.
3. Once we drink, this sensation is suspended for 15 to 30 minutes. If the saltiness is not fully corrected, thirst is stimulated again.

The rate of filtration depends on renal blood flow

1. Blood flow to the renal arteries of the kidneys is generally much more constant than blood flow to other areas of the body.
2. When the concentration of salt or protein wastes in the blood rises, blood flow in the renal arteries increases so the kidneys can process more filtrate.

Angiotensin and renin

1. Angiotensin is the most powerful blood vessel constrictor known.
2. Angiotensin works along with an enzyme—renin—secreted by the kidneys.
3. The kidneys secrete renin in response to a decrease in blood pressure or sodium.
4. Renin then forms angiotensin, which constricts the arterioles which, in turn, increases blood pressure. This increases the filtration rate in the kidneys and causes them to retain water and salt. This, in turn, increases blood volume and, thus, blood pressure.

Aldosterone

1. The hormone aldosterone comes from the adrenal cortex (part of the adrenal gland).
2. Aldosterone stimulates sodium reabsorption in the distal tubules and collecting ducts of the kidneys, as well as in the sweat glands and intestines.
3. Decreased blood sodium concentration causes the adrenal glands to secrete more aldosterone.
4. Aldosterone regulates water loss as well as salt loss.
5. Aldosterone secretion depends not only on the adrenal glands, but also on chemical signals from the liver and electrical signals from the nervous system. This allows regulation of sodium reabsorption in response to changes in blood pressure, salt concentration, water loss, and disease.

Antidiuretic hormone (ADH), or vasopressin

1. Like aldosterone, antidiuretic hormone (ADH, vasopressin) prevents water loss.
2. ADH is secreted by specialized cells in the posterior pituitary (part of the brain) in response to increased salt in the blood.
3. ADH causes the body to retain water by increasing water permeability in the collecting ducts of the kidneys; water then returns to the blood and the urine becomes more concentrated.
4. When ADH reaches the kidneys, it stimulates water reabsorption, diluting the blood and salt.
5. As a result, the total blood volume increases, which also increases blood pressure.
6. Alcohol inhibits ADH release, increasing urine flow.
7. Pain, fear, cold, and stress can all affect ADH levels and the amount of urine.
8. The body controls the composition of the blood and other fluids by regulating the transport of specific substances and altering drinking and eating behavior.

Atrial natriuretic factor

1. Kidney function depends on blood pressure to provide the force for filtration and flow of the filtrate through the tubules.
2. In response to excessive blood pressure, cells in the smaller chambers of the heart (the atria) produce a hormone called atrial natriuretic factor (ANF).
3. ANF:
    - inhibits secretion of renin, aldosterone, and ADH,
    - relaxes smooth muscle,
    - reduces thirst, and
    - increases the kidneys' elimination of sodium ions and water by closing channels in the collecting ducts.
4. ANF thus increases water loss and lowers blood pressure.

HOW DO INVERTEBRATES SOLVE PROBLEMS OF WATER, SALTS, AND WASTES?

1. Animals in different environments must solve different problems of salt and water.

Flatworms

1. The excretory organs of flatworms consist of two or more tubes that run the length of their bodies and empty through tiny pores.
2. Leading into the tubes are many bell-shaped chambers, each formed from a single cell.
3. These bell-shaped cells (flame cells) are lined with cilia that propel liquids into the tubes.
4. Flatworm excretory systems are mainly for excreting water—most nitrogenous wastes are eliminated through the gut.

Earthworms

1. Like mammalian kidneys, earthworm nephridia both filter and reabsorb.
2. Each body segment contains a pair of nephridia.
3. Pressurized fluid in the earthworm's body cavity pushes through an opening in the tubular part of the nephridium.
4. Inside the nephridium, ions and molecules re-enter the circulatory system, and the dilute urine left behind passes from each nephridium through a pore in the body wall.

Marine invertebrates

1. Most marine invertebrates' body fluids have about the same salt concentration as seawater.
2. In many of these animals, the concentration of ions and molecules in the body fluids fluctuates with that of their environment.
3. In marine invertebrates, kidneys work by filtration, absorption, and secretion.
4. They eliminate excess nitrogen in the form of ammonia.
5. Some invertebrates have arrangements of excretory organs very different from humans. Crustaceans, for example, excrete through green glands in their heads.

Insects

1. Insect excretory organs, called Malpighian tubules, are blind outpocketings of the gut, with their blind ends bathed in the insect's body fluids.
2. Fluid flows into the tubules because the tubules actively pump potassium ions, then water and dissolved nitrogen wastes passively flow into the tubules by osmosis.
3. The contents then flow into the gut, where the residues of digestion (feces) mix with the products of excretion (urine).

VOCABULARY BUILDING

In your own words, first write a brief definition, then a full explanation for each of the following terms. Include examples where appropriate. Complete this section from your memory—you will not learn the terms by simply copying definitions from the textbook. Once you have finished, check your responses against the information in the chapter and make any necessary corrections.

homeostasis —

kidneys —

osmolarity —

isotonic —

hypertonic —

hypotonic —

dehydration —

nitrogenous wastes —

excretion —

elimination —

ammonia —

urea —

uric acid —

renal artery —

renal vein —

bladder —

ureters —

urethra —

cloaca —

cortex —

medulla —

glomerulus —

Bowman's capsule —

renal tubule —

nephron —

filtrate —

reabsorb —

proximal tubule —

loop of Henle —

distal tubule —

epithelium —

reabsorption —

secretion —

angiotensin —

renin —

aldosterone —

adrenal glands —

antidiuretic hormone (ADH, vasopressin) —

diuretic —

posterior pituitary —

hypothalamus —

atrial natriuretic factor (ANF) —

flame cells —

nephridia —

CHAPTER TEST

The following test has four parts. Complete as much of the exam as you can from memory. If you cannot answer a question, skip it. Once you complete all that you can, try to answer any questions you skipped. If you still cannot answer them, consult your textbook for the answers. Once you have completed all sections of the test, check your answers for Parts 1 - 3 against those in the back of this book. Highlight any incorrect answers then review that material in your textbook. Correct your answers for future reference.

Part 1: Multiple Choice

For each of the following, select all correct responses—more than one may be correct.

1. The average American consumes what portion of the recommended daily intake of salt?
   a) a quarter of the recommended daily intake
   b) half of the recommended daily intake
   c) one and a quarter of the recommended daily intake
   d) twice the recommended daily intake

2. By weight, water makes up what percent of the body?
   a) 25%
   b) 42%
   c) 50%
   d) 90%

3. If solution A is saltier than solution B, solution A is said to be:
   a) isotonic to solution B
   b) hypertonic to solution B
   c) hypotonic to solution B
   d) more dilute than solution B

4. Which of the following is most likely to excrete large amounts of dilute urine?
   a) marine animals
   b) freshwater animals
   c) terrestrial (land) animals
   d) unicellular organisms

5. Land animals lose water through which of the following routes?
   a) urine
   b) feces
   c) metabolism
   d) evaporation

6. Breakdown of which of the following produces nitrogenous wastes?
   a) proteins
   b) cellular components
   c) carbohydrates
   d) nucleic acids

7. Organisms rid themselves of nitrogenous wastes primarily through:
   a) dehydration
   b) elimination
   c) excretion
   d) evaporation

8. Aquatic organisms rid themselves of nitrogenous wastes in what form?
   a) ammonia
   b) urea
   c) uric acid
   d) none of these

9. Which of the following is the correct path taken by substances during excretion?
   a) blood, ureter, kidney, urethra, bladder
   b) kidney, ureter, bladder, urethra, blood
   c) blood, kidney, ureter, urethra, bladder
   d) blood, kidney, ureter, bladder, urethra

10. In most birds and amphibians, the cloaca serves as a common exit for which systems?
   a) digestive
   b) excretory
   c) reproductive
   d) all of these

11. The functional unit of a kidney is the:
   a) glomerulus
   b) nephron
   c) tubule
   d) Bowman's capsule

12. How is water moved out of the tubule?
    a) by osmosis
    b) by the pumping of ions, which creates osmotic gradients
    c) by diffusion
    d) Water is not moved out of the tubule.
13. The process by which substances are moved from the filtrate to the blood is called:
    a) filtration
    b) reabsorption
    c) secretion
    d) excretion
14. Antidiuretic hormone has which of the following effects?
    a) increases urine output
    b) increases water reabsorption
    c) raises blood pressure
    d) increases the blood's salt concentration
15. Nephridia in earthworms do not perform which of the following functions?
    a) filtration
    b) reabsorption
    c) secretion
    d) They perform all of these.

Part 2: Matching

For each of the following, match the correct term with its definition or example. More than one answer may be appropriate.

reabsorption **secretion**

1. ____________________ Substances move from the tubule into the blood.
2. ____________________ Potassium ions and ammonia are moved by this process.
3. ____________________ Substances moved by this process become more concentrated in the urine.
4. ____________________ Glucose is moved by this process.
5. ____________________ Removes antibiotics and other drugs from the body.

aldosterone **ADH** **angiotensin**

6. ____________________ This is formed in response to renin.
7. ____________________ This is released by the adrenal glands.
8. ____________________ This increases blood pressure.
9. ____________________ This constricts the arterioles.
10. ____________________ This is released in response to decreased blood sodium.

***Part 3:* Short Answer**

Write your answers in the space provided or on a separate piece of paper.

1. What would happen if a hypotonic cell is placed into a hypertonic solution?

2. Land animals derive water from what three sources?

3. What condition results when an organism takes in more water than it loses?

4. Name three compounds that can be used to rid the body of excess nitrogen.

5. What are the three parts of a nephron?

6. What substances do not pass into Bowman's capsule?

7. List, in order, the parts of the tubule after Bowman's capsule.

8. Where in the tubule are microvilli found, and what do they do?

9. By what mechanism does urea help to move water out of the tubule?

10. Briefly explain how renin and angiotensin work together.

11. What triggers the secretion of aldosterone?

12. By what mechanism does antidiuretic hormone cause the body to retain water?

13. List three effects of atrial natriuretic factor.

14. What is the function of flame cells?

15. In insects, the excretory system is linked by the Malpighian tubules to what other system?

Part 4: Critical Thinking: Using Your Knowledge

Answer each of these in essay form, using complete sentences and paragraphs. Provide as much information as you can. (For extra essay practice, write out answers to the Review and Thought Questions in your textbook.)

1. The beginning of this chapter in your book discusses the controversy about the role of a low-sodium diet in battling hypertension. Armed with your new knowledge about how the excretory system regulates salt and water levels, explain specifically how a high-sodium diet might contribute to hypertension in some individuals, and why it apparently has little or no effect in most people. Discuss the specific mechanisms involved, including hormonal controls and renal blood flow.

2. While dining in an upscale restaurant, you notice that patrons in the bar are enjoying free snacks, such as pretzels, peanuts, and popcorn. Armed with your current knowledge of excretory regulation, explain how these free snacks are really meant to entice patrons to buy more drinks.

3. Explain the mechanism by which alcohol can lead to dehydration, increasing thirst, and perhaps also increasing beverage sales (and excessive alcohol consumption).

4. Considering your answer to the last question, explain some of the reasons people often suffer from terrible "hangovers" the day after a drinking binge.

5. Why is alcohol consumption during very hot weather or during strenuous activity a dangerous practice?

6. Thoroughly discuss the movement of substances through the mammalian kidney, tracing their path from the blood to the outside world. Discuss filtrate formation, reabsorption, and secretion, and provide examples of substances moved in each of these processes.

7. A patient with damaged kidneys often experiences swelling in various parts of the body. Explain how the two conditions are related.

8. You show tremendous aptitude during your biology class, and are hired by a major pharmaceutical company that asks you to develop a new diuretic medication. What areas of water metabolism might you target with your new drug to prevent people from retaining water?

9. You set up a beautiful freshwater aquarium and want to fill it with beautiful fish. The brightest ones you find in the pet store are labeled "marine fish." You buy some, put them in your aquarium, and they die. Explain what happened to them.

10. What would happen if you were stranded on a remote island with no fresh drinking water and you began drinking the seawater?

Chapter 40
Defense: Inflammation and Immunity

KEY CONCEPTS

1. The body defends itself against invading pathogens with three lines of defense:
   - the skin and populations of microorganisms on it and in the gut;
   - blood clots and inflammation; and
   - the immune system.
2. Lysozyme in saliva and tears kills most foreign microorganisms; those that enter the intestinal tract must compete with millions of benign microorganisms that live there.
3. Broken skin is closed by wound healing. Broken vessels are closed by hemostasis.
4. Because blood clots are dangerous, synthesis of fibrin is carefully regulated and a system of anticlotting enzymes ensures that clots are rapidly dissolved.
5. Complement proteins and mast cells react nonspecifically to damaged body tissues. Mast cells release histamine, which promotes inflammation.
6. Inflammation helps the body resist infection by opening blood vessels, increasing delivery of white blood cells and plasma proteins. Cytokines, from macrophages, regulate inflammation.
7. The immune response demonstrates specificity, memory, diversity, and the ability to distinguish self from nonself.
8. B lymphocytes, T lymphocytes, and natural killer cells conduct the immune response.
9. Antibody molecules recognize antigens by the specific shape of the antigen's epitopes.
10. Thousands of genes can produce millions of antibodies by:
    - mixing thousands of polypeptides in different combinations, and
    - splicing genes into thousands of combinations not originally specified by the genome.
11. When an antigen binds to B-cell receptors, the B cells divide and differentiate into effector cells (plasma cells), which make and secrete antibodies, or into memory cells, which make more effector cells if the antigen ever appears again.
12. T cells only recognize altered self-antigens that are bound to MHC proteins; T-cell receptors recognize linear segments of polypeptide chains (not epitopes).
13. MHC genes specify membrane proteins that bind and present antigens to T cells.
14. During maturation in the thymus, T cells undergo heavy selection so that only T cells that recognize altered self survive.
15. MHC proteins and other cell surface antigens make the cells of every individual unique.

EXTENDED CHAPTER OUTLINE

THE DANGER MODEL

1. Since a series of experiments in the 1940s, biologists have believed that the immune system protects us by distinguishing self from nonself, and that the system learns what molecules are "self" around birth. This is the self-nonself model.
2. Polly Matzinger has begun to argue that newborn immune systems do not differ much from those of adults, and that the self-nonself model is wrong.
3. By Matzinger's model—the danger model—the immune system distinguishes, instead, between cells and molecules that are dangerous and those that are not.

Jumping off the cliff

1. Early immunology studies focused on organ transplants between individuals, which are often rejected because the cells of the immune system attack and destroy the transplanted organ.
2. English biologist Peter Medawar conducted experiments which showed that he could "teach" newborn mice to tolerate foreign cells.
3. Austrian virologist Frank MacFarlane Burnet proposed that the immune system's main task is to distinguish self from nonself, and that it learns this distinction just once—during fetal development. In 1960, Medawar and Burnet shared a Nobel Prize for their work.
4. This theory does not explain why pregnant women's immune systems do not attack their fetuses, nor how we tolerate bacteria in our mouths, noses, and throats.
5. When a normal cell dies, macrophages are attracted and they eat the dead cell, leaving no trace, in a process called apoptosis.
6. A cell under attack by viruses usually bursts, releasing the cell's cytoplasm, organelles, and virus particles—a mess that must surely attract attention from the immune system.
7. T cells recognize the body's own cells through a series of MHC (major histocompatibility complex) proteins on the cell surfaces.
8. T cells that bind to the body's own healthy cells die, but those that bind to infected or damaged cells are triggered by a second signal which activates the T cells, causing them to move to a lymph node and activate other immune cells.
9. By the self-nonself model, no T cells are activated by a second signal during fetal development—they all die or become inactive. No one knows what the second signal is.
10. The danger model says that the immune system constantly redefines what is dangerous—if no cells are damaged, no second signal is given, and the T cells die.
11. By this model, cells in tumors and warts would not trigger an immune response because the cells are healthy and not dying "messy" deaths, so no second signal is given.
12. This textbook uses the self-nonself model because it is currently the most accepted model.

THE CAST OF CHARACTERS

1. Mammals use three lines of defense to protect themselves against pathogens:
   - the skin;
   - blood clots and inflammation; and
   - the immune system.
2. The immune system alone has specificity—it distinguishes different pathogens.
3. All cells involved in the body's defense are white blood cells (leukocytes), which come from stem cells in the bone marrow.
4. Four white blood cells—neutrophils, eosinophils, basophils, and mast cells—participate mainly in the inflammation response.
5. The two kinds of lymphocytes (B cells and T cells) and natural killer cells (NK cells) participate in the immune response.
6. Macrophages, giant cell-eating white blood cells, participate in inflammation and immunity.

HOW DOES THE SKIN KEEP PATHOGENS OUT?

1. In humans, the skin makes up about 15 percent of the body's weight.
2. The skin epithelium consists of a thick pile of cell layers which together form the epidermis.
3. The outermost cells of the epidermis are nondividing cells that are mostly dead and dying, which protect the dividing cells beneath them.
4. Beneath the epidermis are the dermis (a thick layer of living cells) and subcutaneous tissue.
5. Oil and sweat secreted by glands in the skin create a slightly acidic environment that is hostile to many microorganisms; nonetheless, the skin swarms with microorganisms.
6. Most microorganisms cannot enter the body unless they do so through the mouth, nose, or eyes, and few survive exposure to lysozyme, an enzyme in saliva and tears.
7. Cilia in the respiratory tract sweep mucus and trapped microorganisms toward the esophagus, to be swallowed into the stomach, where acid kills most of them.
8. The gut contains more microorganisms than there are cells in the body, but most are benign and compete with foreign microorganisms which usually cannot survive the competition.

HOW DOES THE BODY DEFEND ITSELF WHEN THE SKIN IS BROKEN?

1. Any significant wound damages blood vessels as well as other tissues.
2. Damaged blood vessels pose two risks:
   - blood loss, and
   - rapid dispersal of microorganisms throughout the body via the circulation.

How does the body control blood loss and keep pathogens out of the circulatory system?

1. In wound healing, normal cells in both the epidermis and the dermis divide when they lose contact with other cells and stop dividing once they have filled the gap.
2. In hemostasis, several mechanisms simultaneously hinder blood flow from the body and passage of microorganisms into the body.
3. Hemostasis involves at least three interconnected mechanisms:
   - vasoconstriction—contraction of smooth muscles in damaged vessels;
   - formation of a platelet plug—clumping of small platelets at the site of injury; and
   - formation of a blood clot (coagulation).
4. Injury to a blood vessel instantly triggers contraction of the smooth muscle in the vessel wall, which may be sufficient to close the vessel.
5. Injury to a blood vessel wall exposes collagen, the major structural protein in connective tissue, and exposed collagen attracts platelets, which adhere first to the collagen and then to each other, forming a temporary plug.
6. Platelets stimulate further vasoconstriction and initiate clotting.
7. Clotting is strictly a property of the plasma—the liquid, noncellular part of the blood.

How does the body prevent damaging blood clots?

1. A single blood clot can block an arteriole in the brain (stroke), heart (heart attack), or lungs (thrombosis), causing permanent damage or death, so clot formation occurs only after a series of molecular checks.
2. A clot is a network of fibers made of the protein fibrin. It is derived from the precursor fibrinogen, which becomes fibrin when the enzyme thrombin removes part of the fibrinogen.
3. At the site of the injury, an enzyme called Factor X (ten) converts inactive prothrombin to active thrombin, initiating the clotting mechanism.
4. The clotting system is so dangerous that it requires a powerful antagonistic system that destroys fibrin; this anticlotting system also consists of multiple enzymes, the last of which is plasmin, which dissolves the clot by breaking bonds in fibrin molecules.
5. Plasmin derives from plasminogen, a precursor protein, and the activation of plasminogen depends on other plasma enzymes or tissue plasminogen activators (tPA), which are enzymes that activate the anticlotting system.

6. Physicians now treat heart attack victims with synthetic tPA, which stimulates production of plasmin, which breaks up the clots blocking the arteries.
7. Mammalian blood contains anticoagulants, which interfere with clotting. Heparin is used to prevent clotting in blood samples and in the body; warfarin is used as a drug in humans and also as a rodent poison, causing them to die from internal bleeding.

How do plasma proteins contribute to nonspecific defense?

1. Complement is a set of blood proteins that attack microbiological invaders.
2. Some of these proteins stimulate special white blood cells called mast cells.
3. Damage to tissues stimulates mast cells to release their granules, which are loaded with histamine, the major stimulus for the inflammation response.

How does inflammation help the body resist infection?

1. During the inflammatory response:
   - capillaries dilate in the injured area, increasing delivery of substances that contribute to the inflammatory and immune responses, and promote healing;
   - neutrophils, attracted by histamine from the mast cells, attach to the capillary walls;
   - neutrophils move into the damaged tissue; and
   - neutrophils phagocytose (engulf) microorganisms in the damaged tissue.
2. Macrophages are large phagocytic cells that consume microorganisms and cellular debris and have a major role in initiating inflammation.
3. Macrophages produce at least three small proteins, called cytokines, which regulate cell division and protein synthesis in other cells of inflammation and immune response:
   - tumor necrosis factor-a (TNF-a),
   - interleukin 1 (IL-1), and
   - interleukin 6 (IL-6).
4. IL-6 stimulates stem cells in the bone marrow to produce more macrophages.
5. TNF-a acts locally to increase the phagocytic activity of macrophages and neutrophils.
6. IL-1 has many effects, including:
   - causing a fever, which interferes with microbial growth and aids the immune response;
   - attracting more phagocytes to the area; and
   - stimulating lymphocytes.
7. Viral infections stimulate production of interferons—another group of cytokines—which interfere with replication of viruses by limiting protein synthesis in virus-infected cells.

HOW DOES THE IMMUNE SYSTEM RECOGNIZE MOLECULES?

1. The immune system produces two distinct responses to invading organisms:
   - production of antibodies—proteins that mark foreign molecules by binding to them; and
   - production of cells that recognize and attack body cells that are infected.

What distinguishes the immune system from other defenses?

1. Four attributes characterize immunity: specificity, memory, diversity, and self-nonself recognition.
2. Immune system cells recognize specific invaders and not others.
3. Once the immune system develops defenses against a particular invader, it can attack it again whenever it appears. Subsequent appearances trigger a faster secondary immune response.
4. The immune system can recognize and attack a wide variety of foreign invaders.
5. The immune system tolerates other cells in the body unless they are infected by a pathogen.
6. The immune system's apparent ability to distinguish the body's own components from others is called self-nonself recognition.

What gives the immune system its specificity?

1. The immune system produces a huge array of proteins, called antibodies, that bind to nonself molecules, which are called antigens.
2. Each antibody recognizes a characteristic shape and charge distribution, not a specific molecule.

Which cells mediate the immune response?

1. Immune cells are a special class of white blood cells called lymphocytes, which develop within the lymphoid tissues.
2. The three main classes of lymphocytes are:
   - B lymphocytes (B cells), which make antibodies that recognize and bind to bacteria, fungi, and protists, and cells that do not belong there, marking but not killing them;
   - T lymphocytes (T cells), which recognize and destroy body cells that have become infected or damaged; and
   - natural killer cells (NK cells), which attack tumor cells and cells infected by a pathogen.

How do antibodies recognize cells and molecules?

1. Each line of B cells makes just one kind of antibody.
2. Every antigen contains one or more epitopes—the specific shapes and charge distributions recognized by antibodies. Epitopes are often only small parts of molecules.
3. All antibodies are immunoglobulins—a group of globular proteins called globulins.
4. All immunoglobulins contain antigen-recognition sites made from two kinds of polypeptide chains—light chains and heavy chains. The five classes of immunoglobulins differ in their heavy chains, but all use the various light chains.
5. Antibodies have at least two binding sites that allow them to bind to more than one antigen. This allows them to link together bacteria, for example, and the clumped mass becomes an easy target for destruction by phagocytes.
6. IgG and IgM activate complement proteins, which destroy the marked cells; the other antibodies do not activate complement proteins.
7. IgM molecules are the first to appear after antigen exposure, while IgG are made in greater amounts in secondary responses.
8. IgA molecules are especially abundant in tears, saliva, and milk.
9. IgE antibodies circulate through the blood and attach to mast cells, stimulating release of histamine, which triggers the inflammatory response.
10. An IgE response to a harmless antigen is called an allergy. Typical symptoms include congestion, sneezing, runny nose, and trouble breathing; extreme responses can be fatal.
11. Some allergic responses occur rapidly and involve antibodies, macrophages, and other cells of inflammation. Others involve T cells and the response appears days after the exposure.

HOW DOES THE IMMUNE SYSTEM GENERATE SO MANY ANTIBODIES?

1. Different antibodies form because each has a unique amino acid sequence, producing a distinct three-dimensional structure that can bind to a distinct epitope.
2. Each line of B cells that produces each antibody is a clone—a group of identical cells.
3. The body generates up to 100 million kinds of B cells.
4. From these, the immune system selects the set of antibodies it needs in response to the presence of a particular infection.

How can thousands of genes produce millions of antibodies?

1. Every antibody is made from two different polypeptides, which can be combined in ways to multiply their differences.
2. Two genes can be spliced together to specify one polypeptide with surprising results.
3. Lymphocytes have genes that other cells do not.

4. Every light chain of an antibody is formed from three different gene segments—one segment has three alleles, one has four, and the third has approximately 300.
5. Mammals apparently can make as many as a billion different kinds of antibody molecules.
6. Immunoglobulin genes mutate at high rates during lymphocyte production, increasing antibody diversity by a factor of ten.

The selection theory explains memory and self-nonself recognition

1. Every B lymphocyte contains just two rearranged genes—one that constitutes an active heavy chain gene and one that constitutes an active light chain gene.
2. The particular gene arrangements within the B cell determine the structure of that antibody.
3. During development, precursors of B cells form a lymphocyte library, with each lymphocyte in the library containing a unique random arrangement of immunoglobulin genes.
4. Each member of the B-lymphocyte library and all its descendants can make only one kind of antigen recognition site.
5. B cells have B-cell receptors—cell surface proteins with the same antigen recognition sites as the antibody that they produce.
6. When an antigen binds to B-cell receptors, the B cell starts to divide and differentiate:
   - some become plasma cells, which make and secrete antibodies; and
   - others become memory cells, which make more effector cells if the antigen reappears.
7. A second exposure to an antigen causes a rapid and very strong response due to memory cells, which rapidly divide and produce plasma cells.
8. Multiplication of memory cells allows lifelong immunity from one exposure to a pathogen.
9. Antigens that mimic disease-causing bacteria or viruses are as effective at producing immunity as the pathogens themselves, forming the basis for immunizations.

HOW DOES THE BODY DEFEND AGAINST GOOD CELLS GONE BAD?

1. B cells cannot respond to body cells that have become infected by pathogens.
2. T cells recognize body cells that have become infected or are growing out of control.
3. T cells are also involved in rejecting transplanted organs and skin grafts.
4. T cells recognize altered self cells and kill "bad" cells—a property called cytotoxicity—by boring holes in their cell membranes.

How do T cells recognize self?

1. T cytotoxic (T_C) cells kill target cells by boring holes in cell membranes.
2. T helper (T_H) cells send messages to nearby B cells and T cytotoxic cells. These messages are cytokines, most of which are paracrine signals—they activate only nearby cells.
3. Some T_H promote B-cell production of antibodies, some promote T-cell immunity, and some also bind antigens to present them more effectively to B cells.
4. Antigens recognized by T cells differ from those recognized by (B cell) antibodies because:
   - antibodies recognize foreign antigens in solution as well as on cell surfaces, but T-cell receptors bind foreign antigens only on the surfaces of cells recognized as self; and
   - antibodies recognize the three-dimensional shapes of proteins and carbohydrates, but T-cell receptors only recognize linear segments of polypeptides.
5. T cells recognize body cells by proteins encoded in a set of 40 to 50 genes called the major histocompatibility complex, or MHC.
6. MHC Class I molecules appear on the surface of essentially every cell in the body except sperm. T cells recognize these as endogenous antigens—proteins made by the body's cells.
7. MHC Class II molecules are on surfaces of specialized antigen-presenting cells and alert T cells to presence of a virus or bacterium inside a cell. The specialized cell phagocytoses the pathogen, then combines antigens from the pathogens with MHC II molecules, creating exogenous antigens—proteins that alert T cells to the presence of agents infecting body cells.

How do T cells recognize altered self?

1. T cells recognize only cells like others in the body but somehow different—altered self.
2. T cells recognize altered self antigens with the help of T cell receptors—molecules on the surface of T cells that recognize foreign antigens as altered self.
3. Rearrangements of T-cell receptor genes produce a T-lymphocyte library, each member of which contains one antigen recognition site.
4. T cells are even more diverse than immunoglobulins.

How does the body eliminate T cells that might attack healthy cells?

1. During the maturation of T_C cells in the thymus, the main survivors are precursor cells that can recognize antigens within the individual's own MHC Class I molecules.
2. T cells that do not distinguish between self and altered self must go, leaving only T cells that can recognize nonself antigens within self MHC molecules.
3. T cells that leave the thymus are highly selected to recognize altered self antigens.

Individuals have distinctive antigens on blood cell surfaces

1. Antigens on blood cell surfaces are important because of blood transfusions.
2. Individuals differ genetically in the kinds of antigens they produce.
3. The most important antigen classification is that of the ABO blood group:
   - Individuals of blood type A have the A antigen on the surfaces of their red blood cells;
   - Individuals of blood type B have the B antigen on the surfaces of their red blood cells;
   - Individuals with type AB have both A and B antigens on their red blood cells; and
   - Individuals with type O lack the A and B antigens.
4. Individuals who do not make A antigen (types O and B) have antibodies against the A antigen; individuals who do not make the B antigen (types O and A) make antibodies against the B antigen.
5. Another antigen on the surface of red blood cells—the Rh factor—determines if the blood is "positive" (has the Rh antigen) or "negative" (lacks the Rh antigen).
6. Individuals who lack the Rh antigen do not generally make antibodies against it unless they are exposed to Rh^+ blood, such as during a transfusion.
7. If an Rh^- woman carries an Rh^+ fetus, the mother may be exposed to the Rh antigen during birth. This sensitization causes the mother to start making antibodies against the Rh antigen, and, in subsequent pregnancies with an Rh^+ child, her antibodies may attack the fetus' red blood cells and the fetus may die.
8. Physicians can prevent this immune response by treating the mother with anti-Rh antibodies, which remove memory cells for the Rh factor.

VOCABULARY BUILDING

In your own words, first write a brief definition, then a full explanation for each of the following terms. Include examples where appropriate. Complete this section from your memory—you will not learn the terms by simply copying definitions from the textbook. Once you have finished, check your responses against the information in the chapter and make any necessary corrections.

macrophages —

apoptosis —

blood clots —

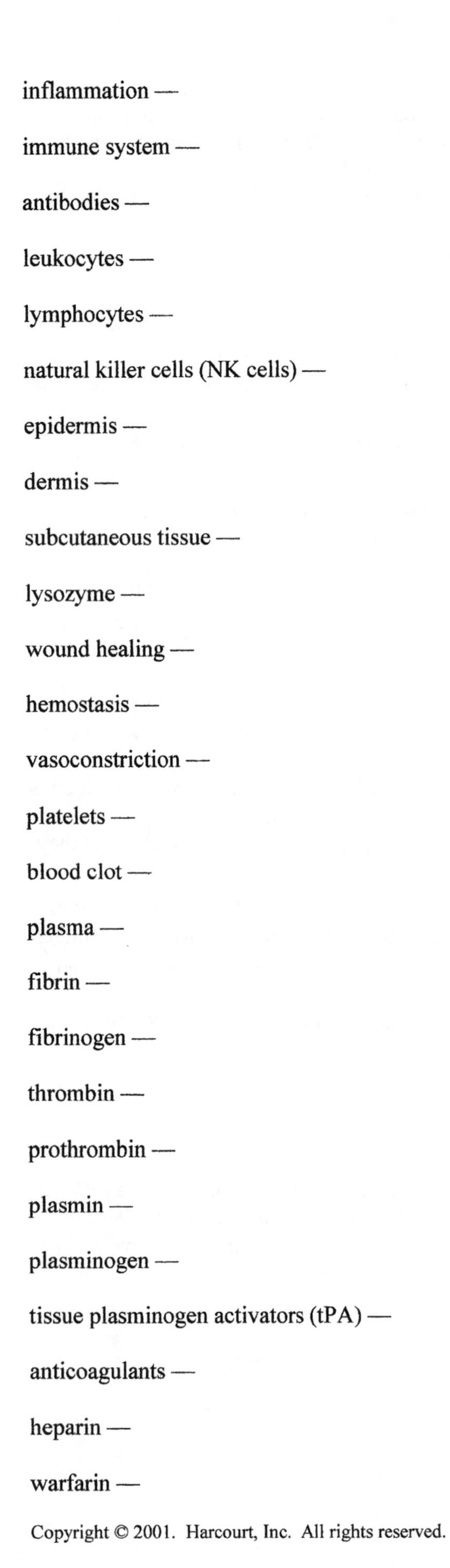

inflammation —

immune system —

antibodies —

leukocytes —

lymphocytes —

natural killer cells (NK cells) —

epidermis —

dermis —

subcutaneous tissue —

lysozyme —

wound healing —

hemostasis —

vasoconstriction —

platelets —

blood clot —

plasma —

fibrin —

fibrinogen —

thrombin —

prothrombin —

plasmin —

plasminogen —

tissue plasminogen activators (tPA) —

anticoagulants —

heparin —

warfarin —

complement —

mast cells —

histamine —

neutrophils —

cytokines —

tumor necrosis factor-a (TNF-a) —

interleukin 1 (IL-1) —

interleukin 6 (IL-6) —

interferons —

specific —

nonspecific —

memory —

diversity —

self-nonself recognition —

secondary immune response —

B lymphocytes (B cells) —

T lymphocytes (T cells) —

antigen —

epitopes —

immunoglobulins —

light chains —

heavy chains —

IgG —

lymphocyte library —

B-cell receptors —

plasma cells —

memory cells —

T cytotoxic (T_C) cells —

T helper (T_H) cells —

major histocompatibility complex (MHC) —

MHC Class I molecules —

endogenous antigens —

MHC Class II molecules —

exogenous antigens —

altered self —

T-cell receptors —

CHAPTER TEST

The following test has four parts. Complete as much of the exam as you can from memory. If you cannot answer a question, skip it. Once you complete all that you can, try to answer any questions you skipped. If you still cannot answer them, consult your textbook for the answers. Once you have completed all sections of the test, check your answers for Parts 1 - 3 against those in the back of this book. Highlight any incorrect answers then review that material in your textbook. Correct your answers for future reference.

Part 1: Multiple Choice

For each of the following, select all correct responses—more than one may be correct.

1. Which of the following is not a leukocyte (white blood cell)?
   a) lymphocyte
   b) antibody
   c) neutrophil
   d) mast cell

2. Which of the following participates in immune responses?
   a) neutrophils
   b) basophils
   c) lymphocytes
   d) natural killer cells

3. Which layer of the skin is primarily composed of dead and dying cells?
   a) dermis
   b) subcutaneous tissue
   c) epidermis
   d) none of these

4. A blood clot consists of fibers made of what protein?
   a) fibrinogen
   b) thrombin
   c) plasminogen
   d) fibrin

5. Which of the following is involved in specific defense of the body?
   a) complement
   b) inflammation
   c) mast cells
   d) T helper cells

6. Which of the following is a function of macrophages?
   a) phagocytosis
   b) initiation of inflammation
   c) disrupt replication of viruses
   d) release histamine

7. Which of the following is not a cytokine?
   a) interleukin
   b) heparin
   c) tumor necrosis factor-a
   d) interferon

8. Antibodies are made by which of the following?
   a) plasma cells
   b) natural killer cells
   c) mast cells
   d) epitopes

9. Which antibody is usually involved in allergic reactions?
   a) IgE
   b) IgM
   c) IgG
   d) IgD

10. Which of the following hasten and strengthen a secondary immune response?
    a) T helper cells
    b) natural killer cells
    c) interferon
    d) memory cells

11. Which of the following activates both B cells and T cells?
    a) T cytotoxic cells
    b) plasma cells
    c) T helper cells
    d) natural killer cells

12. MHC Class II proteins are found:
    a) in blood plasma
    b) on all body cells
    c) on specialized antigen-presenting cells
    d) on all body cells except sperm

13. Epitopes are part of:
    a) antigens
    b) antibodies
    c) tumors
    d) plasma cells

14. Which of the following characterizes a person with type O blood?
    a) presence of both the A and B antigens
    b) presence of antibodies against both the A and B antigens
    c) lack of either the A or B antigen
    d) lack of antibodies against either the A or B antigen

15. Which of the following is a function of the complement system?
    a) kill foreign organisms by putting holes in their cell membranes
    b) coat microbes to target them for phagocytes
    c) stimulate mast cells
    d) all of these

Part 2: **Matching**

For each of the following, match the correct term with its definition or example. More than one answer may be appropriate.

specific response	**nonspecific response**
1. ____________________	Inflammation is one of these.
2. ____________________	This involves T cells and B cells.
3. ____________________	Subsequent exposure causes a quicker reaction.
4. ____________________	Hemostasis is an example of this.
5. ____________________	Allergies are examples of this.

B cells	**T cells**
6. ____________________	Destroys body cells that are damaged or diseased.
7. ____________________	This is a type of white blood cell.
8. ____________________	These become plasma cells that make antibodies.
9. ____________________	These may be involved in allergies.
10. ____________________	Some of these possess cytotoxicity.

Part 3: Short Answer

Write your answers in the space provided or on a separate piece of paper.

1. By Matzinger's danger model, how do immune cells recognize when they are needed?

2. What is the first barrier against invasion of our bodies by pathogens?

3. List the three steps of hemostasis.

4. Starting with prothrombin, trace the chemical reactions that lead to blood clot formation.

5. Why are antihistamine drugs beneficial in treating many allergies?

6. What does the name "interleukin" say about how these cytokines work?

7. List two white blood cells that are phagocytes.

8. List the four attributes that characterize immunity.

9. List three specific areas in which lymphocytes develop.

10. When B cells differentiate, what are the two possible outcomes?

11. Structurally, the five classes of immunoglobulins differ from each other in what way?

12. List two mechanisms by which only thousands of genes can produce millions of antibodies.

13. How can a minor infection with a mild disease, like cow pox, provide immunity against a severe disease, like the deadly smallpox?

14. List and describe the general function of the two kinds of T lymphocytes.

15. Differentiate between the functions of MHC Class I and MHC Class II molecules.

Part 4: Critical Thinking: Using Your Knowledge

Answer each of these in essay form, using complete sentences and paragraphs. Provide as much information as you can. (For extra essay practice, write out answers to the Review and Thought Questions in your textbook.)

1. Thoroughly discuss the basic ideas of the self-nonself model and the danger model of immunity. What phenomena are not easily explained by the self-nonself model? How does the danger model explain these situations?

2. Thoroughly explain the processes involved in inflammation. How does this lead to pain, swelling, heat, and redness? Why is this often painful process necessary?

3. Differentiate between a B-cell response and a T-cell response. How are the two linked together?

4. Explain why successful organ transplantation is so difficult. What factors must be taken into account when finding a match?

5. Explain the basis of blood types, and explain what is meant by saying that people with Type O blood are universal donors, and those with type AB blood are universal recipients.

6. Explain what happens during a transfusion reaction. For example, what happens when a person with type A blood receives a transfusion of type B blood.

7. Sperm do not possess MHC antigens. What is the significance of this in terms of sperm function?

8. HIV (human immunodeficiency virus) specifically targets T helper cells. Explain how this accounts for the numerous complications that lead to death in full-blown AIDS (acquired immune deficiency syndrome).

9. Explain how vaccines, such as the one for tetanus, work, and why you must occasionally receive boosters for several of the vaccines.

10. Explain autoimmune diseases, such as rheumatoid arthritis and lupus, in which an individual's immune system starts to attack the body's own cells and tissues. What happens in these diseases, and why are they so difficult to treat?

Chapter 41
The Cells of the Nervous System

KEY CONCEPTS

1. The nervous system consists of neurons, which carry electrical signals, and glial cells, which provide metabolic and structural support to the neurons.
2. Neurons carry energy impulses that are always the same size and do not dissipate during transmission.
3. A neuron transmits an impulse only when stimulated by a minimum threshold signal.
4. Nerve cells maintain electrochemical gradients that give them a net charge leading to a resting potential of –70 millivolts.
5. The resting potential of a neuron arises because of the selectivity of:
   - ion pumps that concentrate sodium ions outside the cell and potassium ions inside the cell, and
   - channels within the cell membrane, which allow different ions to pass at different rates.
6. The resting potential of a neuron depends on the difference in the concentration of potassium ions between the inside and outside of a cell. The action potential depends on the temporary opening of channels that admit sodium ions into the cell.
7. A sodium channel is a membrane protein that allows sodium ions to pass when the outside of the membrane has a positive charge.
8. An action potential propagates because a charge moves along the cell membrane, opening sodium channels and leading to further depolarization.
9. Because each channel has a refractory period, the charge can move only in one direction.
10. Nerves wrapped in myelin conduct action potentials rapidly because nerve impulses jump from node to node (saltatory conduction).
11. A synapse may be electrical or chemical.
12. Electrical synapses are always excitatory; chemical synapses may be excitatory or inhibitory.
13. At neuromuscular junctions, an increase in the concentration of calcium ions near the synapse triggers release of the neurotransmitter acetylcholine.
14. The acetylcholine channel is a ligand-gated ion channel; when acetylcholine binds to the channel protein, it opens a pore that allows passage of sodium and potassium ions.
15. The firing pattern of the postsynaptic cell depends on the summing of the responses to both excitatory and inhibitory inputs.
16. Receptors for neurotransmitters respond either by opening ion channels or by triggering intracellular biochemical events, such as cyclic AMP production.
17. Neurotransmitters only act briefly because they are removed by enzymes and transporters.
18. In a simple reflex, input from a sensory receptor directly stimulates a motor neuron. More complicated neural circuits involve interneurons, which carry information among neurons with parallel functions.
19. Researchers classify learning into nonassociative learning (habituation and sensitization) and associative learning (the pairing of stimuli).
20. Learning depends on changes in synaptic signaling between individual neurons.

EXTENDED CHAPTER OUTLINE

RITA LEVI-MONTALCINI

1. During development, nerves of the peripheral nervous system appear to grow out from the spinal cord toward the limb.
2. Biologist Viktor Hamburger proposed that the limbs supply some special substance that tells unspecialized nerve cells in the spinal cord what kind of cells they should turn into and in which direction they should grow.
3. Rita Levi-Montalcini's research suggested that embryonic spinal cord nerves specialize but soon die without some special substance from the limb buds to sustain their growth.
4. In the 1950s she and biochemist Stanley Cohen isolated the substance, which they called nerve growth factor. In 1986, they shared the Nobel Prize in Medicine and Physiology.
5. Their discovery has led to possible use of nerve growth factor and similar substances to stimulate growth of damaged neurons, such as in cases of diseases such as Alzheimer's.

HOW ARE NERVE CELLS SPECIAL?

1. In the 1880s, Camillo Golgi developed a method—the Golgi method—for staining only a few cells in a sample.
2. Ramon y Cajal provided clear evidence that the nervous system consists of cells.
3. Nerve cells have exactly the same kinds of organelles as other cells.
4. Nerve cells can be immense.
5. Neurons are nerve cells that carry information in the form of electrical and chemical signals.
6. Glial cells provide metabolic and structural support for the neurons.
7. Some glial cells serve as guides and scaffolding during neural development; others (oligodendrocytes and Schwann cells) wrap around neural cell extensions forming a myelin sheath, which speeds transmission of signals.
8. Most neurons have a distinct center—the cell body (soma).
9. All neurons have extensions out from the cell body:
   - dendrites are numerous short extensions that usually relay signals from other cells to the cell body; and
   - axons are longer, thicker extensions that carry signals from the cell body to other cells.
10. Synapses are highly specialized contacts between axons and dendrites.

How do nerve cells carry information?

1. Nerves respond to electrical impulses.
2. Unlike household electric current:
   - an electrical impulse carried by ions moving back and forth across cell membranes is slower than electrical flow in home wiring,
   - an electrical impulse passing through a nerve is constantly regenerated so the energy at the end of a nerve is the same as that at the beginning; and
   - nerve cells always carry the same-sized pulses of energy.
3. A nerve cell has an all-or-none, or threshold, response. Any stimulus above a certain level triggers the same response, and any stimulus below that level yields no response.
4. As the intensity of a nerve stimulus increases:
   - the frequency of pulses generated and transmitted also increases; and
   - the total number of responding cells increases.
5. The combined effect of these two responses is that the greater the intensity of the stimulus, the greater the number of impulses that reach your brain.

How does a neuron generate an electrical pulse?

1. Membrane proteins that serve as channels allow specific molecules and ions to pass; some of these have gates that open and close in response to environmental stimuli; others actively pump ions or molecules using energy from ATP.
2. The concentration of ions is different on each side of the cell membrane, establishing a concentration gradient, which is a source of potential energy.
3. The electrical gradient that also exists is called voltage, or electrical potential.
4. Voltage is the driver of electrical current—the number of electrons flowing per minute.
5. The greater the voltage, the greater the work that can be performed, and the greater the tendency for a reaction to occur spontaneously.
6. At rest, cells have a net negative charge that results in a potential of about –70 millivolts relative to the extracellular fluid, resulting from both negative and positive ions. This –70 mV charge is the resting potential.
7. Nerve cells contain much less sodium than does the extracellular fluid, but more potassium.
8. When a nerve cell is stimulated, sodium channels open and the cell membrane suddenly becomes permeable to sodium ions, which rush across the cell membrane, creating a slight positive charge inside the cell.
9. A few milliseconds later, the membrane's sodium permeability is lost, but potassium permeability increases, and these ions rush out of the cell, driven by the concentration gradient and the repulsion of the positive charge.
10. The resting potential represents potential energy that can be used to pass an impulse; the sudden electrical change that occurs during an impulse is called an action potential.

How does a neuron maintain its resting potential?

1. The sodium-potassium pump uses energy from one molecule of ATP to push three sodium ions out while moving two potassium ions into the cell.
2. The membrane maintains the uneven ion concentration by restricting passage of sodium and potassium.

How does a neuron generate an action potential?

1. The tendency for electrons to flow is measured in volts: one millivolt is 1/1000 of a volt.
2. The concentration of potassium ions determines the level of the resting potential.
3. Hodgkin and Huxley showed that an action potential, unlike a resting potential, depends on the flow of sodium ions into the cell.

How does a membrane change its permeability to sodium ions?

1. At the normal resting potential, sodium channels are closed. But when the cell is depolarized, or made less negative, the channels open briefly and admit sodium ions.
2. Sodium channels are voltage-gated—they open or close according to the voltage across the membrane.
3. When threshold depolarization occurs, sodium channels open and sodium ions flow inward for about 2 milliseconds, then the channels close and are less likely to open than before.
4. During this time—the refractory period—potassium ions flow out to restore the resting potential.
5. The membrane's sodium-potassium pump soon restores the original distribution of both ions.

How do voltage-dependent sodium channels open and close?

1. Biochemists have isolated the protein that forms the voltage-gated sodium channel.
2. A group of Japanese scientists isolated a gene encoding the channel protein.
3. Four polypeptides assemble to form a pore through which sodium ions can flow when the gate is open, and that pore is controlled by a gate that opens only when the outside of the membrane has a positive charge—when the cell is depolarized.

How does an action potential move down an axon?

1. Six consecutive events mark production of an action potential:
   - something causes a local depolarization (change in ion distribution) of the membrane that exceeds threshold;
   - sodium channels open and sodium ions flow into the cell;
   - as sodium ions enter, the inside of the membrane becomes locally positive;
   - the decreased polarization of the membrane causes more channels to open, increasing the positive charge of the membrane still further;
   - finally, after less than a millisecond, the sodium channels close spontaneously; and
   - potassium ions flow outward, restoring the resting potential.
2. The sodium-potassium pump is not directly involved in the action potential—its purpose is to maintain the correct distribution of ions across the cell membrane.
3. As one spot of the membrane becomes more positive, the charge spreads to the adjacent spot, which then also becomes depolarized and opens its sodium channels, and the process continues the length of the axon.
4. The action potential moves in just one direction because the sodium channels cannot reopen immediately. Thus an impulse cannot pass through again until after the refractory period.

How do nerves speed the transmission of action potentials?

1. Most vertebrate nerves are wrapped in myelin—a specialized, glistening sheath that allows very rapid conduction of nerve impulses.
2. The myelin sheath insulates most of the axon's membrane, so depolarization and ion flow are only possible in the nodes of Ranvier—gaps between the myelin wrappings.
3. The nerve impulses in myelinated axon's move by saltatory conduction—jumping from node to node, down a myelinated axon at rates up to 100 times faster than in unmyelinated axons.
4. An ordinary nerve conducts at about 1.2 m per second, or 2.5 mph; myelinated nerves can conduct at up to 120 m per second, or about 250 mph.

HOW DO NEURONS COMMUNICATE WITH ONE ANOTHER:

1. The synapse is a distinct boundary between most communicating neurons.
2. The presynaptic neuron sends the information; the postsynaptic neuron receives it.
3. The electrical synapse is the simplest type, and joins the neurons through gap junctions—channels through the membranes of adjacent cells (here, the presynaptic and postsynaptic neurons) that allow ions and small molecules to pass freely between the cells.
4. An electrical synapse allows an action potential to continue traveling to the postsynaptic neuron at about the same rate as it traveled down the presynaptic axon.
5. All electrical synapses are excitatory—action potentials in the presynaptic cell stimulate action potentials in the postsynaptic cell.
6. The chemical synapse is the most common, and in this case the two neurons do not join—they are separated by a small space called the synaptic cleft.
7. Communication across a synaptic cleft requires diffusion of one or more neurotransmitters—small signaling molecules made in and released from the presynaptic cell that affect the electrical charge of the postsynaptic membrane.

8. Chemical synapses may excite or inhibit, but the effect is always delayed by the time needed for the neurotransmitter to diffuse across the gap (electrical synapses lack this delay).
9. Biologists have identified at least 20 different neurotransmitters.
10. Chemical synapses have two important advantages over electrical synapses:
    - they can be either excitatory in inhibitory; and
    - they can greatly amplify the signal from a small presynaptic neuron.
11. Most synapses in vertebrate central nervous systems are chemical. Electrical synapses occur in the vertebrate heart and in a variety of invertebrate and vertebrate nerve circuits.

How does a chemical synapse work?

1. Essentially all medications that affect the functioning of the brain act on synapses.
2. The best understood chemical synapse is the vertebrate neuromuscular junction—the synapse between a motor neuron and a voluntary muscle cell.
3. The presynaptic neuron contains tens of thousands of tiny vesicles that are full of the neurotransmitter acetylcholine, which is released by exocytosis.
4. The trigger for acetylcholine release appears to be an increase in intracellular calcium ions.
5. During an action potential, calcium ions flow into a presynaptic neuron locally—through the membrane next to the synapse—by moving through voltage-gated calcium channels.

What does the neurotransmitter acetylcholine do?

1. Once released, acetylcholine diffuses across the synapse to the postsynaptic membrane, where it binds to a receptor molecule, which is, itself, a gated ion channel allowing passage of sodium and potassium ions.
2. This gated receptor is unlocked by acetylcholine.
3. A ligand is a molecule that specifically binds to another molecule, so the acetylcholine receptor is a ligand-gated channel.

Neurotransmitters may produce either excitatory or inhibitory effects in postsynaptic neurons

1. When sodium ions pass through the channel of the acetylcholine receptor, they depolarize the membrane, and this voltage change is called a postsynaptic potential.
2. Acetylcholine excites the cell; the potential is an excitatory postsynaptic potential (EPSP).
3. If an EPSP is sufficiently depolarizing, the postsynaptic cell fires an action potential; if the postsynaptic cell is a muscle cell, the muscle contracts, and if the postsynaptic cell is a neuron, the action potential travels through the cell body and axons to the next synapse.
4. A single EPSP may not depolarize a cell sufficiently to fire an action potential.
5. Many presynaptic neurons can converge on a single postsynaptic neuron, and some may be excitatory while others may be inhibitory.
6. Inhibitory neurotransmitters open channels that allow potassium or chloride ions through, increasing the negative charge inside the cell so that an action potential is less likely to fire.
7. This temporary increase in negative voltage is an inhibitory postsynaptic potential (IPSP).
8. The most common inhibitory neurotransmitter in the brain is gamma-aminobutyric acid (GABA), and when it binds to its receptors, a channel opens that admits negative chloride ions, increasing the negative charge inside the cell.
9. Postsynaptic cell activity depends on the summation of the effects of EPSPs and IPSPs.

Neurons may respond to the same neurotransmitters in different ways

1. The same neurotransmitter can have different effects in different postsynaptic neurons due to the presence of more than one type of receptors on each postsynaptic neuron.
2. Nicotine mimics the effects of acetylcholine in skeletal muscle and in some neurons but has no direct effect on cardiac muscle, whereas the mushroom toxin muscarine stimulates cardiac muscle but not skeletal muscle.

3. Acetylcholine receptors are grouped into two varieties; nicotinic and muscarinic. The effects of acetylcholine are determined by which type of receptor it binds to.
4. Neurotransmitters work in the same way as hormones and other signaling molecules.
5. Postsynaptic receptors are either:
   - ionotropic receptors, which are ligand-gated ion channels, or
   - metabotropic receptors, which interact with another membrane protein.
6. Many metabotropic receptors change the concentration of second messengers.

How are neurotransmitters cleared from the synapse?

1. A new signal cannot be transmitted until the neurotransmitter clears from the synaptic cleft.
2. Termination usually involves reuptake—pumping of neurotransmitter from the synaptic cleft either into the presynaptic neuron or surrounding glial cells, to be recycled or degraded.
3. GABA, for example, is inactivated by an enzyme called GABA transaminase; acetylcholinesterase destroys acetylcholine soon after it is released.

Many psychoactive drugs act on chemical synapses

1. Not all molecules that bind to neurotransmitter receptors are neurotransmitters, and many compounds that bind to these receptors are useful medications.
2. Atropine, for example, blocks acetylcholine's effects in smooth muscle and intestines.
3. We generally call a substance a medicine or a drug when it helps, and a poison when it hurts.
4. Benzodiazepenes, commonly prescribed anti-anxiety drugs, bind to GABA receptors and increase the effectiveness of GABA in opening chloride channels.
5. Other substances influence the release, reuptake, or inactivation of specific neurotransmitters.
6. Selective serotonin reuptake inhibitors (SSRIs) are among the most widely prescribed medications. They include fluoxetine (Prozac) and sertraline (Zoloft), which have mood-elevating effects by blocking the reuptake of serotonin from certain synapses.
7. Many drugs that act on the brain are highly addictive.

HOW DO NEURAL NETWORKS MEDIATE BEHAVIOR?

1. Connections within neural networks determine how an animal will respond to a stimulus.

How do neurons mediate a reflex?

1. Movements and behavior ultimately depend on coordinated signals from motor neurons, which represent the final common pathway of the instructions that direct behavior.
2. Behavior mediated by the spinal cord tends to be the most automatic—least influenced by thought or previous experience.
3. A reflex is the most automatic behavior pattern—motor activity directly responds to a sensory stimulus.
4. A familiar reflex is the knee-jerk reflex:
   - the hammer hits the tendon of the quadriceps muscles as the tendon passes over the knee cap (patella);
   - striking the tendon stretches the muscle, which contains stretch receptors—modified muscle cells whose electrical output depends on how much they are stretched;
   - information from these receptors flows to sensory neurons in the dorsal part of the spinal cord; and
   - these sensory neurons directly contact the motor neurons in the ventral part of the spinal cord, which in turn command the muscle to contract, thus jerking the knee.
5. This stretch reflex is monosynaptic—the information passes through a single neuron.
6. Information from the stretch receptors also travels to a second group of spinal cord neurons that inhibit the hamstrings muscles, which are antagonistic to the quadriceps. This prevents the hamstrings from resisting the action of the quadriceps.

7. Yet another set of neurons in the spinal cord carries information about the status of the leg muscles to brain areas that coordinate movements and posture.
8. Other circuits are more complicated, with at least one interneuron—a neuron that connects one or more neurons to others with parallel functions.
9. Interneurons may be inhibitory, decreasing the likelihood of an action potential in the neurons it connects to.
10. Interneurons can integrate information from many sources, and circuits with such integration are called convergent.
11. An interneuron can also be part of a divergent circuit, in which a single interneuron may coordinate the action of many motor neurons.
12. Input from the brain can modify activity of spinal cord neurons.
13. The behavior that follows the activation of a neural circuit may be quite complex.
14. Once the appropriate connections develop, the proper stimulation of a "trigger" neuron can evoke a standard pattern of behavior. The trigger for action is called a releaser.

HOW DOES EXPERIENCE MODIFY NEURAL NETWORKS?

1. Many neural networks are capable of learning, which is the modification of neural activity and behavior as a result of experience.
2. In nonassociative learning, an individual's sensitivity to a stimulus changes with repeated exposure.
3. The two main types of nonassociative learning are:
   - habituation—a decreased behavioral response following repeated stimulation.
   - sensitization—an increased behavioral response to a noxious stimulus.
4. Associative learning refers to the pairing of seemingly unrelated stimuli.
5. The classic example of associative learning is with Pavlov's dog experiments, in which he trained dogs to salivate at the sound of a bell. He rang the bell (conditioned stimulus) while presenting dinner (unconditioned stimulus), and eventually the dogs salivated at the sound of the bell alone.

Experience can modify the gill withdrawal reflex of the sea hare, Aplysia

1. The neurons responsible for the gill withdrawal reflex lie in the abdominal ganglion.
2. A sensory neuron from the siphon makes a single synaptic connection to the motor neuron (L7) that drives gill withdrawal.
3. A single stimulation of the sensory neuron produces an action potential in L7, but repeated stimulation leads to a decrease, called synaptic depression, in the size of the action potential in L7.
4. Synaptic depression underlies habituation.
5. The depression results from decreased release of the neurotransmitter serotonin.

How does an Aplysia learn?

1. Stimulating the facilitating interneuron reverses the synaptic depression and L7 is facilitated.
2. Synaptic facilitation explains sensitization.
3. When the facilitating interneuron is stimulated, it releases serotonin, which has the effect of lengthening the period during which the axon terminal is depolarized.
4. During this extended depolarization, more calcium enters the axon, leading to release of more neurotransmitter than usual and, thus, a stronger response by the postsynaptic neuron.
5. A single stimulus to the tail (or facilitating neuron) sensitizes the gill withdrawal reflex for several minutes, but repeated stimulation of the tail leads to a much longer sensitization period, lasting days or weeks.
6. This long-term sensitization is a form of associative learning.
7. Long-term sensitization results from a change in gene transcription and protein synthesis.

VOCABULARY BUILDING

In your own words, first write a brief definition, then a full explanation for each of the following terms. Include examples where appropriate. Complete this section from your memory—you will not learn the terms by simply copying definitions from the textbook. Once you have finished, check your responses against the information in the chapter and make any necessary corrections.

neurons —

glial cells —

cell body —

dendrites —

axons —

synapses —

threshold —

channels —

gates —

pump —

concentration gradient —

voltage (electrical potential) —

current —

resting potential —

action potential —

volts —

depolarized —

voltage-gated —

refractory period —

nodes of Ranvier —

saltatory conduction —

presynaptic neuron —

postsynaptic neuron —

electrical synapse —

chemical synapse —

synaptic cleft —

neurotransmitters —

neuromuscular junction —

acetylcholine —

ligand-gated channel —

excitatory postsynaptic potential (EPSP) —

inhibitory postsynaptic potential (IPSP) —

gamma-aminobutyric acid (GABA) —

reuptake —

acetylcholinesterase —

motor neurons —

reflex —

interneuron —

learning —

nonassociative learning —

habituation —

sensitization —

associative learning —

conditioned stimulus —

unconditioned stimulus —

ganglia —

synaptic depression —

serotonin —

CHAPTER TEST

The following test has four parts. Complete as much of the exam as you can from memory. If you cannot answer a question, skip it. Once you complete all that you can, try to answer any questions you skipped. If you still cannot answer them, consult your textbook for the answers. Once you have completed all sections of the test, check your answers for Parts 1 - 3 against those in the back of this book. Highlight any incorrect answers then review that material in your textbook. Correct your answers for future reference.

Part 1: Multiple Choice

For each of the following, select all correct responses—more than one may be correct.

1. Which of the following is a function of glial cells?
   a) forming myelin sheaths
   b) serving as scaffolding
   c) providing metabolic support
   d) all of these

2. Which of the following usually carry signals to the neuron's cell body?
   a) axons
   b) dendrites
   c) myelin sheaths
   d) all of these

3. An electrical gradient, or potential, is called:
   a) voltage
   b) current
   c) threshold
   d) a gate

4. When compared to the extracellular fluid, a neuron's resting potential is:
   a) –30 mV
   b) 0 mV
   c) +50 mV
   d) –70 mV

5. When a neuron is stimulated, what is the first event that changes the membrane potential?
   a) Potassium channels open.
   b) Sodium channels open.
   c) Negative ions rush into the cell.
   d) The sodium-potassium pump moves sodium out of the cell.

6. When a neuron is at rest, which of the following is more concentrated outside the cell?
   a) potassium
   b) sodium
   c) negative ions
   d) all of these

7. Which channels are voltage-gated?
   a) sodium channels
   b) potassium channels
   c) calcium channels
   d) all of these

8. What is the primary function of the sodium-potassium pump in neurons?
   a) causes depolarization
   b) opens sodium channels
   c) maintains the ion distribution that provides the resting potential
   d) transmission of an action potential

9. Electrical synapses allow two neurons to communicate through:
   a) gap junctions
   b) sodium channels
   c) nodes of Ranvier
   d) neurotransmitters

10. In the vertebrate neuromuscular junction, acetylcholine is released in response to increased:
    a) intracellular potassium
    b) extracellular sodium
    c) intracellular calcium
    d) all of these

11. Which of the following is true of electrical synapses?
    a) They are always excitatory.
    b) They may be either excitatory or inhibitory.
    c) No delay occurs when an impulse is sent from one neuron to the next.
    d) A signaling molecule must diffuse across the synaptic cleft.

12. An interneuron is:
    a) a neuron that collects sensory information
    b) a neuron that stimulates a muscle to contract
    c) a neuron that connects other neurons
    d) any of these

13. Habituation refers to:
    a) a decreased response resulting from repeated exposure to a stimulus
    b) an increased response resulting from repeated exposure to a stimulus
    c) a type of associative learning
    d) a type of nonassociative learning

14. Which of the following is an example of associative learning?
    a) A child who used to cry when hearing thunder loses its fear of the noise in time.
    b) After realizing pizzas are delivered by people in cars, a child expects pizza each time a car pulls into the driveway.
    c) An adult grows increasingly irritated the longer the neighbor's stereo is blaring.
    d) A patient's leg jerks each time the doctor strikes just below the kneecap with a hammer.

15. Synaptic depression is associated with:
    a) habituation
    b) sensitization
    c) associative learning
    d) unconditioned reflexes

Part 2: Matching

For each of the following, match the correct term with its definition or example. More than one answer may be appropriate.

electrical synapse **chemical synapse**

1. ______________ The impulse can travel in either direction.
2. ______________ Can have both a presynaptic and postsynaptic neuron.
3. ______________ These may be excitatory or inhibitory.
4. ______________ These involve gap junctions.
5. ______________ Common type in vertebrate central nervous systems.

nonassociative learning **associative learning**

6. ______________ Habituation and sensitization are examples.
7. ______________ This involves pairing of seemingly unrelated stimuli.
8. ______________ Pavlov's dogs are a classic example of this.
9. ______________ Response changes with repeated direct stimulation.
10. ______________ Neural activity and behavior are modified.

Part 3: Short Answer

Write your answers in the space provided or on a separate piece of paper.

1. List the two glial cells that are involved in forming the myelin sheath.

2. Explain the relationship between dendrites, axons, and the cell body.

3. Explain what is meant by neurons responding in an all-or-none and threshold manner.

4. Why does a neuron at rest have fewer positive ions inside than out?

5. What event causes depolarization of the neuron's membrane?

6. Explain the refractory period and why it is needed.

7. Why can an action potential only move in one direction?

8. Why does myelin speed the transmission of the nerve impulse?

9. What are the two components of a neuromuscular junction?

10. By what mechanism does acetylcholine cause an effect in the postsynaptic neuron?

11. What is the most common inhibitory neurotransmitter in the brain?

12. Explain how neurotransmitters are cleared from the synaptic cleft.

13. List three ways in which drugs can have an effect at the synapse.

14. What are the two neural components of a stretch reflex?

15. List two changes in synaptic function that are involved in learning.

Part 4: Critical Thinking: Using Your Knowledge

Answer each of these in essay form, using complete sentences and paragraphs. Provide as much information as you can. (For extra essay practice, write out answers to the Review and Thought Questions in your textbook.)

1. Explain the events that occur in the production of an action potential.

2. Discuss the events that occur at a chemical synapse, and how the nerve impulse is transmitted from one neuron to the next.

3. Review the knee jerk reflex. Now, discuss the events that occur during a withdrawal reflex, for example, when you touch a very hot object.

4. Differentiate between nonassociative and associative learning, and give examples of each from your own life. Discuss how learning is accomplished at the neural level.

5. Explain ways in which the electricity in your home and the electrical impulses in the nervous system are both similar and different.

6. Saltatory conduction occurs in myelinated nerves, and the nerve impulse rapidly jumps from node to node. Not all neurons are myelinated, so some neurons do not use saltatory conduction. Why would this type of conduction be slower? Why does the body use two types of conduction? Myelin never covers an entire neuron, so which parts of a nerve cell do you expect to not have a myelin sheath, and why?

7. A reflex is the simplest and most automatic behavior pattern. What are some examples of reflexes you can think of? What are the advantages to having many activities controlled by reflex behavior instead of conscious control?

8. In many cases, depression can be traced to a deficiency of the neurotransmitter serotonin. Explain precisely how drugs like Prozac and Zoloft alleviate the symptoms of depression.

9. Differentiate between electrical and chemical synapses. What are the advantages and disadvantages of each, and under what circumstances would one be a more effective option than the other?

10. Consider the design and function of the chemical synapse. If you were asked to design a drug that would alter how that synapse performs, what targets might you consider? What information would you need before you could design a specific drug?

Chapter 42
The Nervous System and the Sense Organs

KEY CONCEPTS

1. Nervous systems respond to external and internal environment and coordinate homeostasis.
2. The vertebrate central nervous system consists of the brain and spinal cord. Both are surrounded (and protected) by bones, membranes, and cerebrospinal fluid.
3. The spinal cord receives somatic sensory input about the position of body parts and sends motor instructions to voluntary muscles; it also coordinates many autonomic functions.
4. The hindbrain integrates and processes both sensory information and motor instructions; the midbrain relays sensory information, helps organize movements, and directs attention; and the forebrain plays roles in sensation, perception, movement, learning, memory, and mood.
5. Cerebral hemispheres consist of the cerebral cortex and underlying structures (hippocampus, the amygdala, and the basal ganglia). The human cerebral cortex is divided into four lobes.
6. Memories form in stages. The hippocampus helps form long-term memories.
7. The peripheral nervous system connects the CNS to sense organs and muscles. It includes the autonomic nervous system, which instructs smooth and cardiac muscle, and organs.
8. Receptors detect characteristics of the external world and generate graded receptor potentials. Relay neurons convert this information into patterns of action potentials. Relays of neurons in each sensory pathway convey information to the brain.
9. Sensation involves signal transduction, signal transmission, and signal integration.
10. Sensory information from each sensory modality passes separately through the thalamus and converges separately in different regions of the cerebral cortex.
11. Channels in the ear carry sound to the eardrum, whose vibrations cause vibrations of fluid in the inner ear. Hair cells in the organ of Corti detect the fluid motion. The brain interprets the pattern of stimulation in the organ of Corti as different patterns of sound.
12. Hair cells within the semicircular canals of mammals and within the lateral line system of fishes and frogs help establish the position and the movement of the body.
13. In vision, light energy is captured by a light-sensitive pigment within a photoreceptor cell.
14. In a vertebrate eye, the lens focuses an image on the surface of the retina.
15. Photoreceptor cells contain light-absorbing pigments. Light absorption alters the cell's electrical properties, initiating a series of events that transmit visual information to the brain.
16. The retina's photoreceptor cells transmit information to the bipolar cells, and bipolar cells to the ganglion cells. Connections among photoreceptor cells, bipolar cells, horizontal cells, ganglion cells, and amacrine cells integrate different aspects of the visual world.
17. Axons of the ganglion cells form the optic nerve, whose fibers end in the lateral geniculate nucleus of the thalamus, which integrates visual information and acts as a relay station.
18. Sensory receptors throughout the body detect four kinds of somatic information—touch, position, temperature, and pain. Somatosensory information passes to the spinal cord or medulla, through the thalamus, and to the primary somatosensory area of the parietal lobe.
19. In taste and smell, a small molecule binds to a specific protein, which changes nerve activity in the brain. Information from taste receptors passes to the medulla, through the thalamus, then to the cerebral cortex. Olfactory receptors lie within the olfactory bulb and olfactory signals do not directly reach the cerebral cortex.

EXTENDED CHAPTER OUTLINE

DOES A FALLING TREE MAKE A SOUND WHEN NO ONE IS THERE TO HEAR IT?

1. Perception is the recognition and interpretation of an outside stimulus.
2. Franz Joseph Gall, physician and neuroanatomist, speculated that the size of each brain area varied with use, then tried to correlate an area's size with a particular trait, which he estimated from ridges and bumps on the subject's skull—the practice of phrenology.
3. The human cerebral cortex is far larger than in other animals.
4. Pierre Paul Broca found that damage to a specific area of the cerebral cortex resulted in a patient who could understand words but could not speak.
5. Carl Wernicke found that damage in a different area of the brain allowed speaking, but sacrificed the ability to understand what was being said.
6. Contemporary neuroscientists believe individual brain regions perform specific functions and complex brain functions require cooperation and information flow between regions.
7. Sensation is the detection of external stimuli by vision, hearing, touch, taste, and smell.

HOW DOES THE NERVOUS SYSTEM RESPOND TO ITS ENVIRONMENT?

1. The nervous system receives reports on the state of both the exterior and the interior world.
2. The nervous system is the ultimate homeostatic organ.
3. For the nervous system to maintain homeostasis, it must be able to:
    - detect changes in its internal and external environment;
    - interpret the meaning of these changes; and
    - respond by sending appropriate signals to the muscles and organs of the body.
4. Nerves carry signals from sense organs to the brain or spinal cord, and from the brain or spinal cord to the muscles, organs, and endocrine system.

Do invertebrate animals have brains?

1. Cnidarians are the simplest organisms with nervous systems, but they have no brains.
2. The simplest animals that can learn are flatworms. Their nervous system has two parallel nerve cords with nerves extending out to the muscles. The nerve cords converge at the front to form the simplest possible brain.
3. More complex animals have increasing degrees of cephalization (the concentration of sense organs and ganglia in the front, or head, end).

How is the vertebrate central nervous system organized?

1. Vertebrate nervous systems show the most cephalization and sophisticated sense organs.
2. Most neurons are concentrated in the central nervous system (brain and spinal cord) which is dorsal in vertebrates, but ventral in invertebrates.
3. The brain lies within the thick, bony cranium; the spinal cord passes through a bony canal formed by the vertebrae.
4. The brain and spinal cord are also protected by the three membranes called the meninges.
5. Cerebrospinal fluid, derived from the blood, cushions and protects the brain and spinal cord.
6. Cells of the capillary walls are tightly connected to restrict passage of molecules, forming what is called the blood-brain barrier (blood-nerve barrier in the spinal cord).
7. Only small, nonpolar molecules can pass through the blood-brain barrier, but this includes alcohol, nicotine, heroin, and cocaine can penetrate the blood-brain barrier.

The spinal cord

1. The spinal cord, a long bundle of specialized nerve fibers running from the brain to the tail:
   - receives sensory information and sends motor information;
   - sends motor instructions to the muscles;
   - controls involuntary autonomic functions; and
   - contains sensory and motor neurons responsible for stretch reflexes (and others).
2. Motor neurons (motoneurons) directly stimulate muscle contraction. Motor fibers from the upper brain (cortex) cross in the lower brain (medulla) so the left side of the brain controls muscles on the right side of the body.
3. The inside of the spinal cord is gray matter, reflecting its lack of myelin, and is arranged into:
   - the ventral (anterior) horn, which contains motor neurons, and
   - the dorsal (posterior) horn, which contains the cell bodies of sensory neurons.
4. The outer part of the spinal cord is white matter, composed mostly of myelinated nerve tracts carrying information from sensory neurons to the brain and from the brain to motor neurons.
5. The spinal cord itself can direct even complex behaviors without a connection to the brain.

The vertebrate brain

1. Brains of jawless fish, half a billion years old, show the same divisions seen in today's fish and embryos of contemporary vertebrates—the hindbrain, the midbrain, and the forebrain.

The hindbrain

1. The hindbrain has three parts—the pons, the medulla oblongata, and the cerebellum.
2. The pons and medulla link information passing between the brain and spinal cord.
3. The medulla also contains centers that regulate breathing and blood circulation.
4. The cerebellum coordinates incoming sensory information and outgoing motor instructions so the limbs move smoothly and the body maintains its upright posture.

Midbrain

1. Each side of the midbrain contains a center that processes visual information; another center processes auditory information.
2. Another center, the substantia nigra, is particularly important in initiating movement. Parkinson's disease involves destruction of many cells of the substantia nigra, seriously impairing the person's control of movement.
3. Together, the midbrain, pons, and medulla are called the brainstem.
4. A network of nerves—the reticular formation—runs through the brainstem; it receives information from all of the brain and regulates arousal, attention, sleeping, and dreaming.

Forebrain

1. The forebrain is, in humans, the locus of almost all conscious activity, and it consists of two divisions—the diencephalon and the cerebral hemispheres.
2. The diencephalon has two parts:
   - the thalamus is the major integrator of sensory information about the external world, and
   - the hypothalamus is the major integrator of sensory information from the internal world.
3. The hypothalamus is the master regulator of homeostasis.
4. The cerebral hemispheres consist of a thin surface layer (the cerebral cortex) and underlying structures and, unlike the spinal cord, have their gray matter toward the outside.
5. In humans, the cerebral cortex contains more than 10^{11} neurons.
6. The human cerebral cortex has four areas—the frontal, parietal, temporal, and occipital lobes.
7. The cerebral cortex, like the rest of the brain, has two almost symmetrical halves, which are directly connected only by a thick bundle of nerve fibers called the corpus callosum.

8. Each lobe of the cerebral cortex receives sensory signals (mostly via the thalamus), processes that information, and transmits instructions to some other center within the CNS.
9. The rest of the cortex—the associative cortex—is involved in processing inputs and outputs.
10. The frontal lobe deals principally with movement and smell; the parietal lobe with somatic sensation; the occipital lobe with vision; and the temporal lobe with hearing and formation of memory (as is the underlying hippocampus).
11. The amygdala, underlying the cerebral cortex, coordinates autonomic (involuntary) responses to emotional states, especially anxiety.
12. The underlying basal ganglia are intimately involved in establishing patterns of movement.

How do humans learn?

1. Memory refers to the storage of knowledge about the world, and it forms in stages.
2. Short-term memories can be lost easily, but long-term memories are stabilized so they typically survive head trauma.
3. A reflexive memory has a reflex-like quality—automatic and not dependent on awareness; a declarative memory is a recalled thought or experience.
4. Some skills may first be conscious, declarative memories, then later emerge as unconscious, reflexive memories.
5. Different brain structures may be involved in reflexive and declarative memories.
6. The hippocampus has an important role in the formation of long-term memory, but it also is involved in epileptic seizures.
7. Studies all point to the hippocampus as a crossroads of short-term and long-term memory.
8. We are still far from understanding the basis of memory.

How is the vertebrate peripheral nervous system organized?

1. The peripheral nervous system has two types of nerve cells:
   - afferent nerves carry sensory information to the CNS; and
   - efferent nerves carry motor instructions from the CNS to muscles and organs.
2. Efferent nerves carrying information to striated skeletal muscles compose the somatic, or voluntary, nervous system.
3. Other efferent nerves, forming the autonomic, or involuntary, nervous system, carry instructions to smooth muscle and cardiac muscle.
4. The autonomic nervous system has two systems of nerves, with generally opposite effects:
   - the sympathetic nervous system uses norepinephrine as a final neurotransmitter, and triggers the "fight or flight" reaction; and
   - the parasympathetic nervous system uses acetylcholine as its neurotransmitter, and conserves energy by slowing the heart and increasing intestinal absorption.

How does information move through the nervous system?

1. The sensory receptor cells' information is in the form of receptor potentials—changes in the electrical voltages across their plasma membranes.
2. Relay neurons convert information from one or more receptor cells into a series of electrical discharges, or action potentials.
3. Relay neurons carry information (as action potentials) to particular sets of neurons in the brain.
4. Receptor potentials come in a range of intensities, and these graded potentials are then interpreted as action potentials of different frequencies.
5. Each part of the nervous system and almost every nerve cell receives information from many sources; however, ultimately, all new information comes from the sensory organs.

HOW DO ANIMALS SENSE THEIR ENVIRONMENT?

1. Sensation is the detection of external stimuli; perception is the conscious recognition and interpretation of those stimuli.

What senses contribute to sensation and perception?

1. Neuroscientists refer to the senses as five independent modalities of sensation—audition, vision, gustation, olfaction, vestibular, and somatic sensation.
2. Somatic sensation includes four independent modalities: touch, pain, temperature, and sensing limb motion and position.
3. Some animals detect additional qualities of the external world, such as electric or magnetic fields, or infrared light.
4. Sensation begins in sensory receptors, which are specialized cells that detect a particular property of the external world.
5. Each receptor contributes to only one sense modality.
6. Sensory receptors for hearing, vision, taste, and smell are in specialized organs: ears, eyes, taste buds, and olfactory epithelium.
7. Every sensation except olfaction involves three stages:
   - transduction, which translates the energy of a stimulus into a change in the chemistry of the receptor cell;
   - transmission of a signal from the receptor cells to the CNS; and
   - integration of signals from many receptors to form a mental image of the outside world.
8. Each sensory system responds quantitatively to three characteristics of a stimulus—intensity, duration, and location.

What path does sensory information take to the brain?

1. Sensory receptors feed information to relay neurons.
2. Some relay neurons simply transmit the information from a single sensory receptor to the next step of the neural chain; others integrate information from several sensory receptors or from several other relay neurons.
3. Relay neurons usually send information first to the thalamus, then to particular regions of the cerebral cortex.
4. Information from the olfactory system takes a different route, going first to the other parts of the brain before arriving at the thalamus and cortex.
5. Each sensory receptor feeds information into a characteristic route—a labeled line—so the brain recognizes it as corresponding to a particular sense. Stimulation of a particular receptor always evokes the same sensation.
6. After the neural response to a sensation first arrives in the cortex, the primary sensory region further processes information and sends it to another region of the cortex.
7. The brain is an active agent in sensation—it can actively modify even the start of sensation.

HOW DO ANIMALS HEAR?

1. Sound is actually pulsations of air molecules—waves of pressure.
2. Frequency of vibration is perceived as pitch; amplitude is perceived as loudness.
3. Any sound is the sum of many tones of different frequencies and amplitudes.

How does the mammalian ear detect frequency and amplitude?

1. The mammalian ear consists of three chambers—the outer ear, middle ear, and inner ear.
2. The outer ear collects sound waves and channels them through the auditory canal to the tympanic membrane (eardrum), which vibrates in response to vibrations in air pressure.
3. The middle ear is filled with air, which enters through the Eustachian tubes from the throat.

4. Swallowing, chewing, and yawning can all open the Eustachian tubes and allow the air pressure in the middle ear to equalize with the outside pressure.
5. Inside the middle ear, three small bones (the malleus, incus, and stapes) transmit the movement of the tympanic membrane to the fluid-filled inner ear and amplify the vibrations.
6. The organ of Corti, located in the cochlea within the inner ear, transduces the vibrations into a signal that can be understood as sound.
7. When the fluid in the cochlea vibrates, the organ of Corti moves from side to side.
8. The inner ear interprets this movement as sound by means of hair cells, which are embedded in the membrane of the cochlea and organ of Corti. Movement of the organ of Corti, relative to the cochlear membrane, produces shear forces that cause the hair cells to bend, and this bending alters the membrane's permeability to ions, producing receptor potentials.
9. Each frequency stimulates a different set of hair cells.
10. Signals pass from the organ of Corti to relay neurons in the rear of the brain; relay neurons, in turn, connect to several other regions of the brain.
11. One pathway leads to the thalamus, then to the primary auditory cortex in the temporal lobe.
12. The brain simultaneously processes several separate "images" of sound.

How does the inner ear help maintain balance?

1. Hair cells can also bend in response to gravity.
2. In fish and frogs, hair cells are part of the lateral line system, which consists of two canals running just under the skin on each side of the body.
3. The hair cells detect changes in water movements through the canals.
4. Within the mammalian inner ear are the semicircular canals—three fluid-filled tubes curved into circles lying at right angles to one another.
5. As an animal moves its head, fluid within the canal tends to stay in place due to inertia, and consequently pushes against the hair cells.
6. The pattern of electrical signals from the hair cells tells the cerebellum the head's orientation and movement. The cerebellum integrates this information with other sensory information to determine what muscle activity will maintain posture or execute a certain movement.

HOW DO ANIMALS SEE?

1. The primary event in vision is the action of light energy on a pigment molecule that changes forms, and this event occurs within specialized light-sensitive cells—photoreceptors.
2. Planarians have photoreceptors located within an eye cup that detect a shadow at the cup's edge; these receptors sense the direction of a light source but do not form an image.
3. The vertebrate eye forms a distinct image projected against the back of the eye onto a sheet of cells called the retina.
4. The insect eye is a collection of thousands of individual eyes, none of which forms a true image but, together, the parts of this compound eye compose an accurate image.
5. Insects are amazingly good at seeing movement.

How does a vertebrate's eye form an image of the outside world?

1. The vertebrate eye is encased and protected by a tough connective tissue called the sclera.
2. Light enters the front of the eye through a transparent region of the sclera called the cornea.
3. A pigmented layer inside the sclera is called the choroid which, behind the cornea, forms a muscular donut-shaped disk, called the iris, that gives the eye its color.
4. The iris surrounds the pupil and controls the amount of light that enters the eye.
5. Just behind the pupil lies the clear lens, which focuses light (images) on the retina.
6. The shape of the mammalian lens changes according to the distance of the viewed object. The lens thickens when viewing close objects to bend the light more quickly.
7. The focusing of objects at different distances is called accommodation.

Why must many of us wear glasses?

1. A person who is nearsighted has eyes that are elongated front to back; images of faraway objects focus too soon, in the space in front of the retina, so the image on the retina is out of focus. A concave lens corrects this condition by increasing the focal length.
2. A person who is farsighted has eyes that are too short—the image would focus beyond the retina. A converging (convex) lens corrects this condition by shortening the focal length.
3. Cataracts are cloudy spots that form in the lens and can obscure vision and lead to blindness.
4. The large space between the lens and the retina contains a jellylike substance—the vitreous humor; the space between the iris and cornea is filled with the more watery aqueous humor.
5. Glaucoma results from too much pressure in the aqueous humor.

How does the retina detect an image?

1. The retina is a thin sheet of cells behind the vitreous humor, and consists of five cell types.
2. Photoreceptors are the most numerous of the retina's cells, and they lie in the deepest layer.
3. Photoreceptors convert the light image into electrical signals that the brain can interpret.
4. Rod cells are photoreceptors that:
   - are most dense around the retina's edge,
   - are extremely light-sensitive, and
   - produce colorless, poorly defined images.
5. Cone cells are photoreceptors that:
   - are made for full-color vision in bright light, and
   - are concentrated in the central portion of the retina—the fovea.
6. Rods and cones contain rhodopsins—special pigments that absorb light and transmit a signal; rods contain one pigment; cones contain three, which respond to blue, red, and green light.
7. The visual pigments are called opsins, and each binds to a molecule called retinal.
8. Color blindness results from lack of one or more of the cone opsins (pigments).
9. Because the genes for these opsins are located on the X chromosome, men, with only one X chromosome, are more likely to lack one of the genes than are women.

How does information travel from the retina to the brain?

1. In the visual system, integration actually begins during transduction and transmission.
2. Transduction occurs in the rods and cones (in the photoreceptors).
3. The major pathway of transmission of visual information is from the photoreceptor cells to bipolar cells, to ganglion cells, to the brain.
4. Ganglion cells receive a processed, or integrated, version of the visual world.
5. Horizontal and amacrine cells of the retina are interneurons, and they inhibit nearby cells in a way that increases contrast in the image.
6. A ganglion cell seems to be specialized for detecting changes in light intensity between the center and the outside of its field of photoreceptors.

There is much more to vision than meets the eye

1. Axons of all ganglion cells of the retina form the optic nerve; the two optic nerves converge in the optic chiasma, where some fibers continue on while others cross to the opposite side.
2. Most optic nerve fibers end in a region of the thalamus called the lateral geniculate nucleus.
3. Each lateral geniculate neuron receives input from several ganglion cells and integrates it.
4. Signals that enter the lateral geniculate nucleus continue on to a special region—the primary visual cortex, and information from both of these areas also flows to other brain areas.
5. The pattern of electrical activity in the neurons represents a "mental image" that reflects the spatial organization of the visual image.
6. Ultimately, the interneural connections within the retina and the brain establish not only what we see, but how we respond to it.

OTHER SENSORY PATHWAYS

1. For each of the other senses, the mental image consists of a pattern of neural activity.
2. Perception refers to the conscious experience of objects and events in the external world.
3. Mental "images" are nothing more than electrochemical patterns.

Somatic sensations

1. The somatic sensory system reports on information from three sources:
   - the external world as it impinges on the body's surface;
   - the positions of body parts with respect to one another; and
   - the interior of the body.
2. The somatic sensory system has four types of receptors, for touch (pressure), position, temperature, and pain; each converts information into electrical signals that travel to the central nervous system.
3. These nerve cells connect to neurons in the spinal cord or the medulla, and these, in turn, connect to others in the thalamus.
4. Thalamus neurons connect to neurons in the cerebral cortex's primary sensory area (cortex).
5. The pathways of information roughly maintain the spatial organization of the receptors so that the sensory cortex has a complete, though distorted, map of the whole body.

How do we detect smells and tastes?

1. Perception of smell and taste depend on sensory receptors that respond to molecules from the external environment.
2. These receptors are proteins that bind to small molecules responsible for smell or taste.
3. Taste sensation depends on taste buds on the tongue and on the lining of the top of the digestive tract.
4. A single taste bud consists of about 50 modified epithelial cells—some respond to dissolved chemicals while others provide support.
5. Taste is classified as sweet, sour, salty, and bitter—a single taste bud may detect all four.
6. Taste buds at different locations are more sensitive to some tastes—the human tongue senses:
   - sweetness and saltiness near the tip,
   - bitterness at the back, and
   - sourness at the sides.
7. Information from taste buds flows to the medulla through a distinct part of the thalamus before reaching the cerebral cortex.
8. What we refer to as the "taste" of a food is actually both taste and smell.
9. Smelling, or olfaction, is far more sensitive than taste.
10. Researchers suspect that humans have from 100 to 1000 distinct types of odor receptors.
11. Smell receptors lie in the olfactory epithelium of the nose and are true neurons that connect directly to the brain's olfactory bulb.
12. The olfactory bulb does not reflect the spatial organization of the outside world. Our sense of smell provides no clues about the location of an odor.
13. Electrical signals that represent smells do not reach the cerebral cortex directly, as they do from the other senses; instead, smell information flows to the parts of the brain that are most involved with emotions—the limbic system.

VOCABULARY BUILDING

In your own words, first write a brief definition, then a full explanation for each of the following terms. Include examples where appropriate. Complete this section from your memory—you will not learn the terms by simply copying definitions from the textbook. Once you have finished, check your responses against the information in the chapter and make any necessary corrections.

perception —

sensation —

nerve net —

central nervous system (CNS) —

spinal cord —

motor neurons (motoneurons) —

hindbrain —

pons —

medulla oblongata —

cerebellum —

midbrain —

substantia nigra

brainstem —

reticular formation —

forebrain —

diencephalon —

thalamus —

hypothalamus —

cerebral hemispheres —

cerebral cortex —

gyri (gyrus) —

sulci (sulcus) —

association cortex —

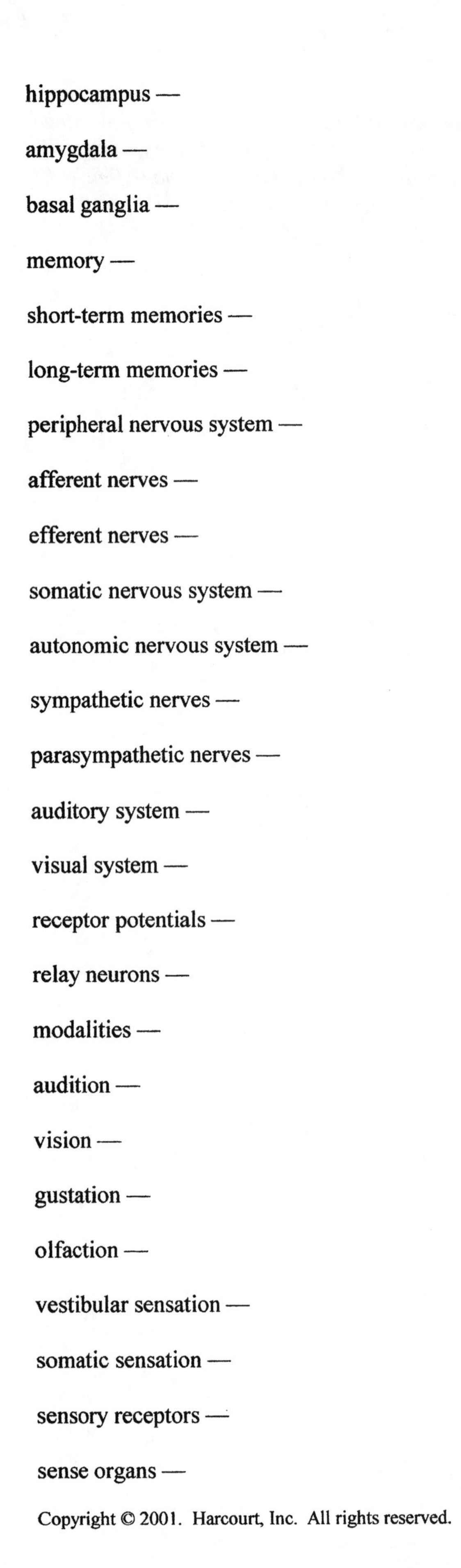
hippocampus —

amygdala —

basal ganglia —

memory —

short-term memories —

long-term memories —

peripheral nervous system —

afferent nerves —

efferent nerves —

somatic nervous system —

autonomic nervous system —

sympathetic nerves —

parasympathetic nerves —

auditory system —

visual system —

receptor potentials —

relay neurons —

modalities —

audition —

vision —

gustation —

olfaction —

vestibular sensation —

somatic sensation —

sensory receptors —

sense organs —

transduction —

transmission —

integration —

transduce —

pitch —

loudness —

outer ear —

middle ear —

inner ear —

auditory canal —

tympanic membrane —

organ of Corti —

cochlea —

hair cells —

lateral line system —

semicircular canals —

photoreceptor —

retina —

cornea —

iris —

pupil —

lens —

accommodation —

nearsighted —

farsighted —

cataracts —

glaucoma —

rod cells —

cone cells —

fovea —

rhodopsin —

opsins —

retinal —

bipolar cells —

horizontal cells —

ganglion cells —

amacrine cells —

optic nerve —

olfactory bulb —

limbic system —

CHAPTER TEST

The following test has four parts. Complete as much of the exam as you can from memory. If you cannot answer a question, skip it. Once you complete all that you can, try to answer any questions you skipped. If you still cannot answer them, consult your textbook for the answers. Once you have completed all sections of the test, check your answers for Parts 1 - 3 against those in the back of this book. Highlight any incorrect answers then review that material in your textbook. Correct your answers for future reference.

Part 1: Multiple Choice

For each of the following, select all correct responses—more than one may be correct.

1. Which of the following is part of the central nervous system?
   a) brain
   b) sensory receptor
   c) spinal cord
   d) all of these

2. Where in the spinal cord are the motor neurons located?
   a) white matter
   b) dorsal horn of gray matter
   c) central canal
   d) ventral horn of gray matter

3. Which of the following is not part of the hindbrain?
   a) pons
   b) cerebellum
   c) thalamus
   d) medulla oblongata

4. What is the major integrator of sensory information about the internal world?
   a) hypothalamus
   b) midbrain
   c) thalamus
   d) medulla oblongata

5. Which of the following is true about the cerebral cortex?
   a) It is made of gray matter.
   b) It is very thin.
   c) It is divided into four lobes.
   d) All of these are true.

6. Which of the following is associated with formation of long-term memory?
   a) hypothalamus
   b) hippocampus
   c) cerebellum
   d) pons

7. Gustation refers to the sense of:
   a) smell
   b) taste
   c) vision
   d) hearing

8. The translation of the energy of a stimulus (such as a sound wave) into an electrochemical event in a receptor cell (such as depolarization) is called:
   a) transmission
   b) integration
   c) perception
   d) transduction

9. The tympanic membrane is located between the:
   a) inner ear and middle ear
   b) middle ear and eustachian tube
   c) outer ear and middle ear
   d) inner ear and cochlea

10. Which of the senses involves bending hair cells to produce a signal in the neurons?
    a) vision
    b) hearing
    c) balance
    d) smell

11. Sensory receptors for hearing are located in the:
    a) olfactory epithelium
    b) retina
    c) semicircular canals
    d) cochlea

12. What part of the eye is involved in accommodation?
    a) iris
    b) lens
    c) cornea
    d) retina

13. Which of the following is true about rod cells?
    a) They are more numerous around the edges of the retina.
    b) They are extremely light sensitive.
    c) They provide our color vision.
    d) All of these are true.

14. What is the function of the iris?
    a) It senses colors.
    b) It moves the eyeball.
    c) It controls the amount of light entering the eye.
    d) It contains photoreceptor cells.

15. What other sense contributes greatly to our sense of taste?
    a) vision
    b) somatic sense
    c) olfaction
    d) audition

Part 2: Matching

For each of the following, match the correct term with its definition or example. More than one answer may be appropriate.

hindbrain **midbrain** **forebrain**

1. ____________________ Includes the pons and medulla oblongata.
2. ____________________ Includes the thalamus and hypothalamus.
3. ____________________ Is a major processor of visual information.
4. ____________________ Parts of this are included in the brainstem.
5. ____________________ Almost all conscious activity occurs here in humans.

audition **vision** **olfaction**

6. ____________________ Receptors are in the organ of Corti.
7. ____________________ Involves the process of accommodation.
8. ____________________ Neural pathways do not reflect spatial organization.
9. ____________________ The information goes to the limbic system.
10. ____________________ Sensation involves bending of hair cells.

Part 3: Short Answer

Write your answers in the space provided or on a separate piece of paper.

1. What three functions must the nervous system fulfill in order to maintain homeostasis?

2. What are the parts of the central nervous system?

3. List three protections that surround the brain.

4. Differentiate between the spinal cord's ventral horn, dorsal horn, and white matter.

5. List the three components of the hindbrain.

6. List the three components of the brainstem.

7. What are the names and functions of the two parts of the diencephalon?

8. What are the four lobes of the cerebral cortex?

9. Give an example of a reflexive memory and of a declarative memory.

10. Differentiate between the two divisions of the autonomic nervous system.

11. Differentiate between perception and sensation.

12. What are the three stages involved in sensation?

13. Receptors for what two senses are located in the inner ear?

14. Explain the process of accommodation.

15. What four types of receptors are part of the somatic sensory system?

Part 4: **Critical Thinking: Using Your Knowledge**

Answer each of these in essay form, using complete sentences and paragraphs. Provide as much information as you can. (For extra essay practice, write out answers to the Review and Thought Questions in your textbook.)

1. Discuss the general organization of the vertebrate nervous system, and compare that to the nervous system of invertebrates.

2. Make a table in which you list the parts of the brain discussed in your textbook, their locations, and their functions.

3. Trace the path of sound as it moves through the ear, and the path of light as it moves through the eye. Discuss each structure that is encountered, and problems that could occur at each location that could impair that sense.

4. For each of the senses discussed in your textbook, trace the path of the signal from the receptor cells through the nervous system, tracking the signal to its final destination in the brain.

5. Many toxic substances have a bitter taste. What is the advantage to having taste receptors for bitter located at the *back* of the tongue, instead of, for example, at the tip or on the sides?

6. When you overheat, your body responds by decreasing blood flow to the skin and increasing sweating. Explain, in detail, the involvement of the nervous system in temperature regulation.

7. Explain what is meant by a child being "cross-eyed." In these cases, it is not uncommon for the child to stop seeing through one of the eyes. What causes this loss of visual perception in an eye capable of normal functioning?

8. Explain how a sore throat can easily lead to an ear infection. Which part of the ear is involved? How might an ear infection disturb your hearing?

9. When you enter someone's home, you often notice a distinct aroma, but after a few minutes you no longer perceive it. When you get dressed, you are keenly aware of your clothing against your skin, but later in the day you barely notice it. Sometimes when you are in bed and can't sleep you clearly hear your refrigerator running, but after awhile you "tune it out." Explain how this is possible, and what advantage this "sensory adaptation" offers.

10. Some animals are active by day (diurnal) while others are active and hunt for food by night (nocturnal). What differences would you expect in their senses? Which senses might be more important by day, and which by night? How would you expect their eyes to differ?

Chapter 43
Sexual Reproduction

KEY CONCEPTS

1. A diploid oocyte produces one haploid ovum and two or three polar bodies.
2. The secondary oocyte moves from the ovary, through the oviduct, then into the uterus.
3. Sperm cells form in seminiferous tubules in the testes, then move through the male genital tract (epididymis, vas deferens, urethra), and out through the penis.
4. Human sperm rarely live more than four days after ejaculation, eggs live only about 12 hours after ovulation, so conception usually requires intercourse within four days before ovulation.
5. Much of the female and male tracts develop from the same embryonic tissues.
6. The sexual response cycle has four phases: excitement, plateau, orgasm, and resolution.
7. Enzymes in the female tract must capacitate sperm before they can fertilize the egg.
8. Progesterone causes the sperm's acrosome to release enzymes that digest the outer layers of the egg, allowing the sperm to enter. Once the sperm is inside, the zona pellucida forms an impenetrable barrier to prevent polyspermy.
9. In men, the hypothalamus secretes GnRH, which stimulates the pituitary to produce LH and FSH. LH stimulates production of testosterone and other androgens, while FSH stimulates the Sertoli cells in the testes to produce sperm.
10. The menstrual cycle's three phases are menstrual, preovulatory, and postovulatory. The postovulatory is nearly always 14 days; the other two are variable.
11. In women, the hypothalamus secretes GnRH, which stimulates secretion of LH and FSH, which then stimulate oocyte and follicle maturation and production of estrogen and other steroid hormones. Growth of the follicles causes buildup of estrogen, triggering release of more GnRH, which causes a surge of LH and FSH.
12. The LH surge triggers ovulation and development of the corpus luteum, whose progesterone stimulates thickening of the endometrium and cessation of GnRH production in preparation for possible pregnancy.
13. If no embryo implants, the endometrium breaks down and sloughs, producing menstruation.
14. Secretion of HCG by the embryo causes the corpus luteum to enlarge and continue progesterone secretion, suppressing menstruation.
15. Between conception and birth, hormones balance the fetal and maternal needs.
16. At parturition, cortisol, oxytocin, and prostaglandins create a positive feedback cycle that leads to expulsion of the baby.
17. After birth, babies first are fed colostrum, but their suckling causes oxytocin to help release milk, and prolactin ensures a continued supply of milk while nursing continues.
18. Some people accept only abstinence or periodic abstinence (rhythm method) for birth control. Barrier methods prevent sperm from reaching the oocyte. Hormonal contraceptives suppress ovulation and implantation (IUDs do the latter). Sterilization is highly effective but usually irreversible.
19. An induced abortion is the removal and destruction of the embryo or fetus; spontaneous abortion is natural death of an embryo or fetus, often one with chromosomal abnormalities.
20. Modern biotechnology offers a wealth of choices to infertile people.

EXTENDED CHAPTER OUTLINE

WHY DID SEX EVOLVE?

1. Textbooks tend to say that the purpose of sex is to recombine genes to increase the genetic diversity in each generation, but this idea is not universally accepted.
2. Richard E. Michod argues that sex evolved to repair damaged genes.
3. Organisms can reconstruct a damaged genome using spare DNA.
4. Michod's research showed that bacteria can survive DNA damage by swapping their own DNA with that of dead bacteria of the same species, and that only bacteria with damaged DNA actively scavenge for spare DNA.
5. As multicellular eukaryotes evolved from bacteria, says Michod, they used the same kind of recombination when making sperm and eggs, so when the sperm and egg join, their chromosomes align, duplicate, and swap DNA. Such recombinations are most common in gaps of DNA, often the site of damage.
6. Sex has a cost to the individual—each parent passes on only half of its genes to its offspring.
7. Sex keeps harmful recessive mutations masked.
8. Michod's ideas are well known, but not yet accepted.

HOW DO MAMMALS FORM GAMETES?

1. The gonads are paired organs where the gametes are formed and mature.
2. Most sexually reproducing organisms form two kinds of gametes—eggs and sperm.
3. Eggs, or ova, form in the ovaries; sperm form in the testes (testicles).
4. Both eggs and sperm are haploid, and when they join at fertilization (syngamy, contraception) their nuclei fuse and the resulting zygote is diploid.
5. When the zygote implants in the uterine lining, it is renamed an embryo; when the embryo develops the basic features of the organism it will become, it is a fetus (in humans, this is at nine weeks).

How do female mammals produce ova?

1. The ovaries lie toward the side walls in the female's pelvis.
2. Much of the process of oogenesis—production of eggs—occurs before birth.
3. Diploid primordial germ cells divide thousands of times to produce up to a million oogonia—diploid cells that, when they grow larger, are renamed primary oocytes.
4. Primary oocytes divide by meiosis to form haploid ova, but they do not complete meiosis unless they are eventually fertilized.
5. A girl is born with about 750,000 primary oocytes; by puberty only about 200,000 remain.
6. Each primary oocyte, with the surrounding granulosa cells, forms a follicle.
7. Follicles begin developing, a few each month, at puberty. One expands and matures faster than others and will be released at ovulation.
8. Before ovulation, the primary oocyte completes meiosis I, forming two haploid cells—a secondary oocyte and a polar body.
9. Just before ovulation, the secondary oocyte enters meiosis II, dividing again to produce the ovum and another polar body.
10. Meiosis II is not completed in humans until after fertilization.
11. Polar bodies reduce the chromosome number but they do not participate in fertilization, so for each primary oocyte that begins to mature, the female produces one ovum.
12. An ovum contains a single set of chromosomes, all organelles and structures contained in most cells, and an abundance of high-fat yolk particles—energy for a developing embryo.
13. Of the 200,000 oocytes present at puberty, only about 400 to 500 will complete meiosis between the start of ovulation, at about age 13, and the end of ovulation, around age 50.

Where do oocytes go after they leave the ovary?

1. Once released from the follicle, the oocyte enters the abdominal cavity then finds its way into one of the oviducts (fallopian tubes in humans) which carry it to the uterus. Cilia at the opening of the oviduct help sweep the oocyte into the tube.
2. Oocytes enter the oviduct reliably, about 99% of the time.
3. Inside the oviduct, more cilia propel the oocyte toward the uterus.
4. The uterus has a thick, muscular wall lined with the endometrium, which has a rich blood supply to provide nutrients to any early embryos.
5. The cervix is the lower part of the uterus and acts as a passage from the vagina to the uterus.
6. The vagina:
   - connects the genital tract to the outside;
   - holds and stimulates the penis;
   - receives ejaculated sperm; and
   - expands at birth for passage of the baby.

How do male mammals produce sperm?

1. In mammals, the male gonads—the testes—are housed within the scrotum, a pouch that lies outside the body.
2. The scrotum keeps the testes at a lower temperature, which is needed for sperm production.
3. A testis consists of hundreds of separate chambers filled with tightly coiled ducts called seminiferous tubules, which are where the sperm mature.
4. Each sperm cell, or spermatozoan, consists of:
   - a head, which contains the nucleus with the chromosomes;
   - a midpiece, which contains a microtubule organizing center and dense concentrations of mitochondria; and
   - a tail, which consists of a long flagellum that provides the sperm with movement.
5. Sperm derive from dividing diploid cells called spermatogonia. Two meiotic divisions of each spermatogonium produce four haploid spermatids, then each spermatid develops into a mature sperm.
6. The front of the sperm contains the acrosome, which is a lysosome whose enzymes allow the sperm to enter the egg.
7. In the seminiferous tubules, developing sperm are surrounded and nourished by Sertoli cells, and the interstitial cells secrete testosterone.
8. After puberty, spermatogonia constantly produce more sperm—about 200 million each hour.

How do sperm cells travel from the testes to the penis?

1. As spermatids mature, they pass into the lumen of the seminiferous tubules, then they pass into the epididymis—an interconnected network of coiled ducts with a single exit tube.
2. Sperm complete their maturation in the epididymis then, during ejaculation, smooth muscle contractions push the sperm toward the vas deferens, which are long tubes that go up into the abdominal cavity then down toward the bladder, carrying the sperm to the urethra.
3. In many birds and some reptiles, sperm (or eggs) and urine pass through a single opening called the cloaca.
4. In all male mammals except monotremes, and in some birds, reptiles, and invertebrates, the penis delivers sperm directly into the female genital tract during sexual intercourse.
5. Just before each vas deferens empties into the urethra, it receives sugars and other nutrients from one of the two seminal vesicles.
6. The two vas deferens join and pass through the prostate gland, which secretes a thin, milky, alkaline fluid into the urethra; then bulbourethral glands add a small amount of mucus.
7. The sperm combined with secretions from the seminal vesicles, prostate gland, and bulbourethral glands make up the semen.

SEXUAL INTERCOURSE: HOW DO THE EGG AND SPERM RENDEZVOUS?

1. Fertilization in mammals and many other animals occurs in the body of the female.

When and where does fertilization occur?

1. Oocytes survive only about 24 hours after being released from the ovaries; sperm can survive in the female tract for up to a week. Thus the *maximum* period of fertility for a woman is eight days—seven before and one after fertilization.
2. In reality, though, the viability and functioning of the gametes decreases rapidly, so conception rarely happens unless intercourse occurs in a few days before ovulation or on the day of ovulation.
3. Fertilization usually occurs in the oviduct.
4. Only after fertilization does the oocyte complete the second meiotic division to form a mature ovum.
5. The chromosomes of the ovum and sperm then align and become a diploid zygote.

How do the external genitalia facilitate sexual intercourse?

1. The human penis has a single exit tube—the urethra—surrounded by spongy cylindrical tissue called corpus spongiosum, which is expanded at the end to form the glans.
2. The penis has two additional cylinders of tissue, called corpora cavernosa, running its length.
3. The foreskin, a flap of skin that covers the glans, is the most sensitive part of the penis, and is often removed surgically in circumcision.
4. The female's external genitalia are collectively called the vulva.
5. The labia minora surround the opening into the vagina and the end of the urethra, and join at the front to form the clitoris, which has the same general structure as the penis.
6. The clitoris includes two corpora cavernosa, a bulb of the vestibule, and a glans (the latter two correspond to the penis' corpus spongiosum).
7. Two outer skin folds—the labia majora—enclose fatty tissue.
8. Bartholin's (vestibular) glands in the female correspond to the male's bulbourethral glands, and secrete mucus at the mouth of the vagina.

How are the male and female genitalia alike?

1. Testes and ovaries develop from the same embryonic structures, as do the labia majora and scrotum, which are also supplied by the same nerves.
2. Male external genitalia are best described as enlarged, closed versions of female genitalia.
3. All embryos develop two kinds of ducts:
   - in the male, the vas deferens and its attachments develop from paired Wolffian ducts;
   - in the female, the oviducts, uterus, and vagina develop from the paired Müllerian ducts.

The male and female sexual responses each consist of four stages

1. Sexual physiologists divide the sexual response cycle into four phases: excitement, plateau, orgasm, and resolution.
2. The first phase—excitement—is marked by increased blood flow to the clitoris, labia minora, and breasts in the female, and to the penis and testes in the male, causing erection of the clitoris, nipples, and penis.
3. Erection in both sexes results from blood filling the tiny spaces within the corpus spongiosum and corpora cavernosa. In females, excitement causes erection of the clitoris and tightening of the tissues around the base of the vagina that squeeze the penis.
4. The male's bulbourethral glands release a small amount of mucus which eases entry of the penis into the vagina, but most of the lubrication comes from the female's Bartholin's glands.
5. During the second phase—plateau—breathing and heart rate increase in response to continued stimulation, the end of the vagina dilates, and the uterus pulls upward creating a space near the cervix where the sperm can pool.

6. The third stage—orgasm—is marked by smooth muscle contractions in the genital tract, skeletal muscle contractions throughout the body, and feelings of intense pleasure.
7. In females, orgasm increases the likelihood of conception but is not required; however, sperm delivery requires ejaculation—the forceful expulsion of sperm out of the penis.
8. Ejaculation has two phases: smooth muscle contractions in the prostate, vas deferens, and seminal vesicles force semen into the urethra in the penis, then smooth muscle contractions in the urethra force the semen out.
9. A human male typically releases about 3.5 mL of semen, containing an average of 200 million sperm (although these numbers are highly variable).
10. In the female, smooth muscle in the uterus and vagina contracts rhythmically and the cervix drops down into the pool of sperm.
11. It is believed that the orgasm-mediated contractions of the uterus and oviducts propel the sperm upward toward the descending ovum.
12. In the last phase—resolution—blood flow and muscle tension return to normal.

How do the sperm reach the egg and penetrate its surface?

1. Once inside the vagina, sperm pile up against a mucus plug in the cervix, which keeps all but the healthiest, most active sperm out of the uterus.
2. The few hundred that get through the cervix fan out into the uterus and up into the oviducts.
3. In many species of aquatic animals and other organisms, the egg releases a chemical that attracts the sperm—it is unknown if mammals use such mechanisms.
4. Sperm are incapable of fertilizing an egg until they have been in the reproductive tract for five to six hours because the head of each sperm is covered with a glycoprotein coat which must be dissolved by enzymes in the female reproductive tract (the process is called capacitation)—sperm not exposed to fluids from the female tract cannot fuse with an egg.
5. Under the influence of the female's progesterone, the acrosome releases its enzymes, which break down the outer layers of the egg—the corona radiata and zona pellucida.
6. Once the sperm passes through the zona pellucida, the layer becomes impenetrable to other sperm, thus avoiding polyspermy.

HOW DO HORMONES CONTROL GAMETE PRODUCTION?

1. The anterior pituitary secretes two gonadotropins—LH and FSH—which act on the gonads.

How do hormones control sperm production?

1. The testes secrete testosterone, which is the most potent of the androgens (male hormones).
2. Testes also secrete estrogen, and the male brain seems to convert some testosterone to estrogen, so estrogen levels rise and fall in synchrony with testosterone levels.
3. The hypothalamus secretes gonadotropin-releasing hormone (GnRH), which stimulates the pituitary gland to produce two hormones that males share with females—luteinizing hormone (LH) and follicle-stimulating hormone (FSH).
4. LH stimulates the interstitial cells of the testes to produce testosterone and other androgens; FSH stimulates Sertoli cells to produce sperm.
5. Testosterone increases the sex drive and aggressive behavior in general in both sexes.
6. Human males do not begin substantial testosterone production until about age 10; by about age 13, they are usually sexually mature and testosterone production increases until about age 20, when it begins a steady decline.
7. By the late 40s or 50s, most men begin to experience some decrease in sexual function similar to that experienced by women (but more gradual), and some experience symptoms typical of women going through menopause.

What are the main events of the menstrual cycle?

1. Beginning at puberty, human females undergo monthly reproductive cycles which simultaneously produce a mature oocyte and prepare the wall of the uterus for implantation.
2. If a zygote does not implant or dies, the uterine wall is sloughed off in menstruation.
3. The cycle is divided into three phases: the menstrual phase, the preovulatory phase, and the postovulatory phase.
4. The average length of the menstrual cycle is 28 days.
5. Menstruation ceases at menopause, which usually occurs in the late 40s or early 50s.
6. The menstrual phase is usually 3 to 7 days, but the period between ovulation and menstruation is nearly always 14 days.

How do hormones regulate the ovarian cycle?

1. Hormones regulate the ovarian cycle—the production of oocytes—and the state of the endometrium.
2. As in the male, the hypothalamus secretes GnRH, which in turn regulates secretion of LH and FSH from the pituitary.
3. During the menstrual phase, increased FSH levels stimulate 5 to 12 follicles in the ovaries to grow and mature. FSH causes growth and development of the oocyte and, with LH, causes the follicle cells to release increasing amounts of estrogen, which causes follicle growth.
4. This positive feedback causes a buildup of estrogen in the days preceding ovulation.
5. Around the time of ovulation, high levels of estrogen stimulate the hypothalamus to release a surge of GnRH, which causes a surge of LS and FSH. The LH triggers ovulation.
6. After the oocyte bursts out of the follicle, the empty follicle becomes the corpus luteum—a structure that secretes progesterone and a little estrogen.
7. Progesterone stimulates the endometrium to thicken in preparation for implantation, and it signals the hypothalamus to stop producing GnRH, which, in turn, reduces LH and FSH.
8. The corpus luteum also produces inhibin, which inhibits the pituitary secretions.
9. If fertilization does not occur, the corpus luteum degenerates and the pituitary is free to again produce FSH and LH.
10. If the egg is fertilized, the zygote produces (human) chorionic gonadotropin (HCG) which maintains the corpus luteum, which can then expand and continue progesterone production.
11. HCG in the blood or urine is the basis of most common pregnancy tests.
12. Estrogen participates in both negative feedback and positive feedback; progesterone only promotes negative feedback.

How do hormones regulate the menstrual cycle?

1. While the follicles are growing, the uterine lining (endometrium) thickens.
2. After ovulation, the corpus luteum and its progesterone stimulates the endometrium to develop further.
3. Estrogen also stimulates the cervix to secrete a clear, fluid mucus to ease passage of sperm.
4. Progesterone causes secretion of thick cervical mucus to prevent entry of sperm or bacteria.
5. If no fertilization occurs, blood vessels supplying the endometrium weaken and break; blood, vaginal secretions, and endometrial tissue flows out of the vagina—this is menstruation.

WHAT HAPPENS DURING PREGNANCY AND LACTATION?

1. Internal reproduction is the rule among land animals and occurs in many aquatics as well.
2. Giving birth to live young, called viviparity, occurs among sharks, snakes, all marsupials, and true mammals.

How can a woman tell if she is pregnant?

1. Pregnancy is the state in which a woman is carrying a fertilized egg, embryo, or fetus inside her body and starts at the moment of conception—when the zygote forms.
2. By tradition, doctors count the pregnancy from the last period, so they refer to a 40-week pregnancy, although biologically the woman is pregnant, on average, 38 weeks—2 weeks less than by the doctor's method.
3. About a week after fertilization, the ball of cells implants in the endometrium of the uterus and starts secreting human chorionic gonadotropin (HCG), which causes the corpus luteum to persist, double in size, and secrete enough progesterone to suppress menstruation.
4. The embryo develops the same four membranes found in all amniotes: an amnion, yolk sac, allantois, and chorion.
5. Mammals also develop a placenta—a flat organ rich in blood vessels that conduct nutrients to the embryo and wastes from it to the mother. An umbilical cord attaches the embryo and placenta.
6. The most obvious early sign of pregnancy is lack of menstruation, but other conditions can cause menstrual irregularities so some women don't realize they are pregnant until late in the pregnancy.
7. From 50 to 75% of all pregnancies end in spontaneous abortion, or miscarriage, most due to chromosomal abnormalities, usually in the first two weeks, so the woman may never know she was pregnant.

What happens between conception and delivery?

1. Human pregnancies are divided into three trimesters, each of about three months.
2. In the first trimester, increased estrogen levels cause enlargement of breasts, uterus, and external genitalia. She may also suffer fatigue and intense nausea (morning sickness).
3. The fetal placenta secretes HCS which decreases the woman's insulin sensitivity, so glucose is redirected to the fetus and increases the mother's metabolism of fat for energy.
4. In the second trimester, the corpus luteum degenerates and HCG levels drop (making pregnancy tests less reliable). The pregnancy is now maintained by estrogen and progesterone from the placenta.
5. The woman feels the fetus kick for the first time around four months.
6. This stage is also called quickening.
7. The mother's abdomen swells noticeably but her energy level increases, morning sickness ends, and she often feels at her best in the middle of the trimester.
8. In the third trimester, the mother's blood volume increases by about 30%.
9. The fetus starts to push on her internal organs. Symptoms of this can include frequent urination, incontinence, leg numbness, cramps, varicose veins, difficulty breathing, heartburn, and restless nights with poor sleep.
10. On average, women gain about 25 pounds during pregnancy.

Parturition: what happens on delivery day?

1. The time of birth (parturition) is likely stimulated by the fetus secreting chemical signals, such as cortisol.
2. Oxytocin from the mother's pituitary stimulates uterine contraction and secretion of prostaglandins from the fetal membranes, which also stimulates contraction.
3. Pressure of the fetus downwards on the cervix triggers a positive feedback situation, as it triggers uterine contractions which stimulate oxytocin release, which in turn stimulates more and stronger contractions.
4. Positive feedback causes contractions to come closer together.
5. The often intense labor pains result at first from lack of oxygen to the uterine muscles, but as labor progresses, tissues start stretching and tearing, causing severe pain.

Nursing the baby

1. After the baby is delivered, the placenta is delivered, and the umbilical cord must also be cut.
2. Within a few hours, the mother's body is reverting to its former state, but her breasts are not.
3. In the first few days, the breasts secrete colostrum, which is similar to milk but has less fat, and also contains high numbers of antibodies.
4. During suckling, stimulation of the nipples causes the mother's pituitary to secrete prolactin, which stimulates milk production.
5. At about three days, the mother's milk "comes in."
6. Human milk is better for the baby than cow's milk because it contains more sugar and less fat and protein, and contains immune cells and antibodies to keep the baby healthy.
7. The amount of milk produced is proportionate to the amount of sucking.
8. Milk ejection ("let-down" reflex) stimulates release of oxytocin, which contracts smooth muscle around the milk glands, shooting the milk out.

HOW DO HUMANS CONTROL REPRODUCTION?

1. Humans, more than any other animal, have sexual contact for reasons other than reproduction; a hypothesis is that sex strengthens social bonding, increasing the chance that a couple will work together to nurture their young.
2. Unlike most other mammals, the human female is sexually receptive even at times when fertilization is impossible.
3. In many countries, nursing continues for about three years, and nursing suppresses ovulation, but in developed countries, weaning occurs very early and reproductions are timed much more closely.
4. Traditions, lifestyles, religion, and even government policies can affect reproductive rates.
5. Birth control (contraception)—the conscious regulation of reproduction—is an important issue for many sexually mature humans.

Abstinence

1. The most certain contraceptive method is to abstain from sexual intercourse.
2. The rhythm method is a form of modified abstinence, in which a couple refrains from sexual intercourse at times when conception is likely to occur. This method is allowed by the Catholic Church, which forbids most forms of birth control.
3. If a couple abstains from four days before to one day after ovulation, no conception should occur, but, because the preovulatory phase can vary enormously, the exact date of ovulation is hard to predict.
4. Of every 100 women practicing the rhythm method, as many as 20 become pregnant in one year. By comparison, in women who use no birth control method, 90 out of 100 would become pregnant.
5. Several factors can increase the success of the rhythm method, such as:
   - having regular and predictable cycles;
   - knowing when ovulation occurs each month;
   - monitoring for the 0.5 to 1° increase in body temperature that accompanies ovulation;
   - noting thinning and stringy quality of mucus in the days just prior to ovulation; and
   - noting mild pain (mittelschmerz) with ovulation.

Withdrawal

1. Coitus interruptus refers to the method in which the man attempts to avoid conception by withdrawing his penis from the vagina before ejaculation.
2. This is one of the least reliable forms of birth control for two main reasons:
   - some sperm usually enter the female tract before ejaculation, and
   - withdrawal before orgasm requires tremendous restraint in an effort to defy the usual pattern of the sexual response.

Barrier methods prevent the union of sperm and ovum

1. A condom is a thin rubberlike sheath that covers the penis and prevents sperm from entering the vagina (a female condom is also available).
2. Condoms prevent not only pregnancy, but also sexually transmitted diseases (STDs), such as acquired immunodeficiency syndrome (AIDS), chlamydia, syphilis, and gonorrhea.
3. The major problem with condoms is that they can break or come off; they must also be timed correctly—if put on too late, sperm may have already entered the vagina, and if left on too long, sperm may leak out from around the flaccid penis.
4. A condom must be put on as soon as the penis is erect, and it must be removed immediately after ejaculation.
5. A cervical cap is a thimble-shaped cap that fits tightly over the cervix; a diaphragm is larger but works in the same way, serving as a barrier to sperm entrance.
6. A spermicide—a cream or jelly that kills sperm—is an essential part of the effectiveness of both cervical caps and diaphragms. Spermicides are also often used with condoms, and one active ingredient found in many spermicides seems to reduce the risk of HIV transmission.
7. When used properly, these barrier methods are effective, producing only two pregnancies per 100 each year, but improper usage and breakage increase this to actual rates of 10 to 13 pregnancies per 100 women per year.
8. The contraceptive sponge also provides a barrier to entry of sperm into the uterus and, unlike the cervical cap and diaphragm, it does not require special fitting. The sponge also releases a spermicide, but, when used alone, gives pregnancy rates of about 17 per 100 women per year. They are most effective used with another method, such as condoms.
9. Some couples attempt to prevent fertilization with spermicides alone, which produces a pregnancy rate of about 15 per 100 per year.
10. Some women attempt to prevent pregnancy with a douche—the washing of the vagina immediately after intercourse—but this method is ineffective, with a pregnancy rate of about 40 per 100 per year.

Hormonal contraceptives and IUDs

1. Chemical birth control can prevent ovulation, fertilization, or implantation.
2. The birth control pill comes in many formulations, including versions that consist of a combination of estrogen and progesterone which prevents the anterior pituitary from secreting LH, thus stopping ovulation and resulting in a regular cycle without ovulation.
3. If used correctly, these formulations are almost entirely effective in preventing pregnancy, but the relatively high hormone doses used sometimes cause undesirable and even dangerous side effects. These risks are greater for smokers than for nonsmokers.
4. Another type of birth control pills ("minipills") contain only progesterone and have fewer side effects than the combination pills but carry a slightly higher pregnancy risk.
5. Another chemical method that interferes with ovulation involves release of a substance from a capsule surgically implanted under the skin. One—Norplant—may last up to ten years.
6. A progesterone derivative called DMPA, or Depo-Provera, is injected and appears to suppress ovulation and inhibit implantation for three months at a time. The risks associated with DMPA are still the subject of debate.
7. The morning-after pill is an oral contraceptive taken in higher than normal doses within 72 hours of unprotected intercourse to interfere with both conception and implantation.
8. The intrauterine device (IUD) is a small piece of plastic that is placed into the uterus, where it interferes with implantation.
9. One type of IUD that was sold in the United States between 1971 and 1974, the Dalkon shield, produced serious side effects, including inflammation of the pelvis, spontaneous abortions, and some deaths.
10. IUDs must be fitted and ultimately removed by a physician.

Sterilization provides effective but generally irreversible birth control

1. Vasectomy—cutting and tying the vas deferens—prevents sperm from entering the urethra.
2. Vasectomy is safe, usually done on an outpatient basis, and takes less than an hour under local anesthesia.
3. Tubal ligation—cutting or blocking the oviducts—prevents eggs from reaching the uterus.
4. Tubal ligation is an invasive procedure and requires abdominal surgery under full anesthesia, although it may be done through two small incisions. Full recovery takes several weeks.
5. Although both procedures may occasionally be reversed, the success rate is poor, so both procedures should be considered permanent.

Induced abortions terminate pregnancies after implantation

1. An induced abortion is the deliberate removal of an embryo or fetus from the uterus.
2. In the first trimester, the embryo is either removed by gentle suction or by dilation and curettage (D & C). In some countries, the drug RU 486 (mifepristone) is used to induce early abortions.
3. Induced abortions in the second trimester are performed only in a hospital; the most common method is to induce uterine contractions by injecting a solution of salt or prostaglandins.
4. Induced abortions in the third trimester are quite rare.
5. In the United States, abortion before quickening (the second trimester) was legal until 1873, when President Ulysses S. Grant enacted a federal law prohibiting all contraception and abortion, as well as any publication or discussion of information about contraception.
6. Similar laws were passed by states in the next decades.
7. In 1973, all of these laws were overturned and declared unconstitutional by the U.S. Supreme Court. The decision (Roe vs. Wade) declared that abortions in the early stages of pregnancy are legal in the United States, but states can restrict abortions in the third trimester as long as the mother's life is not at risk.
8. Currently, abortion is legal in all states but remains deeply controversial.

What can couples do to overcome infertility?

1. About one couple in six in the United States are infertile, meaning they do not conceive after a year of sexual relations without contraception; but about half of infertile couples can achieve pregnancy.
2. About a third of the time, infertility results because the male does not produce and deliver sufficient viable sperm; another third results because the female does not produce fertilizable ova or maintain a pregnancy once it begins; infertility in the remaining third result from combined factors.
3. If the woman is fertile but the man does not produce adequate sperm, the woman can conceive if donor sperm are placed in her vagina at the right time.
4. In *in vitro* fertilization (IVF), an ovum is fertilized in the laboratory, followed by the implantation of the embryo back into the uterus.
5. In IVF, the woman usually is treated with gonadotropins to increase the number of mature follicles, then the surgeon removes the ova from the ovary, places them in a dish, and exposes them to sperm.
6. When the woman is infertile, the man's sperm can be used to fertilize the ovum of a surrogate mother—a woman who will conceive and also carry the resulting embryo to term.
7. Surrogacy is controversial, and surrogate mothers sometimes want to keep the babies.
8. Fertility clinics are increasingly using egg donors—healthy young women willing to donate eggs for a price.

VOCABULARY BUILDING

In your own words, first write a brief definition, then a full explanation for each of the following terms. Include examples where appropriate. Complete this section from your memory—you will not learn the terms by simply copying definitions from the textbook. Once you have finished, check your responses against the information in the chapter and make any necessary corrections.

gonads —

ova (eggs) —

ovaries —

sperm —

testes (testicles) —

syngamy (fertilization, conception) —

zygote —

uterus —

oviducts (fallopian tube) —

fetus —

oogenesis —

primordial germ cells —

oogonia —

primary oocytes —

follicle —

ovulation —

secondary oocyte —

polar body —

endometrium —

cervix —

full term —

scrotum —

seminiferous tubules —

spermatozoan —

spermatogonia —

spermatid —

flagellum —

acrosome —

Sertoli cells —

interstitial cells —

epididymis —

vas deferens —

cloaca —

seminal vesicles —

prostate gland —

bulbourethral glands —

semen —

urethra —

corpus spongiosum —

glans —

corpora cavernosa —

foreskin —

circumcision —

vulva —

labia minora —

clitoris —

bulb of the vestibule —

labia majora —

hymen —

excitement phase —

erection —

Bartholin's glands —

plateau phase —

orgasm —

ejaculation —

resolution phase —

capacitation —

anterior pituitary gland —

hypothalamus —

gonadotropins —

gonadotropin-releasing hormone (GnRH) —

luteinizing hormone (LH) —

follicle-stimulating hormone (FSH) —

implantation —

ovarian cycle —

corpus luteum —

human chorionic gonadotropin (HCG) —

menstruation —

viviparity —

placenta —

spontaneous abortion —

human chorionic somatomammotropin (HCS) —

parturition —

cortisol —

oxytocin —

positive feedback —

colostrum —

prolactin —

milk ejection —

birth control —

abstinence —

rhythm method —

mittelschmerz —

coitus interruptus —

condom —

cervical cap —

diaphragm —

spermicide —

douche —

birth control pill —

Depo-Provera (DMPA) —

morning-after pill —

intrauterine device (IUD) —

vasectomy —

tubal ligation —

induced abortion —

dilation and extraction (D & E) —

RU 486 —

in vitro fertilization —

surrogate mother —

CHAPTER TEST

The following test has four parts. Complete as much of the exam as you can from memory. If you cannot answer a question, skip it. Once you complete all that you can, try to answer any questions you skipped. If you still cannot answer them, consult your textbook for the answers. Once you have completed all sections of the test, check your answers for Parts 1 - 3 against those in the back of this book. Highlight any incorrect answers then review that material in your textbook. Correct your answers for future reference.

Part 1: Multiple Choice

For each of the following, select all correct responses—more than one may be correct.

1. Which of the following are haploid?
   a) primary oocyte
   b) oogonia
   c) secondary oocyte
   d) ovum

2. Where are sperm produced?
   a) seminiferous tubules
   b) epididymis
   c) vas deferens
   d) seminal vesicles

3. Which of the following secretes testosterone?
   a) Sertoli cells
   b) interstitial cells
   c) acrosome
   d) sperm

4. Where do sperm mature?
   a) seminiferous tubules
   b) epididymis
   c) vas deferens
   d) seminal vesicles

5. Primary oocytes undergo two meiotic divisions to produce how many ova (eggs)?
   a) 1
   b) 2
   c) 4
   d) 8

6. Secretions from which of the following are part of the semen?
   a) bulbourethral glands
   b) seminal vesicles
   c) prostate gland
   d) all of these

7. In reality, conception is not likely to occur unless intercourse occurs at what time?
   a) during the two weeks before ovulation
   b) within a few days prior to ovulation
   c) at least two days after ovulation
   d) on the day of ovulation

8. Where does fertilization usually occur?
   a) vagina
   b) uterus
   c) ovary
   d) oviduct
9. What tissue is usually removed from the penis during circumcision?
   a) glans
   b) foreskin
   c) corpora cavernosa
   d) corpus spongiosum
10. What female structure develops from the same embryonic structures as the male's scrotum?
    a) labia minora
    b) clitoris
    c) hymen
    d) labia majora
11. What is the role of the acrosome during fertilization?
    a) It propels the sperm cells through the female's tract.
    b) Its enzymes break down the outer layers of the egg, allowing the sperm to enter.
    c) It forms a mucus plug to block other sperm.
    d) It contains the father's chromosomes.
12. In females, luteinizing hormone (LH) is secreted by which of the following?
    a) ovary
    b) corpus luteum
    c) anterior pituitary
    d) hypothalamus
13. Presence of which of the following is used as a basis for pregnancy tests?
    a) GnRH
    b) LH
    c) progesterone
    d) HCG
14. Which of the following is a barrier method for birth control?
    a) Depo-Provera
    b) condom
    c) diaphragm
    d) IUD
15. Birth control pills (combination pills) work by preventing:
    a) ovulation
    b) menstruation
    c) implantation
    d) sperm entry into the oviducts

Part 2: Matching

For each of the following, match the correct term with its definition or example. More than one answer may be appropriate.

male reproductive system	**female reproductive system**
1. ____________________	Gonads include Sertoli cells.
2. ____________________	External genitalia are called the vulva.
3. ____________________	External genitalia have spongy, erectile tissue.
4. ____________________	Produces more gametes from puberty on.
5. ____________________	Gonadotropin levels fluctuate on a regular cycle.

LH	**FSH** **estrogen**
6. ____________________	Secreted by both males and females.
7. ____________________	Stimulates secretion of sex hormones.
8. ____________________	Stimulates maturation of oocytes.
9. ____________________	Stimulates development of the corpus luteum.
10. ____________________	Secreted from the anterior pituitary.

Part 3: Short Answer

Write your answers in the space provided or on a separate piece of paper.

1. In humans, when do oocytes complete the second meiosis?

2. Why are testes housed in the scrotum, outside the body?

3. List the components of semen.

4. What structures are derived from the Wolffian and Müllerian ducts?

5. List, in order, the four stages of the sexual response.

6. What causes erection?

7. Ejaculation involves smooth muscle contractions in what structures?

8. What mechanism prevents polyspermy once fertilization occurs?

9. What is the relationship between the hypothalamus, anterior pituitary, and gonads?

10. What role does the corpus luteum play in preparing the uterus for a pregnancy?

11. What is the biggest drawback to relying on the rhythm method to prevent pregnancy?

12. What percentage of women who practice no birth control become pregnant in a year?

13. Why is coitus interruptus not reliable for preventing pregnancy?

14. What are two contraceptive methods that do not prevent pregnancy, but, instead, prevent implantation if pregnancy occurs?

15. Why is tubal ligation a more risky and complicated procedure than vasectomy?

Part 4: Critical Thinking: Using Your Knowledge

Answer each of these in essay form, using complete sentences and paragraphs. Provide as much information as you can. (For extra essay practice, write out answers to the Review and Thought Questions in your textbook.)

1. Trace the path of sperm from where they are produced, in the testes, to where fertilization occurs, in the female's oviduct. Discuss sperm production, maturation, formation of semen, and the events that occur at each site through which the sperm pass on their journey.

2. Explain in detail hormonal regulation of the following: sperm production, ovarian cycle, and the menstrual cycle.

3. Sex education in public schools has been hotly debated for years. One side argues that teaching children about sex and contraception encourages them to become sexually active. Many people also maintain that abstinence is the only method of contraception that should be discussed. The other side argues that teen pregnancy rates prove children are sexually active with or without sex education, and that they need to know facts about how the male and female reproductive systems work, how pregnancy occurs, and the risks of sexually transmitted diseases. Many people maintain that contraceptives should be available for those who cannot go to their parents for such assistance. State your position on this issue, and support your position by facts and offer arguments against your opponent's position.

4. Make a chart and include each of the hormones discussed in this chapter. For each, list where it is produced, what triggers its release, and what it does in both males and females (where appropriate).

5. In vitro fertilization, donor eggs, donor sperm, and other modern treatments to bypass infertility can be quite expensive. If a married couple has tried unsuccessfully to conceive, what moral and ethical issues should be considered before they try some of these technologies? Are these issues any different than those faced by adoptive parents?

6. Should insurance companies pay for infertility treatments, fertility drugs, and technologies like in vitro fertilization? If so, should they pay for unmarried couples to conceive? Should they pay for single women to conceive? Should they pay for homosexual couples to have a child? How would you make these decisions?

7. A couple have tried unsuccessfully to conceive for two years. They find a woman willing to be a surrogate mom, and she is impregnated with the husband's sperm. At birth, the surrogate decides she cannot give the child to the married couple. The case goes to court. If you were the judge, how would you decide this case?

8. Some men hope that someday, with medical advances, male pregnancy might be possible. After all, the male and female systems develop from the same tissues and are very similar. If you were assigned to make that happen, what would you have to consider? What approaches might you use? Do you think this may be possible?

9. If a couple is having trouble conceiving, the doctor often recommends that the man switch his underwear selection from briefs to boxers and wear loose pants instead of tight jeans. In many cases, this strategy is successful. Explain how this works.

10. The male reproductive system delivers sperm through the same passageway as urine (the urethra), but the female has completely separate reproductive and urinary systems. Explain the need for this difference.

Chapter 44
How Do Organisms Become Complex?

KEY CONCEPTS

1. Fertilization is a genetic and a developmental event. Protein synthesis and development in early embryos is directed by maternal mRNA—mRNA transcribed from the mother's genes.
2. The eight stages of vertebrate development are gamete formation, fertilization, cleavage, germ layer formation, organ formation, growth, metamorphosis, and aging.
3. Cleavage changes embryos from one large cell to many small cells operating independently.
4. Embryos divide at different rates and in different ways, depending on species and the amount of yolk in the egg. Sea urchins and amphibians divide to form a blastula. Mammals divide to form a blastocyst, which consists of a trophoblast and an inner cell mass.
5. During gastrulation, a blastula (or blastocyst) invaginates, forming a two-layered cup. The inside of the cup is the archenteron, the inside of the future gut.
6. Amphibians invaginate smoothly through the blastopore. The cells of amphibian gastrula can spontaneously organize themselves into three layers of tissue.
7. Although mammalian eggs have little yolk, the embryos gastrulate like their relatives whose eggs are heavy with yolk. The formation of the mammalian embryo takes place in a flattened disk, and the cells invaginate individually through the primitive streak, forming the gastrula.
8. The notochord induces formation of neural plate tissue, which folds into a hollow neural tube that forms the different regions of the brain and the spinal cord. Neural crest cells from the neural tube's surface migrate to distant parts of the embryo to form nerve and other tissue.
9. Programmed cell death (apoptosis) is essential to embryonic development.
10. An embryo forms the bare outlines of all its organs long before those organs are functional. At nine weeks, a human embryo possesses rudiments of all the major organs and is said to be a fetus, but only the heart is functional.
11. Cells' fate is the tissue they will become during normal development; cells' potency is the tissues that they could become under varying environmental influences.
12. Cloning a Scottish mountain sheep in 1997 suggested that differentiated nuclei are totipotent, but Dolly may have come from a stem cell nucleus—totipotency remains in question.
13. Researchers that cloned Dolly, the sheep, induced cultured udder cells into a dormant state in which gene expression was shut down, then transplanted the nuclei into sheep oocytes.
14. The primary organizer establishes the entire organization of an embryo, apparently through chemical signals which are used by many species.
15. Pattern formation seems to result from morphogens, such as retinoic acid.
16. Almost all animals use the same gene products to establish position and identify body parts.
17. Metamorphosis is the process by which larvae are converted to adults. In complete metamorphosis, no adult tissues come from the larva; they derive from imaginal discs—cell groups that do not differentiate in the larva, but are saved for later development.
18. In gradual metamorphosis, embryos hatch into immature nymphs that are like mini-adults.
19. Multicellular organisms lose their ability to reproduce as they age, and their risk of death increases. Effects of aging are apparent even at the cellular level. Like other aspects of development, aging depends on complex interactions between genes and environment.

EXTENDED CHAPTER OUTLINE

HOW DO WE BECOME COMPLEX?

1. According to the theory of preformation, each egg contains a complete embryo, and each embryo contains more complete embryos inside itself, etc. Some microscopists even claimed to have seen a tiny creature—a homunculus—curled up inside the sperm head.
2. Preformation left little room for evolution and only worked as a theory if people believed that Earth was relatively young.
3. In the 1820s, Karl Ernst von Baer described the gradual development of a mammalian zygote. The information about fertilization undermined the theory of preformation.
4. Totipotent cells are those that are capable of developing into any kind of cell, whereas differentiated cells have specialized and lost some potential for developing as other types.
5. Wilhelm Roux did an experiment in which he destroyed one cell of a two-cell frog embryo, and noted that the remaining cell produced only half an embryo. He concluded that embryonic cells are not totipotent.
6. Hans Driesch performed a similar experiment with sea urchins, but separated the cells, and watched a complete sea urchin develop from each, indicating each cell may be totipotent.
7. Driesch's work conclusively discredited the theory of preformation.

HOW DOES FERTILIZATION INITIATE EMBRYONIC DEVELOPMENT?

1. Protein synthesis increases dramatically after fertilization but requires mRNA.
2. The mRNA that directs early protein synthesis is produced in the egg from maternal genes, and is called maternal mRNA.
3. The early development of a zygote is entirely controlled by maternal proteins.

HOW DO CELLS OF THE EMBRYO GIVE RISE TO CELLS OF THE ADULT?

1. The vertebrate body plan is that of a tube within a tube:
   - outermost tube—ectoderm—becomes epidermis, nervous system, and sense organs;
   - innermost tube—endoderm—becomes the gastrointestinal tract, pancreas, and liver; and
   - middle tube—mesoderm—becomes connective tissue, the heart, and kidneys.

Biologists separate vertebrate development into seven stages

1. The eight stages of vertebrate development are:
   - gamete formation—production of egg and sperm through meiosis and specialization;
   - fertilization—the union of the haploid egg and sperm to form the diploid zygote;
   - cleavage—the division of the zygote into many smaller cells;
   - germ layer formation—the movement of embryonic cells to form the three germ layers (this process is called gastrulation);
   - organ formation—the movement and specialization of cells to form functioning organs;
   - growth—the increase in size of an organism after organs and body plan are established;
   - metamorphosis—the series of changes that convert a larval form to an adult form; and
   - aging—further development, which inevitably leads to death.
2. Some developmental biologists divide human development into four stages:
   - gamete formation;
   - embryonic development—an eight-week period from fertilization to organ development;
   - fetal development—the 30-week period of growth and development, through birth, and
   - postnatal development—further growth and development, leading to aging and death.

Why does a zygote divide?

1. A zygote is much larger than the typical body cell, creating two problems—it has too little surface area and too little nucleus for the amount of cytoplasm.
2. A single nucleus cannot meet the enormous demand for mRNA that an embryo has, nor can the nucleus of a large cell adequately distribute the mRNA throughout the cell.
3. Cleavage—rapid divisions in which the cells do not grow—increases the surface area and potential mRNA production, and allows each nucleus (and its cell) to specialize.

How do zygotes cleave?

1. Cleavage patterns and pace vary between different species, and this variation depends on both the orientation of the mitotic spindle and the distribution of yolk within the egg.
2. Eggs with a lot of yolk have slow cleavage divisions or do not completely divide the yolk.
3. Cleavage in mammals occurs slowly, each division taking 12 to 24 hours.
4. After a mammalian embryo divides to eight cells, the cells form a compact morula, which soon forms a blastocyst—a modified blastula with cells enclosing internal cavity.
5. A blastocyst contains two types of cells—a prominent outer layer called the trophoblast, and an inner cell mass which will eventually grow into the embryo itself.
6. Outer trophoblast cells (the chorion) attach the embryo to the uterine wall and become part of the placenta, which is a highly vascularized organ through which mother and fetus exchange nutrients and wastes via chorionic villi. Later in development, the inner cell mass forms the amnion, which is the membrane that lines the space between the inner cell mass and the trophoblast.

Germ-layer formation: how does gastrulation set up the three-layered structure?

1. The three germ layers are set up during gastrulation, a process in which the most visible sign is formation of the archenteron, or primitive gut—the space inside the innermost tube—which is lined with endoderm and will become the digestive tract.
2. Gastrulation is driven by the changing properties of individual cells.

How do amphibians gastrulate?

1. Gastrulation in amphibian embryos also depends on coordinated behavior of sheets of cells and interactions of individual cells with extracellular matrices.
2. Amphibian gastrulation is more complex than that of the sea urchin because of yolk.
3. Amphibian cells gastrulate at the embryo's equator, rather than at the vegetal pole.
4. Amphibian gastrulation starts when an indentation (the blastopore) forms on the side of the blastula due to a shape change in bottle cells.
5. Next, adjacent cells follow the bottle cells into the invagination, forming a hollow archenteron lined with endodermal cells from the surface of the blastula.
6. Future mesodermal cells crawl over the extracellular matrix on the roof of the blastocoel.
7. Individual cells arrange themselves into tissues.

How do mammals gastrulate?

1. In reptiles and their descendants—birds and mammals—germ layer formation depends on separation of parallel sheets of cells followed by movement of individual cells.
2. Separation of cells in the inner cell mass produces two layers—the hypoblast and epiblast. The hypoblast forms the blastodisc—like a blastula—from which adult structures develop.
3. Blastodisc cells move toward the central line, forming the primitive streak—the functional equivalent of the blastopore.
4. Individual cells now move through a groove in the streak into the blastocoel, and eventually separate into endoderm and mesoderm.

Organ formation: how does the nervous system form?

1. Organ formation consists of two major processes:
   - morphogenesis—the creation of form; and
   - differentiation—the specialization of cells.
2. In vertebrates, part of the mesoderm forms the notochord, which is a supportive cord running from head to tail that serves as an organizer for further embryonic development.
3. Above the notochord, ectoderm rearranges to form the nervous system:
   - cells along the central line form the neural plate, which curls along the embryo's length;
   - this gives rise to the hollow neural tube—the precursor for the central nervous system.
4. Cells remaining on the surface become skin ectoderm; those that connected the tube to the surface (neural crest cells) migrate away and give rise to other types of cells.
5. The entire process of forming the neural tube and neural crest cells is called neurulation.
6. A neural tube folds to form beginnings of the different regions of the brain and spinal cord. The front enlarges to form the cerebral hemispheres, and eye bulges appear just behind this.
7. Eye development illustrates some generalizations about organ formation:
   - organ formation precedes differentiation—cells must be in place before they specialize;
   - organ formation often involves complex folding of sheets of cells; and
   - organ formation depends on interactions of cells brought together by movements.

Programmed cell death contributes to normal development

1. Development requires extensive "programmed cell death," or apoptosis—the death and removal of cells.
2. Apoptosis occurs in many developmental pathways in both vertebrates and invertebrates.
3. In apoptosis, cells die quietly, are engulfed by phagocytes, and are digested without a trace.
4. One of the best-studied examples in vertebrates is the programmed death of cells between the digits of embryonic limbs in chickens and other nonaquatic vertebrates, so none of them (including humans) have webbed digits like ducks.
5. Apoptosis in vertebrates depends on gene action, and many of the genes that activate the process are identical to genes that regulate the cell cycle.

HOW DOES HUMAN DEVELOPMENT PROCEED AFTER IMPLANTATION?

1. Mammalian fertilization usually occurs in the oviduct. The zygote begins cleavage and heads toward the uterus.
2. After about a week, the trophoblast of the blastocyst implants in the uterine wall.
3. By 9 to 16 days, gastrulation occurs, producing the three germ layers.
4. By 31 days, the placenta is fully functional. It can also transport harmful substances, such as viruses (rubella and HIV) and teratogens—chemicals that can cause abnormal development.
5. Relatively early in development, each organ appears as a recognizable rudiment (initial stage) from which the final form will develop.
6. In humans, all organs are in place by the end of eight weeks, and the embryo is now a fetus.
7. The final form of each organ requires extensive cell specialization and growth.

8. Different parts of the body grow at different rates. Events in humans include:

9 to 12 weeks:	The fetus doubles in length, external genitalia appear and start to develop, and kidneys excrete their first urine.
12 to 16 weeks:	The length doubles again, ovaries are differentiated and primary follicles contain oogonia.
17 to 20 weeks:	The mother feels the first fetal movements.
20 weeks:	The fetus is about 10 inches long, and the testes have begun to descend.
22 weeks:	Lungs, intestines, and kidneys are sufficiently developed that, with intensive medical care, the fetus might survive outside the womb.
24 weeks:	Lungs begin to secrete surfactant, which allows them to fill with air.
26 weeks:	A prematurely born fetus is likely to survive because the respiratory system is developed and functional.
26 to 36 weeks:	The fetus gains weight.
36 weeks:	Birth.

EMBRYONIC CELLS BECOME INCREASINGLY DIFFERENTIATED

1. The fate of a cell is what it becomes; its potency is what it could become if allowed to develop in another environment.

What kinds of experiments distinguish potency and fate?

1. At early stages of development and for relatively simple organisms, we may follow the fates of cells just by watching and photographing them.
2. In more complex organisms, gastrulation is too complicated, so dyes, radioactive tracers, or carbon particles are used to mark the cells to be followed.
3. Transplanting cells into another embryonic environment can help determine their potential.
4. Cells known to form a particular tissue are described as being "presumptive."
5. Spemann transplanted presumptive neural plate cells into an area of presumptive skin cells. When done in the early gastrula stage, the cells formed skin, demonstrating that their potency was greater than their fate. But when done at the late gastrula stage, the cells developed into neural plate, showing that their fate was now greater than their potency.
6. Spemann concluded that, later in development, cells become committed to their normal developmental fate, gradually losing their potency. The cells' fate is said to be determined.

Are differentiated animal cells totipotent?

1. Modern biologists think of determination as the process that establishes which genes will be expressed and which will not—a process that is influenced by environmental signals.
2. In the 1950s, Robert Briggs and Thomas King devised a method for nuclear transplantation and soon discovered that nuclei from a (frog) blastula are totipotent. But they could not get a complete organism from cells at the neurula stage and concluded that, at least in leopard frogs, nuclei lose their potential to direct complete development.
3. Later, researchers showed that nuclei from intestines of the South African clawed toad tadpoles could generate a normal adult.

Can a differentiated nucleus direct development from egg to adult?

1. Through Briggs and King's work, biologists could produce a group of genetically identical animals—because the nuclei of the blastula cells were identical, the resulting leopard frogs were clones of one another.
2. Gardeners have been cloning plants for decades.

3. In 1997, biologists Ian Wilmut and Keith Campbell at the Roslin Institute in Edinburgh reported that they had successfully cloned an adult sheep, named Dolly, by transplanting the nucleus from a cell in the udder of an adult sheep into an egg cell.
4. Governments hastily passed laws banning the cloning of humans.

How did the Scottish researchers clone Dolly?

1. The trick was to make the DNA of the donor cells behave more like the inactive DNA of an egg or sperm. Researchers grew the udder cells in culture and starved them of essential nutrients, gradually putting the cells into a dormant state in which many genes shut down and replication was not possible.
2. Once the DNA in the donor nuclei was dormant, the nuclei were transplanted into egg cells.
3. During the first three divisions of the sheep embryo, cells replicate their DNA without expressing any nuclear genes—all cell processes remain under control of maternal proteins and RNA. During this time, the nuclear DNA begins losing proteins that it brought in from the udder cell, which at first prevented nuclear genes from being expressed.
4. By the third division, the proteins from the udder are replaced by proteins from the egg, and the maternal (egg) proteins then "reprogram" the nuclear DNA and the embryo starts expressing its own genes.

Did Dolly really come from a differentiated cell?

1. Some biologists question if Dolly was really produced by a differentiated cell because mammary (udder) cells are mixed populations that include stem cells—cells that can make more cells like themselves or undergo differentiation to one or more specialized cell types.
2. Adult vertebrates contain many kinds of precursor stem cells, which are less differentiated than other cells.
3. The developmental fate of stem cells depends on cues from their environment.

HOW DOES DEVELOPMENTAL FATE DEPEND ON CHEMICAL SIGNALING?

1. Since virtually every cell in an animal has the same genes, differences in cell fates must be from selective gene expression, which, in turn, is influenced by the cellular environment.

Can the extraordinary actions of the primary organizer in amphibian development be explained in terms of cells and molecules?

1. Hilde Mangold and Spemann correctly guessed that determination of the ectoderm into neural tissue or skin depended on contact with the underlying mesoderm, which derives from cells that invaginated during gastrulation. These cells are the dorsal lip of the blastopore.
2. Mangold transplanted dorsal lip cells to another region and the cells induced surrounding tissues to form an almost complete second embryo.
3. Spemann termed the dorsal lip the primary organizer—it established the entire organization.
4. The primary organizer's mechanism is chemical—chemical extract from the dorsal lip induces ectodermal differentiation.
5. Researchers do not fully understand embryonic induction, but have found various molecules that help establish the development pattern of whole embryos and individual organs.
6. Three important signaling molecules—nogin, chordin, and follistatin—induce ectoderm to become neural tissue by inhibiting another inducing protein—bmp-4—which prevents neural induction.
7. These same signaling molecules can induce different development patterns in mesoderm.
8. The same signaling molecules are used in different ways by different organisms, thus the same molecules can produce different development patterns.

How do cells "know" where they are and what to do about it?

1. Development requires that individual cells express appropriate genes and that cells be organized into functioning tissues and organs. This requires differentiation of the cells into a characteristic pattern.
2. Pattern formation involves interactions between layers of cells.
3. Special mechanisms specify positional information—chemical cues establish the position of a cell in a pattern.
4. Concentrations of specific substances (morphogens) specify each cell's pattern contribution.
5. The formation of patterns during development depends on each cell "knowing" where it is and also "knowing" what it is.
6. In vertebrates, retinoic acid apparently directly influences gene expression by interacting with a transcription factor (the same way in which steroid hormones exert their influence).
7. Which genes are expressed differs from cell to cell.
8. Homeotic selector genes are genes whose expression affects the overall body plan.

Mammals and flies use many of the same transcription factors to establish patterns during development

1. HOX genes are mammalian counterparts to genes that regulate development in *Drosophila*, and the two sets of genes are remarkably similar in sequences.
2. The biggest surprise in studying HOX genes has been that corresponding genes in flies and mice are arranged in the animals' chromosomes in the same order, and this order matches the pattern of development in the embryo—genes that affect head development lie at one end of the cluster, while those that affect tail development lie at the other end.
3. The most recent common ancestor of *Drosophila* and mice lived about 600 million years ago so, for at least part of the time, flies and mice have used, with little modification, a system of specifying positional information that evolved during or before the Cambrian explosion.

POST-EMBRYONIC DEVELOPMENT

Metamorphosis converts a larva into an adult

1. A larva, the first stage with an independent life, may not resemble the sexually mature adult.
2. Larvae may be so different from the adult form that they appear to be different species.
3. Metamorphosis is more common among invertebrates than vertebrates.
4. Many insects undergo complete metamorphosis—no tissues of the adult are from the larva. Instead, the adult derives from imaginal discs—larval cells set aside for later development.
5. In insects that undergo complete metamorphosis, the larval stage is specialized for feeding and growth, whereas the adult stage is specialized for dispersal and mating.
6. In complete metamorphosis, the wormlike larva may go through several stages of increasing size before entering an inactive phase, called a pupa, which may be enclosed in a cocoon.
7. During the pupal stage:
   - almost all larval tissues are digested;
   - cells of the imaginal discs multiply, move about, and differentiate to form the adult tissues and organs; and
   - the pupa digests away its cocoon and emerges as a sexually mature adult.
8. In gradual metamorphosis, the embryo hatches into an immature form, called a nymph, which is essentially a miniature adult. The nymph undergoes a series of molts, and with each gets bigger and more like the adult.
9. Some scientists consider puberty a variation on the theme of metamorphosis—adolescents undergo a growth spurt and develop various sexual characteristics that enable them to reproduce; these changes, like metamorphosis in insects, are controlled by hormones.

Aging depends on the action of both genes and environment

1. Multicellular organisms lose their ability to reproduce with age, whereas bacteria and unicellular organisms cannot be said to age.
2. From the point of view of natural selection, organisms that can no longer reproduce are dispensable and, for many (including humans), the chance of death increases with time.
3. Many measures of vitality decrease with age, including fertility, cardiac output, brain weight, lung capacity, and muscle mass.
4. Aging connective tissue leads to hardening of the arteries, less elastic skin, and stiff joints.
5. Cells lose their capacity for proliferation.
6. Part of the aging process must depend on genes, but environmental factors, such as disease, predators, parasites, or inadequate nutrition, also contribute to aging.
7. Genetic differences in aging may be from different susceptibilities to environmental factors.

STEM CELLS: HOW CAN SOME CELLS RETAIN THE CAPACITY FOR REGENERATION?

1. Biologists can now grow embryonic stem cells in artificial culture, and these cells can divide to form more stem cells, or differentiate to produce different types of cells.
2. Transgenic mice, knockout mice, and knockin mice are formed by putting specific genes in stem cells then introducing them back into embryo.
3. In late 1998, biologists from the University of Wisconsin reported isolating human ES cells that not only proliferate, but can differentiate into various cell types—cartilage, bone, muscle, and skin, for example.
4. Due to a U.S. ban on support for research on human fetal and embryonic tissue, this work has been supported by private funds.
5. This discovery is viewed by many as a major landmark—a starting point for a revolution in replacement medicine. Other people—biologists and lay people—worried that continued human fetal research would lead to encouragement of abortions. As a result, the U.S. National Institutes of Health (NIH) established stringent guidelines: research can continue on differentiation of ES cells, but new ES cells cannot be harvested.
6. In 1999, more than a dozen labs reported that adult tissue also contains stem cells which can be used for this type of research, avoiding many of the moral concerns.

AND SO WE SAY GOODBYE?

1. Each of our cells and each of our nuclei are the product of not only the genes it carries, but also of the environment in which it lives and develops.
2. Environment of a nucleus partly determines which genes are expressed, and thus phenotype.
3. Similarly, a cell's environment helps determine its fate.
4. Environment also works at the level of determining the phenotype of the whole organism.
5. Finally, your authors conclude with the reminder, "Asking questions and seeking answers is one of the foremost ways of engaging with life."

VOCABULARY BUILDING

In your own words, first write a brief definition, then a full explanation for each of the following terms. Include examples where appropriate. Complete this section from your memory—you will not learn the terms by simply copying definitions from the textbook. Once you have finished, check your responses against the information in the chapter and make any necessary corrections.

preformation —

homunculus —

maternal mRNA —

ectoderm —

endoderm —

mesoderm —

gamete formation —

fertilization —

cleavage —

blastula —

morula —

germ layer formation —

gastrula —

gastrulation —

organ formation —

growth —

metamorphosis —

aging —

blastocoel —

animal pole —

vegetal pole —

trophoblast —

blastocyst —

inner cell mass —

placenta —

archenteron —

blastopore —

hypoblast —

epiblast —

primitive streak —

morphogenesis —

differentiation —

notochord —

neural plate —

neural tube —

neural crest cells —

neurulation —

apoptosis —

teratogens —

rudiment —

trimesters —

miscarriage —

abortion —

fate —

potency —

determined —

stem cells —

primary organizer —

morphogens —

homeotic selector genes —

HOX genes —

metamorphoses —

pluteus —

complete metamorphosis —

imaginal discs —

pupa —

gradual metamorphosis —

nymph —

embryonic stem (ES) cells —

CHAPTER TEST

The following test has four parts. Complete as much of the exam as you can from memory. If you cannot answer a question, skip it. Once you complete all that you can, try to answer any questions you skipped. If you still cannot answer them, consult your textbook for the answers. Once you have completed all sections of the test, check your answers for Parts 1 - 3 against those in the back of this book. Highlight any incorrect answers then review that material in your textbook. Correct your answers for future reference.

Part 1: Multiple Choice

For each of the following, select all correct responses—more than one may be correct.

1. The early development of an embryo is controlled by:
   a) the zygote's proteins
   b) the zygote's genes
   c) the father's genes
   d) the mother's proteins

2. Muscles form from:
   a) ectoderm
   b) mesoderm
   c) endoderm
   d) all of these

3. Gastrulation results in:
   a) a diploid zygote
   b) formation of three germ layers
   c) metamorphosis
   d) a morula

4. Which of the following is a solid ball of cells?
   a) gastrula
   b) blastula
   c) morula
   d) zygote

5. The central nervous system develops from the:
   a) neural tube
   b) neural crest cells
   c) notochord
   d) optic vesicles

6. Which of the following is true about a human fetus?
   a) It has all its organs in place.
   b) It has all its organs developed.
   c) The embryo becomes a fetus around nine weeks.
   d) All of these are true.

7. After what age is a fetus likely to survive if born prematurely?
   a) 9 weeks
   b) 16 weeks
   c) 22 weeks
   d) 26 weeks

8. The programmed death and removal of cells is called:
   a) gastrulation
   b) cleavage
   c) apoptosis
   d) metamorphosis

9. Stem cells are capable of which of the following?
   a) differentiation to more specialized cell types
   b) dividing to provide more cells of the same type
   c) fertilization
   d) all of these

10. Which of the following can affect embryonic development in humans?
   a) HOX genes
   b) morphogens
   c) direct contact between cells
   d) all of these

11. The inactive state of a metamorphosing insect is a:
   a) cocoon
   b) pupa
   c) larvae
   d) pluteus

12. Which of the following is associated with gradual metamorphosis?
    a) pupa
    b) larvae
    c) nymph
    d) imaginal discs

13. Which of the following is not true of insects that undergo complete metamorphosis?
    a) The immature form is more or less a miniature of the adult form.
    b) Imaginal discs do nothing in the larva.
    c) The pupa may be enclosed in a cocoon.
    d) The larval and adult forms may appear to be two unrelated species.

14. The greatest significance of Dolly is that she is the first:
    a) animal to be cloned
    b) adult animal to be cloned
    c) organism of any type to be cloned
    d) animal raised from a transplanted embryo

15. As cells become more differentiated, they tend to lose some of their:
    a) specialization
    b) fate
    c) potency
    d) all of these

Part 2: Matching

For each of the following, match the correct term with its definition or example. More than one answer may be appropriate.

ectoderm **endoderm** **mesoderm**

1. ____________ Middle of the germ layers.
2. ____________ Forms during gastrulation.
3. ____________ Digestive system develops from this layer.
4. ____________ The heart develops from this layer.
5. ____________ The nervous system develops from this.

cleavage **growth** **gastrulation**

6. ____________ The organism increases in size.
7. ____________ The germ layers form.
8. ____________ Cell division occurs to produce smaller cells.
9. ____________ Involves movement of embryonic cells.
10. ____________ This immediately follows fertilization.

Part 3: Short Answer

Write your answers in the space provided or on a separate piece of paper.

1. What are two concepts that led to the demise of the idea of preformation?

2. List, in order, the eight stages of vertebrate development.

3. What are two problems related to the size of a zygote?

4. What two cell types are present in the mammalian blastocyst, and what happens to each?

5. Distinguish between morphogenesis and differentiation.

6. What are two fates of neural crest cells?

7. List three generalizations about organ formation.

8. How does apoptosis differ from ordinary cell death?

9. Differentiate between fate and potency.

10. Outline the procedure used to produce Dolly, the cloned sheep.

11. Why has the Dolly research not necessarily proven that differentiated cells are totipotent?

12. What is the general function of homeotic selector genes and HOX genes?

13. Explain the role of imaginal discs.

14. In what ways do scientists view puberty as a type of metamorphosis?

15. According to natural selection, when in our lives do humans become dispensable?

Part 4: Critical Thinking: Using Your Knowledge

Answer each of these in essay form, using complete sentences and paragraphs. Provide as much information as you can. (For extra essay practice, write out answers to the Review and Thought Questions in your textbook.)

1. Discuss gastrulation as it occurs in sea urchins, amphibians, and mammals.

2. Discuss various chemical signals that help determine developmental fate.

3. The cloning of Dolly caused quite an uproar around the world, but many were shocked at the attention because cloning has been used for years. What are some examples from both plants and animals that illustrate cloning?

4. What are some possible advantages to being able to clone adult animals? Should this research be pursued?

5. In the cloning of Dolly, use of the udder cell has become controversial, but nobody questions the use of the egg cell. Why was an egg cell used instead of two udder cells? Why was an egg cell used instead of a sperm cell?

6. In terms of miscarriage and severe defects, teratogens are a much bigger threat in the first trimester of pregnancy than they are in the later trimesters. Why?

7. When an organism is trying to grow and develop, why is apoptosis—cell death and removal—necessary?

8. What are some potential benefits and applications from research on embryonic stem cells?

9. How do you respond to the argument that research on embryonic stem cells promotes abortion? What kinds of restrictions might be placed on this research, allowing more to happen yet ensuring it does not promote abortion?

10. The President of the United States has selected you to head the new National Bioethics Committee, charged with the tasks of deciding:
    - what new research, such as ES cell and cloning, will be allowed or banned,
    - what restrictions will be placed on these projects, and
    - what the penalty will be for ignoring the restrictions.

    Who would you select to be on your committee, and why? For the two examples given, what would your decisions be, and why?

Answers to Chapter Tests

Chapter 1—The Unity and Diversity of Life

Part 1: Multiple Choice

1. a
2. c
3. d
4. a
5. a
6. b
7. a
8. b
9. b
10. d
11. d
12. b
13. c
14. b
15. b

Part 2: Matching

1. theory
2. hypothesis
3. model
4. negative feedback
5. Homeostasis, adaptation
6. adaptation
7. homeostasis
8. multicellular
9. eukaryotic
10. prokaryotic
11. eukaryotic, multicellular
12. prokaryotic

Part 3: Short Answer

1. a) Focus on a single question or a small set of questions.
   b) Observe.
   c) Develop a model.
   d) Formulate a testable hypothesis.
   e) Design and conduct experiments.
2. Using an adequate control is the only way to be sure the experimental results are due exclusively to the variable you are examining.
3. If experimental results are not as the hypothesis predicts and all other potential errors are ruled out, then the hypothesis is likely incorrect. But even if all experiments conducted support a hypothesis, another experiment not yet conceived might disprove it. It would be impossible to test a hypothesis under every possible circumstance.
4. Good experiments are conducted under carefully controlled conditions, the outcome must depend on only the proposed cause, controls must be included to rule out effects from other variables, statistical analysis must be used to evaluate the results and rule out a chance occurrence, and results should be repeatable.
5. The hypothesis must allow experimentation that can try to disprove it. If it is not testable, it is nothing more than a guess and cannot be scientifically supported.
6. Archaea, Eubacteria, and Eukarya
7. Organisms:
   1. are organized into parts,
   2. perform chemical reactions,
   3. obtain energy from their surroundings,
   4. change with time,
   5. respond to their environments,
   6. reproduce, and
   7. share a common evolutionary history.
8. Eukaryotic organisms are typically more complex and include membrane-bound nuclei and organelles. Prokaryotic cells are usually smaller and simpler, with no membrane-bound organelles. [Examples will vary.]
9. Individual variations in response to an experiment can influence the results. Larger groups make it easier to recognize and rule out effects from individual variations.
10. Statistical analysis allows the investigator to recognize results that occur simply by chance, results that are not really different than the control, and perhaps new lines of inquiry.

11. The mean is an easy value to use for comparisons between different populations. The standard deviation is then used to evaluate how much the data vary from the mean. The standard deviation is only valuable, though, if the data show a normal distribution. All three of these are used together to assess the validity of the data and conclusions drawn from the data.
12. Although two organisms may appear quite different, at the cellular and molecular level they are quite similar, sharing many of the same structures, molecules (such as DNA), and chemical processes.
13. Natural selection is the greater survival and reproduction of individuals with certain inherited traits that less successful individuals lack.
14. Although individuals are unique, all their genes come from the previous generation—half from each parent in organisms that undergo sexual reproduction, and all from one parent in those that undergo asexual reproduction.
15. Homeostasis is maintenance of a relatively constant internal environment. Negative feedback refers to the mechanism used to maintain homeostasis in which a deviation from homeostasis triggers a response that corrects the deviation, then the response ends once the norm is reestablished.

Chapter 2—The Chemical Foundations of Life

Part 1: Multiple Choice

1. c
2. a
3. d
4. a
5. c
6. b
7. c
8. b
9. a
10. c
11. a, b, c, and d
12. c
13. c
14. d
15. a

Part 2: Matching

1. element
2. ion
3. atoms
4. molecules
5. ions
6. compounds
7. isotopes
8. van der Waals interactions
9. hydrophobic interactions
10. ionic bond
11. hydrogen bond
12. covalent bond
13. ionic bond
14. polar molecules
15. polar molecules
16. nonpolar molecules
17. nonpolar molecules

Part 3: Short Answer

1. Elements are composed of atoms.
2. The six major elements in living organisms are carbon, hydrogen, oxygen, nitrogen, phosphorus, sulfur.
3. The molecular weight of glucose is 180 (carbon: 6 x 12 = 72; hydrogen: 12 x 1 = 12; oxygen: 6 x 16 = 96; 72 + 12 + 96 = 180).
4. Radiation treatment can cause nausea, vomiting, diarrhea, hair loss, anemia, and cancer.
5. Sodium has 1 electron in its outer shell that can hold 8; chlorine has 7 electrons in its outer shell that can hold 8. Sodium loses an electron, emptying its outer shell so the next shell in is full. Chlorine gains that electron, filling its outer shell. They are now ions: Na^+ and Cl^-. Their electrical charges hold them together.
6. The electrons shared by the hydrogen and oxygen spend more time near the oxygen. The uneven charge distribution results in polarity—the hydrogen end is slightly positive and the oxygen end is slightly negative.
7. Hydrogen bonds—the polar parts of the molecules form weak attractions with oppositely charged areas on neighboring polar molecules. Hydrophobic interactions—hydrophobic molecules tend to aggregate together in water to avoid water contact. In van der Waals interactions, momentary polarity in one molecule induces polarity in a neighboring molecule.
8. Amphipathic compounds have both hydrophobic and hydrophilic regions. Hydrophilic regions interact with and dissolve in water, but the hydrophobic regions do not.
9. Water has a high density, surface tension, heat capacity, cohesion, and adhesion. It is a powerful solvent, and it forms the basis of pH.

10. Changing heart rate, rate and depth of respiration, and secretion of hydrogen ions by the kidneys all help maintain pH.
11. Daltons are the units of measure for atomic mass.
12. The inner shell can hold only two electrons.
13. Hydrophobic molecules do not interact well with water.
14. Solutions with a high hydrogen ion concentration have a low pH value.
15. An element is a pure substance that cannot be converted to a simpler one. The smallest units of an element are atoms. A stable combination of two or more atoms of an element is a molecule. A compound is a stable assembly of two or more atoms of different elements. A mixture is a combination of compounds that do not affect the properties of each other.

Chapter 3—Biological Molecules Greaat and Small

Part 1: Multiple Choice

1. a
2. d
3. b
4. a and c
5. c
6. a
7. c
8. a
9. d
10. c
11. b
12. b
13. a
14. d
15. a

Part 2: Matching

1. carbohydrates
2. proteins
3. nucleic acids
4. lipids
5. carbohydrates
6. proteins
7. nucleic acids
8. nucleotides and amino acids
9. fatty acids
10. Nucleotides
11. amino acids
12. Sugars
13. nucleotides
14. hydrolysis
15. dehydration
16. both
17. dehydration
18. dehydration

Part 3: Short Answer

1. Proteins, lipids, carbohydrates, and nucleic acids. [Examples will vary.]
2. Amino acids have a carboxyl group, an amino group, and an R group (or side chain) all attached to the α carbon. Variation occurs in the R group.
3. Nucleotides have a sugar (ribose or deoxyribose), one or more phosphate groups, and one of five nitrogenous bases.
4. Phospholipids are made of a glycerol molecule attached to two fatty acids and one phosphate group. They are amphipathic, so their hydrophobic tails avoid water, but their hydrophilic heads interact well with water.
5. Most carbohydrates have hydroxyl groups attached to most of their carbons, and these readily form hydrogen bonds with water molecules.
6. In this process, the equivalent of a molecule of water is removed from the ends of two monomers, and they then join at that spot, forming a larger molecule.
7. It is a carbohydrate; most organisms lack enzymes to break it down.
8. (Answers will vary.)
9. A change in an amino acid's side chain, which is the main determinant of chemical properties, is most likely to have an impact on protein function.
10. The functional groups a molecule contains are most responsible for organic compounds' diverse properties.
11. In hydrolysis, macromolecules are broken apart by adding water, breaking the bonds holding monomers together.
12. [Answers will vary.]
13. Starch is a carbohydrate.
14. Hydroxyl groups promote formation of hydrogen bonds.
15. Fatty acids are amphipathic, so they contain both hydrophobic and hydrophilic regions.

Chapter 4—Why Are All Organisms Made Of Cells?

Part 1: Multiple Choice

1. b
2. a
3. c
4. a, b, and d
5. b
6. c
7. a and c
8. c
9. a, b, and c
10. b
11. c
12. a
13. a
14. c
15. a

Part 2: Matching

1. mitochondrion
2. ribosome
3. ribosome
4. smooth ER
5. Golgi complex
6. rough ER
7. smooth ER
8. pinocytosis
9. phagocytosis
10. exocytosis
11. pinocytosis

Part 3: Short Answer

1. The membrane is mostly composed of phospholipids, which are amphipathic. They naturally arrange with their hydrophobic portions directed inward and their hydrophilic parts aimed outward. Two layers allow the hydrophobic regions to stay away from the watery inside and outside environments.
2. All organisms are made of cells, all cells are alike and are the basic living units of organization, and all cells come from other cells.
3. Rough ER has ribosomes attached to the membrane, giving it a studded appearance.
4. Lysosomes
5. Photosynthesis occurs in chloroplasts. It is the process by which the energy from sunlight is harnessed and converted to chemical energy stored in sugars.
6. The cytoskeleton gives support to the cell, holds organelles in place, directs cell division, produces cell movement, maintains and changes cell shape. [Answers will vary.]
7. Because of the chemical nature of the lipid bilayer, the membrane allows some substances to pass through it easily while others cannot penetrate through.
8. The Golgi complex modifies glycoproteins, packages them in membranes, then alters the carbohydrate to label the glycoprotein, ensuring that it will be transported to the correct place inside the cell or sent to the edge of the cell to be expelled.
9. Phagocytosis is when the plasma membrane extends outward to envelop a rather large amount of material. Pinocytosis is when the membrane invaginates and forms a vesicle around a bit of fluid. Receptor-mediated endocytosis is like pinocytosis except the substance being moved must bind to a specific receptor protein on the membrane before the membrane will take it in.
10. Chromosomes
11. Nuclear pores
12. A nucleus is a membrane-bound organelle in eukaryotes only. Prokaryotes do not have internal membranes, so their DNA is located in the nucleoid, which is a special region in the cytoplasm.
13. [Answers will vary.]
14. The cytoskeleton is composed of microtubules, actin filaments, and intermediate filaments.
15. ATP is produced in the mitochondria

Chapter 5—Directions and Rates of Biochemical Processes

Part 1: Multiple Choice

1. b
2. a
3. c
4. d
5. a
6. b
7. c
8. a
9. b
10. a
11. d
12. c
13. b
14. c
15. c

Part 2: Matching

1. kinetic energy
2. free energy
3. potential energy
4. free energy
5. kinetic energy
6. potential energy
7. endergonic reactions
8. exergonic reactions
9. endergonic reactions
10. endergonic reactions
11. endergonic reaction
12. endergonic reactions

Part 3: Short Answer

1. The calorie is the most common unit of measurement for energy.
2. Heat is a wasted form of energy that is usually not able to do work. As a process occurs, some of the energy is converted to heat, which is then no longer available to do work. Thus, as the Second Law states, the amount of energy available to do work decreases.
3. Heat is a type of kinetic energy.
4. $\Delta G = G_{products} - G_{reactants}$
5. Laws of thermodynamics allow us to predict the direction in which a reaction will occur.
6. Many biochemical reactions are endergonic, so they are coupled to exergonic reactions that release enough energy to allow the endergonic reactions to occur.
7. Catalysts lower the activation energy, allowing a reaction to occur more easily and more quickly.
8. Processes that convert an orderly arrangement to a less orderly arrangement release energy and thus can perform work.
9. Chemical equilibrium occurs when $\Delta G = 0$, meaning the free energy of the reactants equals that of the products. Although the reaction will continue, there is no net change once equilibrium is established.
10. The substrate attaches to the active site of the enzyme by weak noncovalent bonds—hydrogen bonds, hydrophobic interactions, ionic bonds, and van der Waals attractions.
11. Both temperature and pH value affect the rate of a chemical reaction.
12. Enzyme and substrate molecules interact one molecule at a time. The more of each that is present, the more interactions that can occur at one time. If either is present in a low concentration, the rate of the reaction will decrease.
13. Enzymes bring substrates together in positions that favor a reaction. Binding of a substrate to an enzyme may create a transition state in the substrate that has higher free energy, which is then used to drive the reaction. Enzymes may contribute to the reaction, perhaps lending or temporarily accepting an atom.
14. Most human enzymes perform best at pH values between 6 and 8.
15. A competitive inhibitor is one that binds to the active site of an enzyme, thus competing with the substrate molecules.

Chapter 6—How Do Organisms Supply Themselves With Energy?

Part 1: Multiple Choice

1. b and d
2. a
3. c
4. b
5. c
6. a
7. a
8. a
9. c
10. b
11. b
12. d
13. d
14. a
15. B

Part 2: Matching

1. anabolism
2. catabolism
3. catabolism
4. catabolism
5. anabolism
6. anabolism
7. fermentation
8. glycolysis and fermentation
9. citric acid cycle and oxidative phosphorylation
10. citric acid cycle
11. glycolysis
12. oxidative phosphorylation
13. citric acid cycle
14. glycolysis and fermentation

Part 3: Short Answer

1. a) sodium
   b) chloride
   c) chloride
   d) sodium
2. Autotrophs are organisms that can harness energy from their environments and use it to make the organic molecules they need for energy metabolism. Photosynthetic organisms are autotrophs. Heterotrophs must obtain their organic molecules from other organisms.
3. Vitalism is the belief that living systems have powers or qualities beyond those of non-living systems.
4. $C_6H_{12}O_6 + 6O_2 \longrightarrow 6CO_2 + 6H_2O$
   glucose + oxygen $\longrightarrow$ carbon dioxide + water
5. Anaerobic organisms can live without oxygen; aerobic organisms require oxygen to stay alive.
6. a) amino acids
   b) monosaccharides
   c) fatty acids
7. The energy is stored in the chemical bonds that hold the organic molecules together, for example in the carbon-to-carbon bonds in glucose.
8. Electron transport occurs within the mitochondria.
9. NAD^+ and FAD
10. With proton pumping, electron transport sets up an electrical gradient and a chemical or pH gradient.
11. 2 ATP by glycolysis alone
12. A facultative anaerobe can live with or without oxygen. Obligate anaerobes can live only in the absence of oxygen.
13. More ATP is not necessary, so the acetyl-CoA tends to be converted into fats, for energy storage.
14. During anabolic reactions, small organic building blocks are chemically joined to make macromolecules. [Examples will vary.]
15. Fat is mostly nonpolar and hydrophobic, so it binds little water. Carbohydrates may bind up to twice their weight in water.

Chapter 7—Photosynthesis: How Do Organisms Get Energy From the Sun?

Part 1: Multiple Choice

1. b
2. b
3. d
4. d
5. b
6. c
7. b
8. d
9. b
10. a
11. c
12. a
13. a
14. d
15. c

Part 2: Matching

1. photosystem II
2. photosystem I
3. photosystem I
4. photosystem II
5. carbon dioxide and water
6. water
7. carbon dioxide
8. oxygen and water
9. C_4 plants
10. C_3 plants
11. C_4 plants
12. C_4 plants

Part 3: Short Answer

1. Plants must take in water, carbon dioxide, and light.
2. $6CO_2 + 6H_2O \longrightarrow C_6H_{12}O_6 + 6O_2$
   carbon dioxide + water ⟶ glucose + oxygen
3. Like a particle, it travels in straight lines and in particle-like bundles called photons. Like a wave, it has a wavelength between its troughs and crests.
4. Chlorophyll appears green because it absorbs the other colors of visible light—red and blue.
5. Photosystem II best absorbs light with wavelengths of approximately 680 nm.
6. Oxygen is generated by photosystem II, which is part of the light-dependent reactions.
7. The energy stored in these chemical bonds is used to drive the reactions of the Calvin cycle, which synthesize glucose.
8. Sucrose, which stays in solution, can move from cell to cell.
9. The last step of the Calvin cycle regenerates the initial carbon dioxide acceptor—ribulose biphosphate.
10. The total cost for making one molecule of glucose is 18 molecules of ATP and 12 molecules of NADPH, or the energy from 48 photons.
11. Environmental factors that can affect productivity include light intensity, day length, angle of the sunlight, pollution, fog, and overhead shade, such as from trees in a forest.
12. The excited P_{680} from photosystem II can act as an electron donor to replace the electron lost by P_{700} from photosystem I during the noncyclic route of photophosphorylation.
13. We need the oxygen plants release during photosynthesis. Like other heterotrophs, we cannot harness energy from sunlight to make ATP, so we rely on consuming the carbohydrates made by plants during photosynthesis and, in turn, use the energy stored in the chemical bonds to make ATP.
14. Plants produce large amounts of Rubisco.
15. C_3 plants are best adapted for temperate zones.

Chapter 8—Cell Reproduction

Part 1: Multiple Choice

1. b
2. c
3. b
4. d
5. b
6. c
7. d
8. b
9. b
10. a
11. b
12. c
13. a
14. d
15. b

Part 2: Matching

1. G_1, S, and G_2
2. M
3. M
4. G_1
5. S
6. M
7. G_1
8. G_2
9. interphase
10. telophase
11. anaphase
12. telophase
13. anaphase and telophase
14. metaphase
15. prophase

Part 3: Short Answer

1. Single-celled organisms undergo cell division to reproduce—to create new organisms.
2. Mitosis is division of the DNA and nucleus; cytokinesis is division of the rest of the cell (cytoplasm).
3. The materials for cell division are small molecules that provide energy or act as building blocks. The machinery is macromolecular structures that carry out cellular processes. The memory is the DNA.
4. The memory (DNA) comes from both parents, but the materials and machinery are solely from the mother.
5. Binary fission is the relatively simple process by which prokaryotic cells divide by pinching in two.

6. The major stages of the cell cycle are mitosis, cytokinesis, and interphase.
7. Most work for reproduction is done during the G_1 phase of interphase.
8. The phases of mitosis, in order, are prophase, metaphase, anaphase, and telophase.
9. Cytokinesis overlaps mitosis, usually beginning in anaphase and ending after telophase. Thus, it occurs during the M phase of the cell cycle.
10. The mitotic apparatus disappears during telophase.
11. Histones are small, specialized, positively charged proteins that associate with DNA and help it fold so that it will fit in the nucleus.
12. The contractile ring is composed of a bundle of actin filaments.
13. Cell senescence is a limit on the number of times that a cell can divide.
14. When cells contact their neighbors, their cell division is inhibited to prevent overgrowth. When the contact ends, meaning there is room for the cells to spread, their division will resume.
15. Start is the point of no return in cell division. It occurs in G_1. Once it is reached, the cell will complete its division. If Start is not reached, cell division will not occur.

Chapter 9—From Meiosis to Mendel

Part 1: Multiple Choice

1. c
2. a and c
3. a
4. b
5. b
6. a
7. a
8. a and d
9. c
10. b
11. c
12. b
13. a
14. c
15. c and d

Part 2: Matching

1. mitosis
2. meiosis I and II
3. meiosis I
4. mitosis
5. meiosis I and II
6. meiosis II
7. mitosis and meiosis II
8. mitosis and meiosis I
9. sexual reproduction
10. asexual reproduction
11. asexual reproduction
12. sexual reproduction
13. sexual reproduction
14. sexual reproduction

Part 3: Short Answer

1. By the blending model, a population will become more homogeneous in time, eventually showing no variation. Natural selection works on variation, so if blending occurred, natural selection could not.
2. Genotype is the actual genetic makeup of an individual or cell. Phenotype encompasses all the characteristics of that individual. Genotype determines what phenotypes are possible, and the interaction between genotype and environment determines what phenotype is expressed.
3. A gene is a region of DNA that either carries instructions for making a certain protein or regulates the expression of those structural genes.
4. In humans, the gametes—sperm and egg—are haploid.
5. Meiosis produces haploid gametes which then unite during fertilization to produce a diploid zygote. The offspring gets half of the chromosomes from each parent.
6. During synapsis, which occurs in prophase I, homologous chromosomes, each composed of two sister chromatids, align with each other to form a four-chromatid tetrad.
7. In anaphase I, sister chromatids move to the same pole. In anaphase II or in mitosis, the sister chromatids separate and go to opposite poles.
8. In nondisjunction, either the homologous chromosomes (anaphase I) or the sister chromatids (anaphase II) fail to separate correctly. Because they stay together, some of the gametes produced from that meiotic division will have too many chromosomes, and some will have too few. Zygotes produced from these gametes will inherit an incorrect number of chromosomes.
9. Humans have a total of 46 chromosomes—44 autosomes (22 pairs) and 2 sex chromosomes (1 pair).
10. For a recessive trait to be expressed, the individual can have no dominant alleles, so the organism must be homozygous recessive.

11. Females have two X chromosomes, so they will likely carry a normal dominant allele that will prevent expression of the recessive allele. Males have only 1 X chromosome, so they cannot also carry a normal allele to block expression.
12. Egg development is suspended in prophase I before a female is born. The older a woman is at fertilization, the older the egg is and the longer meiosis has been "on hold." Nondisjunction is more likely to occur.
13. Parents' genotype is Ff.
14.

	F	f
F	FF	Ff
f	Ff	ff

15. One-fourth of their children would likely be plastic.

Chapter 10—The Structure, Replication, and Repair of DNA

Part 1: Multiple Choice

1. d
2. b
3. a
4. b
5. c
6. a and c
7. b
8. d
9. a
10. b
11. d
12. c
13. a and d
14. b
15. a

Part 2: Matching

1. DNA ligase
2. DNA polymerase
3. DNA polymerase and DNA ligase
4. DNA polymerase and DNA ligase
5. point mutation
6. point mutation
7. chromosomal mutation
8. chromosomal mutation
9. chromosomal mutation
10. point mutation

Part 3: Short Answer

1. In semiconservative replication, each resulting DNA molecule consists of one original strand and one new strand.
2. Each strand of DNA is arranged in an opposite manner. The 3' end of one is at the 5' end of the other, so the two strands run in opposite directions. The bases are matched so that a certain base on one strand is always aligned with one specific base on the other strand (A with T, C with G) so that each strand contains the same information but in different versions. Thus, each strand can direct the synthesis of the other strand.
3. A virus lacks organelles that carry out cell functions—it is merely a molecular assemblage of protein surrounding nucleic acids.
4. The four bases are adenine, cytosine, guanine, and thymine.
5. Chargaff's rules showed that the proportions of the bases in the DNA varies widely between species, but the proportion of A plus T to C plus G is constant within a species.
6. The bases form the "rungs" on the DNA ladder. They always align so that a purine and pyrimidine are together, giving a constant width to the molecule.
7. Franklin was socially isolated, whereas Watson and Crick were in constant communication with many great scientists with whom they could discuss their work and the problems they encountered, and gain new insights that helped them solve the structure of DNA.
8. Nucleotides are always added to the 3' end.
9. All genetic differences ultimately come from mutations.
10. Mutagens include chemicals, such as alcohol and dioxin, radiation, and ultraviolet light.
11. Mutation hot spots are DNA sequences that have higher mutation rates than other sequences.

12. A nucleotide consists of a sugar, a phosphate, and a base (either a purine or pyrimidine).
13. The sugar and phosphate groups are linked together alternately, forming a sugar-phosphate backbone, from which the bases are hung. Two such polynucleotide strands are aligned, running in opposite directions, with the bases aimed inward. The two strands twist around each other in a double helix.
14. Point mutations include:
    a) base substitutions—one base is replaced by another,
    b) insertions—one or more nucleotides are added, and
    c) deletions—one or more nucleotides are removed
15. Special enzymes recognize common errors and cut out the bad section of DNA. DNA polymerase then rebuilds that section and DNA ligase seals it in place.

Chapter 11—How Are Genes Expressed?

Part 1: Multiple Choice

1. b
2. d
3. c
4. b
5. b
6. a
7. b
8. c
9. a
10. b
11. d
12. a
13. c
14. b
15. c

Part 2: Matching

1. transcription
2. transcription
3. translation
4. translation
5. translation
6. translation
7. DNA
8. mRNA
9. tRNA
10. rRNA
11. mRNA
12. mRNA and tRNA

Part 3: Short Answer

1. There are only 20 amino acids, but 61 codons code for amino acids, so more than one codon can specify the same amino acid.
2. Nonsense, or stop, codons are UAA, UAG, and UGA.
3. The central dogma says that DNA specifies RNA, which specifies proteins (polypeptides), and the information can only flow in that direction.
4. Transcription is when the information in the DNA is transmitted to mRNA.
5. Introns, which are intervening sequences, do not code for amino acids so they are removed, leaving only the exons, which do code for amino acids.
6. RNA polymerase conducts transcription.
7. The reading frame is the grouping of nucleotide triplets—each specifying one amino acid.
8. The large subunit aligns tRNA molecules so peptide bonds can form between adjacent amino acids.
9. Initiation of translation begins when the mRNA is attached to the small ribosomal subunit.
10. The start site is always the codon AUG, which specifies methionine.
11. The three steps of elongation are:
    a) putting the next amino acid in position,
    b) forming a peptide bond between adjacent amino acids, and
    c) moving the ribosome to the next codon.
12. Amino acids are added at the A site, and the growing polypeptide chain is anchored at the P site.
13. Bacteria regulate gene expression so they can respond to changes in their environment and conserve energy.
14. The sense strand contains so many nonsense codons that it would produce only very small chains of amino acids, not the desired functional polypeptide.
15. A conditional mutation is one in which a missense mutation results in a protein that is functional under some conditions but not under others.

Chapter 12—Jumping Genes and Other Unconventional Genetic Systems

Part 1: Multiple Choice

1. b
2. a
3. c
4. a and d
5. b
6. c
7. b
8. a
9. c
10. b
11. d
12. c
13. b
14. d
15. a

Part 2: Matching

1. virus
2. plasmid
3. virus
4. transposon
5. transposon
6. plasmid
7. provirus
8. provirus
9. lytic cycle
10. lytic cycle

Part 3: Short Answer

1. Genes in unconventional genetic systems are either not in the regular genome of the cell nucleus, or they are part of the regular genome but they move around in it.
2. Plasmids and viruses that can replicate independently of their host's DNA contain origins of replication.
3. In autonomous replication, the genes can replicate independently of the host's DNA; in integrated replication, the genes can only replicate when they are integrated into the host's DNA, so they replicate only when it does.
4. Mobile genes include plasmids, viruses, and transposons.
5. A lytic virus is one that causes its host cell to rupture in order to release newly made viruses.
6. An episome is a virus or plasmid that can alternate between autonomous and integrated replication depending on the circumstances.
7. A complex transposon is a stretch of DNA that has two simple transposons located near each other and they move together, carrying along with them any of the host's DNA that is between them.
8. Lytic viruses force the host cell to rapidly produce more viruses, then the cell is destroyed so the viruses are released. Tumor viruses multiply only when their hosts do, and their proteins disrupt the cell's normal control of cell division, resulting in tumor growth.
9. The lysogenic cycle includes the provirus and the lytic cycle.
10. Viral oncogenes likely evolved from proto-oncogenes, which are a cell's normal genetic control of cell division. Biologists believe oncogenes simply provide an overdose of normal cellular proteins that stimulate cell division, speeding up the process.
11. First, most viruses do not cause disease. But of those that do, many can exist in the provirus form for extended periods. During this time, the viral genes replicate but the cell is not lysed, so the virus is not free to cause harm.
12. A retrovirus is one that contains RNA as its genetic material and uses reverse transcriptase to form a DNA version of the RNA that can be integrated into the host cell's DNA.
13. Restriction enzymes, found in many bacteria, cut foreign DNA sequences out of the host's DNA, preventing viral DNA from becoming integrated.
14. Virus cycles are very rapid. Viral regulatory genes often are recognized by the host over its own genes. Viral nucleic acids are streamlined—they lack the "junk" DNA found in eukaryotic cells.
15. (Answers will vary—refer to Box 12.1.)

Chapter 13—Genetic Engineering and Recombinant DNA

Part 1: Multiple Choice

1. d
2. b
3. b
4. a
5. d
6. c
7. d
8. b
9. a
10. a and d
11. c
12. a, b, and c
13. c
14. d
15. d

Part 2: Matching

1. recombinant DNA technology
2. polymerase chain reaction
3. polymerase chain reaction
4. recombinant DNA technology
5. recombinant DNA technology
6. polymerase chain reaction
7. restriction enzyme
8. DNA ligase
9. reverse transcriptase
10. DNA ligase
11. restriction enzyme
12. reverse transcriptase

Part 3: Short Answer

1. Early farmers, cheese makers, and brewers learned to select the organisms with which they worked—early biotechnology.
2. Genetic engineers can damage the gene that controls ethylene production, preventing softening until the distributor treats them.
3. Each end of a restriction fragment has a restriction site—a sequence recognized by the restriction enzyme that cut the original DNA.
4. Restriction fragments are linked together by DNA ligase.
5. Antibiotic resistance is included as a marker—when bacteria are grown in an antibiotic, only those with the antibiotic resistance, and thus also the recombinant DNA, will survive, so researchers know the gene is in those colonies.
6. Vectors carry the genes into a cell or organism where they will be mass-produced (cloned).
7. Bacteria cannot use actual eukaryotic genes (DNA) because the bacteria cannot remove the introns. The cDNA is made from mature RNA (from which introns have been removed), then the cDNA is joined with the vector.
8. Recombinant DNA can be mass-produced through cloning, such as in bacterial colonies, or through the polymerase chain reaction.
9. Genes can be located by using a hybridization probe or antibodies (which locate the protein made by the gene, but that tells the researchers which bacterial colony contains the gene.
10. "Designer genes" refers to recombinant DNA, or engineered genes, which are often "designed" for a specific purpose.
11. A genetic knockout is an organism in which a particular gene has been inactivated.
12. A genetic knockin is an organism in which an allele has been replaced, not just added.
13. Oligonucleotides—short pieces of single-stranded DNA—serve as primers for DNA polymerase.
14. Recombinant DNA consists of two or more DNA segments that are not found together in nature.
15. The plasmid that the gene is joined to can force the bacterium to produce up to a thousand copies of the gene, and researchers can induce the cell to undergo rapid cell division.

Chapter 14—Human Genetics

Part 1: Multiple Choice

1. d
2. b
3. d
4. c
5. a
6. b and c
7. a
8. c
9. b
10. b
11. a and b
12. c
13. d
14. d
15. d

Part 2: Matching

1. coding sequences of DNA
2. noncoding sequences of DNA
3. noncoding sequences of DNA
4. noncoding sequences of DNA
5. coding sequences of DNA
6. amniocentesis
7. chorionic villus sampling
8. amniocentesis and chorionic villus sampling
9. chorionic villus sampling
10. amniocentesis and chorionic villus sampling

Part 3: Short Answer

1. RFLPs are DNA fragments that have been cut by restriction enzymes. Because individuals have different amounts of noncoding DNA, each person's RFLPs will be of unique lengths.
2. Humans differ genetically by only 1%.
3. DNA for fingerprinting can be extracted from any cell with a nucleus.
4. Probability of shared results depends on number of markers examined and population definition.
5. Defense attorneys have argued that the DNA may have deteriorated or the sample may have been contaminated, but neither of these would result in a DNA fingerprint match with an innocent client.
6. There are always some groups who do not neatly fit into any of the defined races; "races" are not pure because of blending; genetic variation between the defined races is only 7%.
7. A completely penetrant dominant allele will always be expressed.
8. Pleiotropy means that a single gene affects many traits. All genes are believed to be pleiotropic.
9. Polygenic means the trait is determined by more than one gene, and redundant refers to the genome's back up system by which the same trait can be accomplished in multiple ways, such as by having multiple genes code for the same effect.
10. The blood types are as follows:
    a) type A
    b) type AB
    c) type O
    d) type A
11. The Human Genome Project began in 1990 under the direction of James Watson.
12. I^A and I^B are codominant to each other, and both of these are dominant over I, so the ABO blood group is polygenic and demonstrates both dominance and codominance.
13. In pedigree analysis, DNA fingerprinting is done to look for RFLP alleles that are consistently associated with inheritance of the disease.
14. Most women who develop breast cancer do not carry the abnormal alleles; knowledge of presence of the abnormal alleles has done nothing to improve prevention or treatment of breast cancer. In addition, insurance companies do not want to insure people who carry the abnormal alleles.
15. Currently, the only options are to carry the child to term or end the pregnancy.

Chapter 15—What is the Evidence for Evolution?

Part 1: Multiple Choice

1. c
2. b
3. c
4. c
5. a
6. b
7. c
8. c
9. a
10. b
11. a
12. c
13. d
14. d
15. d

Part 2: Matching

1. creationism
2. creationism
3. evolution
4. creationism
5. evolution
6. evolution
7. evolution
8. creationism
9. Mesozoic Era
10. Paleozoic Era
11. Precambrian Era
12. Precambrian Era
13. Paleozoic Era
14. Mesozoic Era
15. Mesozoic Era
16. Cenozoic Era

Part 3: Short Answer

1. The prosecution's job is to get a conviction; the defense, supported by the ACLU, wanted a conviction so the ACLU could appeal the decision to a higher court, hoping to strike down the law.
2. Darwin's theory does not address the origin of life.
3. An adaptive trait is one that increases an organism's chance of passing on its genes.
4. In artificial selection, the breeder selects the desired trait for which to breed. In natural selection, forces at work in nature do the selecting—organisms that are best suited for surviving in their particular environment are most likely to reproduce.
5. Humans are classed as follows:
   class: Mammalia
   order: Primates
   family: Hominidae
   genus: *Homo*
   species: *sapiens*
6. Cuvier invented comparative anatomy and paleontology.
7. Cuvier's catastrophism idea said that the organisms in different layers had perished from various catastrophes, the most recent of which was Noah's flood; after each catastrophe, new species were created.
8. Lyell's idea of uniformitarianism, which states that the forces that now mold the Earth's surface always have, and that geologic change is gradual, rejected the idea of multiple catastrophes.
9. Lamarck coined the term "biology" for considering plants and animals as related organisms.
10. Lamarck argued that species pass on acquired characteristics that make them better able to survive.
11. Darwin presented such detailed and thorough evidence, from many different areas of study, that biologists could not refute it.
12. The four eras are Precambrian, Paleozoic, Mesozoic, and Cenozoic.
13. Most dead organisms are consumed.
14. Dead organisms become buried in a bog, sea, or lake, then more layers of mud and sediment cover it; in time, with building pressure, the layers become rock.
15. The rate of change in a species depends on
    a) the intensity of selection,
    b) the extent of inherited variation, and
    c) the extent of variation within a population.

Chapter 16—Microevolution: How Does a Population Evolve?

Part 1: Multiple Choice

1. a
2. c
3. a and d
4. c
5. a
6. a
7. c
8. b
9. d
10. a
11. d
12. a and b
13. c
14. d
15. b

Part 2: Matching

1. Lamarckism
2. blending inheritance
3. natural selection
4. Lamarckism
5. natural selection
6. blending inheritance
7. stabilizing selection
8. disruptive selection
9. directional selection
10. directional selection and disruptive selection

Part 3: Short Answer

1. According to the theory of inheritance of acquired traits:
   a) environmental changes create changed needs in the organisms;
   b) changed needs lead to changed behavior;
   c) behavior changes result in increased use of certain parts, which develops those parts; and
   d) those changes are passed to offspring.

2. Weismann's theory stated that the germ plasm was isolated from the rest of the body and thus was not subject to influence from whatever happened elsewhere; only the germ plasm is passed on so acquired traits cannot be.
3. The modern synthesis joined genetics and evolutionary theory.
4. Population genetics focuses on microevolution.
5. Most genetic variation results from mutation.
6. The rate at which a population evolves depends on the rate at which new mutations spread through the population.
7. Genetic variation spreads through a population primarily by recombination and sexual reproduction.
8. In this equation, p and q represent the frequencies of the two alleles.
9. The five conditions of Hardy-Weinberg equilibrium are:
   a) random mating,
   b) large population,
   c) no mutations,
   d) no breeding with other populations, and
   e) no selection
10. Assortative mating is when individuals choose their mates based on the mate's genotype.
11. Populations can evolve without natural selection through the founder effect or bottleneck effect.
12. Evolution would cease if genetic variation ended.
13. Many deleterious alleles are recessive and "hide" in the heterozygote; some even offer the heterozygote an adaptive advantage. Mutations also ensure that some harmful alleles always persist.
14. Stabilizing selection is common in stable environments, so the average individual is better-suited.
15. Sexual selection includes female choice and male competition.

Chapter 17—Macroevolution: How Do Species Evolve?

Part 1: Multiple Choice

1. b
2. b and c
3. a
4. a
5. b
6. d
7. d
8. a
9. b
10. c
11. c
12. d
13. c
14. d
15. d

Part 2: Matching

1. sympatric speciation
2. allopatric speciation
3. parapatric speciation
4. parapatric speciation
5. allopatric speciation
6. gradualism
7. saltationism
8. saltationism
9. punctuated equilibrium
10. gradualism
11. punctuated equilibrium
12. punctuated equilibrium

Part 3: Short Answer

1. More than half of the known species are insects.
2. Paleontologists study fossils, so they cannot know with certainty who interbred with whom. They also must classify species solely on the basis of morphology of the remaining hard parts.
3. Reproductive isolation means individuals cannot interbreed.
4. A hybrid is an offspring from the mating of two different species, and they are often sterile so they cannot pass on the new genetic combination.
5. Divergent evolution is the separation of one species into two or more. Convergent evolution is the independent development of similar features in separate groups.
6. Vicariance can occur through extinction of populations in between or through creation of geographic barriers.
7. Most of the United States' mammals are placental mammals.
8. Developmental constraints are rules of embryological development that determine the general body form of the individual; they are believed by some to prevent evolution of new body plans.
9. Overall, the extinction rate of species remains relatively constant; with the exception of five mass extinctions.

10. Diversity has increased dramatically after each mass extinction.
11. Gaps in the fossil record may be explained by:
    a) intermediate forms not appearing because they were not completely preserved;
    b) intermediate forms either never existed or existed very briefly; and
    c) paleontologists generally do not recognize intermediate forms as being different.
12. Speciation can occur in small, isolated populations through the founder effect, the bottleneck effect, allopatric speciation, and sympatric speciation.
13. Punctuated equilibrium states that:
    a) species change very little most of the time, and
    b) most change in individual species occurs during a geographically brief period at the time of speciation.
14. Physical anthropologists study humans.
15. The first hominids are believed to have evolved in Africa.

Chapter 18—How Did the First Organisms Evolve?

Part 1: Multiple Choice

1. c
2. a
3. b
4. c
5. a and d
6. d
7. a and c
8. d
9. a and b
10. c
11. b
12. d
13. c
14. d
15. a and c

Part 2: Matching

1. spontaneous generation and prebiotic evolution
2. spontaneous generation
3. colonization from space
4. prebiotic evolution
5. current atmosphere
6. primitive atmosphere
7. primitive atmosphere
8. current atmosphere

Part 3: Short Answer

1. The origin of life is not an ongoing, testable process.
2. [Examples and explanations will vary.] People saw living organisms show up where life had not been before, and didn't realize the organisms had come to the area from elsewhere.
3. The earliest fossils resemble modern prokaryotes.
4. a) Life arose spontaneously from nonliving organic molecules in the primitive environment.
   b) Life arrived from elsewhere in the universe.
5. We can't say that there is not life on Mars, but we do not have enough evidence to prove that there is. A 3.6-billion-year-old Martian meteor contains what appear to be tiny fossils and also complex organic molecules.
6. When the Earth's crust formed, cracks and volcanoes released hot gases from the interior which became the atmosphere.
7. The modern atmosphere is about 78% nitrogen, 21% oxygen, 1% argon, and about 0.1% other gases (carbon monoxide, carbon dioxide, sulfur, nitric oxide, and others).
8. Chemical reactions could have been driven by UV radiation, heat from Earth's interior, or lightning.
9. Protobionts were able to concentrate organic molecules and maintain a separate internal environment.
10. The first true cells would have needed:
    a) a boundary,
    b) enzymes for extracting energy from their environment,
    c) energy storage in ATP,
    d) RNA that specified the structure of specific enzymes,
    e) RNA replication,
    f) specification of each amino acid by a triplet codon, and
    g) primitive tRNAs to connect amino acids to nucleotides.
11. Fermentation evolved first, followed by photosynthesis, and finally respiration.
12. Water is the most abundant source of electrons.

13. Photosynthesis led to accumulation of oxygen. This would have destroyed many organic molecules, which were the food source for many heterotrophs. Many of these organisms would have starved.
14. Mitochondria evolved after photosynthesis, because photosynthesis provided the oxygen the mitochondria rely on for generating useful energy.
15. Endosymbiosis provides an internal source of food for the host and mobility and protection for the inner organism.

Chapter 19—Classification: What's In a Name?

Part 1: Multiple Choice

1. b
2. a
3. b and d
4. d
5. b and c
6. c
7. a
8. b
9. c
10. b
11. c
12. c
13. b
14. a and c
15. a

Part 2: Matching

1. typological species concept
2. biological species concept
3. typological species concept
4. biological species concept
5. typological species concept
6. biological species concept
7. classical taxonomy and phenetics
8. cladistics
9. phenetics
10. cladistics

Part 3: Short Answer

1. By comparing mitochondrial DNA, Wayne and Jenks discovered that the red wolf had no unique DNA sequences—only sequences found in the gray wolf and the coyote.
2. Humans are in the domain Eukarya, kingdom Animalia.
3. The typological species concept overlooks two facts:
   a) All species change over time; and
   b) The individual members of species vary.
4. Species are reproductive units because they are groups of potential mates; they are genetic units because they form a gene pool; and they are ecological units because they interact with other species in their environment in predictable ways.
5. Bacteria often exchange genes with unrelated species.
6. Reproductive isolation (in biologically defined species) tends to lead to evolution of different and divergent morphological characteristics.
7. Homologous similarities are those that are inherited from a common ancestor—they indicate relatedness. Analogous similarities are those that share the same function but different evolutionary paths—they indicate similar adaptations, but in unrelated organisms.
8. Classical taxonomy and phenetics both rely on comparing the maximum number of morphological characteristics.
9. The choice of unit characters is subjective, analogous characteristics suggest a relationship when there is none, some characteristics seem to be more important than others, and environmental conditions affect some characteristics more than others.
10. Cladists reject reptiles as a taxon because they believe all taxons should be monophyletic—including a single ancestral species and all its descendants. Reptiles are paraphyletic—some descendants have evolved into non-reptiles (birds and mammals).
11. Morphological features are largely determined by the genes—the base sequences—so differences in base sequences frequently translate into differences in morphology.
12. Cladistics is based on evolutionary history.
13. The rate of change can be affected by:
    a) chance appearance of new mutations,
    b) geographical changes, and
    c) natural selection.

14. Archaea differ from other bacteria both anatomically and metabolically.
15. The textbook uses these six kingdoms: Monera (prokaryotes), Archaebacteria, Protists, Fungi, Plants, and Animals.

Chapter 20—Prokaryotes: How Does the Other Half Live?

Part 1: Multiple Choice

1. b
2. c
3. d
4. a and c
5. d
6. b and c
7. d
8. b and c
9. a
10. c
11. a
12. a
13. b
14. b
15. c

Part 2: Matching

1. gram-negative bacteria
2. gram-positive bacteria
3. gram-positive bacteria
4. gram-negative bacteria and gram-positive bacteria
5. gram-negative bacteria
6. Archaea
7. Archaea
8. Eubacteria
9. Archaea
10. Eubacteria

Part 3: Short Answer

1. Epidemiologists study the incidence and transmission of diseases in populations.
2. Exposure to antibiotics kills non-resistant bacteria, reducing the competition for the antibiotic-resistant bacteria so that they can thrive.
3. Pathogens usually must invade the host and multiply there, and release a toxin that disrupts the host's normal functioning.
4. Koch's postulates are:
   a) The same pathogen must be found in all people with the disease.
   b) The pathogen must be capable of being grown in a pure culture.
   c) The pure-cultured pathogen must cause the disease when introduced into an experimental animal.
   d) The same pathogen must be present in the infected animal after the disease develops.
5. Antibiotics kill bacterial cells but do not harm eukaryotic cells, such as ours.
6. Most antibiotics are produced naturally by soil bacteria and fungi.
7. Antibiotics may prevent synthesis of the bacterial cell wall, prevent expression of bacterial genes, interfere with the folding of bacterial DNA, inhibit folic acid synthesis, or block protein synthesis.
8. Bacteria exchange genes readily through conjugation, and plasmids can pass genes around rapidly via jumping genes. The resistance genes can be spread through several types of bacteria.
9. The most common prokaryotic shapes are rod-shaped bacilli, spiral-shaped spirilla, and spherical cocci.
10. Gram-positive cells retain gram dyes; gram-negative bacteria can be washed free of the dyes.
11. Capsules protect pathogenic bacteria from attack by the body's white blood cells.
12. Bacteria can replicate their DNA continuously, which allows more rapid cell division. They also have a very streamlined metabolism.
13. Bacterial adaptations include spore formation, production of antibiotics to kill competitors, easy transfer of resistance between bacteria, and the ability to obtain energy from other species' waste products.
14. Archaea differ in the composition of their ribosomes, construction of their cell walls, and the kinds of lipids in their cell membranes.
15. Obligate aerobes require oxygen for respiration; facultative aerobes can grow with or without oxygen; and obligate anaerobes are poisoned by oxygen.

Chapter 21—Classifying the Protists and Multicellular Fungi

Part 1: Multiple Choice

1. b
2. d
3. a
4. d
5. b
6. d
7. a and c
8. c
9. c
10. d
11. a
12. b
13. c
14. c
15. c

Part 2: Matching

1. fungi
2. algae
3. algae and protozoa
4. fungi and algae
5. fungi
6. protozoa
7. ascomycetes
8. myxomycota
9. euglenophytes
10. ciliophora
11. euglenophytes
12. myxomycota

Part 3: Short Answer

1. Protists are grouped as algae, protozoa, and funguslike protists.
2. Algae are photoautotrophs because they produce their own energy by photosynthesis.
3. Gametophytes are haploid; sporophytes are diploid.
4. The rhodophytes (red algae) deposit calcium carbonate.
5. Most algae differ from plants in lacking chlorophyll *b*, structure of the cell wall, and the arrangement of flagella.
6. Chrysophytes can reproduce asexually by swarmer cells swimming away from the colony to form a new one, or by a colony dividing and the parts floating away to form new colonies.
7. Pseudopodia are temporary extensions of the cytoplasm.
8. Sporozoans are nonmotile, spore-forming, parasitic protozoa.
9. Funguslike protozoa resemble other protists in cellular organization and modes of reproduction, but they resemble fungi in general appearance.
10. Saprophytic organisms, like the oomycota, derive their nourishment from dead organisms.
11. Decomposers recycle nutrients from dead organisms.
12. [Examples will vary.]
13. True fungi have rigid cell walls, like plants, but are also heterotrophic, like animals.
14. Fungi extend hyphae into food sources, secrete digestive enzymes to break down the food, then absorb the nutrients.
15. Lichens are a symbiotic association of a fungus (most often an ascomycete) and a photosynthetic partner—either a green algae or a cyanobacterium.

Chapter 22—How Did Plants Adapt to Dry Land?

Part 1: Multiple Choice

1. d
2. d
3. b
4. c
5. a
6. a and d
7. c
8. d
9. c
10. a
11. b
12. c
13. b
14. b and c
15. b

Part 2: Matching

1. equisetophytes and gymnosperms
2. equisetophytes and gymnosperms
3. all three
4. bryophytes
5. gymnosperms
6. flower
7. fruit
8. fruit
9. flower
10. Fruit

Part 3: Short Answer

1. Plants provide oxygen, all the energy heterotrophs get from food, and fossil fuels that humans rely on.
2. For life on land, plants developed a waxy cuticle; the ability to absorb water from dew, rain, or groundwater; enclosed reproductive organs; and enclosed sporangia.
3. The first plants most likely evolved from an ancestor resembling green algae.
4. The Carboniferous period is also called the Age of Ferns.
5. Bryophytes need free-standing water for photosynthesis and reproduction.
6. Rhizoids anchor plants in the ground, but are not specialized for water absorption like true roots.
7. The transport system in vascular plants is composed of xylem and phloem.
8. Psilotophytes are the simplest of the vascular plants.
9. Seed plants are heterosporous because they produce two kinds of gametes from two kinds of spores on two kinds of sporangia.
10. A seed usually has the new sporophyte embryo, a female tissue that nourishes the developing embryo, and the nucellus and integuments (now the seed coat) from the previous sporophyte generation.
11. Gymnosperms have seeds but lack flowers or fruit; angiosperms have all three components.
12. The gymnosperm divisions are conifers, cycads, gingkoes, and gnetae.
13. The gnetae have naked seeds, like gymnosperms, but some of their strobili resemble flowers and the vessels of their leaves and stems are more like those in angiosperms.
14. Pollinators get nectar, some of the high-protein pollen, and nutritious nibblings from petals or other flower parts.
15. After fertilization, the ovary becomes the fruit.

Chapter 23—Protostome Animals: Most Animals Form Mouth First

Part 1: Multiple Choice

1. d
2. b
3. c
4. c
5. b and d
6. a
7. c
8. a
9. c
10. b
11. a
12. b
13. c
14. b
15. c

Part 2: Matching

1. endodern surrounded by mesoderm
2. endoderm
3. endoderm
4. mesoderm
5. ectoderm
6. Cestoda
7. Cnidaria
8. Chelicerata
9. Chelicerata
10. Cnidaria

Part 3: Short Answer

1. All animals are classified as Parazoa or Eumetazoa.
2. A tissue is made of a group of similar cells and the matrix they secrete; an organ is made of two or more tissues.
3. In protostomes, the mouth develops first; but in deuterostomes, the anus develops first.
4. Parazoa develop from embryos and produce sperm and eggs.
5. Ctenophores use their cilia to move through the water.
6. Flatworms are shaped to minimize diffusion distances—flattened bodies and a branched gut cavity extending throughout the body.
7. Parasite adaptations found in tapeworms and flukes include (1) rapid reproduction, (2) distinct stages allowing passage and dispersal through multiple hosts, (3) attachment organs, (4) specialized digestion, and (5) reduced sense organs.
8. Ribbon worms have digestive tracts with a mouth and an anus, and a circulatory system.
9. A pseudocoel is a distribution route for moving, gases, nutrients, and wastes, and it serves as a supportive hydrostatic skeleton.
10. A coelom (1) prevents muscle action from disrupting vital functions, (2) can provide a hydrostatic skeleton against which the muscles can work, and (3) provides protection during gamete production.
11. In most mollusks, the mantle secretes a shell.

12. Cephalopods use their tentacles to catch prey, pull themselves, and steer through the water.
13. Annelids have long, segmented bodies composed of repeating segments called metameres.
14. The Hirudinea, or leeches, are parasites.
15. All arthropods have exoskeletons and jointed limbs.

Chapter 24—Deuterostome Animals: Echinoderms and Chordates

Part 1: Multiple Choice

1. c
2. b and c
3. b
4. d
5. d
6. b
7. c
8. c
9. a and c
10. d
11. a
12. d
13. a and d
14. b
15. c

Part 2: Matching

1. mammals
2. amphibians
3. reptiles
4. amphibians
5. reptiles and mammals
6. mammals
7. reptiles and mammals
8. ectotherms
9. ectotherms
10. endotherms
11. endotherms
12. endotherms

Part 3: Short Answer

1. The phyla of deuterostomes are Echinodermata, Chaetognatha, Hemichordata, and Chordata.
2. All chordates, at some point in life, have a notochord; a dorsal, hollow nerve cord; pharyngeal slits; and a segmented body with a postanal tail.
3. Vertebrates have a segmented vertebral column and a distinct head, with a cranium and brain.
4. The skeleton of all vertebrate embryos is made of cartilage.
5. Today, the only living jawless fish are the lampreys and hagfish.
6. A swim bladder is an air-filled sac that helps bony fish control their buoyancy.
7. Amphibians begin life as aquatic larvae, but most undergo a dramatic change, called metamorphosis, to become terrestrial adults.
8. The reptile orders are Chelonia (turtles and tortoises), Crocodilia (crocodiles and alligators), Squamata (lizards and snakes), and an order that includes only the tuatara.
9. The Mesozoic Era was the Age of the Reptiles.
10. Birds are distinguished from reptiles by having feathers, flight, and endothermy.
11. Flight-enhancing adaptations in birds include loss of teeth, hollowing of the bones, and reshaping of the breastbone.
12. Mammals are distinguished from other vertebrates by having mammary glands, two sets of teeth, and hair.
13. Mammals fall into three subclasses: eutherians (placental mammals), metatherians (marsupials), and protherians (monotremes). [Examples will vary.]
14. Placoderms were armored fish and the first vertebrates with a jaw. No placoderms are alive today.
15. The only protherians alive today are the platypus and spiny anteaters.

Chapter 25—Ecosystems

Part 1: Multiple Choice

1. c
2. b
3. a
4. b and d
5. a
6. a
7. c
8. b and c
9. a and b
10. c and d
11. d
12. b
13. c
14. d
15. b and c

Part 2: Matching

1. producers
2. decomposers
3. consumers and decomposers
4. producers
5. consumers
6. carbon
7. water
8. carbon
9. nitrogen
10. nitrogen
11. carbon
12. nitrogen

Part 3: Short Answer

1. Ecology is the study of interactions between organisms and between them and their environment.
2. As little as 1% of the sunlight shining on Earth may be available for primary consumers to use.
3. Molecular nitrogen is the most abundant element in the Earth's atmosphere.
4. Producers harness energy, usually from the sun, and use it to make their own food. Consumers eat producers or other consumers to get energy. Decomposers live off the energy stored in molecules of dead organisms.
5. A food chain is a linear relationship showing who eats whom, but each ecosystem has numerous food chains and a single organism may be in more than one of them. So, a food web is the collection of all the food chains in an ecosystem.
6. Producers process the most energy.
7. Protists may be producers, consumers, or decomposers.
8. Omnivores eat from multiple levels—producers, herbivores, and other carnivores.
9. The ten percent law states that organisms of any trophic level provide the next trophic level with only 10% of the energy that they have assimilated.
10. Oxygen accounts for about 20% of the atmosphere.
11. Animals consume oxygen to burn organic compounds, releasing carbon dioxide in the process of respiration. Plants, in turn, take in carbon dioxide to make sugars, then release oxygen back to the atmosphere.
12. Plants take up water from the soil and release it as vapor from their stomata during transpiration.
13. Fossils are associated with the carbon cycle.
14. Nitrogen is a key component in proteins and nucleic acids.
15. The largest single source of nitrogen is industrially manufactured fertilizers.

Chapter 26—Biomes and Aquatic Communities

Part 1: Multiple Choice

1. a and b
2. d
3. a
4. d
5. d
6. c
7. c
8. c
9. a
10. b
11. d
12. b and c
13. a
14. b
15. b and c

Part 2: Matching

1. tropical rain forest
2. tundra
3. temperate deciduous forest
4. savanna
5. tropical rain forests
6. tundra
7. eutrophic lake
8. eutrophic lake
9. eutrophic lake
10. oligotrophic lake
11. oligotrophic lake
12. eutrophic lake

Part 3: Short Answer

1. Biomes are mostly determined by climate.
2. Rules that govern air movements are:
   a) hot air rises and cold air falls;
   b) hot air holds more moisture than cold air; and
   c) Earth's rotation twists the moving air.

3. Most of the great deserts are at 30° latitude.
4. As altitude increases, the climate changes similar to when you move through increasing latitude.
5. Periodic fires clear the way for fresh growth and prevent trees from growing to shade the chaparral. But big fires burn so hot that they destroy the roots, killing the plants that would otherwise grow back.
6. Sagebrush has very small leaves to minimize evaporation; cacti store water in fleshy tissue.
7. The African plains are an example of the savanna.
8. Heavy rainfall leaches nutrients from the soil.
9. Lianas root in the soil but grow up into the canopy. Epiphytes grow entirely on other plants.
10. The taiga is characterized by coniferous trees.
11. Tundra soil is wet because permafrost prevents the water from draining, and the cold air minimizes evaporation.
12. Oceans resist temperature changes because the waters are constantly mixing, and water has a high heat-retaining capacity.
13. The benthic division consists of all organisms that live on the ocean bottom; the pelagic division consists of all organisms that live in open water.
14. As organisms die, their remains fall out of the oceanic zone, taking their minerals with them, and they reach the ocean's bottom and are not returned upward.
15. Estuaries receive a constant flow of nutrients from rivers and tides.

Chapter 27—Communities: How Do Species Interact?

Part 1: Multiple Choice

1. c
2. b and c
3. a
4. c
5. d
6. c
7. a
8. c
9. d
10. c
11. a
12. a and b
13. c
14. c
15. a and d

Part 2: Matching:

1. mutualism
2. parasitism
3. commensalism
4. parasitism, mutualism, and commensalism
5. mutualism
6. pioneer community
7. pioneer community
8. climax community
9. pioneer community
10. climax community

Part 3: Short Answer

1. A community is both an integrated set of interacting species and a group of organisms that occur together because their ranges happen to overlap.
2. Interspecific interactions include those between the consumers and the consumed, between competitors, and between species that associate with one another.
3. Consumers include scavengers, decomposers, herbivores, predators, and prey.
4. In Batesian mimicry, a harmless species mimics a harmful species; in Müllerian mimicry, equally dangerous species have similar coloration.
5. No—only in parasitism is any harm done, and just to one species.
6. Coevolution is when two or more species evolve together, as in mutualistic relationships.
7. The competitive exclusion principle says that when two species compete directly for the same resources, ultimately one will prevail and the less well-adapted species will be eliminated.
8. A habitat is the place the organism lives; a niche is how it functions within its habitat.
9. Resource partitioning—splitting up the niche—allows more species to share the same niche by eliminating direct competition.
10. Species richness means the number of different species in an ecosystem, but it counts all species, even if only one is present. Relative abundance measures how common each species is in that ecosystem.

11. Predators typically consume more than one prey species. If the population of one of the prey species grows, competition between it and other prey species could lead to the demise of some species. But, as the population of one prey species increases, predators usually consume more of that species, maintaining a balance between the different populations.
12. [Examples will vary.]
13. Climax communities do change, but very slowly compared to earlier stages.
14. Resilience refers to how quickly an ecosystem will return to a particular form after a major disturbance.
15. Competition may force an organism into the smaller realized niche.

Chapter 28—Populations: Extinctions and Explosions

Part 1: Multiple Choice

1. b
2. c
3. a and b
4. a and c
5. c
6. d
7. b
8. d
9. b
10. a
11. c
12. a, b, and d
13. a
14. c
15. d

Part 2: Matching

1. *r*-selected species
2. *r*-selected *and* *K*-selected species
3. *r*-selected species
4. *K*-selected species
5. *K*-selected species
6. diagonal survivorship curve
7. convex survivorship curve
8. concave survivorship curve
9. convex survivorship curve
10. concave survivorship curve

Part 3: Short Answer

1. Exponential growth means the population doubles in some constant time period.
2. Population size is determined by birth, death, immigration, and emigration.
3. [Examples will vary, but may include competition, predation, parasitism, disease, and emigration.]
4. [Examples will vary, but may include fire, weather, natural disasters, and human-made disasters.]
5. Population size of *r*-selected species is often controlled by density-independent factors.
6. [Examples will vary.]
7. Future age structure can be predicted by knowing a) the present age structure, b) mortality of each cohort, c) age structure of immigrants and emigrants, and d) fertility of each cohort.
8. Demography is the statistical study of populations.
9. Completed family size refers to the total number of children who reach reproductive age that are born to a family, but replacement reproduction means that only two offspring are born to each couple—just enough offspring to replace the parents.
10. The world's population is over 6 billion people.
11. With population momentum, population size can continue to increase even if completed family size is below replacement level. This is possible if a large portion of the population has yet to reach reproductive age. As they reproduce in large numbers, population size will increase.
12. People in cities and suburbs rely on agriculture for the food consumed to provide energy.
13. [Examples will vary.]
14. In biomagnification, toxic chemicals accumulate to the highest degree, and are thus amplified, at the higher levels of the food chain. Toxins often accumulate in tissues, especially fat, and since each level of the food chain eats more than the previous level, the top levels consume the most of these toxins.
15. Automobiles operate by the combustion of fossil fuels, which emits carbon dioxide and sulfur and nitrogen compounds. Carbon dioxide traps heat within the atmosphere, leading to global warming. Sulfur and nitrogen oxides are carried up to thousands of miles away from the source by wind currents, then combine with water and fall to the earth as acid rain.

Chapter 29—The Ecology of Animal Behavior

Part 1: Multiple Choice

1. a
2. b
3. c
4. b
5. a
6. c
7. c
8. d
9. b
10. d
11. c
12. d
13. a
14. b
15. a, c, and d

Part 2: Matching

1. innate behavior
2. learned and innate behavior
3. innate behavior
4. innate behavior
5. learned behavior
6. kin selection
7. sexual selection
8. sexual selection
9. kin selection

Part 3: Short Answer

1. Experimental psychologists believed most behavior is learned; ethologists believed most is innate.
2. The adaptive value of behavior is determined by ecology—the individual's relationship to its environment.
3. Modal action patterns are stereotyped patterns of innate behavior that are triggered by sign stimuli.
4. Most behaviors, like most other traits, are governed by complex interactions between more than one gene and the environment.
5. A sensitive period is the time during which a complex behavior can be learned. If it is not learned at that time, it will not be learned.
6. Mating rituals allow the female to assess the health and vigor of potential mates so she can choose the best mate. Sexual selection is the differential ability of individuals to acquire mates—in other words, sexual selection increases the likelihood of being chosen as a mate.
7. Agonistic behavior includes aggression, aggressive displays, appeasement, and retreat.
8. A dominance hierarchy is a ranking of individuals that determines who consistently has first access to resources or mates. [Human examples will vary.]
9. Natural selection can only work on inherited aspects of behavior.
10. Animals are most likely to expend energy to defend territories if that action increases their reproductive potential.
11. With social behavior, larger groups can more easily spot predators and other dangers, defend themselves, and work together to kill larger prey. However, larger groups are more easily spotted by predators, must compete for resources, breeding site, and mates, and can more easily spread diseases.
12. All social insects cooperate in raising the young, only a few members of the group reproduce while sterile workers assist and defend the fertile queen, and generations overlap.
13. [Examples will vary.]
14. Inclusive fitness is the sum of an individual's genetic fitness, and it is determined by adding the genetic fitness of the individual's direct offspring and the individual's influence on the fitness of its close relatives.
15. Reciprocal altruism refers to altruism based on the expectation that the recipient will reciprocate in the future if the opportunity arises.

Chapter 30—Structural and Chemical Adaptations of Plants

Part 1: Multiple Choice

1. b
2. c
3. c
4. a
5. d
6. b
7. d
8. d
9. a
10. a
11. d
12. b
13. c
14. c
15. a and d

Part 2: Matching

1. leaves
2. stems
3. roots
4. roots, stems, and leaves
5. leaves
6. monocot or dicot
7. dicot
8. monocot
9. monocot
10. dicot

Part 3: Short Answer

1. Ames concluded that the production of free radicals, during normal metabolism, is the cause of these degenerative diseases.
2. Biologists believe these plants evolved from green algae, or chlorophytes.
3. Land plants had to develop new mechanisms for obtaining water and carbon dioxide from the environment, and for supporting themselves against the force of gravity.
4. Xylem carries water and minerals, phloem transports energy-rich products of photosynthesis.
5. Primary cell walls are immediately outside the plasma membrane. Some cells also have an additional secondary cell wall, which is thicker, outside of the primary cell wall. The secondary cell wall provides additional mechanical support.
6. Secondary growth thickens the plant's roots or shoots.
7. The fluid-conducting cells in the phloem are the sieve tube members.
8. Monocots and dicots differ in many ways, including the number of flower parts, the arrangement of their vascular bundles, and the structure of their root systems.
9. Taproots are found in many dicots.
10. Nutrients dissolved in water are absorbed from the soil by the plant's roots.
11. Roots have three concentric rings—epidermis on the outside, cortex in the second ring, and the stele (vascular system) in the center.
12. The root's cortex is mostly made of parenchymal cells.
13. Leafs are typically arranged in one of three patterns: alternate, opposite, or whorled.
14. [Answers will vary; examples include waxy cuticle, hairy or sticky leaves, spines, thorns, and prickles.]
15. Lignin provides structural support and also repels herbivores.

Chapter 31—What Drives Water Up and Sugars Down?

Part 1: Multiple Choice

1. c
2. d
3. a
4. c
5. a and c
6. a and d
7. b and c
8. d
9. c and d
10. d
11. c
12. a
13. b
14. a and b
15. d

Part 2: Matching

1. pressure flow hypothesis
2. cohesion-tension theory
3. cohesion-tension theory
4. transpiration-photosynthesis compromise
5. transpiration-photosynthesis compromise
6. cohesion-tension theory
7. stomata open
8. stomata close
9. stomata open
10. stomata close

Part 3: Short Answer

1. The state had an interest in protecting both the nurseries and the grape growers.
2. The xylem transports water and minerals from the soil.
3. Guttation is caused by root pressure.
4. Adhesiveness refers to water's tendency to cling to surfaces, such as those in the tubes of xylem. Cohesiveness refers to water's tendency to stick to itself, moving in continuous columns. Both properties prevent water from slipping back downward due to gravity.

5. The cohesion-tension theory states that transpiration from the leaves pulls the water upward through the plant in continuous columns, held together by water's cohesiveness.
6. In the spring, woody plants undergo a burst of secondary growth, forming new xylem in which water can flow continuously.
7. The greater the transpiration rate, the greater the tension in the xylem. As tension increases, the xylem's walls are pulled inward (contract), which decreases the diameter of the tree trunk.
8. Open stomata allow carbon dioxide in for photosynthesis, but also allow water to leave. The more carbon dioxide brought in, the more water is lost, so a balance must be maintained between providing carbon dioxide and conserving water.
9. CAM plants open their stomata to collect carbon dioxide only at night, when the air is cooler so less water is lost. They store it as an organic acid compound that can then release CO_2 for photosynthesis during the day, when the stomata remain closed.
10. Humus is the residue of dead organisms that are broken down by bacteria and fungi.
11. Sands have the largest particles, followed by silts, then clays (with the smallest particles).
12. Roots use active transport to pump ions into their cells, and active transport requires energy (ATP), which is provided through oxygen-requiring respiration.
13. Carnivorous plants typically live in areas where acidic water interferes with the plant's uptake of dissolved nitrates. These plants digest animals to obtain the needed nutrients.
14. After girdling, bark above the girdle remains healthy, but bark below it shrivels and dies.
15. Plant sap contains sugars, minerals, carbohydrate derivatives, and plant hormones.

Chapter 32—Growth and Development of Flowering Plants

Part 1: Multiple Choice

1. b
2. d
3. b
4. a
5. c
6. c
7. d
8. a
9. b
10. a
11. b
12. d
13. d
14. b or c
15. b

Part 2: Matching

1. megaspore
2. microspore
3. microspore
4. megaspore
5. microspore
6. all three
7. stem cutting
8. grafting
9. tissue culture
10. grafting

Part 3: Short Answer

1. Pollen grains are a generative cell and a vegetative cell enclosed in a common wall.
2. Double fertilization occurs when two sperm fuse with two nuclei in the embryo sac. One sperm fuse with the egg to form the zygote; the other sperm fuses with the two central nuclei, forming the triploid endosperm.
3. The suspensor attaches the embryo to the surrounding tissues.
4. The cotyledons' attachment divides the axis into the epicotyl and hypocotyl.
5. Imbibition, the first stage of germination, is when the seed swells with water so that the dormant embryo is hydrated and awakens.
6. The radicle is the first part of the embryo to emerge from the seed coat.
7. Premature germination is prevented in some plants by a hard seed coat that requires scarification; other plants use chemical inhibitors that must be washed away by rain.
8. Plants that germinate in the absence of sufficient light keep their hooks instead of straightening, and they grow thin and spindly, with poor leaf development and little or decreased chlorophyll.
9. The apical meristem generates the same plant parts repeatedly by building the same module many times.
10. Cells expand by taking water into their central vacuoles, and cellulose in the wall allows the cell to expand only at one end, causing elongation.

11. Ethylene causes cellulose fibers to align longitudinally, running parallel to each other, which promotes wider but shorter growth patterns.
12. The root's growth zone has the root cap, the apical meristem, and the elongation zone.
13. Apical dominance means that presence of terminal buds suppresses development from axillary buds, leading to taller plants.
14. The four whorls, from the outside, are:
    a) sepals
    b) petals,
    c) stamens, and
    d) carpels.
15. Corn is monoecious, meaning both male and female flowers are on a single plant. Corn's flowers are incomplete because they do not have all four whorls; they are imperfect because the individual flowers have either stamens or carpels, but not both.

Chapter 33—How Do Plant Hormones Regulate Growth and Development?

Part 1: Multiple Choice

1. c
2. a
3. a and b
4. c
5. d
6. d
7. c
8. c
9. a
10. c
11. a
12. a and b
13. c
14. c and d
15. d

Part 2: Matching

1. auxin and cytokinin
2. cytokinin
3. auxin
4. auxin and cytokinin
5. ethylene
6. cytokinin
7. gravitropism and phototropism
8. gravitropism
9. phototropism and photoperiodism
10. phototropism and photoperiodism
11. gravitropism
12. phototropism

Part 3: Short Answer

1. Agent Orange, a synthetic auxin, was used in the Vietnam War to exfoliate forests so ground activity was detectable from the air.
2. Plant hormones move to their targets by diffusion and by transport through the vascular tissues.
3. Auxin moves from the lit area to the shaded area, and auxin promotes cell elongation. Cells in the shaded area, where auxin accumulates, elongate more—the plant turns toward the light.
4. Gibberellins have the greatest overall effect on the growth of intact plants.
5. The breakdown of starches and sucrose increases the solute concentration inside the cells and this, in turn, draws more water into the cells, increasing their expansion.
6. Auxin inhibits development of axillary buds; cytokinin stimulates cell division and growth when applied directly to an axillary bud.
7. Senescence refers to the breakdown of cellular components, ultimately leading to cell death. Abscission does not imply death—it is the process by which plants drop their leaves and fruit.
8. Of the plant hormones, ethylene is the simplest, having just two carbons and four hydrogens.
9. Ethylene, which promotes ripening, is present in smoke.
10. Ripened fruit releases ethylene, which causes other fruits to ripen and release still more ethylene, promoting over-ripening.
11. Abscisic acid opposes the actions of gibberellins and auxin.
12. Salicylic acid promotes flowering, inhibits ethylene production, and prevents plant infections.
13. [Examples will vary.]
14. Long-day plants require short nights for flowering; short-day plants require long nights for flowering.
15. During the day, both phytochromes exist, with P_{fr} predominating. But at night, P_{fr} levels drop because it is converted to P_r.

Chapter 34—Form and Function in Animals

Part 1: Multiple Choice

1. c
2. a and c
3. d
4. d
5. a
6. b
7. a and b
8. c
9. c
10. b
11. c
12. a and b
13. b
14. b and c
15. d

Part 2: Matching

1. small animals
2. small animals
3. small animals
4. large animals
5. large animals
6. smooth muscle and cardiac muscle
7. skeletal muscle and cardiac muscle
8. skeletal muscle
9. smooth muscle and cardiac muscle
10. smooth muscle

Part 3: Short Answer

1. An organ's form depends on its function.
2. Animals exclusively adapted to be fast swimmers are typically oblong.
3. Snakes can move by undulation, rectilinear movements, and sidewinding.
4. Most vertebrate flyers walk on two legs because their forelimbs are modified into wings.
5. Stride length can be increased by increasing the length of the leg in proportion to the rest of the body, increasing shoulder flexibility, or alternately flexing and extending the spine.
6. Tendons attach muscles to bones; ligaments attach bones to bones.
7. Exoskeletons can support more body weight than endoskeletons, but exoskeletons cannot grow.
8. Tension is the pulling action of two opposing forces; shear is the twisting action created by forces that are not opposite to each other.
9. Bone is also strengthened by calcium phosphate crystals.
10. Exercise causes bones to thicken and strengthen.
11. Muscles can only pull, so two muscles are usually involved in movements, pulling in opposite directions.
12. Smooth and cardiac muscles are called involuntary, because they cannot be consciously controlled.
13. Glycolytic fibers produce power in short bursts; oxidative fibers power muscles over long periods.
14. Myoglobin is an iron-containing muscle protein that binds oxygen.
15. Cross-bridges are extensions of the myosin molecules.

Chapter 35—How Do Animals Obtain Nourishment From Food?

Part 1: Multiple Choice

1. b
2. b
3. b
4. a
5. b and c
6. a and c
7. c and d
8. b and c
9. b
10. c
11. c
12. d
13. c
14. a
15. b and c

Part 2: Matching

1. amylase
2. amylase and lipase
3. pepsin
4. lipase
5. pepsin
6. amylase
7. stomach
8. small intestine
9. large intestine
10. stomach
11. small intestine
12. large intestine

Part 3: Short Answer

1. Undernourishment refers to taking in too few calories; malnourishment refers to a deficiency of one or more of the essential nutrients.
2. Animal cells are protected from digestive enzymes because digestion is, for most animals, extracellular.
3. Pinocytosis is the process by which extracellular material, such as food, is trapped into a vesicle when the plasma membrane folds inward. Phagocytosis is the same process, but the membrane expands outward and larger particles are taken in, usually for removing cellular debris instead of for feeding.
4. Two-ended digestive tracts allow specialization of the different regions, and thus sequential processing.
5. Spiders secrete digestive enzymes into their prey to begin hydrolyzing the food, then the spider ingests the liquefied prey.
6. The digestive system performs movement, secretion, digestion, and absorption.
7. Incisors bite and cut, canines tear, premolars and molars grind and crush.
8. Mucus coats food particles and lubricates their movement; salivary amylase hydrolyzes carbohydrate.
9. The pharynx is a shared entryway for the digestive system and the respiratory system.
10. Hydrochloric acid unfolds pepsinogen and cuts part of it away, forming the active enzyme pepsin.
11. The gizzard is a bird's unspecialized stomach which contains abrasives and stones that grind the food.
12. The villi increase the surface area to increase absorption of nutrients.
13. Diarrhea (watery feces) results when the contents of the colon move through too quickly, decreasing the amount of water absorbed. Constipation occurs when the contents move through the colon too slowly, so too much water is absorbed.
14. Some of the enzymes in the digestive system include amylase, pepsin, proteases, peptidases, lipases, and nucleases.
15. The nerve network receives information from receptors in the epithelium of the gut, and from the brain.

Chapter 36—How Do Animals Coordinate Cells and Organs?

Part 1: Multiple Choice

1. a and c
2. d
3. c
4. d
5. a
6. b
7. d
8. b
9. c
10. a and b
11. c
12. d
13. d
14. a and b
15. b

Part 2: Matching

1. hypothalamus and posterior pituitary
2. anterior pituitary and posterior pituitary
3. hypothalamus
4. anterior pituitary
5. posterior pituitary
6. lipid-soluble signals
7. water-soluble signals
8. lipid-soluble signals
9. water-soluble signals
10. lipid-soluble signals

Part 3: Short Answer

1. Estrogen and testosterone trigger development of the secondary sex characteristics and of germ cells.
2. Synthetic estrogenics include atrazine, chlordane, DDT, DES, and PCBs.
3. Pheromones are secreted by one organism and elicit a response in another organism of the same species; hormones are secreted by endocrine cells, enter the circulatory system, and travel to their target cells; paracrine signals affect only cells in the immediate vicinity of the cells that release them.
4. During the postabsorptive state, glycogen is broken down to increase the amount of glucose circulating in the blood, and cells derive energy more from internal sources—glycogen, fats, and proteins.
5. In IDDM, the immune system attacks and destroys the insulin-making cells so the person cannot make insulin; most people with NIDDM make insulin, but their cells do not respond to it.

6. Alpha cells secrete glucagon, beta cells secrete insulin, and delta cells secrete somatostatin.
7. Cortisol from the adrenal cortex, and epinephrine from the adrenal medulla are released during stress.
8. Hyperthyroidism is overproduction of thyroid hormone, which leads to uncontrolled, rapid metabolism.
9. The posterior pituitary consists of the ends of nerves that originate in the hypothalamus.
10. In females, FSH stimulates estrogen production and maturation of the follicle during the menstrual cycle; in males, it stimulates testosterone production. In females, LH induces ovulation and stimulates estrogen production; in males, it increases testosterone production.
11. In negative feedback, information about the output of the system is used to reduce further output. When there is a change away from homeostasis, the body responds in such as way as to correct the deviation, and once the correction is made, the response ends.
12. The hypothalamus integrates information from both of the control systems.
13. Steroids include the glucocorticoids, mineralocorticoids, and the sex steroids.
14. For signaling to occur between cells, a cell must make the signal molecule and release it; the molecule must travel to the target cells; the target cells must detect the signal and respond to it; and something must end the signaling process.
15. The most widely used second messenger is cyclic adenosine monophosphate (cAMP).

Chapter 37—How Do Animals Move Blood Though Their Bodies?

Part 1: Multiple Choice

1. c
2. a
3. c
4. b and c
5. d
6. b
7. a and c
8. b
9. d
10. c
11. d
12. a and b
13. d
14. a
15. c

Part 2: Matching

1. pulmonary circulation
2. systemic circulation
3. systemic circulation
4. systemic circulation
5. pulmonary circulation
6. veins
7. capillaries
8. arterioles
9. capillaries
10. capillaries

Part 3: Short Answer

1. The pulmonary artery carries deoxygenated blood from the right ventricle to the lungs.
2. The atria receive blood that is returning to the heart, then send it into the ventricles; the ventricles pump the blood out of the heart.
3. Blood follows this path: heart, arteries, arterioles, capillaries, venules, veins, heart.
4. Invertebrate blood lacks blood cells.
5. Erythrocytes carry oxygen; leukocytes defend against infections; and platelets help blood clot.
6. Hemoglobin and myoglobin both bind oxygen.
7. In open circulatory systems, blood and extracellular fluid mix freely and bathe the organs. In closed circulatory systems, the blood is contained within vessels.
8. Arthropods and most mollusks have open circulatory systems.
9. Capillaries have a thin wall—just a single cell thickness—and they are leaky.
10. The lymphatic system returns extracellular fluid, proteins, and large particles to the blood; filters the lymph; and fights infections.
11. Elastic allows the vessels to narrow or expand in response to fluctuations in blood pressure.
12. The "lub" sound is from the atrioventricular valves snapping shut; the "dup" sound is from the semilunar valves snapping shut.
13. Ventricular fibrillation occurs when the ventricles undergo continuous disorganized contractions. It is fatal because the ventricles do not pump blood, including their own supply, so the heart muscle dies from lack of oxygen.

14. Cardiac output is the amount of blood that leaves the heart in one minute. It is determined by:
    a) heart rate—the number of beats per minute; and
    b) stroke volume—the amount of blood that leaves the ventricles with each contraction.
15. Yes, histamine is beneficial because it increases blood flow to the affected area, bringing in nutrients and materials needed for tissue repair, and white blood cells to fight infection.

Chapter 38—How Do Animals Obtain and Distribute Oxygen?

Part 1: Multiple Choice

1. b
2. a and b
3. d
4. d
5. b
6. d
7. b
8. d
9. b and c
10. b
11. d
12. d
13. b
14. d
15. a

Part 2: Matching

1. gills
2. lungs
3. lungs
4. lungs
5. lungs and gills
6. oxygen
7. oxygen
8. carbon dioxide
9. oxygen
10. carbon dioxide

Part 3: Short Answer

1. Cigarette sales leveled in 1964—the Surgeon General issued a report that cigarette smoking is hazardous to health; sales declined two years later—Congress required labeling cigarettes hazardous.
2. Exercise uses oxygen, so the partial pressure of oxygen around the cells decreases, which increases the oxygen gradient. Exercise also releases carbon dioxide, which lowers the pH, and that reduces hemoglobin's affinity for oxygen.
3. Tar paralyzes the cilia that help keep the lungs clean. Tar and debris accumulate in the lungs and the only way to clear this debris, since the cilia are not working, is to cough it out.
4. Mucus traps foreign particles; cilia sweep mucus to the throat to be swallowed; and macrophages remove small particles and debris.
5. In emphysema, the walls of the alveoli are broken down through chronic coughing, which develops in an attempt to keep the lungs cleared when the normal mechanisms fail.
6. People with untreated lung cancer live only eight months; 90% in which the disease is treated aggressively die within five years.
7. The total pressure of a gas mixture is equal to the sum of the partial pressures of each component gas.
8. The rate at which gas molecules can enter a solution is determined by temperature, surface area, and concentration gradient. An increase in any of these causes more molecules to enter the solution.
9. Spiders breathe with a book lung—actually a set of internal gills.
10. The epiglottis prevents food from entering the larynx (air tube) during swallowing.
11. Dead space is the air in respiratory passages that does not reach the alveoli, allowing no gas exchange.
12. During inspiration, the diaphragm and rib muscles contract to expand the thoracic cavity, decreasing pressure in the lungs relative to outside the body, so air is drawn in to reduce the pressure difference.
13. Respiratory gases dissolve in the liquid present, then move from one area to the other by diffusion, moving along a concentration gradient from high concentration to the area of low concentration.
14. Hyaline membrane disease affects primarily premature babies, whose immature lungs lack surfactant, making it difficult to expand the alveoli and making breathing difficult.
15. An increase in BPG, which binds to hemoglobin, increases its oxygen delivery efficiency; increased secretion of the hormone erythropoietin stimulates red blood cell production.

Chapter 39—How Do Animals Manage Water, Salts, and Wastes?

Part 1: Multiple Choice

1. d
2. c
3. b
4. b
5. a, b, and d
6. a, b, and d
7. c
8. a
9. d
10. d
11. b
12. a and b
13. b
14. b and c
15. c

Part 2: Matching

1. reabsorption
2. secretion
3. secretion
4. reabsorption
5. secretion
6. angiotensin
7. aldosterone
8. aldosterone, ADH, and angiotensin
9. angiotensin
10. aldosterone

Part 3: Short Answer

1. The cell would lose water by osmosis and shrink.
2. Land animals get water from drinking, eating, and metabolism.
3. Water accumulates if more is taken in than is lost, causing edema (swelling) and possibly water intoxication.
4. Excess nitrogen can be removed as ammonia, urea, or uric acid.
5. A nephron consists of a glomerulus, Bowman's capsule, and tubule.
6. Blood cells and large molecules do not become part of the filtrate.
7. After Bowman's capsule, a tubule is composed of the proximal tubule, loop of Henle, and distal tubule.
8. Microvilli are found in the proximal tubule, ascending limb of the loop of Henle, and collecting duct. They increase the surface area and pump ions across the epithelium.
9. Urea diffuses out of the collecting duct, increasing the urea concentration outside of the tubule which, due to osmosis, pulls water out of the tubule.
10. A drop in blood pressure or sodium causes the kidneys to release renin. Renin then triggers formation of angiotensin in the blood. Angiotensin constricts blood vessels and causes the kidneys to retain water and salt, which increases blood volume. The combined effects increase blood pressure.
11. A decrease in blood sodium triggers an increase in aldosterone secretion.
12. ADH increases water permeability in the collecting ducts of the kidney, so more water is reabsorbed.
13. ANF inhibits secretion of renin, aldosterone, and ADH; relaxes smooth muscles in blood vessels; reduces thirst; and increases loss of sodium and water by closing channels in the collecting ducts.
14. In flatworms, flame cells line the excretory tubes and their cilia direct fluid into the tubes.
15. In insects, the excretory and digestive systems are linked, so excretory and digestive products mix.

Chapter 40—Defense: Inflammation and Immunity

Part 1: Multiple Choice

1. b
2. c and d
3. c
4. d
5. d
6. a and b
7. b
8. a
9. a
10. d
11. c
12. c
13. a
14. b and c
15. d

Part 2: Matching

1. nonspecific response
2. specific response
3. specific response
4. nonspecific response
5. specific response
6. T cells
7. B cells and T cells
8. B cells
9. B cells and T cells
10. T cells

Part 3: Short Answer

1. By the danger theory, immune cells are stimulated into action when damaged cell contents are spilled.
2. The first barrier against pathogen invasion is the skin.
3. Hemostasis begins with vasoconstriction, followed by formation of a platelet plug, and finally clotting.
4. Factor X converts inactive prothrombin to thrombin; thrombin digests away part of fibrinogen to form fibrin, and fibrin strands form the clot.
5. Most allergies involve stimulation of mast cells, which dump large quantities of histamine. Histamine is the main stimulus for inflammation, so antihistamine drugs block the inflammatory effects of histamine.
6. Interleukins act as chemical messengers between different kinds of white blood cells (leukocytes).
7. Neutrophils and macrophages (immature form is the monocyte) are phagocytes.
8. Immunity is characterized by specificity, memory, diversity, and self/nonself recognition.
9. Lymphocytes develop in lymphoid tissue, such as the spleen, lymph nodes, thymus, and tonsils.
10. Differentiated B cells become effector (plasma) cells that make antibodies, or become memory cells.
11. Immunoglobulin classes differ in their heavy chains.
12. First, each antibody is made of two polypeptide chains, so different polypeptides can be combined to produce numerous arrangements. Second, two separate genes can be spliced together to specify different polypeptides. Also, immunoglobulin genes mutate at high rates during lymphocyte production.
13. B lymphocytes respond to epitopes—often small parts of the whole molecule—so exposure to the epitope for one disease establishes immunity against any other diseases that have the same epitope.
14. T cytotoxic cells kill cells recognized as altered self; T helper cells send chemical messages to nearby B cells and T cytotoxic cells, stimulating their response.
15. MHC I molecules identify cells as belonging in the body. MHC II molecules appear on the surface of antigen-presenting cells and alert T cells to the presence of foreign (nonself) cells.

Chapter 41—The Cells of the Nervous System

Part 1: Multiple Choice

1. d
2. b
3. a
4. d
5. b
6. b
7. d
8. c
9. a
10. c
11. a and c
12. c
13. a and d
14. b
15. a

Part 2: Matching

1. electrical synapse
2. electrical and chemical synapse
3. chemical synapse
4. electrical synapse
5. chemical synapse
6. nonassociative learning
7. associative learning
8. associative learning
9. nonassociative learning
10. both types of learning

Part 3: Short Answer

1. Oligodendrocytes and Schwann cells are involved in forming the myelin sheath.
2. Dendrites carry signals to the cell body; axons carry signals away from the cell body and to the synapse.
3. A minimum amount of stimulation (threshold) must occur before a neuron transmits a signal. If that threshold is met, the neuron responds completely, but if the threshold is not met, there will be no response at all.
4. A resting neuron has fewer positive ions inside because
   a) the sodium-potassium pump moves 3 sodium ions out for each 2 potassium ions it moves in; and
   b) the membrane's selective permeability to these two positive ions keeps them segregated.
5. Depolarization occurs when the sodium channels open, allowing sodium to rush inside the neuron.
6. During the refractory period, potassium ions flow back out of the cell to restore the resting potential. Until this is done, another impulse could not be transmitted.

7. After a patch of the membrane is depolarized, the sodium gates cannot immediately reopen, so depolarization cannot occur there again for a brief period, but it does occur in the next adjacent patch of the membrane where the sodium gates can open.
8. Ions cannot flow through myelinated sections, so depolarization only occurs at the bare patches.
9. A neuromuscular junction is the synapse between a motor neuron and a voluntary muscle cell.
10. The acetylcholine receptor on the postsynaptic neuron is an ion channel. Acetylcholine binds to the receptor and opens the channel, allowing ions to flow into the postsynaptic neuron.
11. Gamma-aminobutyric acid (GABA) is the most common inhibitory neurotransmitter in the brain.
12. Neurotransmitters are usually removed by reuptake—pumped into the presynaptic neuron or glial cells; then they are recycled or inactivated by enzymes.
13. Drugs may bind to neurotransmitter receptors, either mimicking or blocking the effects of the neurotransmitter; or they may influence the release, reuptake, or inactivation of the neurotransmitter.
14. A stretch reflex is monosynaptic, involving only a sensory and a motor neuron, directly linked.
15. Learning may involve synaptic depression, which underlies habituation, or synaptic facilitation, which underlies sensitization.

Chapter 42—The Nervous System and the Sense Organs

Part 1: Multiple Choice

1. a
2. d
3. c
4. a
5. d
6. b
7. b
8. d
9. c
10. b and c
11. d
12. b
13. a and b
14. c
15. c

Part 2: Matching

1. hindbrain
2. forebrain
3. midbrain
4. hindbrain and midbrain
5. forebrain
6. audition
7. vision
8. olfaction
9. olfaction
10. audition

Part 3: Short Answer

1. To maintain homeostasis, the nervous system must be able to detect changes in its environment, interpret the meanings of the changes, and respond appropriately to those changes.
2. The central nervous system is composed of the brain and the spinal cord.
3. The brain is protected by the cranium, the meninges, and the cerebrospinal fluid.
4. In the spinal cord, the ventral horn (gray matter) contains motor neurons, the dorsal horn (gray matter) contains the bodies of sensory neurons, and the white matter contains myelinated nerve tracts.
5. The hindbrain contains the pons, medulla oblongata, and cerebellum.
6. The brainstem consists of the pons, medulla oblongata, and midbrain.
7. The diencephalon contains the thalamus, which integrates sensory information about the external world, and the hypothalamus—the master regulator of homeostasis—which integrates sensory information about the internal world, and links the endocrine and nervous systems together.
8. The lobes of the cerebral cortex are the frontal, parietal, occipital, and temporal.
9. [Answers will vary.]
10. The sympathetic nervous system triggers the "fight or flight" reaction, mobilizing energy stores in preparation for vigorous activity; the parasympathetic nervous system conserves energy.
11. Sensation is the mere detection of external stimuli; perception is the conscious recognition and interpretation of the stimuli.
12. Sensation involves:
    a) transduction—translation of a stimulus into an electrical event at the receptor cell;
    b) transmission—sending a signal from the receptor cells to the central nervous system; and
    c) integration—combining the signals to form a mental image of the outside world.
13. Receptors for both audition and the sense of balance are located within the inner ear.

14. Accommodation is when the lens of the eye changes its shape, becoming more rounded or flattened, to alter the focal length (bending) of light when viewing objects at different distances.
15. The somatic sensory system has receptors for touch, position, temperature, and pain.

Chapter 43—Sexual Reproduction

Part 1: Multiple Choice

1. c and d
2. a
3. b
4. b
5. a
6. d
7. b and d
8. d
9. b
10. d
11. b
12. c
13. d
14. b and c
15. a

Part 2: Matching

1. male reproductive system
2. female reproductive system
3. male reproductive system and female reproductive system
4. male reproductive system
5. female reproductive system
6. LH, FSH, and estrogen
7. LH and FSH
8. FSH
9. LH
10. LH and FSH

Part 3: Short Answer

1. The second meiosis is not completed until after an oocyte is fertilized.
2. Sperm production requires a lower temperature than is found inside the body.
3. Semen is composed of:
   a) sperm, for fertilization,
   b) sugars and nutrients from the seminal vesicles that nourish the sperm,
   c) alkaline secretions from the prostate gland that regulate pH, and
   d) mucus from the bulbourethral glands, for lubrication.
4. Wolffian ducts give rise to the vas deferens and its attachments; Müllerian ducts give rise to the oviducts, uterus, and vagina.
5. The stages of the sexual response are excitement, plateau, orgasm, and resolution.
6. Erection results from blood filling the spongy spaces in the corpora cavernosa and corpus spongiosum.
7. Ejaculation involves contraction of muscle in the prostate, vas deferens, seminal vesicles, and urethra.
8. Once the sperm enters the oocyte, the zona pellucida becomes impenetrable to other sperm.
9. The hypothalamus secretes GnRH, which causes the anterior pituitary to release LH and FSH, which both act on the gonads.
10. The corpus luteum secretes progesterone, which:
    a) stimulates thickening of the endometrium in preparation for implantation, and
    b) blocks GnRH secretion, and thus LH and FSH secretion, to prevent a new cycle from starting.
11. The rhythm method requires that the woman knows when her ovulation occurs so she can avoid intercourse in the days before and on the day of ovulation and, in theory, conception could occur during a week-long window, which precludes intercourse for about one fourth of each month.
12. About 90% of women not using birth control would become pregnant in a year.
13. Some sperm enter the vagina prior to ejaculation, and withdrawal before ejaculation requires tremendous self-control a time when it goes against the normal behavior pattern.
14. Both the IUD and "morning after" pill block implantation, but do not necessarily prevent conception.
15. The vas deferens is cut and tied before it enters the abdomen; tubal ligation requires abdominal surgery.

Chapter 44—How Do Organisms Become Complex?

Part 1: Multiple Choice

1. d
2. b
3. b
4. c
5. a
6. a and c
7. d
8. c
9. a and b
10. d
11. b
12. c
13. a
14. b
15. c

Part 2: Matching

1. mesoderm
2. ectoderm, endoderm, and mesoderm
3. endoderm
4. mesoderm
5. ectoderm
6. growth
7. gastrulation
8. cleavage
9. gastrulation
10. cleavage

Part 3: Short Answer

1. Preformation left little room for evolution, the Earth is too old to allow all embryos to be present, housed within each other, and genetic material from each parent combines during fertilization.
2. The stages of development are fertilization, cleavage, germ layer formation, organ formation, growth, metamorphosis, and aging.
3. The large size of a zygote provides too little surface area for adequate material exchange, and too little nucleus for the amount of cytoplasm that it must regulate and direct.
4. The blastocyst has trophoblast cells, some of which become part of the placenta, and the inner cell mass, which eventually grows into the embryo.
5. Morphogenesis is the creation of form; differentiation is the specialization of cells.
6. Neural crest cells can become many types of cells, including those of nerve tissue and pigment cells.
7. Organ formation precedes differentiation, often involves complex folding of sheets of cells, and depends on interactions among cells that are brought together by movements.
8. In apoptosis, cells die quietly and are engulfed without initiating an immune attack.
9. Fate refers to what a cell becomes through development; potency is what it could become in a different environment.
10. Udder cells from an adult sheep were cultured in a way to induce a dormant state, shutting off many genes and replication. The dormant nuclei were then transplanted into egg cells.
11. Udders contain mixed populations of cells, some of which are undifferentiated stem cells, and Dolly may have developed from one of these.
12. Homeotic selector genes and HOX genes determine the formation of the organism's overall pattern.
13. Imaginal discs are groups of unspecialized cells in the larva that wait to develop until after metamorphosis—they give form to the adult.
14. Like metamorphosis, puberty involves a burst of growth, and development of various sexual characteristics that enable them to reproduce. It is controlled by hormones.
15. By the view of natural selection, we become dispensable when we can no longer reproduce.